SLon 立环脉动高梯度磁选机研制与应用论文集

熊大和 著

北京
冶金工业出版社
2021

图书在版编目(CIP)数据

SLon 立环脉动高梯度磁选机研制与应用论文集/熊大和著. —北京：冶金工业出版社，2021.8
ISBN 978-7-5024-8847-5

Ⅰ.①S… Ⅱ.①熊… Ⅲ.①高梯度磁选机—文集 Ⅳ.①TD457-53

中国版本图书馆 CIP 数据核字（2021）第 116991 号

出版人　苏长永
地　址　北京市东城区嵩祝院北巷 39 号　邮编 100009　电话（010）64027926
网　址　www.cnmip.com.cn　电子信箱　yjcbs@cnmip.com.cn
责任编辑　张熙莹　美术编辑　彭子赫　版式设计　禹　蕊
责任校对　李　娜　责任印制　禹　蕊

ISBN 978-7-5024-8847-5
冶金工业出版社出版发行；各地新华书店经销；北京捷迅佳彩印刷有限公司印刷
2021 年 8 月第 1 版，2021 年 8 月第 1 次印刷
787mm×1092mm　1/16；38.25 印张；923 千字；599 页
380.00 元

冶金工业出版社　投稿电话　（010）64027932　投稿信箱　tougao@cnmip.com.cn
冶金工业出版社营销中心　电话　（010）64044283　传真　（010）64027893
冶金工业出版社天猫旗舰店　yjgycbs.tmall.com

（本书如有印装质量问题，本社营销中心负责退换）

作者简介

熊大和，男，1952年7月生，汉族，江西宜丰人，中共党员，教授级高级工程师；1982年毕业于江西冶金学院（现江西理工大学），获学士学位；1988年毕业于中南工业大学（现中南大学）选矿机械专业，获博士学位；SLon立环脉动高梯度磁选机第一发明人，我国著名的磁电选矿专家。他从事磁选技术与设备研究40余年，实现了立环脉动高梯度磁选技术从理论研究到大规模工业应用的飞跃，研发了一系列拥有自主知识产权的先进选矿设备，带动了国内外氧化铁矿、钛铁矿、锰矿、钨矿、钽铌矿、稀土矿等弱磁性矿石及非金属矿矿产资源的综合利用技术水平与装备的进步，创造了显著的社会效益和经济效益，为我国选矿事业的发展和科技进步作出了重要贡献。

熊大和在国内外发表论文及编写研究报告百余篇；获国家科技奖励3项，均排名第一；获中国专利优秀奖1项，排名第一；获省部级科技进步奖特等奖1项，发明专利金奖1项、一等奖6项、二等奖4项，技术发明奖二等奖1项。先后被国家和各级组织授予"庆祝中华人民共和国成立70周年纪念章""国家有突出贡献专家""享受国务院政府特殊津贴""全国优秀科技工作者""全国五一劳动奖章""江西省劳动模范""有色总公司特等劳动模范""江西省十佳杰出青年""江铃科技精英奖"等荣誉。

笔 者 的 话

亲爱的读者，感谢你抽空阅读此论文集，但愿它对你的工作有所启发和帮助。

笔者从1980年大学三年级开始学习高梯度磁选技术，就立志研制出最好的强磁选设备，至今从事了40年高梯度磁选技术和设备的研究与应用。在总结国内外高梯度磁选和强磁选设备的优缺点之后，笔者与同事们站在前人的肩膀上，在国内外同行的大力支持和帮助下研制了SLon立环脉动高梯度磁选机系列产品。从1985年开始成功研制第一台SLon-1000立环脉动高梯度磁选机开始至2000年SLon磁选机达到成熟的工业应用经历了15年。从2001年至今，SLon立环脉动高梯度磁选机成为国内外最先进的高梯度磁选设备或强磁选设备，占据了国内90%以上的市场份额，国外70%以上的市场份额。作为该研究成果的最高成就，"弱磁性矿石高效强磁选关键技术与装备"获得了2014年度国家科技进步奖二等奖。

笔者作为一个出自寒门的农村青年，致力于选矿工业40年并有所作为，想出版一部文集，以此方式回报国家的培养和国内外同行的支持与帮助。

本书收录了笔者曾发表的78篇学术论文，其中大部分是中文论文，小部分是英文论文。论文的内容都是笔者在研制SLon立环脉动高梯度磁选机及其应用方面的心得体会。SLon立环脉动高梯度磁选机从实验室研究到大规模工业应用，经历了很多曲折和艰辛。好比高楼大厦平地而起，每篇论文中如有新的内容，都是表示在给SLon立环脉动高梯度磁选机的研制添砖加瓦。

笔者写的论文力求做到通俗易懂，大部分论文是在工作繁忙之余所写，因此很多地方有刀痕斧迹，缺少精雕细琢，若有不妥之处，还请读者指正。

笔者希望每篇论文都为读者呈现新的内容，但是论文前后又有很多重复的地方，为了使读者更容易找到每篇论文的新内容，笔者在每篇论文之后均给出了写作背景。笔者出版此论文集有以下三个目的：

(1) 把SLon立环脉动高梯度磁选机研制与应用的心路历程记录下来呈献社会；

（2）向国内外选矿同行宣传普及 SLon 磁选机的知识，当新人进入选矿行业，他们并不一定能看到笔者以前发表的所有文章，因此每篇论文都尽量做到通俗易懂，前后连贯；

（3）不管是陌生的人还是熟悉的人，我们都要经常性地向他们报道和推介我们产品的优点、研制与应用的最新进展，我们在努力与他们保持亲切关系，就如同可口可乐不断地在做广告，重复它的优点，保持与顾客的亲切关系一样。

最后要说明的是，笔者所发表的论文是由笔者手写起草，绝大部分是由同事打印、制图、编辑和校稿完成的。如同 SLon 立环脉动高梯度磁选机研制成果一样，笔者起了一牵头作用，而大部分工作是其他同事和同行完成的，在此谨向对该研制事业有贡献的同事、同行和 SLon 立环脉动高梯度磁选机的使用客户表示衷心的感谢和崇高的敬意！

熊大和

2020 年 12 月

目　录

振动高梯度磁选 …………………………………………………………………………… 1
高梯度磁选机发展的新动向 ……………………………………………………………… 7
脉动高梯度磁选细粒氧化铁矿的研究 …………………………………………………… 14
SLon-1000 立环脉动高梯度磁选机的研制 ……………………………………………… 20
脉动高梯度磁选中冲程冲次对选矿指标的影响 ………………………………………… 26
SLon-1500 立环脉动高梯度磁选机的研制 ……………………………………………… 33
SLon 型立环脉动高梯度磁选机及其应用的新发展 …………………………………… 38
SLon 型立环脉动高梯度磁选机的改进及其在红矿选矿的应用 ……………………… 43
SLon-2000 立环脉动高梯度磁选机的研制 ……………………………………………… 50
SLon 立环脉动高梯度磁选机的发展与应用 …………………………………………… 55
弱磁性矿粒在棒介质高梯度磁场中的动力学分析 ……………………………………… 61
SLon-1750 立环脉动高梯度磁选机的研制与应用 ……………………………………… 67
应用 SLon 立环脉动高梯度磁选机提高资源利用率 …………………………………… 72
提高姑山赤铁矿生产指标的工业试验研究 ……………………………………………… 78
新一代强磁选设备——SLon 立环脉动高梯度磁选机 ………………………………… 84
上厂铁矿尾矿复选赤铁矿新技术 ………………………………………………………… 89
SLon 立环脉动高梯度磁选机发展与应用现状 ………………………………………… 94
SLon 型磁选机在齐大山选矿厂的应用 ………………………………………………… 100
SLon 立环脉动高梯度磁选机的应用研究 ……………………………………………… 105
SLon 立环脉动高梯度磁选机在提高红矿质量的应用 ………………………………… 114
The Development of A New Type of Industrial Vertical Ring and Pulsating
　　HGMS Separator …………………………………………………………………… 121
New Development of the SLon Vertical Ring and Pulsation HGMS Separator ……… 127
Development and Commercial Test of SLon-2000 Vertical Ring and Pulsating
　　High-gradient Magnetic Separator ………………………………………………… 134
Development and Applications of SLon Vertical Ring and Pulsating High Gradient
　　Magnetic Separators ………………………………………………………………… 142
New Technology of Pulsating High Gradient Magnetic Separation …………………… 150
Research and Commercialisation of Treatment of Fine Ilmenite with SLon Magnetic
　　Separators …………………………………………………………………………… 164
A Large Scale Application of SLon Magnetic Separator in Meishan Iron Ore Mine …… 171
SLon 磁选机分选东鞍山氧化铁矿石的应用 …………………………………………… 177
SLon 立环脉动高梯度磁选机选别鞍山红矿的研究与应用 …………………………… 186

SLon 磁选机在钛铁矿选矿工业中的研究与应用 ………………………………………… 191
SLon Magnetic Separator Applied to Upgrading the Iron Concentrate ……………… 196
SLon-2000 磁选机在调军台选矿厂的工业试验与应用 ………………………………… 202
SLon 磁选机研究与应用新进展 …………………………………………………………… 208
SLon 磁选机在淮北煤系高岭土除铁中的应用 ………………………………………… 212
脉动高梯度磁选垂直磁场与水平磁场对比研究 ……………………………………… 217
SLon 型磁选机在红矿选矿工业中的应用 ……………………………………………… 224
SLon Magnetic Separators Applied in the Ilmenite Processing Industry …………… 230
SLon 磁选机在红矿强磁反浮选流程中的应用 ………………………………………… 238
SLon 立环脉动高梯度磁选机分选红矿的研究与应用 ………………………………… 245
SLon 磁选机在黑色金属选矿工业应用新进展 ………………………………………… 256
SLon 立环脉动高梯度磁选机与《金属矿山》共发展 ………………………………… 267
SLon Magnetic Separator Promoting Chinese Oxidized Iron Ore Processing Industry ……… 273
SLon 脉动与振动高梯度磁选机新进展 ………………………………………………… 280
SLon 磁选机在大型红矿选厂应用新进展 ……………………………………………… 287
SLon 磁选机分选锰矿的研究与应用 …………………………………………………… 296
SLon Magnetic Separators Applied in Various Industrial Iron Ore Processing
　　Flow Sheets …………………………………………………………………………… 302
SLon 磁选机在包钢选厂的试验与应用 ………………………………………………… 311
SLon 磁选机分选氧化铁矿研究与应用新进展 ………………………………………… 318
SLon Magnetic Separators Applied to Beneficiate Low Grade Oxidized Iron Ores ……… 324
SLon 强磁机在低品位氧化铁矿中的新应用 …………………………………………… 332
Application of SLon Magnetic Separators in Modernising the Anshan Oxidised Iron
　　Ore Processing Industry …………………………………………………………… 338
SLon 磁选机在鞍山氧化铁矿选矿工业现代化的应用 ………………………………… 347
SLon 磁选机与离心机组合技术分选氧化铁矿 ………………………………………… 354
SLon 立环脉动高梯度磁选机新技术 …………………………………………………… 361
SLon 磁选机在云川贵选矿工业中的应用 ……………………………………………… 371
SLon-2500 立环脉动高梯度磁选机的研制与应用 ……………………………………… 381
SLon 磁选机与离心机组合流程分选氧化铁矿新进展 ………………………………… 389
SLon 立环脉动高梯度磁选机技术创新与应用 ………………………………………… 399
A New Technology of Applying SLon-2500 Magnetic Separator to Recover Iron
　　Concentrate from Abandoned Tails ……………………………………………… 407
SLon 磁选机分选氧化铁矿工业应用新进展 …………………………………………… 415
The Integrative Technology of SLon Magnetic Separator and Centrifugal Separator for
　　Processing Oxidised Iron Ores …………………………………………………… 423
SLon 磁选机分选氧化铁矿十年回顾 …………………………………………………… 431
SLon 立环脉动高梯度磁选机大型化研究与应用 ……………………………………… 442

应用 SLon 磁选机提高选钛回收率的回顾与展望 ·················· 449
Recent Development and Applications of SLon VPHGMS Magnetic Separator ·················· 463
SLon 立环脉动高梯度磁选机大型化研究与应用新进展·················· 473
应用 SLon 磁选机提高弱磁性金属矿回收率 ·················· 481
The Creative Technologies of SLon Magnetic Separators in Beneficiating Weakly
　Magnetic Minerals ·················· 495
SLon 磁选机在大型氧化铁矿选矿厂的应用·················· 505
SLon 立环脉动高梯度磁选机研制与应用新进展·················· 515
SLon 磁选机提高弱磁性金属矿回收率的研究与应用·················· 524
SLon 立环脉动高梯度磁选机研制与应用新进展·················· 535
SLon 立环脉动高梯度磁选机在多种金属矿选矿中的应用·················· 546
SLon-3000 立环脉动高梯度磁选机的研制与应用 ·················· 554
SLon 磁选机与离心机分选氧化铁矿新技术·················· 562
SLon 磁选机在安徽分选弱磁性矿石的研究与应用·················· 571
SLon 磁选机提高选钛回收率的研究与应用·················· 581
新型 SLon 立环脉动高梯度磁选机应用实践 ·················· 591

振动高梯度磁选

熊大和　刘永之　刘树贻

（中南工业大学）

高梯度磁选是20世纪60年代末期发展起来的技术。其基本原理是利用铁磁性丝状介质在磁场中产生的高梯度强磁力捕收微细的弱磁性粒子。该技术首先在高岭土提纯、钢铁厂废水处理和动力厂循环水净化等方面获得应用。近年来开始用它分选金属矿物。由于金属矿物的粒度比高岭土粗，尽管高梯度磁力分选金属矿物时一般能够获得很高的磁性物料回收率，但存在着磁性精矿品位低和磁介质堵塞的问题，在一定程度上阻碍了该技术在金属矿山的推广应用。本研究的目的是用振动的办法解决这些问题。

1　机械式振动高梯度磁选机

试验所用的振动高梯度磁选机是在螺线管高梯度磁选机上装置一套机械振动系统，如图1所示。一根非导磁不锈钢制成的振杆穿过上磁极头深入到分选腔的内部，振杆上钻有一些小圆孔，穿过小圆孔可插入一些不锈钢丝，可使钢毛、钢丝网等磁介质固定在振杆上，并随振杆一起振动。振杆上端装有一个滚动轴承，以减少传动摩擦阻力，振杆可在凸轮和弹簧的驱动下作强迫振动，凸轮的转速由直流调速电机控制。该机的一些可调参数及调节方法见表1。

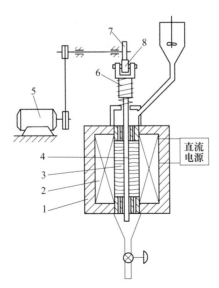

图1　机械式振动高梯度磁选机

1—铁铠；2—激磁线圈；3—磁介质；4—振杆；5—直流调速电机；6—弹簧；7—凸轮；8—滚动轴承

---

❶ 原载于《有色金属》，1986（04）：13~17。

表 1 振动高梯度磁选机的可调参数

可调参数	调节范围	调节方法
振幅（1/2 冲程）	0~1.4mm	改变凸轮的偏心距
振次	0~4500 次/min	调节凸轮的转速
振动曲线形状	正弦、非正弦	更换不同轮廓的凸轮
磁介质堆高度	0~24cm	安装不同长度的磁极头
分选腔直径	3.2~8cm	安装不同直径的分选筒
背景场强	0~12000Oe	调节激磁电流
矿浆流速	0~8cm/s	控制排矿口阀门

该机除了具有多种参数可调的优点外，振动参数的调节比较方便，且不受背景磁场的影响，冲洗磁性产品时，磁介质可做强烈的振动，不用压力水就可迅速地将磁性产品冲洗出来；该机能够分选多种金属矿和非金属矿，是进行振动高梯度磁选工艺和理论研究的较为理想的设备。

本文所述的振动曲线均是正弦曲线，磁介质堆高度为 17.5cm，选矿效率用道格拉斯公式计算，即：

$$E = \frac{\varepsilon - \gamma}{100 - \gamma} \cdot \frac{\beta - \alpha}{\beta_{\max} - \alpha} \times 100\%$$

式中，黑钨矿的 $\beta_{\max} = 76.5\% WO_3$，三氧化二铁的 $\beta_{\max} = 70.0\% Fe$。

2 纯矿物试验

2.1 分级试验

分级试验所用纯矿物是人工挑选的天然黑钨矿和石英。黑钨矿的纯度为 97.37%，比磁化系数实测为 $33 \times 10^{-6} cm^3/g$。黑钨矿和石英分别分成四个粒级，混合入选时，试料近似地按自由沉降等降比配比，见表 2。试验的固定条件见表 3。一次选别的试验结果如图 2 所示。

表 2 混合料的搭配

试料编号	黑钨矿粒度/μm	石英粒度/μm	混合料品位（WO_3）/%
A	20~40	40~74	14.10
B	10~20	20~40	12.82
C	5~10	10~20	11.91
D	0~5	0~10	11.50

表 3 分级试验的固定条件

黑钨粒度/μm	磁介质类型	磁介质丝径/μm	给矿负荷率/g·cm^{-3}	振幅/mm	背景场强/Oe	给矿流速/cm·s^{-1}	给矿浓度
20~40	16 目低碳钢丝网	315	2.92	1.0	10000	3.8	单一黑钨矿试验5%，混合料入选时10%
10~20	16 目低碳钢丝网	315	2.92				
5~10	40 目低碳钢丝网	250	2.92				
0~5	40 目低碳钢丝网	250	1.46				

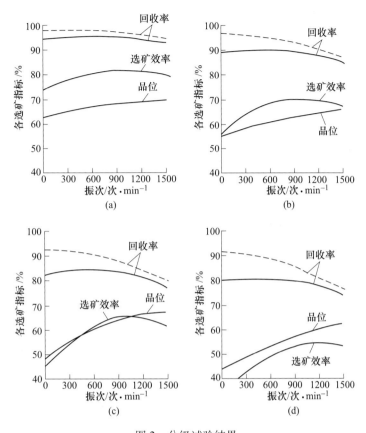

图 2　分级试验结果

(a) 20~40μm 黑钨矿，40~74μm 石英；(b) 10~20μm 黑钨矿，20~40μm 石英；
(c) 5~10μm 黑钨矿，10~20μm 石英；(d) <5μm 黑钨矿，<10μm 石英

（虚线表示单一黑钨矿回收率随振次的变化关系；实线表示混合试料各选矿指标随振次的变化关系）

根据图 2 作如下分析：

（1）石英对黑钨矿回收率的影响。用单一黑钨矿考察回收率试验，磁介质不振动（振次为零），各级别的回收率都能达到 90% 以上。分选混合物料时，由于石英的干扰，黑钨矿的收率普遍下降。但是，随着振次的增加，石英对黑钨矿回收率的影响逐渐减小，在图 2 上表现为虚线与实线的距离随振次的增加而缩小，这种现象对细粒级尤为显著。如 0~5μm 粒级黑钨矿，振次为零时，混合料黑钨矿的回收率比单一黑钨矿的回收率约低 12%，而当振次为 1500 次/min 时，二者仅相差约 2.5%。此外，振次较低时，混合试料中黑钨矿各粒级的回收率随振次的增加而稍有提高，而当回收率曲线到达峰值后，则随振次的增加而下降。因此，在一定的振次范围内，振动高梯度磁选基本上不会降低磁性产物回收率。

（2）磁性物料的品位曲线随振次的增加而上升，细粒级的变化比粗粒级的更显著。

（3）选矿效率曲线存在着随振次的增加而上升，到达峰值后随振次的增加而下降的规律。选矿效率值上升阶段主要是品位值的迅速增加。当振次很高时，品位的增加趋于平缓，回收率值下降较快，这导致了选矿效率值的下降。

在无振动条件下，石英很容易夹杂在磁介质堆中，占据捕集区域。这不但降低磁性精

矿的品位和易造成堵塞,而不能捕获一部分磁性矿粒,降低了磁性矿物的回收率。此外,石英与黑钨矿及石英与磁介质表面也存在着一定的表面作用,粒度越小,表面力越显著。在磁介质的振动从无到有、从弱到强的变化过程中,石英的夹杂和表面力都逐渐下降是次要因素。因此,适当的振动条件对消除非磁性矿物的干扰,提高磁性精矿的品位和选矿效率都有利。

2.2 纯黑钨矿与纯石英混合料宽级别试验

试料由 0~40μm 黑钨矿和 0~76μm 石英混合而成,给矿品位为 12.82%WO₃。试验用的磁介质是 16 目低碳钢丝网,给矿浓度为 10%,给矿负荷率为 0.4,其余固定条件同表 3。试验结果如图 3 所示。

图 3 中各曲线的变化规律和分级试验相同。此外还可看出最佳的选矿效率出现在振次为 1000~1200 次/min 范围内,例如,振次为 1200 次/min 时,精矿品位为 68.03%WO₃,回收率为 93.40%,选矿效率为 79.75%。与振次为零的指标比较,回收率基本不变,而品位提高约 12%,选矿效率提高约 20%。

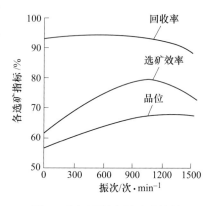

图 3 宽级别混合料试验结果

3 几种矿物的分选试验

3.1 齐大山氧化铁矿的分选

试料采用鞍钢齐大山选矿厂粉矿车间,粒度为 20~0mm。在实验室先用对辊机破碎至小于 2mm,然后用球磨机磨至小于 200 目(74μm)占 80%。试样中有较多的强磁性矿物。采用了两种简单试验流程:一是弱磁选—强磁选流程,弱磁选场强 600Oe,振次 3000 次/min,弱磁选尾矿强磁选,强磁选场强 6000Oe,振次 1800 次/min,弱磁选、强磁选的精矿合并为综合精矿。二是单一强磁选流程,磁场强度 4000Oe,振次 1800 次/min。弱磁选和强磁选都是在图 1 所示的振动高梯度磁选机上进行,分选条件和试验指标见表 4。

表 4 齐大山粉矿分选试验指标

流程	产物名称	产率/%	品位/%	回收率/%	选矿效率/%	振次/次·min⁻¹ 弱磁选	振次/次·min⁻¹ 强磁选
弱磁选—强磁选流程	弱磁选精矿	48.73	50.51	82.89	35.96	0	0
	强磁选精矿	9.08	30.87	9.43			
	综合精矿	57.81	47.41	92.32			
	尾矿	42.19	5.28	7.68			
	给矿	100.00	29.69	100.00			
	弱磁选精矿	27.13	61.62	56.72	55.52	3000	1800
	强磁选精矿	20.27	48.17	33.11			
	综合精矿	47.40	55.86	89.83			
	尾矿	52.60	5.70	10.17			
	给矿	100.00	29.48	100.00			

续表4

流程	产物名称	产率/%	品位/%	回收率/%	选矿效率/%	振次/次·min^{-1}	
						弱磁选	强磁选
一次选别流程	精矿	58.50	44.91	89.70	28.85	0	
	尾矿	41.50	7.27	10.30			
	给矿	100.00	29.29	100.00			
	精矿	45.59	56.42	87.81	51.71	1800	
	尾矿	54.41	6.56	12.19			
	给矿	100.00	29.29	100.00			
	精矿	37.56	59.34	73.59	42.59	3000	
	尾矿	62.44	11.21	26.41			
	给矿	100.00	29.29	100.00			
固定条件	矿浆流速7cm/s；给矿浓度10%；振幅1.4mm；磁介质：16目低碳钢丝网，介质充填率4%；给矿负荷率3.5g/cm³						

由表4可看出，在适当的振动分选与不振动分选相比较，可保证（综合）精矿回收率基本不变，而（综合）精矿品位和选矿效率分别提高约10%和20%。当振次超出一定范围，精矿品位的提高很缓慢，铁回收率的下降却很快，例如，一次选别流程中，振次1800次/min与3000次/min相比较，精矿品位仅从56.42%提高到59.34%，回收率却从87.81%下降到73.59%。此外，弱磁选—强磁选流程中的弱磁选能从品位为29.48%的给矿中一次分选出品位为61.62%、回收率为56.72%（场强为6000Oe，振次为3000次/min）的精矿，富集比如此之高，是任何其他类型磁选机难以实现的。

3.2 锰矿尾泥的分选

某矿的废弃锰矿尾泥，在实验室筛分后得到小于40μm部分作为试料，其中含锰约12%。经一次分选所得结果及其试验条件见表5。

表5 锰矿尾泥分选指标

产品名称	产率/%	品位（Mn）/%	回收率/%	振次/次·min^{-1}	磁介质
精矿	36.83	21.04	60.98	0	4号钢毛
尾矿	63.17	7.85	39.02		
给矿	100.00	12.71	100.00		
精矿	29.03	25.54	61.50	800	
尾矿	70.97	6.54	38.50		
给矿	100.00	12.06	100.00		
精矿	15.27	27.35	32.43	400	40目低碳钢丝网
尾矿	84.73	10.27	67.57		
给矿	100.00	12.88	100.00		
精矿	11.08	30.23	27.60	800	
尾矿	88.92	9.88	72.40		
给矿	100.00	12.14	100.00		
固定条件	场强10000Oe；矿浆流速2~4cm/s；给矿浓度5%；介质充填率5%；给矿负荷率1.6g/cm³				

由表5可见，采用4号钢毛作磁介质，振次800次/min，从含锰12.06%的给矿一次

分选得含锰 25.54%、回收率为 61.50% 的精矿，与不振动比较，回收率基本不变，而品位提高 4.5%；当采用 40 目低碳钢丝网作介质时，一次分选得含锰 27.35%~30.23% 的精矿，但回收率仅 30% 左右。若从品位较高更易出售，又是从尾矿中回收有用金属的观点看，使用网状介质的选矿指标是有意义的。

3.3 低品位黑钨细泥的分选

试料粒度为小于 200 目 69.66%，含 WO_3 为 0.44%，其中还含有一部分褐铁矿、菱铁矿和黄铁矿等与黑钨矿磁性大致相等的弱磁性矿物，而磁选对这些矿物几乎没有选择性，这就限制了磁性产物中 WO_3 的含量，但由于磁选可以抛去大量尾矿，作为一种粗选手段仍是有意义的。经一次分选的指标以及条件见表 6。

表 6　低品位黑钨矿细泥分选指标

产品名称	产率/%	品位（WO_3）/%	回收率/%	振次/次·min^{-1}
精矿	12.50	3.16	89.77	
尾矿	87.50	0.051	10.23	0
给矿	100.00	0.44	100.00	
精矿	9.87	4.12	92.41	
尾矿	90.13	0.037	7.59	800
给矿	100.00	0.44	100.00	
精矿	4.58	8.06	83.94	
尾矿	95.42	0.074	16.06	1500
给矿	100.00	0.44	100.00	
固定条件	场强 10000Oe，矿浆流速 2.5cm/s，磁介质：16 目低碳钢丝网，介质充填率 4%，给矿负荷率 3.5g/cm^3，给矿浓度 10%，振幅 1mm			

从表 6 可知，当振次为 800 次/min 时，精矿品位和回收率都超过了不振动的相应指标，但精矿品位改善不多。当振次为 1500 次/min 时，虽然回收率比不振动分选稍低一些，但精矿含 WO_3 8.06%，是不振动精矿品位的 2.6 倍。

本文列举的纯矿物及 3 种矿石的分选，都做了振动与不振动的高梯度磁选对比试验。结果表明，只要振动条件选择恰当，可在大致保持回收率不变的条件下，大幅度提高精矿品位；此外，振动对消除堵塞，加速磁性产物的排出，节省用水等方面都有益。

写作背景　本文是作者在中南工业大学攻读硕士研究生期间所做的一些实验工作。如图 1 所示的机械式振动高梯度磁选机的振动系统和磁介质是本人设计和制造的，而磁体是实验室原有的。该机是世界上第一台机械式振动高梯度磁选机，其振幅和频率可以定量调节。利用该设备进行了黑钨矿与石英的分离试验以及赤铁矿、锰矿、钨锡细泥等实际矿物的选矿试验，证明了有振动与没有振动比较，有振动的高梯度磁选大大提高了选矿效率。

高梯度磁选机发展的新动向

熊大和　刘树贻

（中南工业大学）

摘　要　本文介绍了两种立环式高梯度磁选机和3种振动高梯度磁选机及其工作原理和选矿性能。

1　前言

高梯度磁选是20世纪60年代末发展起来的新技术，其基本原理是利用铁磁性丝状介质在磁场中产生的高梯度强磁力捕收细粒弱磁性粒子。与齿板类强磁选机比较，其磁场梯度和磁力可高出1~2个数量级、处理细粒弱磁性粒子普遍可获得高得多的回收率。但是，由于密集的铁磁性介质堆、强磁力和颗粒之间的相互作用，高梯度磁选机较易形成介质堵塞和更多的非磁性颗粒被夹杂在磁性物料中，降低磁性产品的纯度及非磁性有用成分的回收率。处理金属矿物时，高梯度磁选存在的另一个问题是分选粒度上限较低，一般要求筛除0.1mm或0.2mm以上的粗料再入选（齿板类强磁选机的粒度上限为0.5mm），因此，适应于高梯度磁选机的物料较少，一般选矿厂要达到如此严格的分级也比较困难。

高梯度磁选存在的问题在一定程度上阻碍了该技术在矿山、选厂的推广应用，使高梯度磁选回收率高的优势难以发挥。针对高梯度磁选精矿品位较低和较易堵塞的问题，近年国内外出现了立环高梯度磁选机和振动高梯度磁选机。立环式高梯度磁选机对防止磁介质堵塞很有效，也有助于提高分选粒度上限，但对提高磁性精矿品位效果不大。振动高梯度磁选机是目前已知分选细粒弱磁性矿物最有效的磁选设备，它既可提高磁性精矿品位和消除介质堵塞，又能提高分选粒度上限和发挥高梯度磁选回收率的优势。因此，振动高梯度磁选机可能会成为一种很有前途的磁选设备。

2　立环高梯度磁选机

以往连续工作的高梯度磁选机为单环式，单环高梯度磁选机的给矿方向与磁性产品排出的方向一致，粗颗粒和木渣草屑均须穿过磁介质堆才能排出去，较易形成堵塞。立环高梯度磁选机的给矿与磁性产品方向相对于磁介质堆反向，粗粒和木渣草屑可循原路冲出，故不易堵塞。

图1是1985年美国发表的一种永磁立环高梯度磁选机[1]，名为铁轮磁选机（ferrous wheel magnetic separator），该机共由16个环组成，处理量为50~80t/h，各环的尺寸一样，可以互换并可根据现场要求增环或减环，最大可增至25环。如图1所示为单环的工作原理，转环下部两边装有一对永久磁铁，磁力线水平穿过转环，在两磁极之间形成磁场区，转环内可充填导磁不锈钢毛、钢板网或其他丝状介质，处于磁场区的磁介质被磁化形成高梯度磁场，矿浆从磁场区上部给入，通过磁介质，磁性颗粒被吸着在磁介质上，非磁性颗

---

❶ 原载于《选矿机械》，1987，01。

粒随矿浆排出。图1转环上方另布置有一对较小的永久磁极,磁性产物随转环到达此区间后,可进行一次清洗以提高磁性精矿品位,磁性产品在转环的顶部被冲出至接矿斗中。

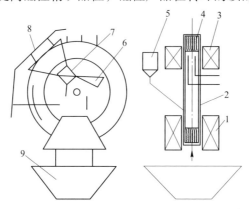

图1　铁轮磁选机

1—永久磁系Ⅰ；2—转环；3—永久磁系Ⅱ；4—磁介质；5—给矿盒；6—磁性产品接矿斗；
7—中矿斗；8—供水管；9—非磁性产品接矿斗

这种设备的优点是造价较低、更省电,25环的设备运转时仅一台4kW传动电机耗电。其缺点是场强较低。

图2是捷克斯洛伐克和苏联联合研制的一种大型电磁双立环高梯度磁选机[2],名为VMS磁选机。该机由激磁线包、铁铠、转环、传动装置、给矿和排矿等部件组成。其磁路保留了马斯顿磁路,即铁铠螺线管磁路的特点。分选区位于磁系的内部,磁力线与给矿方向基本平行。分选区场强最高可达1.7T。两个转环并排布置。转环直径为2m,每个转环宽1m,整机处理能力为100t/h左右,该机有3种规格,其主要技术参数如下：

场强/T	0.3~0.8, 1.0~1.2, 1.5~1.7
转环数	2
转环直径/mm	2000
转环宽度/mm	2×1000
外形尺寸/mm×mm×mm	5960×3302×5000
处理能力（矿浆）/$m^3 \cdot h^{-1}$	300
设备质量/t	80, 120, 210
电耗（当场强为0.5T时）/kW	30.2

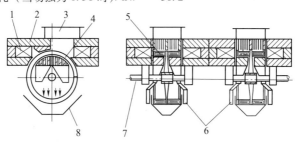

图2　VMS立环高梯度磁选机

1—激磁线圈；2—铁铠；3—给矿盒；4—转环；5—磁介质；
6—非磁性物排矿管；7—转环轴；8—磁物接矿斗

该机在苏联克里沃伊罗格（кривойрог）地区对其储量大的石英岩细粒氧化铁矿进行了工业考核试验。2000h 的考核期末发现磁介质堵塞现象。选矿指标见表1。

表1　VMS 磁选机分选石英岩细粒氧化铁指标

分选阶段	产物	试样1		试样2		备注
		品位/%	回收率/%	品位/%	回收率/%	
Ⅰ	粗精矿	50.7	90.5	49.7	83.9	第一段磨矿细度为小于74μm 占80%～90%，磁选抛尾后，粗精矿进入第二段磨矿，磨矿细度为小于44μm 占93%
	尾矿	11.4	9.5	15.3	16.1	
	给矿	38.3	100.0	36.5	100.0	
Ⅱ	精矿	61.7	71.9	59.0	71.5	
	尾矿	31.5	19.6	26.0	12.4	
	给矿	20.0	29.1	21.7	28.5	

从表1可知，VMS 立环高梯度磁选机在给矿品位为36.5%～38.3%，磨矿细度为80%～90%小于74μm 时，粗精矿品位为50%左右，回收率为83.9%～90.5%，即回收率较高但精矿品位不够理想；第二段磨至小于44μm 占93%时，精选后精矿品位达到60%左右，但回收率降至71%左右。

立环高梯度磁选机虽然可以解决堵塞问题，但分选指标与平环高梯度磁选机大致相同，欲得到较高的精矿质量，都要靠细磨和多次选别才能实现，这容易导致生产成本增高和磁性产品回收率下降。

3　振动高梯度磁选机

为了解决高梯度磁选介质堵塞和精矿品位较低的问题，1974年出现了振动高梯度磁选机。在国内，也已于1982年前后开始了振动高梯度磁选的研究，并获得了初步成果。

自1974年振动高梯度磁选机在美国获得专利以来，已有3种振动高梯度磁选机。

最初的振动高梯度磁选机是在螺线管高梯度磁选机的介质中设置交流线圈使介质振动[3]。如图3所示，这是一种周期式螺线管电磁振动高梯度磁选机，其介质堆被分成若干节，每节介质的下部放置一个交流线圈，交流线圈也可绕在介质堆的外围，交流线圈与背景磁场相互作用产生振动，并带动介质振动。交流线圈在磁场中的振动原理如图4、图5所示。

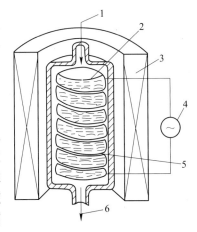

图3　周期式螺线管电磁振动高梯度磁选机

1—进料；2—磁介质；3—激磁线圈；4—交流电源；5—振动线圈；6—出料

图4为交流线圈在均匀磁场中的受力图，线圈受到的电磁力矩为：

$$L = P_m B \sin\theta \quad (1)$$

式中，L 为线圈所受电磁力矩；P_m 为线圈的磁矩；B 为背景磁感强度；θ 为 P_m 与 B 的夹角。

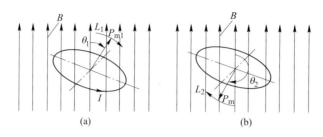

图 4 交流线圈在均匀磁场中的受力图

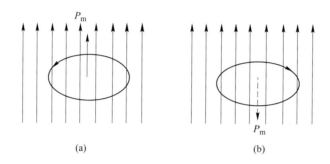

图 5 交流线圈在非均匀磁场中的振动

$$P_m = NAi \quad (2)$$

式中，N 为线圈匝数；A 为线圈包围的面积；i 为交变电流强度。

当 i 的方向发生变化，P_m 和 L 的方向随之做 180°转变，因此，只要 $\sin\theta \neq 0$，交流线圈就可在电磁力矩的作用下发生摇摆振动。

当 $\sin\theta = 0$ 时，只要交流线圈是置于非均匀磁场中，仍能产生沿磁场方向的往复振动（图5），其振动力为：

$$F = P_m \text{grad}B \quad (3)$$

式中，F 为交流线圈受到的磁力；$\text{grad}B$ 为磁场梯度。

置于非均匀磁场中的交流线圈，只要其磁矩 P_m 不与磁力线平行，它就同时受到以上介绍的两种力，即摇摆振动力矩和往复振动力的作用。在磁选机中，背景磁场的非均匀性总是存在的。因此，交流线圈在背景磁场中总是受到交变力的作用，背景磁场越高，交流线圈的安匝数越多，振动强度也越大。不仅周期式高梯度磁选机的介质中可安装振动线圈，而且连续式高梯度磁选机也可安装振动线圈，使介质振动。

发明这种振动高梯度磁选技术的目的是为了克服介质堵塞。但这种振动装置是使导线位于分选腔内，选矿时受到矿浆的浸泡和冲刷，容易磨损漏电而且不便于维修，这可能是这种技术以后没有得到工业应用的原因。

第二种振动高梯度磁选机是横向磁场电磁振动高梯度磁选机（图6），这是一种小型试验设备[4]。研制这种设备的目的是为了解决难选细粒黑钨锡石的分离问题，获得了很好的效果。该设备用线圈—铁芯磁系的间隙作背景磁场，两个交流线圈使分选盒振动。两个交流线圈分别贴近磁选机磁极的上下表面，因而能受到磁极边缘非均匀磁场力的作用，产生振动，图7（a）表示交流线圈的N极贴近磁选机的磁极，因而受到磁选机S极的吸引和N极的排斥，向右摆动；图7（b）表示交流线圈的极性发生了变化，向左摆动。线圈左右摆动时，带动分选盒左右运动，与磁选机的极面交替碰撞，产生强烈振动。振频率为50次/s，振幅以1.5mm左右为宜。

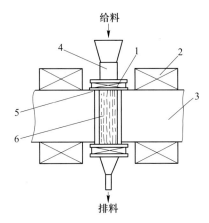

图6 横向磁场电磁振动高梯度磁选机
1—铁芯；2—激磁线圈；3—振动线圈；4—分选盒；5—线架；6—钢毛介质

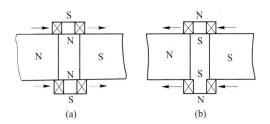

图7 线圈、铁芯磁系振动高梯度磁选机的振动原理

用该试验设备对广西珊瑚锡矿细粒难选黑钨、锡石中矿进行的振动和非振动高梯度磁选比较，试验结果列于表2。结果表明，振动高梯度磁选比非振动高梯度磁选的效果好得多，精矿WO_3品位高15.04%，WO_3回收率只低1.74%；非磁性产品的Sn回收率高15.94%。

表2 珊瑚锡矿细粒黑钨、锡石难选中矿的振动与非振动高梯度磁选试验结果比较 （%）

产品名称	产率	品位		回收率		比较条件
		WO_3	Sn	WO_3	Sn	
精矿	37.31	55.16	0.58	93.89	1.75	振动
尾矿	62.69	2.14	19.71	6.11	98.25	
给矿	100.00	21.58	12.58	100.00	100.00	
精矿	52.13	40.32	4.42	95.63	17.69	非振动
尾矿	47.87	2.01	22.35	4.37	82.32	
给矿	100.00	21.98	13.00	100.00	100.00	

这种设备构造简单，工作可靠，操作方便，省电，但磁极间隙不宜过大，因而应用规

模受到限制。

第 3 种振动高梯度磁选机是我们最近研制的螺线管周期式机械振动高梯度磁选机（图 8）[5]，该机是在原有 φ8mm 螺线管周期式高梯度磁选机的基础上加一套机械振动机构而成。振动机构由传动装置、凸轮、振杆和弹簧等构成。工作时，将分选介质固定在振杆上，由传动件带动凸轮，并在弹簧的配合下，迫使振杆上下运动。该设备的场强、矿浆流速、磁介质充填高度、磁介质类型、振动曲线形状均可在较大范围内调节，其振次在 0~4500 次/min、振幅在 0~1.4mm 内调节。

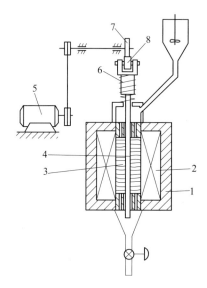

图 8　周期式机械振动高梯度磁选机
1—铁铠；2—激磁线圈；3—磁介质；4—振杆；5—调速电机；6—弹簧；7—凸轮；8—滚动轴承

机械振动高梯度磁选机与上述电磁振动高梯度磁选机相比的优点是，振动不受磁场限制，因而不仅分选时介质可以振动，而且断磁后冲洗磁性物时，介质也可以振动，故既能提高精矿品位，又能缩短冲洗时间；冲程冲次可调，有利于对不同物料选择各自合适的振动条件。用该机对多种细粒弱磁性物料进行试验证明，能显著提高分选效果。例如，用于齐大山铁矿粉矿试验时，分选一次可获得表 3 的结果，振动高梯度磁选比非振动高梯度磁选的精矿品位高 11.15%，回收率只低 1.89%，选矿效率高 22.86%。

表 3　齐大山铁矿粉矿振动与非振动高梯度磁选试验指标比较

产品名称	产率/%	品位/%	回收率/%	效率/%	比较条件
精矿	45.59	56.42	87.81		振动 1800 次/min，振幅 1.4mm
尾矿	54.41	6.56	12.19	51.71	
给料	100.00	29.29	100.00		
精矿	58.50	44.91	89.70		非振动
尾矿	41.50	7.27	10.30	28.85	
给料	100.00	29.29	100.00		
其他固定条件：给矿粒度小于 74μm（200 目）占 80%，给矿粒度上限 0.5μm，矿浆流速 7cm/s，磁场强度 0.4T					

4 结语

(1) 立环高梯度磁选机的给矿方向与排磁性产品的方向相反,有利于消除磁介质堵塞问题,但立环高梯度磁选机的磁性精矿品位仍然较低。

(2) 振动高梯度磁选是高梯度磁选技术的进一步发展,能大幅度提高磁性产品质量和非磁性成分的回收率,避免分选介质堵塞,提高分选粒度上限和缩短冲洗磁性物的时间,因此,振动高梯度磁选机将是一种很有希望的细粒弱磁性矿物分选设备。

(3) 现有的振动高梯度磁选机尚处于小型试验阶段,只能周期性地工作,有必要研制出大型连续设备,以便在铁矿山、锰矿山等金属矿山中推广应用。

参 考 文 献

[1] Arvidson, Bo R, Fritz, A J. New Inexpensive High Gradient Magnetic Separator [C]. Proceedings of 15th Int Mineral Processing Congress, 1985: 317~329.

[2] Cibulka J, et al. A New Conception of High Gradient Magnetic Separators [C]. Proceedings of 15th Int Mineral Processing Congress, 1985: 363~370.

[3] Anonymous U, S. Pat. N, 1974, 3, 838, 773.

[4] 刘树贻,彭世英. 珊瑚锡矿细粒钨锡混合物料的振动高梯度磁选研究 [J]. 中南矿冶学院报,1983 (增刊1): 103~112.

[5] 熊大和. 振动高梯度磁选 [D]. 长沙:中南矿冶学院,1985.

写作背景 本文是作者在中南工业大学攻读博士研究生期间写的一篇论文。介绍了当时高梯度磁选的发展趋势。其中图1所示的铁轮磁选机(水平磁系)和图2所示的VMS立环高梯度磁选机(磁系在上,转环在下)为我们后来发展立环脉动高梯度磁选机提供了参考依据。

脉动高梯度磁选细粒氧化铁矿的研究[1]

熊大和 刘树贻

（中南工业大学）

摘　要　提出了脉动高梯度磁选原理。用处理量为 4~7t 的立环脉动高梯度磁选机在马钢姑山铁矿进行了 1500h 的工业试验，证明了脉动高梯度磁选可大幅度地提高细粒级铁精矿品位和获得较高的回收率。

关键词　脉动高梯度磁选　细粒氧化铁矿　回收率

近年磁选设备发展迅速，磁选工艺在我国铁矿、锰矿、钨矿等选厂所占地位日显重要。齿板类湿式强磁选机及球介质立环湿式强磁选机已广泛使用。高梯度磁选各种细粒弱磁性矿的小型试验和工业试验已有不少报道，并普遍认为高梯度磁选微细粒弱磁性矿物有较高的捕收效率，但至今未被工业应用。其主要原因是：磁介质较易堵塞；磁性精矿品位较低；给料粒度上限（0.2mm 左右）限制过死，导致筛分作业困难。

针对上述问题，中南工业大学从 1980 年开始研究振动、脉动高梯度磁选机[1~4]，先后对赤铁矿、锰矿、黑钨矿和硫化物等试样做了大量的探索试验。结果表明，振动或脉动高梯度磁选能显著提高磁性精矿品位和选矿效率，有助于防止介质堵塞和提高给矿粒度上限。

为了使振动、脉动高梯度磁选获得工业应用，赣州有色冶金研究所投资与中南工业大学联合研制了 SLon-1000 立环脉动高梯度磁选机。该机于 1987 年 10 月安装在马钢姑山铁矿进行工业试验，用于处理 φ350 旋流器溢流等细粒物料。至 1988 年 1 月 27 日，已累计运转 1500h，设备力学性能良好，运转稳定可靠，选矿效率较高。本文介绍前一段的工业试验情况，有关设备构造和性能拟另文发表。

1　脉动高梯度磁选原理

高梯度磁选处理的物料粒度通常较细，影响选矿过程的力除了磁力、流体力、重力之外，磁性矿粒与脉石之间的表面力（包括静电力和范德华力）也是不容忽略的因素。这些力对矿粒的作用都是单向的。如矿浆从上至下流动时，部分脉石被其他矿粒或介质丝架位（图 1）。因流体力 R 方向朝下，故这些脉石不能脱离，导致精矿品位下降，严重时还会堵塞磁介质。此外，附着在磁介质表面的脉石占据了部分有效捕收表面，影响磁介质对磁性矿粒的捕收。

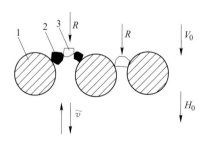

图 1　非磁性矿粒夹杂的形成
1—磁介质；2—磁性矿粒；3—非磁性矿粒

---

[1] 原载于《金属矿山》，1988（07）：34~37。

在脉动高梯度磁选中，除了上述各种力之外，还人为地施加一种脉动流体力，以松散群粒，提高选矿指标和防止堵塞。矿浆的最大脉动速度和平均脉动速度分别为：

$$\tilde{v} = \frac{1}{2}s\omega\sin\omega t \tag{1}$$

$$\tilde{v}_{\max} = \frac{1}{2}s\omega \tag{2}$$

$$\tilde{v} = \frac{1}{2\pi}\int_0^{2\pi}\left|\frac{1}{2}s\omega\sin\omega t\right|d(\omega t) = \frac{s\omega}{\pi} \tag{3}$$

式中，s 为选矿区有效冲程；ω 为脉动波角速度；t 为时间变量。

矿浆的实际流速为给矿速度 v_0 和脉动速度的叠加：

$$v = v_0 + \tilde{v} = v_0 + \tilde{v}_{\max}\sin\omega t \tag{4}$$

在本试验中，实际取值大致为 $v_0=6$ cm/s；$\tilde{v}_{\max}=16$ cm/s，因此 $v=6+16\sin\omega t$（cm/s）。

如图2所示是根据上式绘制的一个脉动周期内矿浆在选矿区的流速图。图中阴影部分表示矿浆的实际流速与给矿方向相反，此时对停留在磁介质上方的矿粒产生一个反向松散力，使图1所示被卡住的脉石受到一个自下至上的冲力，脱离约束状态而进入尾矿。

一般高梯度磁选中，当给矿方向从上至下时，绝大多数被捕集的磁性矿粒停留在磁介质的上表面，下表面捕获的矿粒很少[5,6]。在脉动高梯度磁选中，分选区矿浆不断变换流动方向，磁介质上下表面都能机会大致均等地捕获磁性矿粒。因此，尽管脉动力的存在增大了竞争力，但在适当的冲程冲次范围内，因捕获区增加可使磁性矿粒的捕获得到补偿。

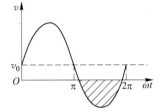

图2　矿浆在选矿区的流速

表1是根据试验所用场强、介质丝径等参数对单颗赤铁矿或石英受力的估算值。从表1中可知，粒度大于 10μm 时，磁力最大，脉动流体力居第二位，成为影响选矿指标的第二要素；进浆流体力为第三要素；静电力、范德华力和重力比前3种力小1~2个数量级，对选矿指标影响较小。当粒度小于 10μm 时，静电力和范德华力越来越接近于磁力，成为不可忽略的因素。

表1　矿粒受力估算值

矿粒直径/μm	磁力（赤铁矿）/dyn	平均脉动流体力/dyn	进浆流体力/dyn	静电力/dyn	范德华力/dyn	重力（赤铁矿）/dyn
2	0.2×10⁻⁴	1.9×10⁻⁴	1.1×10⁻⁴	0.85×10⁻⁵	0.083×10⁻⁵	0.02×10⁻⁶
10	22×10⁻⁴	9.4×10⁻⁴	5.6×10⁻⁴	4.3×10⁻⁵	4.2×10⁻⁵	2.6×10⁻⁶
20	158×10⁻⁴	19×10⁻⁴	11×10⁻⁴	8.5×10⁻⁵	8.3×10⁻⁵	20.5×10⁻⁶
计算方法	磁力公式	斯托克斯公式	斯托克斯公式	异质絮凝理论	DLVO 理论	重力公式

注：1dyn = 10^{-5}N。

2　工业试验

姑山铁矿是马钢主要原料基地之一，年采选能力约100万吨原矿。铁矿石属宁芜式赤铁矿，其构造呈块状、网状、浸染状及角砾状，铁矿物嵌布粒度极不均匀，属难磨难选红矿。现行选矿流程为：原矿经粗碎、中碎、洗矿和细碎后，用跳汰和感应辊式强磁选机预选得出部分精矿并抛弃部分尾矿，产率约占原矿一半的中矿进入主厂房。中矿经一段磨矿至小于 74μm（200 目）50%左右，用 ϕ500 和 ϕ300 旋流器分级两次，沉砂用螺旋溜槽分选，ϕ350 旋流器溢流用离心机选别，但离心机分选指标差，精矿品位和回收率都很低。

本试验设备处理的物料由离心机给料中分出，其细度为小于 74μm（200 目）90%左右，浓度 20%左右，铁品位约 33%，但粒度、浓度、品位波动均较大。由于粒度细，次生泥多，品位低，且其预选作业的跳汰、感应辊强磁机已选出部分易选铁矿物。因此，这是姑山选厂最难选的一部分物料。表 2 列出了曾采用不同方法对该物料选别的生产和试验指标。由表 2 可知，当给矿品位为 30%左右时，浮选和离心机分选的回收率均低于 50%，精矿品位为 50%左右，强磁选一次分选也不能得到好指标。

表 2　姑山铁矿 ϕ350 旋流器溢流历史选别指标

选别方法	日期	规模	铁品位/%			回收率/%
			给矿	精矿	尾矿	
浮选	1982 年 6~9 月	生产	32.14	49.23	24.08	49.09
离心机	1983 年 9~12 月	生产	31.37	49.70	24.26	44.28
离心机	1987 年 1~6 月	生产	34.12	52.36	26.79	43.99
Shp 湿式强磁	1983 年 10 月	小型试验	30.36	53.90	19.52	55.98
平环高梯度[7]	1983	工业试验	25.74	53.20	16.44	52.30

本试验所用脉动高梯度磁选固定参数为：有效脉动冲程 10~15mm；给矿平均流速 6cm/s；导磁不锈钢板网规格 0.3mm×3.2mm×8mm；磁介质堆高度 7cm；磁介质充填率 6%；给矿量 3~5t/h；给矿粒度上限 0.8mm。

2.1　脉动冲次与选矿指标的关系

铁矿物与脉石的分离是一个非常短暂的过程，例如当给矿流速为 6cm/s，磁介质堆高度为 7cm 时，矿浆在介质堆内停留时间仅 1.2s 左右。若脉动次数为 300 次/min，则 1.2s 内仅作用 6 次，这一时间特点决定了脉动高梯度磁选宜采用较小的冲程和较大的冲次。冲次试验结果见表 3。

表 3　脉动冲次试验指标

冲次/次·min^{-1}	给矿品位/%	精矿品位/%	精矿产率/%	尾矿品位/%	回收率/%
0	34.45	48.80	50.23	19.97	71.15
200	33.32	53.42	45.23	16.72	72.52
300	34.45	56.78	44.60	16.47	73.52
350	33.95	58.41	42.37	15.97	72.89
400	33.70	59.53	39.31	16.97	69.44

试验时固定背景场强为 0.92T（9200Gs）。由表 3 可知，当冲次为零（即无脉动）时，精矿品位仅 48.80%，随着冲次增大，精矿品位大幅度上升，冲次为 400 次/min 时，精矿品位高达 59.53%。此外，脉动使磁介质表面的有效捕收区得到充分利用。因此，在适当的冲次

范围内，尾矿品位随冲次的增大而略有下降。试验表明较合适的冲次为300~400次/min。

2.2 背景场强与选矿指标的关系

矿浆脉动加大了作用在矿粒上的流体竞争力，假如忽略重力和表面力不计，磁性矿粒被捕收的必要条件是：

$$F_m > R + \widetilde{R} \tag{5}$$

式中，F_m 为磁性矿粒所受磁力；R、$\widetilde{R}$ 分别为给矿平均流速和脉动流速对磁性矿粒的作用力。

因此，脉动高梯度磁选所需场强宜比无脉动时高一些，以便得到尽可能高的回收率。表4列出了背景场强的试验结果。从表4可知，在0.53~1.1T，精矿品位是随场强的增高而略为下降，但变化甚微，4个试验点的精矿品位均为56%左右；尾矿品位随场强的增高而较明显下降，回收率上升。当场强超过0.92T时，尾矿品位和回收率变化甚微。因此，适合的场强为0.76~0.92T。

表4 背景场强试验指标

场强/T	给矿品位/%	精矿品位/%	精矿产率/%	尾矿品位/%	回收率/%
0.53	28.70	56.66	28.66	17.47	56.57
0.76	28.95	56.28	33.42	15.23	64.98
0.92	30.20	56.16	37.34	14.73	69.44
1.10	28.45	55.79	34.61	13.98	67.87

注：脉动冲次为350次/min；1T=10000Gs。

2.3 给矿粒度与选矿指标的关系

表5为给矿粒度与选矿指标的统计值。由该表可知，脉动高梯度磁选对姑山矿小于74μm（200目）占60%~98%的各个粒级，当给矿品位为（36±2）%时，一次选别均能得到品位高于57%、回收率68%左右的精矿。这说明脉动高梯度磁选机适应的粒度范围较宽，精矿品位在上述粒度范围内基本上不受粒度变化的影响。

表5 给矿粒度试验指标

小于74μm（200目）/%	取样批数	给矿品位/%	精矿品位/%	精矿产率/%	尾矿品位/%	回收率/%
60~70	2	34.57	58.88	37.68	19.87	64.18
70~80	8	37.92	57.60	44.87	21.90	68.16
80~90	36	36.49	57.54	43.91	20.01	69.24
90~98	41	35.18	57.09	42.24	19.16	68.54

2.4 给矿浓度对选矿指标的影响

表6是保持给矿体积12.5m³时给矿浓度与选矿指标的统计值。由该表可知，给矿浓度波动在10%~42%时，精矿品位均保持在57%以上。而给矿浓度以25%~30%最佳，精矿品位高达58.54%。随着浓度增大，尾矿品位略有升高，回收率略有下降。这是因为给

矿体积不变时，浓度增大等于增加给矿量，磁介质负荷量增大，捕收能力有所下降。

表6 给矿浓度试验指标

给矿浓度/%	取样批数	给矿品位/%	精矿品位/%	精矿产率/%	尾矿品位/%	回收率/%
10~20	23	35.07	56.76	42.24	19.21	68.36
20~25	39	35.75	57.38	43.00	19.43	69.02
25~30	16	37.22	58.54	43.07	21.09	67.74
30~42	7	36.75	57.04	41.41	22.41	64.27

2.5 给矿品位对选矿指标的影响

表7为给矿品位与选矿指标的统计值。由表7可知，给矿品位从23%~27%（平均24.53%）增至38%~40%（平均38.91%）时，精矿品位由56.07%增加到58.19%，尾矿品位由15.16%增至21.22%，回收率由52.35%增至71.56%。即给矿品位每增加1%，精矿品位增加约0.15%，回收率增加约1.3%，尾矿品位增加约0.42%。

表7 给矿品位试验指标

给矿品位/%	取样批数	精矿品位/%	尾矿品位/%	回收率/%
23~27	11	56.07	15.16	52.35
27~30	12	56.45	16.53	60.78
30~35	16	56.81	18.32	68.46
35~38	35	57.40	20.72	67.90
38~40	21	58.19	21.22	71.56

3 其他试验

3.1 ϕ350mm 旋流器溢流一粗一精试验

试验目的是为姑山铁矿实现精矿品位60%的远期攻关目标提供参考依据。一粗一精开路流程脉动高梯度磁选试验结果见表8。由表8可知，当给矿品位为38.28%时，精选作业的精矿品位61.90%，回收率为62.49%。试验结果为姑山矿难选细粒氧化铁矿实现精矿品位60%提供了一条易行的途径。

表8 一粗一精试验结果   （%）

产品名称	产率	品位	回收率
给矿	100.00	38.28	100.00
粗选精矿	49.73	57.61	74.84
精选精矿	38.65	61.90	62.49
中矿	11.08	42.65	12.35
粗选尾矿	50.27	19.16	25.16

注：背景场强：粗选1.1T，精选0.76T。

3.2 洗矿尾泥的回收试验

姑山矿洗矿尾泥产率约占采出矿的5%，铁品位约28%，其金属量占总金属量的4%左右，粒度为小于74μm（200目）50%左右。该部分洗矿泥长期作尾矿排放。用立环脉动高梯度磁选机所做一粗一精开路流程的试验结果列于表9。由表9可见，品位为28.42%的洗矿泥，经一粗一精选后可得品位58.58%、作业回收率为56.44%的精矿。若姑山矿应用这一试验成果，其总回收率可提高2%左右，每年可增加产值约70万元。

表9 洗矿泥一粗一精选矿试验指标   （%）

产品名称	产率	品位	回收率
给矿	100.00	28.42	100.00
粗选精矿	35.49	55.23	68.97
精选精矿	27.38	58.58	56.44
中矿	8.11	43.92	12.53
粗选尾矿	64.51	13.67	31.03

注：粗选和精选的背景场强均为0.76T。

4 小结

通过1500h的工业试验表明，脉动高梯度磁选能显著地提高铁精矿品位，保持较高的回收率；有助于防止堵塞和扩大分选粒度范围。试验结果为姑山铁矿今后的流程改造提供了切实可行的途径。

参 考 文 献

[1] 刘树贻，等. 一种高效率的磁选方法——振动加脉动高梯度磁选法 [J]. 中南矿冶学院学报，1983 (s1)：88~105.

[2] 刘树贻，等. 珊瑚锡矿细粒钨锡混合物料的振动高梯度磁选研究 [J]. 中南矿冶学院学报，1983 (s1)：106~111.

[3] 熊大和，等. 有色金属：选矿部分，1986 (4)：13~17.

[4] 孙仲元. 有色金属：选矿部分，1987 (4)：47~53.

[5] Cowen C, et al. High gradient magnetic field particle capture on a single wire [J]. IEEE Trans. On Mag., 1975 (5)：1600~1602.

[6] Luborsky F E. Buildup of particles in a high field-high gradient separator [J]. IEEE Trans. On Mag., 1976 (5)：463~465.

[7] 李明德，等. 金属矿山，1985 (4)：29~33.

写作背景 作者和作者的指导老师刘树贻等人在实验室做了很多振动和脉动高梯度磁选试验，效果虽好，但工业上应用不理想。作者很想将这些新的方法应用到工业上去。1985~1988年作者在中南工业大学攻读博士研究生，得到了赣州有色冶金研究所资助和鼎力帮助，研制出了第一台SLon-1000立环脉动高梯度磁选机，并得到了马钢姑山铁矿的帮助，在该矿进行了分选细粒氧化铁矿的工业试验，获得了良好的选矿指标。由于当时论文投稿时项目尚未鉴定，因此发表时没有公布设备的结构。

SLon-1000 立环脉动高梯度磁选机的研制[1]

熊大和

（中南工业大学）

摘　要　SLon-1000 立环脉动高梯度磁选机利用磁力、脉动流体力和重力等多种力的综合力场进行选矿，其转环立式旋转，反冲精矿，配有脉动机构，具有富集比高、不易堵塞、适应性强，工作稳定等优点。本文着重介绍该机的结构及工作原理。

关键词　脉动高梯度　转环　磁系　激磁线圈

近年国内陆续报道过一些有关振动、脉动高梯度磁选的试验内容，大量的小型试验结果表明，振动或脉动高梯度磁选机可大幅度地提高磁性精矿品位，并保持高梯度磁选对细粒磁性矿物回收率较高的优点。此外，振动或脉动对防止磁介质堵塞也是相当有效的。但是，以往的试验都是在实验室的周期式高梯度磁选机上进行的，这种高效的选矿方法能否在工业生产中获得应用，关键在于研制出适用工业生产的连续振动或脉动高梯度磁选机。

SLon-1000 立环脉动高梯度磁选机是由赣州有色冶金研究所与中南工业大学联合研制的第一台工业型脉动高梯度磁选机。该机于今年3月在马钢姑山选厂完成3000余小时的工业试验，至于6月上旬通过冶金部和中国有色金属总公司联合组织的设备鉴定。该机的诞生，标志着脉动高梯度磁选已达到了工业实用阶段。

1　整机结构与选矿原理

SLon-1000 立环脉动高梯度磁选机结构主要由如图1所示13个部分组成。立环内装有导磁不锈钢板网磁介质（也可以根据需要充填钢毛等磁介质）。选别时，转环作顺时针旋转，矿浆从给矿斗给入，沿上铁轭缝隙流经转环，转环内的磁介质在磁场中被磁化，磁介质表面形成高梯度磁场，矿浆中磁性颗粒吸着在磁介质表面，由转环带至顶部无磁场区，被冲洗水冲入精矿斗；非磁性颗粒沿下铁轭缝隙流入尾矿斗排走。

为了保证脉动选矿，维持矿浆液面的高度是至关重要的。通过调节尾矿斗下部阀门，可使液面保持在液位线以上，液位显示管为透明有机塑料管，操作者随时可观察液位高度及脉动情况。脉动机构驱动装置安装在尾矿斗上的橡胶鼓膜往复运动，只要矿浆液位保持在液位线以上，脉动能量就能有效地传到选矿区。该机采用调速电动机驱动脉动冲程箱，脉动冲次由调速电机的控制器调节，脉动冲程的调节是通过调节冲程箱内的偏心块来实现的。

图1中左图绘出了上下磁轭的分布情况，下磁轭有11道缝与尾矿斗分别通过上磁轭的8条缝和2条缝与分选区沟通，磁性矿和非磁性矿在分选区得到分离。漂洗水的作用是进一步清除未排干净的非磁性颗粒，以提高磁性精矿品位。下磁轭与排水斗沟通的3条缝是供排水用的，其上方称为排干区，在此区间转环内的磁性矿物继续受磁力的作用黏着在

---

[1] 原载于《金属矿山》，1988（10）：37~40。

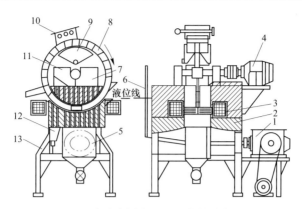

图 1 脉动高梯度磁选机结构示意图

1—脉动机构；2—铁轭；3—激磁线圈；4—转环驱动机构；5—尾矿斗；6—液位显示管；7—给矿斗；8—转环；9—精矿斗；10—精矿冲洗装置；11—漂洗水斗；12—排水斗；13—机架

磁介质上，而水及其夹带的非磁性颗粒流经排水斗排走。上磁轭位于排干区上方有 2 条缝与大气相通，空气及时填补了转环内因水流走而留下的空间，以便转环内的水在重力的作用下迅速排走。转环转出磁系的部分虽然不再受磁场力的作用，但转环内基本上不含流动水，磁性矿依靠表面力附着在磁介质上，被带到转环上方冲洗出来。

分选区和排干区之间没有缝隙的部位称隔断区。无论旋转至哪个部位，转环上至少有一块隔板位于隔断区，这将保证分选区的矿浆不会大量地朝排干区流动和脉动能量的传播集中在分选区。

2 磁系

磁系的结构不仅关系到场强、漏磁、功耗等技术参数，而且关系到选别指标优劣和矿浆通路是否流畅等问题。因此，磁系的最佳化是制造优质磁选机的关键之一。

该机磁系立体图和磁路图分别见图 2 和图 3。它是由一个水平放置的用空心电工矩形铜管绕制的激磁线圈、1 块下铁轭、2 块上铁轭和 2 块月牙板而构成。上下铁轭之间的弧形空间为选别区。磁系包角为 90°，当激磁线圈有直流电通过时，在选别区产生背景磁场，磁力线从下铁轭极头指向上铁轭极头，然后沿铁轭形成闭合磁路。两个上磁极头相邻面相距一段距离（图中已放大），供转环辐板通过。上下铁轭极头上的缝隙结构是矿浆和水的通路。这一磁系的优点是漏磁小，激磁功耗低，平面线圈较易绕制，铜材利用率高。磁系的设计技术参数见表 1。电磁性能实测值见表 2。比较表 1 和表 2。

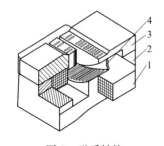

图 2 磁系结构

1—激磁线圈；2—下铁轭；3—上铁轭；4—月牙板

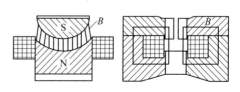

图 3 磁系磁路

表 1 磁系技术参数

额定背景场强/T	额定激磁电流/A	额定激磁电压/V	额定激磁功率/kW	冷却水量/m³·h⁻¹	冷却水压/MPa·cm⁻²	磁场空间尺寸（长×宽×高）/mm×mm×mm	磁系包角/(°)	线圈匝数/匝	线圈工作温度/℃	线圈串联电阻/Ω	铜管规格/mm
1.0	1200	21.2	25.5	1.1	0.2~0.5	706×340×104	90	88	<50	0.0177	22×18×15

表 2 电磁性能实测值

激磁电流/A	激磁电压/V	激磁功率/kW	背景场强/T
300	4.5	1.35	0.30
390	6.0	2.34	0.38
510	8.0	4.08	0.50
720	11.5	8.28	0.70
930	15.0	13.95	0.90
1100	18.0	19.8	1.00
1200	20.0	24.0	1.08
1480	24.5	36.3	1.20

可知，当背景场强为1.0T时，实测电流和激磁功率分别为1100A和19.8kW，低于设计值1200A和25.5kW，这说明磁系设计是成功的。

3 转环

3.1 结构

图4为转环结构图。转环两侧各为一块普通不锈钢环板，环板之间焊有24块梯形普通不锈钢隔板，各隔板中部与一环形普通不锈钢加强圈焊接，用一块纯铁辐板与加强圈和轮毂焊接。两侧环板和梯形隔板围成24个矩形分选室，各分选室靠轴线一侧焊有一些不锈钢内垫条，以阻止磁介质往轴线方向转动；各分选室外侧装有2~4根不锈钢外垫条（现场用2根φ12mm普通不锈钢外垫条），外垫条可拆，以便安装磁介质。

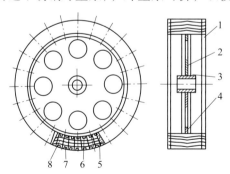

图 4 转环结构示意图

1—环板；2—辐板；3—轮毂；4—加强圈；5—隔板；6—内垫条；7—外垫条；8—磁介质

3.2 磁介质及其固定方式

现用的磁介质为 0.3mm 导磁不锈钢冲制的菱形网，网孔对角尺寸为 3.2mm×8mm，网与网之间用 0.4mm×10mm×25mm 大孔普通不锈钢菱形网隔开，每个分选室的上、下端各放一块 6 目普通不锈钢丝网。安装网堆时用手尽量压紧，以防松动，网堆的安装方式如图 5 所示。

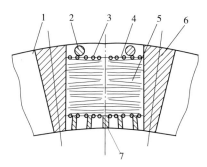

图 5　磁介质堆固定方式

1—环板；2—外垫条；3—6 目不锈钢丝网；4—磁介质；5—不锈钢大孔网；6—隔板；7—内垫条

3.3 转环传动

转环用调速电机驱动，摆线针轮减速器减速，其二级减比速比为 121，电动机皮带轮与减速器输入轴皮带轮的速比约为 2，总减速比为 240，电动机输出转速为 120~1200r/min。因此，转环的调速范围为 0.5~5r/min，可满足多种矿石的选矿需要。

4 脉动机构

在设计中采用脉动机构，对提高磁性精矿品位和选矿效率、防止堵塞都起着重要的作用。有关脉动高梯度磁选过程的实现在前已详述。该机的主要技术参数见表 3。

表 3　SLon-1000 高梯度磁选机技术参数

转环外径，内径 /mm	转环宽度 /mm	转环转速 /r·min^{-1}	磁介质堆尺寸（长×宽×高）/cm×cm×cm	给矿粒度 /mm	最佳分选粒度 /mm	矿浆通过能力 /m^3·h^{-1}	干矿处理量 /t·h^{-1}	额定背景场强 /T	最高背景场强 /T	额定激磁电流 /A
1000,800	300	0.5~5	26×10×7	-0.8~0	0.1~0.01	12.5~20	4~7	1.0	1.2	1200

额定激磁电压 /V	额定激磁功率 /kW	有效脉动冲程 /mm	脉动冲次 /min^{-1}	转环电机功率 /kW	脉动电机功率 /kW	供水压力 /MPa	耗水量 /m^3·h^{-1}	主机质量 /t	最大部件质量 /t	外形尺寸（长×宽×高）/mm×mm×mm
21.2	25.5	0~20	0~400	1.5	1.1	0.2~0.5	10~20	5.34	2.2	2000×1500×2500

脉动机构由 JZT-24-4 型电磁调速电动机驱动。脉动冲次的可调范围为 0~400r/min 冲

程箱输出冲程的可调范围为 0~30mm。选别区的有效冲程和冲程箱的输出冲程换算关系如下：

$$S_e = 0.66S$$

式中，S_e 为有效冲程；S 为冲程箱输出冲程。

例如，在姑山铁矿的工业试验中，该机使用的冲程箱输出冲程多为 20mm，其有效冲程为 $S_e = 0.66 \times 20 = 13.2$mm。

5 工业试验

马钢姑山铁矿属细微嵌布石英质赤铁矿石，铁矿物需磨至小于 0.029mm 才能单体解离，具有难磨难选的特点。近年来球磨选矿工段的生产指标为：给矿品位 35%~40%，精矿品位 55%，尾矿品位 24%。其中细粒级（小于 0.074mm（200 目）占 90% 左右）用离心机分选，1987 年 1~6 月的生产指标为：给矿品位 34.12%，精矿品位 52.36%，尾矿品位 26.72%，作业收率 43.99%。可见，离心机的选矿指标远低于生产平均指标。经查定细粒级的选矿问题是影响姑山铁矿生产的主要因素之一。

脉动高梯度磁选机分选姑山铁矿 ϕ350mm 旋流器溢流（即离心机给料）3000 余小时的生产考核指标见表 4。其平均生产指标为：给矿品位 33.26%，精矿品位 56.78%，尾矿品位 19.31%，作业收率 63.56%。与采用离心机工艺（1987 年 1~6 月的生产指标）比较，精矿品位提高 4% 以上，尾矿品位降低 6%，作业收率提高 20% 左右；与离心机工艺历史上最好月平均指标比较，精矿品位提高 4%，回收率提高 10% 左右。

表 4 脉动高梯度磁选机分选 ϕ350mm 旋流器溢流生产考核指标

年-月	运转时间/h	处理原矿量/t	生产精矿量/t	给矿/%			精矿/%			尾矿品位/%	磁介质充填率/%
				小于0.074mm(200目)	浓度	品位	产率	品位	回收率		
1987-10~11	489	1402	516	94.23	19.98	32.88	35.71	55.92	60.74	20.08	11
1987-12	613	2023	901	87.19	22.42	36.60	43.56	56.87	67.68	20.96	6
1988-01	637	2085	812	84.60	22.26	32.52	37.98	56.32	65.78	17.95	6, 3.5
1988-02	623	2254	763	83.77	24.21	33.02	32.95	57.45	57.33	21.01	3.5
1988-03	690	2613	871	87.91	25.13	30.87	32.55	57.48	60.61	18.02	3.5
合计	3052	10377	3863	87.39	22.85	33.26	37.23	56.78	63.56	19.31	

6 结语

SLon-1000 立环脉动高梯度磁选机在马钢姑山铁矿的工业考核表明，该设备的设计和制造是成功的，其主要设备性能参数和选别性能均优于设计指标，是一种分选细粒弱磁性矿物的新型高效设备。它具有富集比大，不易堵塞，分选粒度范围宽，操作和维护简单，能够在现场生产流程中长期稳定地运转等优点，可广泛用于赤铁矿、锰矿、黑钨矿等金属矿和非金属矿的细粒级选别和尾矿中有用金属的回收。该机将朝着大型化发展，以满足大规模生产的需要。

写作背景 1988年3月，首台SLon-1000立环脉动高梯度磁选机在马钢姑山铁矿完成了3000余小时的工业试验，并于同年6月通过了冶金工业部和中国有色金属工业总公司联合组织的设备鉴定，该机的诞生，标志着脉动高梯度磁选初步达到工业实用阶段。本论文公布了这个研究成果。这里要说明的一点是，为什么第一台SLon-1000磁选机是脉动而不是振动？这是因为脉动在工业上更有把握实现。

脉动高梯度磁选中冲程冲次对选矿指标的影响

熊大和　刘树贻　刘永之　陈　苂

（中南工业大学）

摘　要　本文介绍了周期式脉动高梯度磁选机的工作原理，从理论上分析了冲程冲次对脉动高梯度磁选指标的影响，用黑钨矿和赤铁矿进行了验证试验，证实了脉动对消除脉石的机械夹杂和提高磁性精矿品位有重要作用。通过理论分析和试验研究，提出了脉动高梯度磁选的捕集图。

关键词　分选　黑钨矿　赤铁矿　脉动高梯度磁选　冲程

The Effects of Stroke and Frequency on Pulsating High Gradient Magnetic Separation

Xiong Dahe　Liu Shuyi　Liu Yongzhi　Chen Jin

(*Central South University of Technology*)

Abstract　In this paper, a cyclic pulsating high gradient magnetic separator of laboratory size and its principle are described. The effects of pulsating stroke and frequency on separation response are discussed. Binary Mineral system of wolframite quartz and natural hematite ore are used in this study. It is demonstrated that pulsation can effectively liberate the non-magnetic particles trapped in the matrix and consequently greatly enhance the beneficiation ratio of magnetic concentrate. The process of particle built up in pulsating high gradient magnetic separation is proposed based on the theoretical and experimental studies.

Keywords　Separating; Wolframite; Hematite; Pulsating high gradient magnetic separation; Stroke

我国拥有丰富的红铁矿、锰矿、黑钨矿等弱磁性矿石资源，但多数矿床的原矿品位低、嵌布粒度细，需要经过细磨深选才能满足冶炼要求。现有的选矿设备和工艺尚不能满足选矿工业的要求，细粒选矿的回收率和效率低是一个普遍存在的问题，尾矿中细粒级金属的损失严重，因此有必要研究新型选矿设备和新工艺以提高我国细粒选矿水平。普通高梯度磁选具有对细粒弱磁性矿物回收率高的优点，缺点是磁性精矿品位较低。脉动高梯度磁选可大幅度地提高磁性精矿的品位，并保持了普通高梯度磁选对细粒弱磁性矿石回收率高的优点。这是一种高效的选矿方法，可望在工业上获得大规模应用。至今国内外对脉动高梯度磁选的原理研究甚少，本研究旨在这一方面取得一些进展。

---

❶　原载于《中南矿冶学院学报》1989（05）：497～504。

1 试验设备及其选矿原理

本研究所用的试验设备为图 1 所示的 dia80mm 周期式脉动高梯度磁选机。该机与普通周期式高梯度磁选机的不同之处是它的下部装有一脉动机构,由偏心连杆机构产生的交变力 $\widetilde{F}$ 推动橡胶鼓膜往复运动,使分选腔内的矿浆产生脉动。调节下部阀门可控制矿浆的流速和液位高度,改变流经激磁线圈的直流电强度可调节分选区的背景场强。本试验用的磁介质是规格为 0.3mm×3.2mm×8mm 的导磁不锈钢菱形网,其体积充填率约为 5%。为了简便起见,本文后面的理论分析假定了磁介质为圆柱形铁磁体。

选矿时,先在分选腔内灌满水,然后从给矿盒给入矿浆,分选腔内的磁性矿物和非磁性矿物在磁力、脉动流体力、重力的综合力场作用下得到分离,磁性矿粒被吸着在磁介质表面上。非磁性矿粒随矿浆从下部排走。给矿完毕后,放干水分,切断激磁电流,然后用干净水将磁性物冲洗出来。

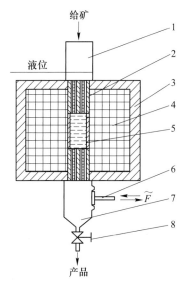

图 1 周期式脉动高梯度磁选机
1—给矿盒;2—磁极头;3—磁轭;4—激磁线圈;
5—磁介质;6—脉动机构;7—接矿斗;8—阀门

2 脉动高梯度磁选的理论分析

2.1 分选区矿浆的流速和加速度

当脉动机构工作时,脉动能量从下磁极头的通孔传入到分选区,驱使分选区矿浆产生脉动。矿浆的瞬时脉动速度、最大脉动速度和平均脉动速度分别为

$$\widetilde{v} = \frac{1}{2} S \cdot \omega \cdot \sin(\omega t) \tag{1}$$

$$\widetilde{v}_{max} = \frac{1}{2} S \cdot \omega \tag{2}$$

$$\bar{v} = \int_0^{2\pi} \left| \frac{1}{2} S \cdot \omega \cdot \sin(\omega t) \right| d(\omega t) = \frac{S \cdot \omega}{\pi} \tag{3}$$

式中,S 为选矿区脉动冲程;ω 为脉动波角速度;t 为时间变量。

矿浆的实际流速为给矿速度 v_0 和脉动速度的叠加:

$$v = v_0 + \widetilde{v} = v_0 + \frac{1}{2} S \cdot \omega \cdot \sin(\omega t) = v_0 + \widetilde{v}_{max} \cdot \sin(\omega t) \tag{4}$$

矿浆的脉动瞬时加速度、最大脉动加速度和平均脉动加速度分别为:

$$\widetilde{\alpha} = \frac{1}{2} S \cdot \omega^2 \cdot \cos(\omega t) \tag{5}$$

$$\widetilde{\alpha}_{max} = \frac{1}{2} S \cdot \omega^2 \tag{6}$$

$$\bar{\alpha} = \frac{1}{2\pi}\int_0^{2\pi} \left| \frac{1}{2} S \cdot \omega^2 \cdot \cos(\omega t) \right| \mathrm{d}(\omega t) = \frac{S \cdot \omega^2}{\pi} \tag{7}$$

2.2 脉动对矿粒行程的影响

图 2（a）和（b）分别为无脉动和有脉动时矿粒穿过磁介质堆的运动轨迹。设磁介质整齐地排 n_0 层，总高度为 L_0，层与层间距为 ΔL；假定矿粒随矿浆从 C_1 点进入磁介质堆，然后从 C_2 点出来，为了直观起见，图 2（b）中有意将矿粒的出来点错开至 C_2' 点，尽管实际出来点可能是 C_2 点。显然无脉动时矿粒所穿过磁介质层的次数（以下简称穿过次数）就等于磁介质的实际层数 n_0；当有脉动时，由式（4）可知，只要满足 $\tilde{v}_{\max} > v_0$，矿浆便不断地改变流向，时而从上至下，

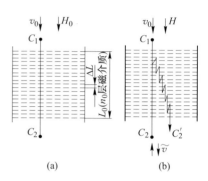

图 2 无脉动和有脉动时的矿粒运动轨迹
（a）无脉动；（b）有脉动

时而从下至上，因此随矿浆运动的矿粒的轨迹是一条上下波动的曲线，见图 2（b）。由于惯性力的作用，矿粒的运动速度可能会滞后于流体速度，为了简便起见，这里暂不考虑这种滞后性。脉动使矿粒可能不止一次地穿过每一层磁介质，即矿粒的穿过次数可能比无脉动时高。矿粒的穿过次数与它在垂直方向上的行程成正比，矿粒在一个脉动周期 T 内的行程为：

$$L_T = \int_0^T | v_0 + \tilde{v}_{\max} \cdot \sin(\omega t) | \mathrm{d}t$$

$$= \begin{cases} v_0 T & \tilde{v}_{\max} \leq v_0 \\ \dfrac{2}{\pi} v_0 T \left[\arcsin \dfrac{v_0}{\tilde{v}_{\max}} + \sqrt{\left(\dfrac{\tilde{v}_{\max}}{v_0}\right)^2 - 1} \right] & \tilde{v}_{\max} > v_0 \end{cases} \tag{8}$$

当 v_0 为常数时，无论有无脉动，矿粒经过磁介质堆的时间都是 $\Delta t = L_0/v_0$，当脉动冲次为 N 次/min 时，在 Δt 时间内的脉动次数为：

$$\Delta N = \frac{N}{60}\Delta t = \frac{\Delta t}{T} = \frac{L_0}{T v_0} \tag{9}$$

由式（8）和式（9）得到矿粒穿过磁介质堆的垂直行程为：

$$L = \Delta N \cdot L_T$$

$$= \begin{cases} L_0 & \tilde{v}_{\max} \leq v_0 \\ \dfrac{2}{\pi} L_0 \left[\arcsin \dfrac{v_0}{\tilde{v}_{\max}} + \sqrt{\left(\dfrac{\tilde{v}_{\max}}{v_0}\right)^2 - 1} \right] & \tilde{v}_{\max} > v_0 \end{cases} \tag{10}$$

式（10）两边除以 ΔL，并代入 $v_{\max} = \dfrac{1}{2} S \cdot \omega = \dfrac{\pi}{60} S \cdot N$，得矿粒的穿过次数为：

$$n = \begin{cases} n_0 & \tilde{v}_{\max} \leqslant v_0 \\ \dfrac{2}{\pi} n_0 \left[\arcsin\left(\dfrac{60 v_0}{\pi S \cdot N}\right) + \sqrt{\left(\dfrac{\pi S \cdot N}{60 v_0}\right)^2 - 1} \right] & \tilde{v}_{\max} > v_0 \end{cases} \quad (11)$$

利用上式可以证明，只有 $\tilde{v}_{\max} > v_0$，就是 $n > n_0$。

2.3 穿过次数对回收率的影响

设磁性矿粒通过单层磁介质被捕获的概率为 A，则它 n 次穿过磁介质被捕获的概率为：

$$P = 1 - (1 - A)^n \quad (12)$$

当 n 较大时，$(1 - A)^n \approx \exp(-A \cdot n)$，代入上式得：

$$P = 1 - \exp(-A \cdot n) \quad (13)$$

设磁性矿粒的理论最高回收率为 $\varepsilon_{\max}$，则实际回收率为：

$$\varepsilon = \varepsilon_{\max} P = \varepsilon_{\max} \left[1 - \exp(-A \cdot n) \right] \quad (14)$$

2.4 脉动对单层捕获概率 A 的影响

弱磁性矿粒通过单层磁介质被捕获的概率近似地与磁力成正比，与竞争力成反比，即

$$A = K \dfrac{F_m}{F_c} \quad (15)$$

式中，K 为比例常数。当磁力 F_m 不变时，A 仅与竞争力 F_c 有关。F_c 为流体力、重力和惯性力之和，假定各力的作用方向一致，则有：

$$F_c = R + \bar{R} + G + F_N \quad (16)$$

式中，R 为给矿速度 v_0 对矿粒产生的黏滞力，$R = 6\pi\mu b v_0$；$\bar{R}$ 为脉动对矿粒产生的平均黏滞力，结合式（3）：

$$\bar{R} = 6\pi\mu b \bar{v} = 6\mu b S \omega$$

式中，μ 为黏滞系数；b 为颗粒半径；G 为作用在矿粒上的重力（忽略浮力），$G = mg$；F_N 为脉动加速度使矿粒产生的平均惯性力，结合式（7）：

$$F_N = m\bar{\alpha} = \dfrac{m}{\pi} S \cdot \omega^2$$

将以上各式代入式（16）并注意 $\omega = 2\pi N/60$ 得：

$$F_c = (6\pi\mu b v_0 + mg) + \dfrac{\pi}{5}\mu b SN + \dfrac{\pi}{900} m SN^2 \quad (17)$$

上式还可写成：

$$F_c = K_1 + K_2 SN + K_3 SN^2 \quad (18)$$

其中，K_1，K_2，K_3 分别对应于式（17）中的常数项。从上式可知 F_c 是冲程 S 的一次函数，是冲次 N 的二次函数，可见当 N 较大时，F_c 随 N 增加的速率是很大的。

将式（15）代入式（14），得回收率公式为

$$\varepsilon = \varepsilon_{\max} \left[1 - \exp\left(-K F_m \dfrac{n}{F_c}\right) \right] \quad (19)$$

式中，n 和 F_c 分别由式（11）和式（18）计算。

2.5 冲程冲次对磁性精矿品位的影响

冲程冲次对磁性精矿品位的影响可用 Oberteuffer[1] 导出的磁性物纯度公式解释，即

$$G_m = \frac{1}{1 + A_{nm} K' \dfrac{F_i}{F_c}} \tag{20}$$

式中，G_m 为磁性精矿的纯度；A_{nm} 为给矿中非磁性物料与磁性物料的重量比；K' 为比例常数；F_i 为磁性矿粒与非磁性矿粒的相互作用力；F_c 为促使非磁性矿粒离开捕集区的竞争力，它也可用式（17）表示。

上式还可写成：

$$\beta = \beta_{\max} G_m = \frac{\beta_{\max}}{1 + A_{nm} K' \dfrac{F_i}{F_c}} \tag{21}$$

其中，β，$\beta_{\max}$ 分别为磁性精矿品位和理论最高品位。

由式（21）易知，β 随 F_c 的增大而单调上升，当 $F_c \to \infty$ 时，$\beta \to \beta_{\max}$，即 β 是以 $\beta_{\max}$ 为渐近线的一条随 F_c 增大而单调上升的曲线。因 F_c 随脉动冲程冲次的增大而增大，因此 β 一般是随冲程冲次的增大而单调上升的。

3 试验结果与讨论

图 3 为细粒黑钨矿与石英混合料脉动高梯度磁选试验结果，其中给矿含 WO_3 约 12%，黑钨矿比磁化系数为 $x = 33 \times 10^{-6} cm^3/g$，$B_0 = \mu_0 H_0 = 1.0T$，$v_0 = 4cm/s$，$L_0 = 10cm$，$S = 1.0cm$。图 4 为安徽省姑山铁矿细粒赤铁矿试验结果。其中给矿中含 Fe 约 33%，给矿粒度为小于 $74\mu m$（200 目）占 90%，赤铁矿的比磁化系数 $\chi = (40 \sim 60) \times 10^{-6} cm^3/g$，$B_0 = 0.9T$，$v_0 = 6cm/s$，$L_0 = 7cm$，$S = 1.3cm$。从图 3（a）和图 4（a）可看出，当冲次不大时，在一定范围内磁性物料的回收率随冲次的增加而略有上升，这一阶段矿粒的穿过次数 n 值随冲次增加较快，即矿浆的脉动使矿粒有更多的机会与磁介质碰撞，尽管竞争力 F_c 也在增加；但它增加的速度稍慢；当冲次 N 增加到一定程度后，由式（18）可知，F_c 随 N 的增

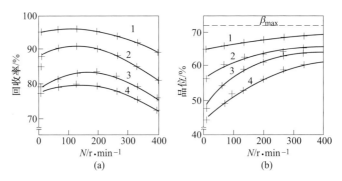

图 3 细粒黑钨矿与石英混合料试验结果
（a）冲次对回收率的影响；（b）冲次对品位的影响
1—40~20μm；2—20~10μm；3—10~5μm；4—5~0μm

大而呈抛物线的关系增加，此外，由式（11）可以证明，$S \cdot N$ 较大时，n 随 $S \cdot N$ 呈线性关系增加，即 n 的增加速度落后于 F_c 的增加速度，由式（19）易知，回收率随 n/F_c 的值减小而下降。

由图 3（b）和图 4（b）可看出，磁性精矿的品位随冲次的增加而单调上升，这主要是作用在非磁性矿粒上的竞争力增大了，减少了非磁性矿粒的机械夹杂和因表面力作用而黏附磁性矿表面的概率，试验结果与式（21）所作出的理论定性分析相符合。

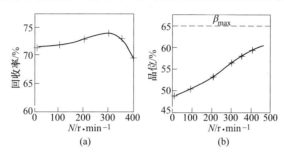

图 4　姑山铁矿细粒赤铁矿试验结果
（a）冲次对铁回收率的影响；（b）冲次对铁品位的影响

4　脉动高梯度磁选过程的捕集图

国外一些学者如英国 Nesset J E[2] 等用细粒焦磷酸锰（$Mn_2P_2O_7$，$x = 83 \times 10^{-6} cm^3/g$）为给料建立了单丝圆柱形捕集图，主要结论有两点：

（1）磁介质捕获磁性物的量随竞争力的增大而减少；

（2）矿浆的流动方向不变时，磁介质主要是迎着给矿方向的一面捕获磁性物，而另一面几乎不捕获矿粒，即使在 $v_0 = 23.3 cm/s$ 时，这种现象依然存在。

根据本文的理论分析、试验结果和上述前人的研究结果，可定性地描绘出脉动高梯度磁选单丝圆柱形磁介质的捕集图如图 5 所示，其中假定 v_0 为常数。由该图可知，当脉动最大速度 $\tilde{v}_{max} = 0$ 时，磁介质主要是上表面捕获矿粒，且非磁性矿粒的机械夹杂比较严重，磁性精矿的品位较低，随着 $\tilde{v}_{max}$ 的增大，非磁性矿粒的机械夹杂越来越少，磁介质下表面因矿浆流速不断改变方向而成为有效捕集区，且捕获的矿粒数量逐渐与上表面趋于一致，因此，在合适的冲程冲次范围内，磁性精矿品位提高而回收率基本不变或略有增加；当冲程冲次超过一定范围后，因作用在磁性和非磁性矿粒上的竞争力都增加很快，磁介质上下表面捕获的矿粒显著减少，从而磁性精矿品位继续上升而回收率显著下降。

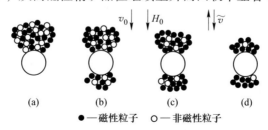

图 5　脉动高梯度磁选单丝捕集图
（a）无脉动；（b）较小脉动；（c）较大脉动；（d）更大脉动

5 结论

（1）脉动增加了矿粒与磁介质的碰撞次数，这是对提高磁性矿回收率有益的一面，脉动也增大了作用在磁性矿粒上的竞争力，这是对磁性矿回收率不利的一面，碰撞次数与竞争力的同时增长使磁性精矿的回收率在一定的冲程冲次范围内基本不变或有所增加。

（2）脉动增大了作用在非磁性矿粒上的竞争力，这有利于消除非磁性矿粒的机械夹杂和因表面力黏附在磁性矿粒上的可能性，因此磁性精矿的品位随冲程冲次的增大而单调上升。

（3）矿粒与磁介质的碰撞次数近似地与冲程冲次呈线性增长关系；竞争力与冲程呈线性增长关系，而与冲次呈抛物线增长关系。增大竞争力有利于提高磁性精矿品位，因此，改变冲次对提高磁性精矿品位比改变冲程更有效果。

参 考 文 献

[1] Oberteuffer J A. IEEE Trans on Magnetics, 1974 (2)：223.
[2] Nesset J E, et al. IEEE Trans on Magnetics, 1981：1506.
[3] 刘树贻，彭世英. 中南矿冶学院学报：选矿论文专辑, 1983 (S1)：98.
[4] 熊大和，刘树贻. 金属矿山, 1988 (7)：34.

写作背景　本文是作者攻读博士学位时做的关于脉动高梯度磁选理论研究工作，探讨了脉动的冲程冲次对选矿指标的影响。

SLon-1500 立环脉动高梯度磁选机的研制[①]

熊大和

(赣州有色冶金研究所)

摘 要 SLon-1500 立环脉动高梯度磁选机利用磁力、脉动流体力分选细粒弱磁性矿物,具有富集比大、不易堵塞、适应性强的优点。本文介绍该机的结构、工作原理及其在马钢姑山铁矿的应用情况。

关键词 SLon-1500 立环脉动高梯度磁选机　细粒弱磁性矿物　赤铁矿

赣州有色冶金研究所与中南工业大学合作曾于 1987 年研制出 SLon-1000 立环脉动高梯度磁选机,并在马钢姑山铁矿选厂分选细粒赤铁矿取得良好指标[1,2]。为了满足大中型选厂需要,赣州有色冶金研究所于 1989 年又研制出 SLon-1500 立环脉动高梯度磁选机(下称 SLon-1500 机),同年 6 月在姑山铁矿选厂投产,用于分选细粒赤铁矿。SLon-1500 较 SLon-1000 机处理量增至 4~5 倍,在延长鼓膜寿命和克服转环抖动等方面做了进一步的改进。至今(1990 年 3 月)该机带矿运转已超过 5000h。较长时间的生产考核表明,其分选效率高,运转稳定。该机的诞生,为我国增添了一种新型高效的磁选设备。

1 整机结构及选别原理

SLon-1500 机的结构见图 1。它主要由脉动机构、激磁线圈、铁轭和转环等组成。选别时,转环作顺时针旋转,矿浆从给矿斗给入,沿上铁轭缝隙流经转环,矿浆中的磁性颗粒

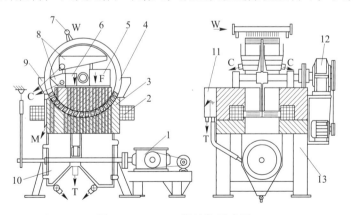

图 1　SLon-1500 机结构示意图

1—脉动机构；2—激磁线圈；3—铁轭；4—转环；5—给矿斗；6—漂洗水；
7—精矿冲洗水管；8—精矿斗；9—中矿斗；10—尾矿斗；11—液面斗；12—转环驱动机构；
13—机架；F—给矿；W—清水；C—精矿；M—中矿；T—尾矿

[①] 原载于《金属矿山》,1990 (07):43~46。

吸着在磁介质表面，由转环带至顶部无磁场区，被冲洗水冲入精矿斗；非磁性颗粒则沿下铁轭缝隙流入尾矿斗排走。

转环立式旋转的好处是：如果给矿中有粗颗粒不能穿过磁介质堆，一般会停留在磁介质堆的上表面，即靠近转环内圆周，当磁介质堆随转环到达顶部时，正好旋转了180°，粗颗粒位于磁介质堆的下部，很容易被精洗水冲入精矿斗中。

SLon-1500机的脉动机构由电动机、冲程箱、双向橡胶鼓膜和中心传动杆组成。冲程箱初步设计的最大冲程为50mm，后经试验和实际使用改为13～20mm就够了，现已改成最大冲程为30mm。现生产上使用冲程为15mm，冲次为300次/min。橡胶鼓膜采用两块ϕ700mm×560mm×95mm 的U形鼓膜，设备出厂时安装的鼓膜至今完好，说明其使用寿命达5000h以上，大约是腰子形鼓膜寿命的10倍。

当鼓膜在冲程箱的驱动下做往复运动时，只要液面高度在转环下部磁介质的上面，分选室的矿浆便做上下往复运动，脉动流体力使矿粒群在分选过程中始终保持松散状态，从而可有效地消除非磁性颗粒的机械夹杂，显著地提高磁性精矿品位。此外，脉动显然对防止磁介质的堵塞也大有好处。

为了保证良好的分选效果，使脉动充分发挥作用，维持矿浆液面高度是至关重要的。该机的液位调节可通过调节尾矿斗下部流量阀、给矿量和漂洗水量来实现。SLon-1500机本身也有一定的液位自我调节能力，当外部因素引起液面升高时，尾矿的排放有尾矿斗下部阀门和液位斗溢流面两种通道；当液面较低时，液位斗不排尾矿，尾矿只能从尾矿斗下部阀门排走。此外因液面至该阀的高差减小，压力降低，尾矿流速自动变慢，液位计的液面与分选区液面基本同样高，它既起到自我调节液面的作用，又可供操作者观察液面位高度。当给矿量有20%波动时，该机靠本身的自我调节能力稳定液面。姑山铁矿在圆筒筛前面安装了一个溢流矿斗，多余的矿浆流回浓缩池，因此给矿量的波动小于20%。

SLon-1500机的分选区大致分为受矿区、排矿区和漂洗区三部分（参见图1左图）。现以转环上的一个分选室为例说明。当它随转环刚进入分选区时，主要接受来自给矿斗的矿浆，其内的磁介质迅速捕获矿浆中的磁性颗粒，并排走一部分尾矿；当它随转环转至分选区的中部时，上铁轭位于此处的缝隙与大气相通，其内的尾矿迅速从尾矿斗中部的尾矿管排走；当它转至左边漂洗区时，室内的尾矿已不多了，脉动漂洗水将剩下的尾矿洗净；当它转出分选区时，室内剩下的水分及其夹带的少量矿粒从中矿斗排走，中矿可酌情排入尾矿或返回给矿；室内磁性矿一小部分靠自重落入精矿小斗，大部分被带至顶部用水冲入精矿大斗。

该机主要技术参数：

转环外径×宽/mm×mm	ϕ1500×600
磁介质堆尺寸（长×宽×高）/mm×mm×mm	(280+280)×100×95
给矿粒度/mm	-1.0
给矿浓度/%	10～50
矿浆通过能力/$m^3 \cdot h^{-1}$	60～100
干矿处理量/$t \cdot h^{-1}$	20～35
额定背景场强/T	1.0
额定激磁功率/kW	38
传动功率/kW	4+4

脉动冲程/mm	0~30
脉动冲次/次·min^{-1}	0~300
供水压力/MPa	0.2~0.5
耗水量/m^3·h^{-1}	50~100
主机质量/t	20
最大部件质量/t	5.0
外形尺寸（长×宽×高）/mm×mm×mm	3500×3000×3200

2 磁系

SLon-1500机磁系立体图和磁路图分别见图2和图3。它是由一个水平放置的矩形激磁线圈、1块下铁轭、2块上铁轭和2块月牙板构成，上下铁轭之间的弧形空间为选别区。为了获得较大的分选空间，磁系包角采用100°；下磁极两侧的尖角改为钝角，这样消除了磁力线过度聚焦的现象，转环进出磁场的部位所受磁场力变化较平缓，从而克服了转环的抖动现象。激磁线圈采用规格为22mm×18mm×5mm空心矩形铜管绕制，水内冷，低电压大电流激磁。当激磁线圈有直流电通过时，在选别区产生背景磁场，磁力线从下铁轭磁极头指向上铁轭磁极头，然后沿铁轭形成闭合回路，两个上磁极头相距一段距离，供转环辐板通过，上下磁极头上的缝隙均为矿浆和水的通路。该磁系具有漏磁小、激磁功耗低、分选空间大、平面线圈较易绕制、铜材和工程纯铁利用率高的优点。其电磁性能实测值见表1。该机的最高背景场强可达1.1T，能满足多数弱磁性金属矿的选矿需要。

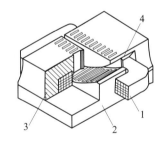

图2　磁系结构

1—激磁线圈；2—下铁轭；3—上铁轭；4—月牙板

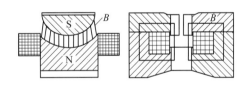

图3　磁系磁路

表1　SLon-1500机磁系电磁性能实测值

激磁电流/A	激磁电压/V	激磁功率/kW	背景场强/T
340	13	4.4	0.38
400	16	6.4	0.45
500	20	10.0	0.56
650	27	17.6	0.75
770	32	24.6	0.86
930	39	36.3	0.99
940	40	37.6	1.02
1100	47	51.7	1.10

3 转环

转环结构见图4。它是用普通不锈钢加工的1块中环板、2块侧环板和74块梯形隔板围成双列共74个分选室。各分选室用1.0mm×4mm×12mm导磁不锈钢板网和0.7mm×10mm×25mm普通不锈钢大孔网交替重叠构成磁介质堆,导磁网的充填率为12%,大孔网充填率为3.2%。各分选室靠轴线一侧焊有一些不锈钢内垫条,以阻挡磁介质堆往轴线方向移动。各分选室外侧装有2根不锈钢外垫条,外垫条可拆下,以便安装磁介质。

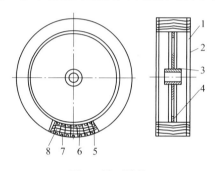

图4 转环结构

1—中环板;2—侧环板;3—轮毂;4—辐板;5—隔板;6—内垫条;7—外垫条;8—磁介质

该机试验初期曾用过0.5mm×4mm×10mm($b=0.3$)这种较细的导磁不锈钢板网,但网丝太细软导致充填工作量大,且在长期生产中受磁力的往复作用易拆断,因此现场使用的磁介质不宜选得太细。较粗的磁介质虽然磁场梯度较低,但易于增加介质充填率,因为适当增加介质充填率可弥补磁场梯度降低所带来的影响。

转环的中环板与辐板相焊接,并通过轮毂与传动机构相连接。

4 选矿生产

姑山铁矿年采选能力100万吨原矿,矿石系宁芜地区红矿,中温热液矿床,铁矿物主要为赤铁矿,少量假象赤铁矿、褐铁矿和菱铁矿,矿石构造以角砾状、致密块状为主,网状和浸染状次之。其现行选矿流程为:原矿经粗碎、中碎、洗矿和细碎后,用跳汰机和感应辊强磁选机得出部分粗精矿和抛弃部分尾矿,产率约占原矿一半的中矿进入主厂房一次磨矿至小于0.074mm(200目)含量占50%左右,然后分两部分。一部分用SQC湿式强磁选机一粗一精选;另一部分用SQC湿式强磁选机粗选和螺旋溜槽精选。

SLon-1500机在姑山铁矿主要用于处理细粒尾矿和一部分难选中矿,其给矿来源及干矿量如下:(1)部分洗矿溢流(原生细泥),约140t/d;(2)SQC强磁精选尾矿经φ350旋流器分级的溢流,约80t/d;(3)跳汰工段沉淀池沉砂,约50t/d;(4)选矿厂总污水池沉淀物,100t/d;(5)原矿细碎底流(次生细泥),约50t/d;(6)主厂房污水沉料,约20t/d。

以上几部分矿料混在一起,每天计440t左右,品位波动在22%~35%Fe,粒度波动在小于0.074mm(200目)50%~95%,其性质变幻莫测,以前因无合适的分矿设备,都作为尾矿排弃,现集中进入φ24m浓缩池,经浓缩后用筛孔为1.0mm(16目)的圆筒筛除渣，筛下物用 1 台 SLon-1500 机分选。选厂生产分选条件大致为：背景场强 0.9T；脉动冲次 300min^{-1}、冲程 15mm；磁介质充填率 12%；转环转速 3r/min；给矿体积 90m^3/h；干矿处理量 22.24t/h（分选条件随矿石性质不同有所变化）。SLon-1500 机自分选这种混合物料以来的平均选矿指标见表 2。

表 2　一次分选姑山选厂混合料指标　　　　　　　　　　（%）

给矿		精矿				尾矿品位
浓度	小于 0.074mm（200 目）含量	品位	产率	品位	回收率	
21	71	28.13	28.28	55.65	55.94	17.28

从表 2 可知，SLon-1500 机可从品位为 28.13%Fe 的给矿中，一次分选得到品位为 55.56%Fe 的合格铁精矿（姑山选厂合格粉矿品位为 5%Fe），作业回收率为 55.94%，每年大约可从过去被抛弃的尾矿中回收约 2 万~3 万吨合格的铁精矿。

5　结语

（1）SLon-1500 立环脉动高梯度磁选机是在 SLon-1000 机的基础上加以扩大和改进研制成功的，其转环立式旋转，反冲精矿，并配有脉动机构，具有富集比大、磁介质不易堵塞、适应性强的优点，是一种新型高效的磁选设备。

（2）该机在马钢姑山铁矿用于从尾矿中回收赤铁矿，当给矿品位为 22%~35%Fe 时，一次分选获得品位为 56% 以上的合格铁精矿，作业回收率平均为 60.67%，每年可从尾矿中多回收 3 万~4 万吨铁精矿，年经济效益百万元以上。

（3）该机在 5000h 运转过程中，因圆筒筛破裂，大量粗粒进入而堵塞过一次，正常情况下没出现过堵塞现象；设备运转稳定，单机月作业率达 98% 以上，长时间的生产考核证明了该机有优异的选矿性能和机电性能，可望在其他弱磁性金属矿选厂推广应用。

参 考 文 献

[1] 熊大和，刘树贻. 脉动高梯度磁选细粒氧化铁矿的研究 [J]. 金属矿山，1988（7）：34~37.
[2] 熊大和. SLon-1000 立环脉动高梯度磁选机的研制 [J]. 金属矿山，1988（10）：37~40.

写作背景　1988 年作者促成赣州有色冶金研究所与马钢姑山铁矿签订了研制 SLon-1500 立环脉动高梯度磁选机的合同，作者在博士毕业前夕完成了该机的图纸设计工作。该机于 1989 年研制成功，在姑山铁矿分选细粒氧化铁矿获得了良好的选矿指标。该机采用双鼓膜脉动，工业试验期间发现双鼓膜不好维修，且对提高选矿指标没什么帮助，因此之后设计的 SLon 立环脉动高梯度磁选机都采用单鼓膜。

SLon 型立环脉动高梯度磁选机及其应用的新发展[❶]

熊大和[1,2]

(1. 赣州金环磁选设备有限公司；2. 赣州有色冶金研究所)

摘　要　SLon 型立环脉动高梯度磁选机是一种新型高效磁选设备，具有不易堵塞、作业率高、分选细粒弱磁性矿物效率较高的优点。本文介绍了该机结构、工作原理及其在姑山铁矿、梅山铁矿和弓长岭选矿厂的新应用。

关键词　脉动高梯度磁选　综合力场　细粒选矿

首台 SLon-1000 立环脉动高梯度磁选机于 1988 年研制成功，以其独特新颖的结构和良好的选矿性能引起了国内外一些选矿专家的兴趣与关注，近年该机正以较快的速度朝着大型化、实用化方向发展，该机的研制者不仅成功地研制了 SLon-1500 立环脉动高梯度磁选机，在其结构方面做了不少改进，使其稳定性和选矿性能有了进一步提高，而且成功地将该机在马鞍山钢铁公司姑山铁矿、上海钢铁公司梅山铁矿、鞍山钢铁公司弓长岭选矿厂等地推广应用，有效地推动了这些厂矿的技术进步。

1　设备结构和工作原理

SLon 型立环脉动高梯度磁选机是一种利用磁力、脉动流体力和重力的综合力场选矿的工业设备。其特点是转环立式旋转、反冲精矿、配有矿浆脉动机构。该机分选细粒弱磁性矿物具有富集比高、不易堵塞和适应性强等优点。

该机主要由图 1 所示的 13 个部分组成。脉动机构、激磁线图、铁轭和转环是关键部件。其脉动冲次和冲程可以调节。激磁线圈用空心铜管绕制并采用水内冷冷却方式。沿转环周边是多个矩形分选室，室内装有导磁不锈钢板网磁介质。当该机工作时，转环顺时针旋转，矿浆从给矿盒给入，沿上铁轭缝隙流经转环，位于分选区的磁介质被磁场磁化，矿浆中的磁性颗粒被吸着在磁介质表面，被转环带至无磁场区，被冲洗水冲入精矿斗；非磁性颗粒在矿浆的脉动力、重力和流体力的作用下穿过转环沿下铁轭缝隙进入尾矿斗。脉动机构推动鼓膜作往复运动，只要将矿浆的液位调至液位斗液流面之上，脉动动能就能有效地传递至分选区。

对于每一组磁介质而言，冲洗磁性精矿的方向与给矿方向相反，粗颗粒不必穿过磁介质堆便可冲洗出来。矿浆的脉动可使磁介质堆中的矿粒群始终保持松散状态。显然，反冲精矿和矿浆脉动可防止磁介质堵塞；脉动分选可提高磁性精矿的质量；此外，这些措施既保证了该机可有效地捕收下限为 10μm 左右的弱磁性矿物，又使其分选上限提高到

[❶]　原载于《江西有色金属》，1993（02）：66~69。

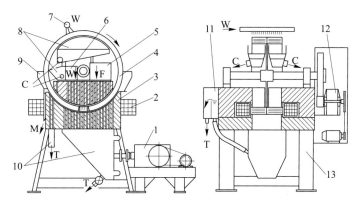

图 1 SLon 型磁选机结构图

1—脉动机构；2—激磁线圈；3—铁轭；4—转环；5—给矿斗；6—漂洗水斗；7—精矿冲洗装置；
8—精矿斗；9—中矿斗；10—尾矿斗；11—液位斗；12—转环驱动机构；13—机架；
F—给矿；W—清水；C—精矿；M—中矿；T—尾矿

1000μm 左右，从而扩大了分选粒度范围和简化了筛分作业。

SLon-1000 和 SLon-1500 型立环脉动高梯度磁选机主要技术参数见表 1。其分选区的平均场强高达 1.0T，对于大多数氧化铁矿来说场强足够。

表 1 SLon-1000 和 SLon-1500 型机技术参数

型号	转环直径 /mm	背景场强 /T	激磁功率 /kW	驱动功率 /kW	脉动冲程 /mm	脉动冲次 /r·min^{-1}	给矿粒度 /mm	处理量 /t·h^{-1}	机重 /t
SLon-1000	1000	0~1.0	25.5	1.1+1.1	0~30	0~400	−1.0	4~7	6
SLon-1500	1500	0~1.0	38	3+4	0~30	0~400	−1.0	20~35	20

2 SLon 型磁选机的新应用

2.1 在姑山铁矿细粒赤铁矿全磁选流程中的应用

姑山铁矿是马鞍山钢铁公司主要矿石原料基地之一，矿石属中温热液矿床，主要有用矿物为赤铁矿，其嵌布粒度细且不均匀，属难选矿石。1978 年建成选矿厂以后，其选矿原则流程是以跳汰、螺旋溜槽、离心机为主的全重选流程。原矿破碎成小于 12mm 以后，用跳汰机选出部分块状铁精矿和丢弃部分块状尾矿。占原矿 1/2 左右的跳汰中矿用球磨机一次性磨细并分成两个粒级，然后分别用螺旋溜槽和离心选矿机分选，多年来细粒选矿生产指标较低，如 1988 年细粒选矿生产指标为：入选品位 37.16%Fe，精矿品位 55.22%Fe，尾矿品位 24.47%Fe，铁回收率仅 61.32%。

为了提高选矿指标，1989~1992 年，细粒选矿流程已改为阶段磨矿—强磁—高梯度全磁选流程，其中采用了 2 台 SLon-1500 型立环脉动高梯度磁选机作为扫选设备。其原则流程见图 2。该流程的综合选矿指标为：给矿品位 32.48%，精矿品位 58.19%，尾矿品位 13.76%，铁回收率 75.49%。与原重选流程比较，铁精矿品位提高了 2.87%，回收率提高了 14.17%。回收率的提高主要依靠了 SLon-1500 磁选机从 SQC 强磁选机尾矿中回收了一部分本来很难回收的铁精矿。

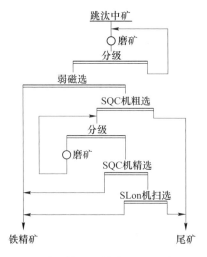

图 2　姑山铁矿细粒全磁选原则流程

2.2　在梅山铁矿高硫高磷铁精矿脱硫脱磷试验流程中的应用

梅山铁矿是一个储量较大的铁矿床，其矿石类型为磁铁矿、菱铁矿、赤铁矿和黄铁矿的混合矿，其铁精矿主要供给上海冶金公司作球团原料，但因含硫和磷偏高铁精矿不能单独作为球团原料，只能作为配料，从而限制了铁精矿的生产量。

为了降低铁精矿中硫和磷的含量，梅山铁矿与有关科研单位合作，于1990年用全磁选流程对其铁精矿进行了降硫降磷的半工业试验，其试验原则流程见图3。该流程中采用了 SLon-1000 型立环脉动高梯度磁选机为扫选设备，该机有效地控制了尾矿品位，对提高回收率起了重要作用。试验结果见表2。

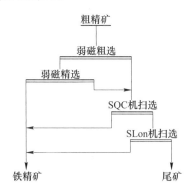

图 3　梅山铁精矿降硫降磷试验原则流程

表 2　梅山铁精矿降硫降磷半工业试验指标　　　　　　　　　　　　（%）

名称	产率	品位			回收率		
		TFe	S	P	TFe	S	P
铁精矿	89.29	57.50	0.144	0.232	95.63	50.01	54.45
尾　矿	10.71	21.91	1.200	1.620	4.37	49.99	45.55
粗精矿	100.00	53.69	0.257	0.381	100.00	100.00	100.00

由表 2 可见，尾矿中硫和磷的回收率分别为 49.99% 和 45.55%，铁的损失率仅 4.37%。铁精矿的品位由 53.69% 升高到 57.50%，且硫和磷的含量显著降低，其质量大幅度提高了。梅山铁矿第二期扩建工程准备采用此流程。

2.3 用于弓长岭贫赤铁矿的分选

弓长岭选矿厂是鞍钢的主要原料基地之一，其红矿选矿车间年处理能力为 280 万吨，采用阶段磨矿阶段选别流程。该厂对小于 0.074mm（200 目）占 80% 左右的细粒级（产率占原矿的 50% 左右）采用 ϕ1600mm×900mm 离心选矿机一次粗选抛尾和二次精选得精矿。生产上突出的问题是粗选离心机尾矿品位太高，严重影响了铁精矿的回收率。

1989 年 9 月开始我们与鞍钢矿山公司及其所属弓长岭选矿厂合作，首先用一台 SLon-1000 立环脉动高梯度磁选机对该厂细粒级红矿进行了连续一个月的现场试验。SLon-1000 机安装在粗选离心机的位置，处理相同给矿。在一个月的对比试验中，共运转 713h，单机作业率达 99%，与离心机对比选矿指标见表 3。SLon-1000 机与粗选离心机选矿指标比较，精矿品位高 4.12%，尾矿品位低 10.16%，回收率高 19.65%，大大优于粗选离心机分选指标。这一试验为 SLon 型立环脉动高梯度磁选机在弓长岭选矿厂的推广应用打下了良好的基础。

表 3　SLon-1000 机与离心机选矿指标对比　　　　　　　　　　　　　　　（%）

设　备	品　位			精矿产率	精矿回收率
	给矿	精矿	尾矿		
SLon-1000	31.72	50.81	9.07	54.26	86.92
离心选矿机	31.82	46.69	19.23	45.85	67.27

为了探索 SLon 型立环脉动高梯度磁选机在生产上全面取代弓长岭选矿厂粗选离心机的可能性，1991 年 3 月至 1992 年 7 月，鞍钢矿山公司、弓长岭选矿厂与赣州有色冶金研究所和赣州有色冶金机械厂合作，采用 5 台 SLon-1500 型立环脉动高梯度磁选机在该厂红矿选矿车间 7、8 系统进行了为期 16 个月的工业试验，5 台 SLon-1500 机取代了 24 台 ϕ1600mm×900mm 粗选离心机。试验流程与原流程（5、6 系统）的选矿指标见表 4。

表 4　试验流程与原流程选矿指标对比　　　　　　　　　　　　　　　　（%）

系　统	原矿品位	精　矿			尾矿品位
		品 位	产 率	回收率	
7、8	28.47	64.31	31.84	71.92	11.73
5、6	28.47	62.78	25.36	55.93	16.81
差　值	0	+1.53	+6.48	+15.99	-5.08

由表 4 可见，7、8 系统采用 SLon-1500 机后，尾矿品位显著下降，铁回收率提高 15.99%，铁精矿品位提高 1.53%。这一试验成果为提高鞍山式赤铁矿的选矿回收率提供了一条有效的途径。目前，7、8 系统的 5 台 SLon-1500 型机已正式作为生产设备。

3　结语

SLon 型立环脉动高梯度磁选机在姑山铁矿、梅山铁矿、弓长岭选矿厂的应用表明：

该机分选细粒弱磁性矿物具有良好的选矿性能。SLon-1500 机用于姑山铁矿和弓长岭选矿厂回收细粒赤铁矿，使铁回收率大幅度提高，SLon-1000 机用于梅山铁矿粗精矿脱硫脱磷的半工业试验中，使其铁精矿品位显著提高，硫和磷含量大幅度降低。SLon 型立环脉动高梯度磁选机在这些厂矿的成功应用，促使了其技术水平和生产水平的提高。

总之，SLon 型立环脉动高梯度磁选机将有关磁选和重选理论和方法有机地结合起来，成为国内外第一种同时利用脉动流体力、重力和高梯度磁场力的工业连续选矿设备。它既继承了传统高梯度磁选机梯度高、磁力大、能有效回收细粒弱磁性矿物的优点，又在提高磁性精矿品位、防止磁介质堵塞、扩大分选粒度范围和改善机械稳定性等方面有较大的突破。它为细粒弱磁性矿物的选矿提供了一种新型高效的选矿设备。这种设备有望在今后老厂的技术改造和新矿建设中得到较大的发展和应用。

写作背景　这是作者 1993 年 2 月在《江西有色金属》期刊上发表论文，介绍了 SLon-1000、SLon-1500 立环脉动高梯度磁选机结构、工作原理以及在马钢姑山铁矿和梅山铁矿的选矿试验流程和选矿指标。从本论文开始，SLon 磁选机逐渐在工业现场推广应用。

SLon 型立环脉动高梯度磁选机的改进及其在红矿选矿的应用

熊大和

（赣州有色冶金研究所）

摘 要 SLon 型立环脉动高梯度磁选机分选细粒弱磁性矿物具有富集比大、分选效率高、不易堵塞、分选粒度范围宽和适应性强的优点，在国内红矿选矿生产中得到较快的推广应用。近年研制者对其结构、磁介质、传动机构、冲程箱、激磁线圈、整流器等作了较多的研究和改进，使其可靠性和分选性能进一步提高。

关键词 立环脉动高梯度磁选机 赤铁矿 褐铁矿 红矿选矿

The Improvement of SLon Vertical Ring and Pulsating High Gradient Magnetic Separator and its Application in Processing Oxidized Iron Ore

Xiong Dahe

(*Ganzhou Nonferrous Metallurgy Research Institute*)

Abstract This separator possesses the advantages of large beneficial ratio, high efficiency, not easy to be blocked, wide separating grain size range and strong adaptability. It has been rapidly applied in processing oxidized iron ore. In recent years, a lot of research work have been done for improving its structure, matrix. driving system, pulsating mechanism, energizing coil and rectifier. Its operational reliability and efficiency have been further raised.

Keywords High gradient vertical ring plusating magnetic separator; Hematite; Iimonite; Red ore processing

赣州有色冶金研究所与中南工业大学、马钢总公司姑山铁矿共同合作，分别于1988年和1990年完成首台SLon-1000和首台 SLon-1500 立环脉动高梯度磁选机的研制，它们在姑山铁矿分选细粒赤铁矿均获得了良好的效果[1~3]，近几年来，为了提高 SLon 磁选机的机电、选矿性能和可靠性，我所对该机又作了较多的研究和改进。迄今为止该机已在马钢总公司姑山铁矿、鞍钢弓长岭选矿厂、合肥钢铁公司钟山铁矿、山西二峰山铁矿、湖北铜

---

❶ 原载于《金属矿山》，1994 (06)：30~34。

绿山铜矿、昆钢八街铁矿、宁夏中卫铁厂等地获得工业应用，其中大部分用户获得较显著的经济效益。但是，其中个别厂矿也曾出现过一些管理、操作上的失误，使该机未能发挥应有的水平。本文结合作者多年研制、试验、改进该设备的体会作一介绍。

1 设备结构和工作原理

如图1所示为经改进的SLon-1500立环脉动高梯度磁选机，与首台SLon-1500磁选机比较[3]，原有双鼓膜改成了单鼓膜，减轻了冲程箱的负荷和维修量。该机主要由脉动机构、激磁线圈、铁轭、转环及各种给料、排料装置组成。其工作原理为：转环内装有导磁不锈钢板网或棒介质并随转环作顺时针旋转，矿浆经给矿盒和上铁轭缝隙流经转环下部磁场区，磁性矿粒被转环内的磁介质吸住，由转环带至顶部无磁场区用水冲入精矿斗；非磁性矿粒在矿浆脉动力、重力和流体力的作用下穿过转环流入尾矿斗排走。

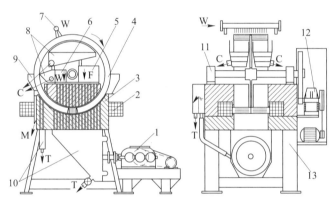

图1 改进后的SLon-1500磁选机示意图

1—脉动机构；2—激磁线圈；3—铁轭；4—转环；5—给矿斗；6—漂洗水斗；7—精矿冲洗装置；
8—精矿斗；9—中矿斗；10—尾矿斗；11—液位计；12—转环驱动机构；13—机架；
F—给矿；W—清水；C—精矿；M—中矿；T—尾矿

该机转环立式旋转，对于每一组磁介质而言，冲洗磁性精矿的方向与给矿方向相反，粗颗粒不必穿过磁介质堆便可冲洗出来。该机的脉动机构驱动矿浆产生脉动，可使位于分选区磁介质堆中的矿粒群保持松散状态，使非磁性矿介质堆中的矿粒群保持松散状态，使非磁性矿粒尽快穿过磁介质堆进入到尾矿中去。显然，反冲精矿和矿浆脉动可防止磁介质堵塞；脉动分选可提高磁性精矿的质量。这些措施保证了该机具有较大的富集比、较高的分选效率和较强的适应能力。

2 在红矿选矿工业中的应用

SLon磁选机最初在马钢总公司姑山铁矿分选细粒赤铁矿试验成功，该机在原重选流程中曾用于分选φ350mm旋流器溢流、洗矿溢流、跳汰作业沉淀池细泥，均获得前所未有的良好指标[1,2]。1989年姑山铁矿、马鞍山矿山研究院、马钢矿山设计研究所合作完成了跳汰中矿弱磁—强磁高梯度阶段磨选工业试验并于1992年完成了该流程的技术改造。流程中采用了两台SLon-1500磁选机扫选SQC湿式强磁选机精选尾矿,有效地控制了小于30μm粒级的回收。1993年全年的生产指标达到铁精矿品位58%、回收率75%的好指标,与原一段磨重选流程比较,精矿品位提高2.9%,回收率提高14%,原矿处理量不变,每年增产铁精矿约3万吨。

1991~1992年鞍钢矿山公司及其所属弓长岭选矿厂与赣州有色冶金研究所和赣州有色冶金机械厂合作采用SLon-1500磁选机对弓长岭细粒贫赤铁矿进行了工业试验[4],采用5台SLon-1500磁选机取代原7、8生产系统中24台$\phi1600mm\times900mm$粗选离心选矿机,使细粒级的尾矿品位从原有的18%左右降低到10%左右,回收率从53%左右提高到70%左右。

近年SLon磁选机还陆续地在其他厂矿用于分选红矿,已正常投产应用和通过半工业试验或现场试验的情况见表1。从表上部可知,SLon磁选机在生产中用于分选红矿一般可使全流程回收率或作业回收率提高10%~20%。据统计,平均每台SLon-1500磁选机每年可增产约1.0万~1.5万吨铁精矿,年效益达100万元以上;从表下部可知,SLon磁选机在一部分红矿选矿厂均存在应用潜力。

表1 SLon磁选机分选红矿已投产应用和半工业或现场试验统计

	矿山名称	设备型号	使用台数	分选矿种	原流程及分选指标 TFe/%	新流程及分选指标 TFe/%	备注
已投产使用单位	姑山铁矿	SLon-1500	2	赤铁矿	一段磨矿螺溜离心机重选,α 34.74,β 55.10,θ 22.24,ε 60.33	阶段磨矿SQC与SLon磁选,α 34.74,β 58.21,θ 15.03,ε 76.49	1. SLon扫选SQC精选尾矿; 2. 1993年生产平均 β 58%,ε 75%
	弓长岭选矿厂	SLon-1500	5	赤铁矿	阶段磨矿强磁重选,α 28.47,β 63.65,θ 17.44,ε 53.36	阶段磨矿强磁高梯度重选,α 28.47,β 64.31,θ 11.73,ε 71.92	1. SLon取代原粗选离心机; 2. 1992年4~7月全流程指标
	钟山铁矿	SLon-1500	1	赤铁矿	二段连磨螺溜重选,α 34,β 60,θ 25,ε 45	二段连磨SQC、SZC、SLon磁选,α 34,β 60,θ 22,ε 55	1. SLon扫选SZC中磁精选尾矿; 2. 所列为全流程指标
	二峰山铁矿	SLon-1500	1	磁、赤铁混合矿	一段磨矿单一弱磁选	原流程增加SLon扫选,SLon指标:α 18.62,β 53.64,θ 10.43,ε 54.60	
	八街铁矿	SLon-1500	1	褐铁矿	一段磨矿强磁一粗一扫,α 38.69,β 51.30,θ 31.82,ε 46.76	一段磨矿强磁粗选SLon扫选,α 38.81,β 52.26,θ 25.55,ε 66.82	1. 强磁机为ϕ1500双立环,球介质; 2. 所列为生产调试指标
	宁夏中卫铁矿	SLon-1500	1	褐铁矿		一段磨矿SLon一次粗选,α 46.36,β 54.17,θ 38.60,ε 58.23	新建选厂调试指标
	鸡冠山金矿	SLon-1000	1	褐铁矿		氰化法选金后SLon一次粗选,α 43.84,β 53.02,θ 33.82,ε 63.03	新建选厂调试指标

续表1

	矿山名称	设备型号	使用台数	分选矿种	原流程及分选指标 TFe/%	新流程及分选指标 TFe/%	备注
工业试验或现场试验单位	江西铁坑铁矿	SLon-1000	1	褐铁矿	二段磨矿强磁浮选，浮选指标：α 29.43，β 44.38，θ 24.73，ε 36.78	二段磨矿 SQC 强磁 SLon 扫选，SLon 指标：α 29.71，β 48.65，θ 23.43，ε 40.78	1. SLon 现场运转 80 个班平均指标；2. 浮选同期指标
	海南铁矿	SLon-1000	1	赤铁矿	强磁—粗—扫	SLon 分选强磁扫选尾矿，SLon 指标：α 20.91，β 49.79，θ 16.95，ε 28.68	1. SLon 现场运转 144h 平均指标；2. 小于 38μm（400 目）β 57.53%，ε 91.86%
	海南铁矿	SLon-1000	1	赤铁矿	强磁—粗—扫	SLon 分选强磁尾矿，α 29.50，β 55.30，θ 24.75	SLon 现场运转 72h 的平均指标
	梅山铁矿	SLon-1000	1	铁精矿除磷	原铁精矿，TFe 53.69，S 0.257，P 0.381	弱磁—强磁—SLon 扫选，TFe 57.50，S 0.144；P 0.232，ε 95.63	1. ε—对原铁精矿 TFe 回收率；2. 现场 104h 中试指标
	东鞍山二选	SLon-1000	1	赤铁矿	浮选	浮尾用 SLon 扫选，SLon 指标：α 24.29，β 35.60，θ 9.02，ε 83.79	SLon 精矿品位 35.60%，与采出原矿品位相当，有待作精选研究
	铜绿山铜矿	SLon-1000	1	赤铁矿	浮铜尾矿用 SQC 强磁回收赤铁矿	SQC 尾矿用 SLon 扫选，SLon 指标：α 22.64，β 54.98，θ 14.30，ε 48.20	已购 1 台 SLon-1500 机建厂

3 SLon 磁选机在应用中的问题和改进措施

3.1 网介质与棒介质的对比

迄今 SLon 磁选机在生产中多采用导磁不锈钢板网作为磁介质（以下简称网介质），网介质具有磁场梯度较高、对微细粒捕收能力较强的优点，在网介质清洁无堵塞时可获得较高的回收率。但是，网介质有几个较难克服的缺点：

（1）维护工作量较大，要求每 3~5 个月拆出清洗一次。

（2）要求操作维护者每天检查一次松紧程度，发现松动必须及时停机添加，否则，网介质在松动的条件下运转，由于受强磁场力的往复作用而容易破碎。据统计，网介质的维护工作量占该机成套设备维护工作量的 75% 左右，它是影响该机作业率提高的主要因素。现在使用网介质的 SLon 磁选机作业率达到 97%~98%（指运转时间与给矿时间之比）后，很难继续提高，平均每天要停机 20~30min 检修网介质。

（3）选矿指标随网介质的清洁程度而变化，如果长期不清洗，选矿指标会显著下降。

为了克服网介质的缺点，赣州有色冶金研究所与弓长岭选矿厂合作用 ϕ4mm 和 ϕ5mm 导磁不锈钢棒为磁介质（以下简称棒介质）已进行了为期 16 个月的运转试验。实践证明棒介质具有运转稳定、永不堵塞、维护工作量小、选矿指标稳定的优点。此外，网介质的固定压杆是靠转环边板固定，边板磨损后就会导致转环报废或需要大修；而棒介质是整盒用螺栓固定在转环隔板上，隔板几乎无磨损。因此，棒介质转环寿命比网介质转环长得多。使用棒介质的 SLon 磁选机停机检修时间大幅度减少，其作业率可达 99% 以上。棒介质的缺点是磁场梯度较低，与清洁的网介质比较，其尾矿品位稍高。但若网介质长期不清洗或堵塞较严重，其尾矿品位反而高于棒介质所选品位。弓长岭选矿厂在 SLon-1500 磁选机上使用棒介质与网介质的对比尾矿品位（注：给矿品位和精矿品位基本相同且随时间波动很小）。网介质运转时间小于 6 个月时，其尾矿品位低于棒介质所选品位，超过 6 个月后，其尾矿品位高于棒介质所选品位。因此，网介质必须 5~6 个月清洗一次，而棒介质不需清洗。

3.2 转环传动机构的改进

首台 SLon-1500 磁选机的转环主轴与摆线针轮减速器输出轴的连接采用硬性联轴节，但因安装误差二者不易同轴线运转，使转环轴承、减速器轴承承受巨大的偏心力，曾出现过轴承被压碎，摆线针轮减速器外壳被胀破的故障。即使采用柔性联轴节或十字滑块联轴节也不易克服转环抖动。因此，从第二台开始，转环主轴安装大齿轮与减速器输出轴安装小齿轮啮合，避免了同轴线多个轴承的缺点；采用耐冲击力较好的圆柱齿轮减速器取代摆线针轮减速器，并在转环主轴与轴承座相交处安装防水卡以阻止矿水进入到轴承内。改进后的传动系统具有运转平稳、作业率高、使用寿命长的优点。

3.3 脉动冲程箱的改进

早期设计的 SLon 磁选机脉动冲程箱参考了工业跳汰机冲程箱结构，其往复杆的重量完全由铜套支承，往复杆与铜套相互滑动摩擦。经应用证明，其箱体内易进矿进水，铜套、轴承和往复杆使用寿命较短，不适应于冲次较高的 SLon 磁选机使用。1991 年以后制造的 SLon 磁选机设计了全滚道式脉动冲程箱，其往复杆的重量全由轴承和平面滚道支承，实现了全滚动摩擦，往复杆输出端没有防水装置。它具有摩擦阻力小、力量大、工作稳定、使用寿命长和适应于较高冲次的优点。

3.4 激磁线圈的改进

早期的 SLon 磁选机激磁线圈涂灌环氧树脂防水防矿，用橡胶包扎后安装在磁选机上，这种绝缘和安装方式不能可靠地阻挡矿水渗到线圈内部，线圈使用时间长了后较易引起绝缘性能下降，严重的会产生短路或使线圈失效。此外，在安装、运输和起吊的过程中也较易碰坏绝缘层。为了克服这些缺点，我所又试制了不锈钢外壳全封闭式激磁线圈，已在弓长岭选矿厂试用了 21 个月，其可靠性明显优于原有线圈。

3.5 整流器的改进

现有 SLon 磁选机整流器的二极管和整流变压器次级线圈采用水冷散热。水冷有结构简单、无噪声的优点，但时间较长或水质较差时易在管路中产生淤泥沉积，严重时会阻塞水路引起故障。此外，连接胶管老化后较易炸裂，引起漏水并危害电器。迄今部分 SLon 磁选机整流器已改为风冷式，其优点是故障率较低，但有一定的噪声。如弓长岭选矿厂于 1993 年 6 月将 5 台 SLon-1500 磁选机整流器全改成风冷，用 800A 风冷二极管取代原有

500A 水冷二极管，整流器顶部安装 1 台 0.25kW 风扇，整流变压器二次线圈撤去冷却水。经改风冷后的整流器运转已达 10 个月，至今未出现故障。至于噪声问题，0.25kW 风扇噪声较小，一般可被选矿厂其他设备声音淹没。权衡利弊，风冷整流器优于水冷整流器。

4　SLon 磁选机在选矿流程中的配置

SLon 磁选机在选矿流程中的配置如下：

（1）若 SLon 磁选机作为粗选抛尾设备且其粗精矿用其他设备精选，精选尾矿（或中矿）连生体含量高，应返回球磨机再磨后再返回粗选作业。若不磨矿返回，则容易造成粗粒和连生体恶性循环。如弓长岭选矿厂细粒级选别流程为 SLon-1500 磁选机粗选抛尾，其粗精矿用离心机精选两次，试验初期离心机第一次精选尾矿直接返回 SLon-1500 磁选机给矿，导致中矿恶性循环，后改为返回二次球磨分级机[4]。

（2）分选褐铁矿应有扫选作业。由于褐铁矿的磁性很弱，用 SLon 磁选机一次选别回收率 60% 左右（见表 1 下部）。为了提高回收率，应用 SLon 磁选机增加一次扫选作业，用该机一粗一扫，铁回收率可达 70% 以上。

（3）SLon 磁选机具有富集比较高的优点，用于精选作业可望比其他强磁选设备获得较高的铁精矿品位。

（4）该机暂不适应于含石灰水较高的矿浆和冲洗水。例如在铜陵鸡冠山金矿，该机用于从选金尾矿中回收褐铁矿。由于矿水中含石灰水太高，在磁介质中易结垢而较难处理，从而曾一度中断选矿生产。此问题有待于进一步研究。

5　结语

（1）SLon 立环脉动高梯度磁选机具有富集比大、分选效率较高的优点。获得良好选矿指标的要点是保证矿浆液位高度使脉动充分发挥作用和保持磁介质清洁整齐无损。

（2）棒介质具有工作稳定、作业率高、寿命长、永不堵塞、维护工作量小的优点，其缺点是磁场梯度较低；网介质的磁场梯度较高但维护工作量较大。凡棒介质能满足选矿生产要求的工艺上应尽量采用以棒为介质，若采用网介质则必须加强管理。

（3）降低棒介质 SLon 磁选机的尾矿品位，优化其排列组合是今后的研究方向之一。

（4）SLon 磁选机在我国多个红矿选厂应用普遍获得较好的效果，尤其是在提高细粒级回收率方面效果比较明显。

（5）该机经过数年的发展、改进，已成为一种适应于分选细粒弱磁性矿物的新型、高效、工作可靠的选矿设备。但该机还需要进一步发展与完善。

参 考 文 献

[1] 熊大和，刘树贻. 脉动高梯度磁选细粒氧化铁矿的研究［J］. 金属矿山，1988（7）：34~37.
[2] 熊大和. SLon-1000 立环脉动高梯度磁选机的研制［J］. 金属矿山，1988（10）：37~40.
[3] 熊大和. SLon-1500 立环脉动高梯度磁选机的研制［J］. 金属矿山，1990（7）：43~46.
[4] 王棣华. SLon-1500 型立环脉动高梯度磁选机分选弓长岭贫赤铁矿的工业试验［J］. 金属矿山，1993（2）：32~36.

写作背景　首台 SLon-1000 和首台 SLon-1500 磁选机虽然获得了较好的选矿指标,但设备故障率比较高。1990~1994 年,作者团队对 SLon 磁选机进行了多方面的改进,例如双鼓膜改为单鼓膜,磁介质由网介质改为棒介质等。改进后的 SLon 磁选机可靠性显著提高,在工业上获得更广泛的推广和应用。

SLon-2000 立环脉动高梯度磁选机的研制[1]

熊大和

（赣州有色冶金研究所）

摘 要 针对红矿选矿工业生产中存在的一些问题研制了 SLon-2000 立环脉动高梯度磁选机，在鞍钢弓长岭选矿厂进行了 6 个月分选赤铁矿的工业试验，证明了该机具有富集比大、分选效率高、分选粒度范围宽、适应性强、运转可靠的优点，尤其是其棒形磁介质长期保持清洁不堵塞，克服了高梯度磁选机网介质较难维护和平环强磁选机齿板较易堵塞的缺点。

关键词 赤铁矿选矿 强磁选机 立环脉动高梯度磁选机

The Development of SLon-2000 Vertical Ring and Pulsating High Gradient Magnetic Separator

Xiong Dahe

(*Ganzhou Nonferrous Metallurgy Research Institute*)

Abstract The SLon-2000 vertical ring and pulsating high gradient magnetic separator is developed to solve some of the problems existing in the oxidized iron ore processing industry. The commercial test of processing hematite for 6 monthes in Gongchangling mineral Processing plant of Anshan Iron and Steel Company has proved that it possesses the advantages of large concentration ratio, high efficiency, wide separating grain size range, strong adaptability and high reliability. Especially its rod matrix always keeps clean and unblocked, which overcome the problems of net matrix which is difficult to maintain in high gradient magnetic separator and of groved plate matrix, which is easy to be blocked in wet high intensity magnetic separator. The successful development of SLon-2000 is a great advance in the development of high intensity magnetic separator.

Keywords Hematite beneficiation; High intensity magnetic separator; High gradient vertical ring pulsating magnetic separator

现代红矿采选工业中遇到的主要问题是：随着采选规模扩大，可采富矿日益减少，贫矿和难选矿入选比例逐年增加，强磁选设备在红矿选矿中的应用占据重要的地位，但是，多数强磁选机均存在着磁介质的堵塞问题。

在鞍钢弓长岭选厂的赤铁矿选矿生产中，上述问题较为突出。20 世纪 80 年代，该厂赤铁矿入选品位为 30% 左右，20 世纪 90 年代降至 28% 左右，即每年品位以 0.2% 左右的

---

[1] 原载于《金属矿山》，1995 (06)：32~34，47。

速度递减。该厂在20世纪80年代安装了5台ϕ2m平环强磁选机用于一次球磨分级机溢流的粗选作业,因其给矿粒度达2mm及强磁性矿物较多,虽齿顶间隙放大到3.5mm,但仍存在着齿板堵塞频繁、浮水严重、选矿指标不稳定和维修工作量大的缺点。

为了解决上述问题,鞍钢弓长岭矿山公司和弓长岭选矿厂决定采用1台SLon-2000立环脉动高梯度磁选机(以下简称SLon-2000机)与ϕ2m平环强磁选机进行工业对比试验。该机由赣州有色冶金研究所设计,赣州有色冶金机械厂制造,合同要求在给矿量、给矿性质、精矿品位相同的条件下,SLon-2000机的尾矿品位应比ϕ2m平环强磁选机低2%以上,并必须做到磁介质不堵塞。

1 设备结构和工作原理

如图1所示为SLon-2000机的结构示意图。它主要由脉动机构、激磁线圈、铁轭、转环及各种给料、排料装置组成。其工作原理为:转环内装有导磁不锈钢棒介质并随转环做顺时针旋转,矿浆经给矿斗和上铁轭缝隙流经转下部位于磁场中的磁介质堆,磁性矿粒被磁介质吸住,由转环带至顶部无磁场区用水冲入精矿斗;非磁性矿粒在重力和流体力的作用下穿过转环流入尾矿斗排走。

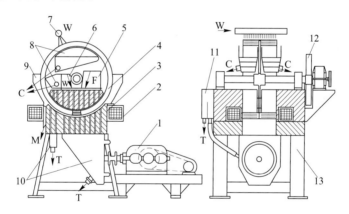

图1 SLon-2000机结构示意图

1—脉动机构;2—激磁线圈;3—铁轭;4—转环;5—给矿斗;6—漂洗水斗;7—精矿冲洗装置;
8—精矿斗;9—中矿斗;10—尾矿斗;11—液位斗;12—转环驱动机构;13—机架;
W—清水;C—精矿;M—中矿;T—尾矿

该机转环立式旋转,对于每一组磁介质而言,冲洗精矿的方向与给矿方向相反,粗颗粒不必穿过磁介质堆便可冲洗出来。该机的脉动机构驱动矿浆产生脉动,使位于分选区磁介质堆中的矿粒群保持松散状态,使非磁性矿粒尽快穿过磁介质堆进入到尾矿中去,反冲精矿和矿浆脉动可防止磁介质堵塞;脉动分选可提高磁性精矿的质量。这些措施保证了该机具有较大的富集比,较高的分选效率和较强的适应能力。

2 设计依据及改进

在总结SLon-1000、SLon-1500机研制与应用[1~5]的基础上,设计SLon-2000机时做了如下主要的改进:

(1) 扩大处理能力,台时处理量达到 50~70t。

(2) 磁系包角由 90°~100°改为 120°,在转环转速不变的条件下,每一磁介质堆在磁场中运行的时间与磁系包角成正比,较大的磁系包角使矿浆得到较长的分选时间。

(3) 采用 ϕ5mm 导磁不锈钢棒为磁介质,实行优化组合排列。与网介质比较,棒介质具有运转稳定、长期不堵塞、维护工作量小、选矿指标稳定、转环使用寿命长的优点。

(4) 采用全滚道脉动冲程箱,其往复杆的质量全由轴承和平面滚道支承,实现了全滚动摩擦。往复杆输出端配有防水装置。与滑动摩擦冲程箱比较,它具有摩擦阻力小、力量大、工作稳定、使用寿命长、能适应较高冲次的优点。

(5) 采用不锈钢外壳全封闭式水冷激磁线圈,可防止矿浆和水浸入绝缘层,提高其工作可靠性和使用寿命。

(6) 配套的整流器由水冷改为风冷,采用感应调压器无级调节激磁电流,提高了整流器的工作可靠性并方便了激磁电流的调节。

(7) 对配套的圆筒筛作了较大的改进,采用特制粗丝不钢筛网,增加传动部件的防水措施和提高滚轮的抗磨能力,提高了其工作稳定性和使用寿命。

采取以上措施后,SLon-2000 机成套设备的工作可靠性、机电性能和选矿性能均得到显著改善。对该机测试各项技术指标均达到设计要求。该机的主要技术参数(其中右侧括号内为该机分选弓长岭贫赤铁矿实际使用值)见表1。表2为该机电磁性能实测值。

表 1 SLon-2000 机主要技术参数

转环外径×宽/mm×mm	ϕ2000×900
转环转速/r·min^{-1}	3
磁介质盒尺寸(长×宽×高)/mm×mm×mm	426×154×144
给矿粒度/mm	0~2.0(0~2.0)
给矿浓度/%	10~40(35~45)
矿浆通过能力/m^3·h^{-1}	100~200(100~170)
干矿处理量/t·h^{-1}	50~70(60~70)
额定背景场强/T	1.0(0.85)
额定激磁电流/A	1080(800)
额定激磁电压/V	76(52)
额定激磁功率/kW	82(41.6)
传动电动机功率/kW	5.5+7.5
脉动冲程/mm	0~30(14)
脉动冲次/次·min^{-1}	0~300(300)
供水压力/MPa	0.3~0.5(0.3)
耗水量/m^3·h^{-1}	100~200(193)
主机质量/kg	50000
最大件质量/kg	14000
外形尺寸/mm	4200×3500×4300

注:设计值,括号内为弓长岭选厂使用值。

表2 SLon-2000机电磁性能实例

项 目	实 测 值										
激磁电流/A	100	200	300	400	500	600	700	800	900	1000	1080
激磁电压/V	8.0	15.2	20.0	26.0	33.8	39.6	46.0	52.0	58.0	64.0	71.0
激磁功率/kW	0.80	3.04	6.0	10.4	16.9	23.8	32.2	41.6	52.2	64.0	76.7
背景场强/T	0.109	0.230	0.342	0.453	0.580	0.694	0.781	0.851	0.918	0.975	1.026

3 SLon-2000机的工业试验

鞍钢弓长岭选矿厂重选车间年处理贫赤铁矿约260万吨,近年入选原矿品位28%左右,氧化亚铁在2%~7%,其中主要有用矿物为赤铁矿、磁铁矿、假象赤铁矿等。脉石矿物以石英为主,次为绿泥石、方解石、角闪石、铁锂云母等。5~8号系统采用阶段磨矿强磁重选工艺流程,一次球磨分级机的溢流和二次球磨机的排矿经滚筒永磁中磁场磁选机分选出强磁性矿物,其底流进入ϕ2m平环强磁选机粗选抛尾,粗精矿再进入螺旋溜槽等设备精选。

SLon-2000机于1994年8月安装在该厂重选车间ϕ2m平环强磁选机旁边,从1994年9月1日至1995年2月28日,与4号平环强磁选机进行了6个月的对比试验,对比选矿指标见表3。

表3 SLon-2000机工业试验对比选矿指标 (%)

试验时间	机型	给矿品位	精矿			尾矿品位
			品位	产率	回收率	
1994年9月	SLon-2000机	25.34	39.39	51.64	80.26	10.34
	ϕ2m平环强磁选机	25.34	32.94	55.29	71.88	15.94
1994年10月	SLon-2000机	27.88	42.03	53.83	81.16	11.38
	ϕ2m平环强磁选机	27.88	34.98	60.02	75.31	17.22
1994年11月	SLon-2000机	27.22	44.48	47.17	77.08	11.81
	ϕ2m平环强磁选机	27.22	36.21	51.74	68.83	17.58
1994年12月	SLon-2000机	26.66	43.23	48.43	78.53	11.10
	ϕ2m平环强磁选机	26.66	36.26	51.07	69.46	16.64
1995年1月	SLon-2000机	23.81	38.06	50.61	80.89	9.21
	ϕ2m平环强磁选机	23.81	30.12	59.76	75.59	14.44
1995年2月	SLon-2000机	24.70	40.05	47.93	77.72	10.57
	ϕ2m平环强磁选机	24.70	33.49	52.64	71.37	14.93
6个月平均值	SLon-2000机	25.94	41.21	49.89	79.25	10.74
	ϕ2m平环强磁选机	25.94	34.00	54.90	71.95	16.13
	差值	0	7.21	-5.01	7.30	-5.39

4 结语

(1) SLon-2000机在6个月的工业试验中,磁介质始终保持清洁无堵塞,从根本上克服了采用网介质的SLon磁选机经常需要添加网介质和定期清洗网介质的缺点,也克服了平环强磁选机齿板堵塞频繁的缺点。

(2) 该机作为弓长岭选矿厂0~2mm贫赤铁矿粗选抛尾设备,在6个月的工业试验中,其粗选精矿品位平均提高7.21%,尾矿品位降低5.39%,铁回收率提高7.30%,多丢弃产率为5.01%的尾矿,为后续螺旋溜槽等设备的精选作业创造了有利的条件,也减轻了二段球磨机的磨矿负荷。

(3) 通过弓长岭选矿厂实验室筛水析检验,该机 9~0μm 粒级的尾矿品位为 11.14%,与大于 9μm 各粒级尾矿品位差别不大,说明该机具有回收粒度范围宽、适应性强的优点。

(4) 在入选原矿品位逐年降低的形势下,该机具有较大的富集比和较高的处理能力,可为降低采矿边界品位,增加矿石工业储量和延长矿山的开采年限作出一定的贡献。

(5) 在 6 个月的工业试验中,该机每月运转 670~700h,对日历时作业率为 94% 左右,扣除因停矿、水、电等外部因素造成的停机外,其实际作业率平均 98.3%,说明该机有很高的运转可靠性。

参 考 文 献

[1] 熊大和. SLon-1000 立环脉动高梯度磁选的研制 [J]. 金属矿山, 1988 (10): 37~40.
[2] 熊大和. SLon-1500 立环脉动高梯度磁选机的研制 [J]. 金属矿山, 1990 (7): 43~46.
[3] 王重渝, 杨庆林, 谢金清. 应用 SLon-1500 立环脉动高梯度磁选机回收姑山细粒赤铁矿尾矿的工业试验 [J]. 金属矿山, 1991 (5): 38~43.
[4] 王棣华. SLon-1500 型立环脉动高梯度磁选机分选弓长岭贫赤铁矿的工业试验 [J]. 金属矿山, 1993 (2): 32~36.
[5] 熊大和. SLon 型立环脉动高梯度磁选机的改进及其在红矿选矿的应用 [J]. 金属矿山, 1994 (6): 30~34.

写作背景 从 1993~1995 年完成首台 SLon-2000 立环脉动高梯度磁选机的研制与工业试验,在 SLon-1000 和 SLon-1500 的基础上经过了多方面的改进,该机的设备作业率达到 98% 以上。在鞍钢弓长岭选矿厂用于分选氧化铁矿获得了良好的选矿指标。该机与现场原有的 ϕ2000 平环强磁选机比较,不仅解决了平环强磁选机磁介质(齿板)频繁堵塞的技术难题,而且选矿指标大幅度提高。

SLon 立环脉动高梯度磁选机的发展与应用[①]

熊大和

（赣州有色冶金研究所）

摘　要　本文介绍了 SLon 立环脉动高梯度磁选机的研制历程及技术特点，该机具有运转可靠、分选效率高、磁介质不易堵塞、分选粒度范围宽、适应性强的优点，近年在马钢姑山铁矿、鞍钢弓长岭选矿厂、昆钢八街铁矿和罗次铁矿、宁夏中卫铁厂、山西二峰山铁矿应用于分选赤铁矿和褐铁矿，显著地提高了铁精矿品位和回收率，获得了良好的技术经济指标。

关键词　高梯度磁选机　赤铁矿　褐铁矿

1　发展概况

周期式高梯度磁选机在 20 世纪 70 年代就已被应用于高岭土提纯和废水处理，高梯度磁选技术已被公认为对微细弱磁性颗粒具有回收率较高的优点，人们很早就想利用这一技术来提高红矿细粒级选矿的回收率。针对铁矿选矿处理量大和磁性物料产率大的特点，20 世纪 70 年代末期和 80 年代初期，国内外均研制出平环高梯度磁选机，其中最有影响的为 Sala HGMS480 平环高梯度磁选机。该机转环外径 7.5m，介质宽达 1.5m，沿转环一周可配置 4 个磁极头。该机 1979 年 8 月在瑞典 Strassa 铁矿进行了从细粒尾矿中回收赤铁矿的工业试验，但此后很少见到应用这种设备的报道。

平环高梯度磁选机的给矿与排矿方向一致，粗粒和木渣草屑都必须穿过磁介质堆才能排走，从而存在磁介质较易堵塞、脉石的机械夹杂较严重、磁性精矿品位不高、对给矿粒度要求苛刻的缺点。磁介质较易堵塞是平环高梯度磁选机难以在红矿选矿工业中推广应用的主要原因。

中南工业大学于 1981 年开始在实验室研究振动、脉动高梯度磁选。大量的小型试验结果表明：振动或脉动高梯度磁选可显著提高磁性精矿的品位，并保持高梯度磁选对细粒弱磁性矿物回收率较高的优点。例如用振动高梯度磁选机分选赤铁矿，与无振动比较，铁精矿品位可提高 8%~12%，铁回收率基本不变。这种高效的选矿方法能否转化为生产力，关键在于研制出适应于工业生产的连续振动或脉动高梯度磁选机。

从 1986 年开始至今的 10 年中，赣州有色冶金研究所先后与中南工业大学、马钢姑山铁矿、赣州有色冶金机械厂、鞍钢矿山公司及弓长岭选矿厂合作研制出了 SLon-1000、SLon-1500 及 SLon-2000 立环脉动高梯度磁选机。该机针对铁矿选矿特点，在技术上采取了如下有力的措施：

（1）将设备工作可靠性和实用性放在第一位，尽可能做到坚固耐用、结构简单、便于操作与维护。将理论与实践相结合，不片面地追求理论上预测的最高选矿指标。

（2）将重选理论与磁选理论相结合，设置脉动机构驱动分选区的矿浆脉动，以利减少

[①] 原载于《全国矿山与冶金设备技术报告会论文集》，1996。

脉石的机械夹杂，提高铁精矿品位。

（3）转环立式旋转、反冲精矿。对于每一组磁介质堆而言，其给矿方向与排矿方向相反，粗粒和木渣草屑不必穿过磁介质堆便可冲洗出来，可有效地防止磁介质堵塞。

（4）扩大分选粒度范围，简化现场筛分作业。目前该机的分选粒度上限可达2mm，其有效捕收粒度下限为10μm左右，比平环高梯度磁选机的适应范围（0~0.2mm）大得多。实际上，该机的给矿粒度上限与齿板类强磁磁选机一致。

（5）为了克服网介质较易松动，经常需要添加和需要定期清洗的缺点，近年来部分SLon磁选机已开始采用棒介质，安装在弓长岭选矿厂SLon-1500磁选机上的棒介质已试用2年多，证明棒介质具有工作稳定、从不堵塞、寿命长、维护工作量小、选矿指标稳定的优点。棒介质的应用彻底解决了磁介质的堵塞问题，为该机的大型化打下了良好的基础。

由于SLon立环脉动高梯度磁选机具有运转可靠、分选效率高、磁介质不易堵塞（网介质）和不堵塞（棒介质）、分选粒度范围宽、适应性强的优点，该机近年在马钢姑山铁矿、鞍钢弓长岭选矿厂、合钢钟山铁矿、昆钢八街铁矿和罗次铁矿、宁夏中卫铁厂、山西二峰山铁矿等地得到应用，主要用于分选赤铁矿和褐铁矿，为各厂矿取得了较好的技术经济指标。目前在国内外有扩大应用之势。

SLon磁选机结构原理图如图1所示，其主要技术参数见表1。

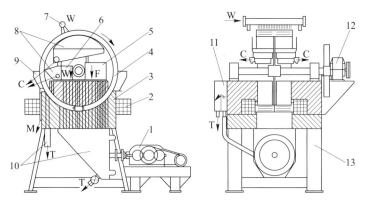

图1　SLon磁选机结构原理图

1—脉动机构；2—激磁线圈；3—铁轭；4—转环；5—给矿斗；6—漂洗水斗；
7—精矿冲洗装置；8—精矿斗；9—中矿斗；10—尾矿斗；11—液位斗；12—转环驱动机构；
13—机架；F—给矿；W—清水；C—精矿；M—中矿；T—尾矿

表1　SLon磁选机主要技术参数

型号	转环直径 /mm	背景场强 /T	激磁功率 /kW	驱动功率 /kW	脉动冲次 /r·min⁻¹	脉动冲程 /mm	给矿粒度 /mm
SLon-2000	2000	0~1.0	0~82	5.5+7.5	0~30	0~400	0~2.0
SLon-1500	1500	0~1.0	0~38	3+4	0~30	0~400	0~1.3
SLon-1000	1000	0~1.0	0~25.5	1.1+1.5	0~30	0~400	0~1.3

型号	处理能力 /m³·h⁻¹	给矿浓度 /%	给矿体积 /%	供水压力 /MPa	耗水量 /m³·h⁻¹	主机质量 /t	最大件质量 /t
SLon-2000	50~70	0~40	100~200	≥0.3	100~200	50	14
SLon-1500	20~35	0~40	50~100	≥0.3	50~100	20	5
SLon-1000	4~7	0~40	10~20	≥0.2	10~20	6	1.5

2 SLon 磁选机在赤铁矿选矿工业应用实例

2.1 在姑山细粒赤铁矿全磁选流程中的应用

姑山铁矿是马钢主要矿石原料基地之一，矿石属中温热液矿床，主要有用矿物为赤铁矿，其嵌布粒度细且不均匀，属难选矿石。1978年建成选矿厂以后，其选矿原则流程是以跳汰、螺旋溜槽、离心机为主的全重选流程。原矿破碎成小于12mm以后，用跳汰机选出部分块状铁精矿和丢弃部分块状尾矿，约占原矿一半的跳汰中矿用球磨机一次磨细并分成两个粒级后分别用螺旋溜槽和离心选矿机分选，多年来细粒选矿生产指标较低，如1988年细粒选矿生产指标为：入选品位37.16%Fe，精矿品位55.22%Fe，铁回收率仅61.32%。

为了提高选矿指标，1989~1992年，细粒选矿流程已改为阶段磨矿强磁高梯度全磁选流程，其中采用了两台SLon-1500立环脉动高梯度磁选机作为扫选设备。

该流程的综合选矿指标为：给矿品位35.51%，精矿品位58.17%，尾矿品位16.32%，铁回收率75.12%，与原流程比较，铁精矿品位提高了2.95%，回收率提高13.80%。回收率的提高主要依靠SLon-1500磁选机从SQC强磁选尾矿中回收了一部分本来很难回收的铁精矿。

2.2 用于弓长岭贫赤铁矿的分选

弓长岭选矿厂红矿选矿车间年处理贫赤铁矿260万吨，采用阶段磨矿阶段选别流程，其小于74μm（200目）占80%左右的细粒级产率占原矿的50%左右，生产用ϕ1600mm×900mm离心选矿机一次粗选抛尾和二次精选得精矿，其突出的问题是粗选离心机尾矿品位太高，严重影响了铁精矿的回收率。

1989年该厂采用一台SLon-1000立环脉动高梯度磁选机对该厂细粒级红矿进行了连续一个月的现场试验，共运转713h，与粗选离心机对比选矿指标见表2。

表2 SLon-1000机与粗选离心机选矿指标对比结果　　　　　　（%）

设备	铁品位			精矿产率	精矿铁回收率
	给矿	精矿	尾矿		
SLon-1000机	31.72	50.81	9.07	54.26	86.92
离心选矿机	31.82	46.69	19.23	45.85	67.27

SLon-1000机与粗选离心机比较，精矿品位高4.12%，尾矿品位低10.16%，回收率高19.65%，显著优于粗选离心机分选指标，这一试验为SLon磁选机在弓长岭选矿厂的推广应用打下了良好的基础。1992年5~7月，弓长岭选矿厂使用5台SLon-1500立环脉动高梯度磁选机取代7~8系统的24台ϕ1600mm×900mm粗选离心机，连续3个月达到的选矿指标与未改造流程（5~6系统）的选矿指标见表3。

表3 SLon-1500机取代粗选离心机后全流程指标 （%）

系统	原矿铁品位	精矿			尾矿铁品位
		铁品位	产率	铁回收率	
7、8	28.47	64.31	31.84	71.92	11.73
5、6	28.47	62.78	25.36	55.93	16.82
差值	0	+1.53	+6.48	+15.99	-5.08

从表3可见，7、8系统采用SLon-1500机后，尾矿品位显著下降，铁回收率提高15.99%，铁精矿品位提高1.53%，这一试验成果为提高鞍山式赤铁矿的选矿回收率提供了一条有效的途径。

2.3 SLon-2000机用于赤铁矿粗选抛尾作业

弓长岭选矿厂80年代安装了5台φ2m平环强磁选机用于一次球磨分级机溢流的粗选作业，因其给矿粒度粗达2mm及强磁性矿物较多，虽齿顶间隙放大到3.5mm，仍存在着齿板堵塞频繁、选矿指标不稳定和维修工作量大的问题。为了解决这些问题，该厂于1994年8月安装了一台SLon-2000立环脉动高梯度磁选机与平环强磁选机进行工业对比试验，分选0~2mm的一次球磨机溢流，运转6个月的平均对比指标见表4。

表4 SLon-2000机与平环强磁机对比结果

设备	给矿/%			精矿/%			尾矿铁品位/%	处理量/t·h^{-1}
	浓度	小于74μm(200目)	铁品位	铁品位	产率	铁回收率		
SLon-2000	35~45	50	25.94	41.21	49.89	79.25	10.74	60
平环强磁	35~45	50	25.94	34.00	54.90	71.95	16.13	40
差值	0	0	0	7.21	-5.01	7.30	-5.39	20

由表4可知SLon-2000机的各项指标均显著优于平环强磁选机，更重要的是该机磁介质从不堵塞，从根本上克服了齿板类强磁机介质板堵塞问题，显著地提高了设备作业率和减少了设备检修工作量。

2.4 用SLon-1500机回收尾矿中赤铁矿

由于SLon磁选机有较大的富集比，可从部分选矿厂的尾矿中一次分选得到较高的铁精矿品位，目前应用SLon-1500立环脉动高梯度磁选机从尾矿中回收赤铁矿的实例有昆钢罗次铁矿和山西二峰铁矿，生产指标见表5。

表5 SLon-1500机从尾矿中回收赤铁矿指标 （%）

厂矿	SLon给矿铁品位	精矿			尾矿铁品位
		产率	铁品位	铁回收率	
罗次铁矿	30.39	32.18	55.01	58.25	18.71
二峰山铁矿	18.62	18.95	53.64	54.60	10.43

3 SLon 磁选机在褐铁矿选矿工业中的应用

3.1 SLon 磁选机在宁夏中卫铁厂的应用

宁夏中卫铁厂于 1994 年建成褐铁矿选矿厂，采用一段磨矿，2 台 SLon-1500 磁选机一粗一扫流程，该流程结构简单、投资少、见效快。其选矿生产指标见表 6。

表 6　SLon 磁选机分选中卫铁厂褐铁矿指标　　　　　　（%）

厂矿	给矿铁品位	精矿			尾矿铁品位
		产率	铁品位	铁回收率	
SLon 粗选	46.36	49.84	54.17	58.24	38.60
SLon 扫选	38.60	42.21	52.43	57.33	28.50
综合指标	46.36	71.01	53.65	82.18	28.50

3.2 SLon 磁选机在昆钢八街铁矿的应用

八街铁矿选矿车间处理的矿石主要是褐铁矿，原采用一段磨矿、用 ϕ1500mm 双立环球介质强磁选机一粗一扫的工艺流程。其突出的生产问题是尾矿品位太高，铁回收率低。1993 年该矿采用了 1 台 SLon-1500 机取代原有的扫选强磁选机。改造前后的对比指标见表 7。

表 7　SLon 磁选机在八街铁矿应用对比指标　　　　　　（%）

	流程或设备	给矿		精矿			尾矿铁品位
		小于 0.074mm（200 目）	铁品位	产率	铁品位	铁回收率	
改造前	强磁机一粗一扫综合	70	38.69	35.27	51.30	46.76	31.82
改造后	强磁精选作业	70	38.81	19.64	51.78	26.20	35.64
	SLon 扫选作业	75	35.64	36.76	52.72	54.38	25.71
	综合指标	70	38.81	49.18	52.34	66.33	25.71

由表 7 可知，SLon-1500 机取代扫选球介质强磁选机后，与粗选球介质强磁选机比较，在给矿品位较低的条件下，获得了较高的铁精矿品位和高得多的作业回收率，使全流程的综合铁精矿品位提高 1.04%，回收率提高 19.57%，尾矿品位降低 6.11%，取得了显著的技术经济效果。

4 结论

（1）在 SLon 立环脉动高梯度磁选机 10 年研制过程中，将设备的可靠性和实用性放在第一位；采用转环立式旋转、反冲精矿并配置脉动机构等措施防止磁介质堵塞和提高选矿指标。现已研制出 SLon-1000、SLon-1500 和 SLon-2000 立环脉动高梯度磁选机，其台时处理量为 4~70t。该机已成为我国红矿选矿工业中应用较为广泛的新型高效磁选设备。

（2）SLon 磁选机具有很强的适应能力，在马钢姑山铁矿、鞍钢弓长岭选矿厂、昆钢八街铁矿、宁夏中卫铁厂等地的应用证明，它既可用于分选粒度很细、磁性很弱的褐铁矿，又可用于分选粒度粗达 2mm（小于 0.074mm（200 目）占 50% 左右）和含有较多强磁性矿物的赤铁矿，并可获得良好的选矿指标。

（3）SLon 磁选机采用棒介质后，彻底克服了普通强磁选机磁介质堵塞这个长期未能解决的问题，标志着我国强磁选设备的发展进入一个新的阶段。

写作背景　本文是作者第一次在全国性的学术会议上发表的论文，向同行介绍了 SLon-1000、SLon-1500 和 SLon-2000 立环脉动高梯度磁选机的研制与应用现状。文中列举了 SLon 磁选机与离心选矿机、平环强磁选机和球介质强磁选机的选矿指标对比数据，表明了 SLon 磁选机可获得较高的选矿指标。

弱磁性矿粒在棒介质高梯度磁场中的动力学分析[1]

熊大和

(赣州有色冶金研究所)

摘 要 本文推导了弱磁性矿粒在单根棒介质高梯度磁场中的动力学方程,计算了弱磁性矿粒在磁力捕获区的运动速度、加速度和捕获时间,绘出了矿粒运动轨迹图,为有关理论研究、设备研究和生产应用提供了一些有参考价值的数据。

关键词 高梯度磁选 弱磁性矿粒 圆柱形磁介质

Dynamic Analysis of Weakly Magnetic Mineral Particles in High Gradient Magnetic Field of Rod Medium

Xiong Dahe

(*Ganzhou Nonferrous Metallurgy Research Institute*)

Abstract A dynamic equation for weakly magnetic mineral particles in high gradient magnetic field of single rod medium is derived. The movement velocity, acceleration and capture time of weakly magnetic particles in the magnetic capture zone are calculated and the movement path of particles is plotted, which provides data of reference value for the related theoretic study, equipment development and practical application.

Keywords High gradient magnetic separation; Weakly magnetic mineral particles; Cylindrical magnetic medium

圆柱形磁介质(或称棒介质)具有工作可靠性高、易于实现优化组合排列和不易堵塞的优点。近年来在工业生产中应用的 SLon 立环脉动高梯度磁选机越来越多地采用棒介质,因此研究棒介质的选矿机理对设备研究和应用具有一定的指导意义。

1 方程的推导

1.1 矿粒所受磁力

根据电磁场理论,可推导出无限长圆柱形铁素体外的磁场强度表达式为:

---

[1] 原载于《金属矿山》1998 (08): 3~5。

$$H_r = \left(1 + \frac{M}{2H_0} \cdot \frac{a^2}{r^2}\right) H_0 \cos\theta$$

$$H_\tau = -\left(1 - \frac{M}{2H_0} \cdot \frac{a^2}{r^2}\right) H_0 \sin\theta$$

由上式得：

$$H^2 = H_r^2 + H_\tau^2 = H_0^2 \left(1 + \frac{Ma^2}{2H_0 r^2}\cos 2\theta + \frac{M^2}{4H_0^2} \cdot \frac{a^4}{r^4}\right)$$

$$\text{grad}H^2 = \frac{\partial(H^2)}{\partial r}\boldsymbol{r}^0 + \frac{1}{r}\frac{\partial(H^2)}{\partial \theta}\boldsymbol{\tau}^0$$

$$= -\left(\frac{2H_0 Ma^2}{r^3}\cos 2\theta + \frac{M^2 a^4}{r^5}\right)\boldsymbol{r}^0 - \left(\frac{2H_0 Ma^2}{r^3}\sin 2\theta\right)\boldsymbol{\tau}^0 \tag{1}$$

根据弱磁性矿粒所受磁力公式：

$$\boldsymbol{f} = xmH\text{grad}B = \frac{1}{2}xm\mu_0 \text{grad}H^2 \tag{2}$$

将式（1）代入上式并注意到 $B_0 = \mu_0 H_0$，$m = \frac{4}{3}\pi b^3 \delta$ 得：

$$\boldsymbol{f} = -(6\pi b\eta v_m)\frac{a^3}{r^3}\left(\cos 2\theta + \frac{M}{2H_0} \cdot \frac{a^2}{r^2}\right) \cdot \boldsymbol{r}^0 - (6\pi b\eta v_m)\frac{a^3}{r^3}\sin 2\theta \cdot \boldsymbol{\tau}^0 \tag{3}$$

式中，v_m 为具有速度量纲，习惯上称为磁速度：

$$v_m = \frac{2}{9}\frac{b^2 \delta x B_0 M}{\eta a}$$

现结合图 1 对式（3）作些分析：磁力 f 至少与 r^3 成反比，这表明磁力随着 r 的增加而迅速减小，当 $r=3a$ 时，磁力小于或等于 $r=a$ 时的 1/27，因而普遍认为磁力的作用范围为 $a<r\leqslant 3a$。圆柱形磁介质的磁场可分为图 1 虚直线所分割的 4 个区域，在磁力线密集的两个区域内，顺磁性矿粒受到磁场引力；在其余两个区域内，矿粒受到的磁场力很小或受到斥力，在分析顺磁性矿粒的捕集情况时，这两个区域将不予考虑。

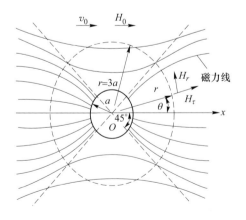

图 1　圆柱形磁介质周围的磁力场

1.2 矿粒所受到的流体力

根据流体力学的推导,当流体为理想流体时,圆柱体外任意一点流体的流速为:

$$u = u_r \boldsymbol{r}^0 + u_\tau \boldsymbol{\tau}^0 = \left(1 - \frac{a^2}{r^2}\right) v_0 \cos\theta \cdot \boldsymbol{r}^0 - \left(1 + \frac{a^2}{r^2}\right) v_0 \sin\theta \cdot \boldsymbol{\tau}^0 \tag{4}$$

在高梯度磁选中,虽然矿浆不是理想流体,但矿浆流速较低,雷诺数较小,圆柱体迎水面的流线与理想流体的流线大致相同(如图2所示)。

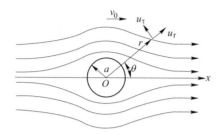

图2 理想流体在圆柱体附近的流场

设矿粒相对于圆柱体的运动速度为:

$$\boldsymbol{v} = v_r \boldsymbol{r}^0 + v_\tau \boldsymbol{\tau}^0 = \frac{\mathrm{d}r}{\mathrm{d}t} \boldsymbol{r}^0 + r \frac{\mathrm{d}\theta}{\mathrm{d}t} \boldsymbol{\tau}^0 \tag{5}$$

将圆柱体视为静止参考系,流体视为牵连参考系,矿粒视为运动物体,矿粒相对于流体的运动速度为:

$$\boldsymbol{v} - \boldsymbol{u} = \left(\frac{\mathrm{d}r}{\mathrm{d}t} - u_r\right) \boldsymbol{r}^0 + \left(r \frac{\mathrm{d}\theta}{\mathrm{d}t} - u_\tau\right) \boldsymbol{\tau}^0 \tag{6}$$

假定微细矿粒所受流体阻力服从斯托克斯阻力公式,即:

$$\boldsymbol{R} = -6\pi b \eta (\boldsymbol{v} - \boldsymbol{u})$$

上式右端的负号表示流体阻力与矿粒相对于流体运动的方向相反,将式(6)和式(4)代入上式得:

$$\boldsymbol{R} = 6\pi b \eta \left[\left(1 - \frac{a^2}{r^2}\right) v_0 \cos\theta - \frac{\mathrm{d}r}{\mathrm{d}t}\right] \boldsymbol{r}^0 - 6\pi b \eta \left[\left(1 + \frac{a^2}{r^2}\right) v_0 \sin\theta + r \frac{\mathrm{d}\theta}{\mathrm{d}t}\right] \boldsymbol{\tau}^0 \tag{7}$$

1.3 矿粒相对于圆柱形磁介质运动的加速度方程

在高梯度磁选中,微细矿粒所受到的重力相对于磁力和流体阻力可忽略不计,例如当一颗直径为10μm的黑钨矿在水中以1cm/s的速度运动时,它所受到的流体阻力大约是它所受重力的250倍,因此以下考虑的竞争力主要是磁力和流体阻力。

根据牛顿第二力学定律,矿粒相对于圆柱形磁介质的加速度为:

$$\boldsymbol{a} = a_r \boldsymbol{r}^0 + a_\tau \boldsymbol{\tau}^0 = \frac{1}{m}(\boldsymbol{f} + \boldsymbol{R})$$

将式(3)、式(7)及 $m = \frac{4}{3}\pi b^3 \delta$ 代入上式,整理得其分量表达式为:

$$a_r = \frac{9\eta}{2b^2\delta} \left[\left(1 - \frac{a^2}{r^2}\right) v_0 \cos\theta - \frac{dr}{dt} - v_m \left(\cos 2\theta + \frac{M}{2H_0} \cdot \frac{a^2}{r^2} \right) \frac{a^3}{r^3} \right]$$

$$a_\tau = \frac{9\eta}{2b^2\delta} \left[\left(1 + \frac{\alpha^2}{r^2}\right) v_0 \sin\theta + r \frac{d\theta}{dt} + v_m \sin 2\theta \cdot \frac{a^3}{r^3} \right] \tag{8}$$

以上便是矿粒相对于圆柱形磁介质的加速度方程，给定初始条件可计算矿粒在任一时刻的加速度、速度、位置及绘出矿粒的运动轨迹。

2 计算机求解及计算结果分析

计算区间的划分如图 3 所示，假定矿浆由上至下流动，矿浆中有 9 颗球形磁性矿粒分别从 1~9 号位置随矿浆进入磁力场范围内。将磁力场用经线和纬线均匀划开，假定 $N=1$ 的纬线与各经线的交点是矿粒的初始位置，矿粒进入磁力场的方向与矿浆流动方向相同，那么位于点 $(R(1), \theta)$ 的矿粒的初速度为：

$$v_r|_{t=0} = v_0 \cos\theta, \quad v_\tau|_{t=0} = v_0 \sin\theta \tag{9}$$

至此，每一矿粒的初始位置和初速度确定了，根据式（8）和式（9），用计算机分别计算了半径为 $1.0\mu m$、$2.5\mu m$、$5\mu m$、$10\mu m$、$20\mu m$ 5 种不同粒度的黑钨矿的五套动力学数据。计算方法采用标准四阶龙格—库塔法，计算步长（时间间隔）是笔者建立的预估—迭代法。表 1 和表 2 列出了计算中必要的基本参数。

表 1 计算不同粒度矿粒的参数值

位置	矿粒半径 b /μm	磁介质半径 a /μm	矿浆背景流速 v_0 /cm	计算角度 θ /(°)	纬线最大规范半径 R_S	相邻纬线间距 R_I
1 号	1.0	25	-1	0~30	2	0.05
2 号	2.5	50	-2	0~30	3	0.1
3 号	5.0	100	-3	0~3	3	0.1
4 号	10	150	-4		3	0.1
5 号	20	200	-6	0~40	3	0.1

注：$R_S = r/a$，最大 $R_S = R(1)$，$R_I = R(1) - 1/20$。

表 2 各程序公用参数名称

黑钨矿比磁化系数 χ /$m^3 \cdot kg^{-1}$	黑钨矿密度 δ /$kg \cdot m^{-3}$	矿浆黏性系数 η /$N \cdot s \cdot m^{-2}$	背景磁场强度 H_0 /$A \cdot m^{-1}$	磁介质磁化强度 M /$A \cdot m^{-1}$
0.4125×10^{-6}	7300	0.0011	79.58×10^4	135.28×10^4

根据计算结果，如图 3 所示绘出了两种不同半径黑钨矿的运动轨迹。其中可看出较粗矿粒向中线靠拢的速度较快，这说明磁场作用在矿粒上的切向力随粒度增长的幅度较流体力更大，较粗矿粒能够被磁力捕获的范围也大一些，例如半径小于或等于 $5\mu m$ 的矿粒从方位角为 40°的经线与 $N=1$ 的纬线交点处进入磁力场，则磁力竞争不过流体力，矿粒不能被捕获。

如图 4 所示绘出了各矿粒沿 $\theta=30°$ 运动的径向速度曲线，容易看出矿粒离磁介质由远而近速度越来越快，矿粒越粗速度变化越显著，反之亦然。

如图 5 所示绘出了各粒度矿粒沿 $\theta=30°$ 运动的径向加速度随位置的变化关系。从中可见磁性矿粒邻近磁介质表面的加速度是很大的。

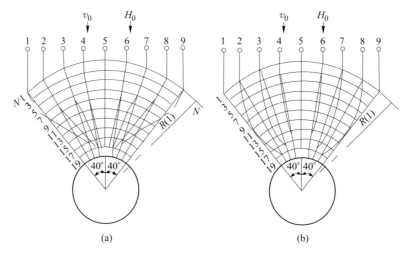

图 3　黑钨矿的运动轨迹

(a) 半径为 10μm 黑钨矿的运动轨迹；(b) 半径为 2.5μm 黑钨矿的运动轨迹

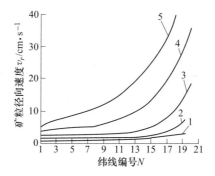

图 4　矿粒径向速度随位置的变化关系

(初始角 $\theta=30°$)

1—$b=1.0$μm；2—$b=2.5$μm；3—$b=5.0$μm；4—$b=10.0$μm；5—$b=20.0$μm

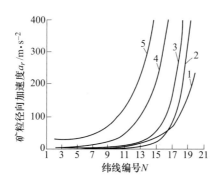

图 5　矿粒径向加速度随位置的变化关系

(初始角 $\theta=30°$)

1—$b=1.0$μm；2—$b=2.5$μm；3—$b=5.0$μm；4—$b=10.0$μm；5—$b=20.0$μm

如图 6 所示绘出了各矿粒沿 $\theta=30°$ 运动的时间随位置的变化关系。从中可看出，在表

1 所定的条件下，各矿粒在磁力场中运动达到终点的时间相差不大，在同一个数量级范围内。

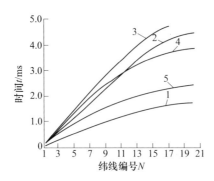

图 6　矿粒运动时间随位置的变化关系
（初始角 $\theta = 30°$）
1—$b = 1.0\mu m$；2—$b = 2.5\mu m$；3—$b = 5.0\mu m$；4—$b = 10.0\mu m$；5—$b = 20.0\mu m$

3　结语

（1）振动（磁介质振动、矿浆不振动）或脉动（矿浆振动、磁介质静止）频率对捕获矿粒有多大的影响？如图 6 所示，磁介质捕获矿粒的时间一般不超过 5ms。假定磁介质以 1500r/min 的频率振动，则振动周期为 40ms；或假定矿浆以 300r/min（工业上常用）脉动时，振动周期为 200ms。因此，无论振动还是脉动，磁介质有充分的时间捕获矿粒，对磁性物回收率影响很小。

（2）振动或脉动频率和振幅适当时，磁介质与矿浆相对运动速度出现时正时负的现象，即磁介质的迎水面和背水面不断交替转化，计算结果表明，这有利于磁介质两面捕获矿粒。

（3）根据计算数据绘出的矿粒运动轨迹图（如图 3 所示）可清楚地看到不同粒度的矿粒从不同角度进入磁场的运动路线，这有助于深入了解高梯度磁选捕收机理。

（4）磁性矿粒邻近磁介质表面的加速度是很大的，例如 $1\mu m$ 和 $10\mu m$ 矿粒到达磁介质表面时的加速度分别为 $238 \sim 334 m/s^2$ 和 $1974 \sim 2151 m/s^2$。当磁介质的振幅为 1mm，振次为 1500r/min 时，其最大加速度为 $24.7 m/s^2$。这说明振动对已被吸附在磁介质表面的磁性矿粒产生的惯性力与它们所受的磁力相比仍是次要的。

写作背景　本文是作者在攻读硕士学位时所做的一些理论性的工作，是硕士论文中一部分的缩写。介绍了弱磁性矿粒在圆柱形磁介质周边所受到的磁力、流体力及其运动轨迹、运动速度、运动加速度和运动时间。为了解高梯度磁选原理有一定的帮助。

SLon-1750 立环脉动高梯度磁选机的研制与应用

熊大和[1]　杨庆林[2]　谢金清[2]　钱士湖[2]　汤桂生[2]

（1. 赣州有色冶金研究所；2. 姑山矿业公司）

摘　要　本文针对马钢姑山铁矿用于赤铁矿粗选作业的 SQC-6-2770A 型湿式强磁选机精矿品位偏低、尾矿品位偏高、齿板较易堵塞的问题，研制了 SLon-1750 立环脉动高梯度磁选机，取代 1 台 SQC-6-2770A 磁选机进行工业对比试验，证明了该机具有富集比较大，选矿效率高，磁介质不堵塞的优点。

关键词　SLon-1750 磁选机　赤铁矿选矿　脉动高梯度磁选

The Development and Application of SLon-1750 Vertical Ring Pulsating High Gradient Magnetic Separator

Xiong Dahe[1]　Yang Qinglin[2]　Xie Jinqing[2]　Qian Shihu[2]　Tang Guisheng[2]

（1. Ganzhou Nonferrous Metallurgy Research Institute；2. Gushan Mining Company）

Abstract　In view of the problems of low concentrate grade, high tailings grade and the easy plugging of the tooth plate of SQC-6-2770A high-intensity wet magnetic separators used in hematite roughing in Gushan Iron Mine, Maanshan Iron & Steel Co., a SLon-1750 vertical ring pulsating high gradient magnetic separator was developed. Its industrial comparative test with the existing SQC-6-2770A magnetic separator has shown that it has the advantages of great concentration ratio, high separation efficiency and unplugging of magnetic medium.

Keywords　SLon-1750 magnetic separator；Hematite separation；Pulsating high gradient magnetic separation

1　SLon-1750 磁选机的研制

SLon 型立环脉动高梯度磁选机是一种适合于分选细粒弱磁性矿物的高效磁选设备，其转环立式旋转、反冲精矿，并配有脉动机构，具有富集比大、分选效率高、磁介质不易堵塞、对给矿粒度、浓度和品位波动适应性强、工作可靠、操作维护方便等优点。作者通过十余年的研究，已研制出 SLon-750、SLon-1000、SLon-1250、SLon-1500、SLon-2000 等型号的立环脉动高梯度磁选机，它们已在工业上广泛应用于分选赤铁矿、菱铁矿、褐铁

● 原载于《金属矿山》1999（10）：23~26。

矿、钛铁矿及长石矿的提纯。

姑山铁矿是马钢主要原料基地之一，主要金属矿物为赤铁矿，矿石硬度大、嵌布粒度细且不均匀，属难选红铁矿。其选矿厂自 1978 年投产以后，选矿工艺流程历经多次改进，生产技术指标稳步提高，但随着矿山服务年限的延长，矿床深部的矿石开采量日益增加，矿石性质逐渐发生变化，其中矿石含磷增高是影响铁精矿质量的主要因素之一。

姑山选厂主厂房现采用强磁-脉动高梯度磁选、阶段磨选流程选别跳汰中矿。其中一段球磨分级溢流采用 3 台 SQC-6-2770A 型强磁选机抛尾。由于给矿粒度较粗（小于 74μm（200 目）占 42%左右），该机齿板经常堵塞，粗精矿品位较低，尾矿品位较高，使全厂的选矿指标受到较大的影响，目前的生产流程和选别设备越来越不能适合生产的需要。

为使姑山选矿技术进一步提高，解决 SQC-6-2770A 强磁选机精矿品位偏低、尾矿品位偏高和齿板较易堵塞的问题，研制出了新一代的 SLon-1750 立环脉动高梯度磁选机，并在生产上取代了 1 台 SQC-6-2770A 型强磁选机。

1.1 设备结构及工作原理简介

SLon-1750 磁选机结构如图 1 所示。该机主要由脉动机构、激磁线圈、铁轭、转环和各种矿斗、水斗等组成。

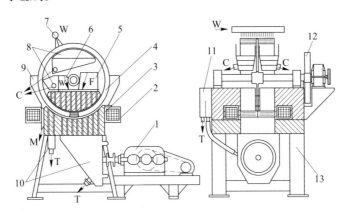

图 1 SLon-1750 磁选机结构

1—脉动机构；2—激磁线圈；3—铁轭；4—转环；5—给矿斗；6—漂洗水斗；
7—精矿冲洗装置；8—精矿斗；9—中矿斗；10—尾矿斗；10—液位斗；
12—转环驱动机构；13—机架；F—给矿；W—清水；C—精矿；M—中矿；T—尾矿

该机的工作原理为：当激磁线圈通以直流电时，在激磁线圈的内腔产生磁场，通过铁轭的导磁作用将磁势降尽可能地压缩在圆弧形分选腔内，使分选腔内获得一个较高的背景磁场。

该机转环内装有导磁不锈钢棒磁介质。选矿时，转环做顺时针旋转，矿浆从给矿斗给入，沿上铁轭缝隙流经转环，转环内的磁介质在磁场中被磁化，磁介质表面形成高梯度磁场，矿浆中的磁性颗粒被吸着在磁介质表面，随转环转动被带至顶部无磁场区，用冲洗水冲入精矿斗中，非磁性颗粒沿下铁轭缝隙流入尾矿斗排走。

该机转环立式旋转，如果给矿中有粗颗粒不能穿过磁介质堆，粗粒一般会停留在磁介质上表面，即靠近转环内圆周；当磁介质被转环带至顶部时，已经旋转了 180°，粗粒位于磁介质下部，不必穿过磁介质堆即可很容易地被冲洗水冲入精矿斗。

该机的脉动机构由鼓膜和冲程箱组成。当鼓膜在冲程箱的驱动下做往复运动时，只要矿浆液面高度能浸没转环下部的磁介质，分选室的矿浆便做上下往复运动，脉动流体力使矿粒群在分选过程中始终保持松散状态，从而有效地消除非磁性颗粒的机械夹杂，显著地提高磁性精矿品位。此外，矿浆脉动显然对防止磁介质堵塞也有好处。

1.2 主要技术参数及电磁性能

主要技术参数、电磁性能见表1、图2和表2。

表1 主要技术参数

转环外径 /mm	转环宽度 /mm	额定背景磁感强度 /T	最高背景磁感强度 /T	额定激磁电流 /A	额定激磁电压 /V	额定激磁功率 /kW	传动功率 /kW	脉动冲程 /mm
1750	750	1.0	1.15	1050	34	36	5.5+4	0~20
脉动冲次 /r·min^{-1}	给矿粒度 /mm	给矿浓度 /%	矿浆通过能力 /m³·h^{-1}	干矿处理量 /t·h^{-1}	供水压力 /MPa	耗水量 /t	主机质量 /t	最大部件质量 /t
0~300	0~1.3	10~50	75~150	30~50	0.2~0.4	80~150	35	11

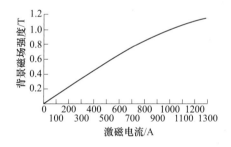

图2 SLon-1750立环脉动高梯度磁选机激磁曲线

表2 SLon-1750立环脉动高梯度磁选机电磁性能测定数据

档 次	I/A	U/V	P/kW	B_0/T
Ⅰ-低	550	18.0	9.9	0.594
Ⅰ-中	600	19.0	11.4	0.638
Ⅰ-高	620	20.0	12.4	0.660
Ⅱ-低	900	28.0	25.2	0.905
Ⅱ-中	940	30.0	28.2	0.931
Ⅱ-高	1000	32.0	32.0	0.970
Ⅲ-低	1180	39.0	46.0	1.089
Ⅲ-中	1220	42.0	51.2	1.109
Ⅲ-高	1300	43.0	55.9	1.144

1.3 技术改进措施

为了提高SLon-1750磁选机的机电性能及选矿性能，研制该机时采用了较多的新技术。与以往的SLon型立环脉动高梯度磁选机比较，研制该机时主要采用了如下技术改进措施：

(1) 采用电子计算机 CAD 辅助设计与制图技术，对所有图纸进行精密设计与装配模拟定位，消除了设计上可能出现的误差，绘制了成套高精度的图纸。

(2) 采用全滚道数值式脉动冲程箱。冲程箱内往复杆的质量全部支承在滚动轴承上，采用数字式偏心套调节方式，具有摩擦阻力小、力量大、寿命长，适应于较高频率的冲次，冲程输出准确且易调节的优点。

(3) 磁系设计采用了水内冷、低电压、大电流、低电流密度的激磁方式；铁轭的设计采用电子计算机优化设计，尽可能减少漏磁通，从图 2 和表 2 可知，该机达到额定背景磁感强度 1.0T 时，激磁电流为 1040A，激磁电压 34V，激磁功率仅 36kW（该机的处理能力为 SLon-1500 磁选机的 1.6 倍，而 SLon-1500 磁选机达到 1.0T 时激磁功率为 42kW），因此，该机具有能耗低、效率高的优点。

(4) 该机采用 φ4mm 导磁不锈钢棒为磁介质，在磁场理论研究、选矿探索试验的基础上采用计算机精确定位和优化组合排列，具有选别效率高、长期不堵塞、高度耐磨、寿命长、磁介质维护工作量极小的优点。

(5) 研制高耐磨尾矿流量控制阀。SLon 磁选机的尾矿斗下部流量控制阀因要承受脉动矿浆的高压冲刷，过去市面上购买的阀门使用寿命仅一个月左右。为了解决尾矿阀寿命短的问题，研制 SLon-1750 磁选机时专门研制了高耐磨的尾矿阀。该阀采用流线型节流及耐磨材料衬里，预计该阀的使用寿命是普通阀门的 10 倍以上。

(6) 给矿斗、精矿斗及各矿浆通道均采用优化设计，易受矿浆冲刷的地方均采用耐磨橡胶衬里或增加缓冲装置，延长其使用寿命。

2 工业应用

姑山选厂主厂房现有的选矿生产流程及 SLon-1750 磁选机安装位置见图 3，其中 SLon-1750 磁选机取代了原有的 4 号 SQC-6-2770A 强磁选机作为粗选抛尾设备。SLon-1750 磁选机自 1999 年 4 月 1 日投产，至 4 月 22 日连续运转 22d 共 528h。该机与 5 号 SQC-6-2770 A 强磁选机平均指标对比见表 3。

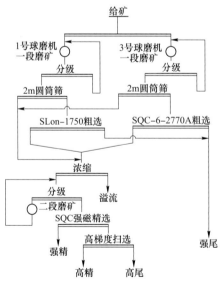

图 3 姑山选矿厂主厂房选矿流程

表 3 磁选机 SLon-1750 与 SQC-6-2770A 对比指标

型号	干矿处理量 /t·h^{-1}	给矿细度（小于74μm（200目））/%	给矿浓度 /%	给矿铁品位 /%	精矿铁品位 /%	尾矿铁品位 /%	精矿产率 /%	铁回收率 /%	磁感强度 /T	激磁功率 /kW	传动功率 /kW	总功率 /kW
SLon	33~35	33~60	20~50	40.45	51.55	14.91	69.72	88.84	1.11（背景）	52	5.5+4	61.5
SQC	33~35	30~60	20~50	42.42	48.48	25.31	73.83	84.39	1.3（齿尖）	60	10	70
差值	0	0	0	-1.97	+3.07	-10.40	-4.11	+4.45		-8	-0.5	-8.5

从表 3 可知，SLon-1750 磁选机与 SQC-6-2770 A 磁选机比较有如下优点。

（1）在给矿品位低 1.97 个百分点的情况下，铁精矿品位高 3.07 个百分点，有利于最终铁精矿品位的提高，体现了该机富集比较高的优点。

（2）尾矿品位低 10.40 个百分点，可大幅度降低全流程的尾矿品位，提高全流程的铁回收率。

（3）精矿产率低 4.11%，降低了二段球磨机的入磨量及其负荷，有利于提高二段磨矿细度及二段磁选机精选指标。

（4）铁回收率提高 4.45%。加上该机精矿品位较高，可显著提高二段磁选机精选和全流程的铁回收率。

（5）该机总电耗低 8.5kW，每年可节电 $5.3×10^4$ kW·h。

3 结语

（1）针对马钢姑山铁矿现有 SQC-6-2770 A 强磁选机精矿品位偏低、尾矿品位偏高和齿板较易堵塞的问题，研制了处理量相当的 SLon-1750 立环脉动高梯度磁选机，采用 CAD 辅助设计与制图技术，全滚道数值式脉动冲程箱，水内冷，低电压、大电流、低电流密度激磁等新技术，使该机具有优异的机电性能和选别性能。

（2）通过工业试验证明，SLon-1750 磁选机各项指标全面超过 SQC-6-2770 A 强磁选机，充分证明了该机具有富集比大、选矿效率高的优点。此外，该机的磁介质具有耐磨和长期不堵塞的优点，从根本上解决了 SQC-6-2770A 强磁选机齿板堵塞频繁的缺点。

（3）该机研制成功，不仅为姑山铁矿选矿生产指标上台阶提供了技术保证，而且为国内外弱磁性矿石的选矿工业提供了又一新型号的高效磁选设备。

写作背景 马钢姑山铁矿原采用的 SQC-6-2770A 平环强磁选矿机存在齿板磁介质容易堵塞的缺点，为了解决这个难题和进一步提高选矿指标，1998 年作者团队研制了首台 SLon-1750 立环脉动高梯度磁选机，这两种机型处理量和安装空间相当。通过工业试验证明了 SLon 磁选机各方面优于 SQC 磁选机。之后的两年中，姑山铁矿 6 台 SQC-6-2770A 平环强磁选机由 6 台 SLon-1750 磁选机取代。

应用 SLon 立环脉动高梯度磁选机提高资源利用率

熊大和

（赣州有色冶金研究所）

摘 要 SLon 立环脉动高梯度磁选机是一种利用磁力、脉动流体力和重力连续分选细粒弱磁性矿物的新型设备。文中介绍了该机结构、工作原理及在昆钢罗次铁矿、上厂铁矿、攀钢选钛厂、安徽来安皖东长石厂应用情况。实践证明，该机可有效地回收细粒红铁矿、钛铁矿及用于非金属矿提纯工业。为充分利用我国的矿产资源，延长矿山服务年限、降低生产成本和增加效益做出了一定的贡献。

关键词 磁选 立环脉动高梯度磁选机 红铁矿 钛铁矿 长石提纯

我国的矿产资源具有储量大、品位低、天然结晶粒度细、难选矿石多等特点，许多矿产资源因无合适的选矿方法和设备暂时得不到利用。细粒选矿回收率较低或精矿质量达不到商品矿的要求，造成了我国大量的矿产资源浪费或得不到合理的开采利用。因此，研究高效的选矿设备和工艺对提高我国矿产资源的利用率具有重要的意义。

针对我国细粒弱磁性矿物选矿效率较低的问题，我们从 1986 年开始研制 SLon 立环脉动高梯度磁选机，采用转环立式旋转、反冲精矿、配置脉动机构松散矿粒群及磁介质优化组合等措施。显著提高了分选效率，成功地解决了国内外平环式强磁选机和平环式高梯度磁选机磁介质容易堵塞的问题。迄今已研制出 SLon-750、SLon-1000、SLon-1250、SLon-1500、SLon-1750、SLon-2000 等型号的立环脉动高梯度磁选机。该系列磁选机分选粒度下限可达 $10\mu m$ 左右，并且具有富集比大、分选效率高、磁介质不易堵塞、对给矿粒度、浓度和品位波动适应性强，且工作可靠、操作维护方便等优点。

1 设备结构和工作原理简介

SLon 立环脉动高梯度磁选机（以下简称 SLon 磁选机）结构如图 1 所示，它主要由脉动机构、激磁线圈、铁轭、转环和各种矿斗、水斗组成。用导磁不锈钢制成的钢板网或圆棒作磁介质。

该机工作原理为：激磁线圈通以直流电，在分选区产生感应磁场，位于分选区的磁介质表面产生非均匀磁场即高梯度磁场；转环作顺时针旋转，将磁介质不断送入和运出分选区；矿浆从给矿斗给入，沿上铁轭缝隙流经转环，矿浆中的磁性颗粒吸附在磁介质棒表面上，被转环带至顶部无磁场区，被冲洗水冲入精矿斗，非磁性颗粒在重力、脉动流体力的作用下穿过磁介质堆，与磁颗粒分离。然后沿下铁轭缝隙流入尾矿斗排走。

该机的转环采用立式旋转方式，对于每一组磁介质而言，冲洗磁性精矿的方向与给矿

---

❶ 原载于《金属矿山》，2000（09）：200~203。

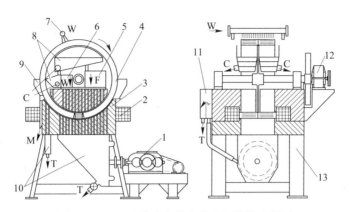

图1 SLon立环脉动高梯度磁选机结构示意图

1—脉动机构；2—激磁线圈；3—铁轭；4—转环；5—给矿斗；6—漂洗水斗；7—精矿冲洗装置；
8—精矿斗；9—中矿斗；10—尾矿斗；11—液位斗；12—转环驱动机构；13—机架；
F—给矿；W—清水；C—精矿；M—中矿；T—尾矿

方向相反，粗颗粒不必穿过磁介质堆便可冲洗出来。该机的脉动机构驱动矿浆产生脉动，可使位于分选区磁介质堆中的矿粒群保持松散状态，使磁性矿粒更容易被捕获，使非磁性矿粒尽快穿过磁介质堆进入到尾矿中去。

显然，反冲精矿和矿浆脉动可防止磁介质堵塞；脉动分选可提高磁性精矿的质量。这些措施保证了该机具有较大的富集比、较高的分选效率和较强的适应能力。SLon系列立环脉动高梯度磁选机主要技术参数见表1。

表1 SLon磁选机主要技术参数

机型	转环外径/mm	背景场强/T	激磁功率/kW	传动电动机/kW	脉动冲程/mm	脉动冲次/r·min⁻¹	给矿粒度/mm	给矿浓度/%	矿浆通过能力/m³·h⁻¹	干矿处理量/t·h⁻¹	主机质量/t
SLon-2000	2000	0~1.0	0~82	5.5+7.5	0~30	0~300	0~2.0	10~40	100~200	50~80	50
SLon-1750	1750	0~1.0	0~62	4+4	0~26	0~300	0~1.3	10~40	75~150	30~50	35
SLon-1500	1500	0~1.0	0~44	3+4	0~20	0~300	0~1.3	10~40	50~100	20~30	20
SLon-1250	1250	0~1.0	0~35	1.5+2.2	0~20	0~300	0~1.3	10~40	20~50	10~20	14
SLon-1000	1000	0~1.2	0~28.6	1.1+2.2	0~20	0~300	0~1.3	10~40	10~20	4~7	6
SLon-750	750	0~1.0	0~20.4	0.55+0.75	0~50	0~400	0~1.3	10~40	1.0~2.0	0.1~0.5	3

2 昆钢罗次铁矿提高赤铁矿选矿效率的应用

罗次铁矿选矿厂于1985年建成投产，随着矿山开采年限延长，可采矿石急剧减少，出矿品位降低，矿山本部铁矿开采量已不能满足选厂生产能力所需，不得不越来越多地收购民间小矿点矿石。由于收购矿石性质多变及自采矿石品位降低，原有的生产设备越来越不适应形式的发展，导致选矿生产指标呈逐年下降的趋势。

罗次铁矿的矿石是以磁铁矿和赤铁矿为主的混合矿，并含部分褐铁矿。原生产流程为两段连续磨矿后，要用弱磁选—强磁选—重选流程，要求保证铁精矿品位大于59%。其改造前的选矿指标：原矿品位42.89%，最终精矿品位59.98%，回收率70.98%。其中磁选部分的尾矿品位高达 26.69%，磁选部分的回收率仅 63.31%，回收率较低，导致生产成本居高不下。

为了提高选矿指标和解决难选矿石增多的问题，罗次铁矿从 1994~1997 年陆续引进 3 台 SLon-1500 磁选机，取代了原有的 8 台 φ1500mm 双立环强磁选机；将原有的一次粗选、一次扫选、一次精选的强磁选流程改造为全高梯度磁选的一次粗选、一次精选强磁选生产流程；用 2 台 SLon-1500 磁选机代替 4 台 φ1500mm 双立环强磁机作粗选，用 1 台 SLon-1500 磁选机代替 2 台 φ1500mm 双立环强磁机作精选，并取消强磁选扫选作业和重选扫选作业，大大简化了生产流程，提高了分选效率。图 2 是改造后的选矿数质量流程，改造前后的强磁选作业对比指标见表 2。

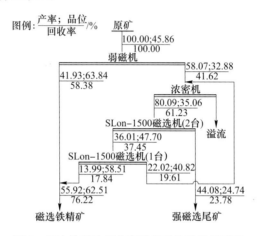

图 2　罗次铁矿选厂全高梯度磁选数质量流程

表 2　罗次铁矿新流程与原流程强磁选指标比较　　　　　　　　　　　　　　（%）

年份	流程	给矿品位	精矿品位	尾矿品位	强磁作业回收率
1997 年	一粗一精新流程	32.88	58.51	24.74	42.89
1994 年	一粗一精一扫原流程	32.58	56.81	29.69	26.99
差　值		+0.3	+1.7	-4.95	+15.9

从表 2 可见，采用 SLon-1500 磁选机所组成的强磁选流程与原强磁选流程比较，在给矿品位相近的条件下，新流程的强磁精矿品位提高 1.7%，尾矿品位降低 4.95%，强磁作业回收率提高 15.9%（折算成全流程回收率为 7.99%）。由此可见采用 SLon-1500 磁选机对全流程回收率提高是非常显著的。由于 SLon-1500 高梯度磁选机具有富集比较大的优点。因此，在保证铁精矿品位大于 59% 的前提下，新流程可降低入选矿的品位，减少富矿的外购量，从而可大幅度降低矿石采购成本。罗次铁矿不仅因回收率增加而获得显著效益，而且每年可节约采矿费用 600 万元。

3　昆钢上厂铁矿回收细粒铁矿的应用

上厂铁矿 1972 年投产，经 28 年的开采，矿山资源已趋于枯竭。该矿尾矿库已堆积含铁 22% 左右的尾矿近千万吨。为了充分利用尾矿资源，延长服务年限，该矿从前几年开始回采尾矿，并建立了尾矿分选系统，采用跳汰机等设备从尾矿中回收粗粒铁精矿，每年可从尾矿中回收约 8 万吨铁品位为 60% 左右的粗粒铁精矿，而细粒部分未回收。

上厂铁矿的铁矿石主要是有赤铁矿和褐铁矿组成，另有少量的磁铁矿、碳酸铁等，其特点是结晶粒度粗，含泥量大，铁矿物粒度越粗品位越高，粒度越细品位越低，且含泥量大。故上厂铁矿采出原矿的分选系统中，仅有破碎分级设备而无磨矿设备，选别设备以感应辊强磁选机和跳汰机为主回收粗粒铁矿物，而粒度小于 1mm、铁品位为 22% 左右的细粒尾矿未经选别就排入尾矿库中。

因此，上厂铁矿原矿选矿系统和尾矿回采选矿系统中的细粒铁矿都有回收价值。上厂铁矿于 1999 年 5 月在其选厂安装了 3 台 SLon-1500 磁选机，并于当年 7 月投入试生产。该流程中，从尾矿库采出的尾矿用高压水枪打散，然后用 φ1200mm 单螺旋分级机分级，分级机的沉砂进入跳汰系统，分选较粗的铁精矿。溢流与原生产系统（处理原矿）的 φ1500mm 双螺旋分级机的溢流混合后进入图 3 所示细粒尾矿回收生产流程。至今生产流程稳定，在给矿品位 TFe 22% 时，精矿品位 TFe 55% 以上，精矿产率 15% 以上。该系统每小时处理 40t 细粒尾矿，平均每小时可生产铁精矿 6t 以上，现已达到年产铁精矿 3 万吨的水平。

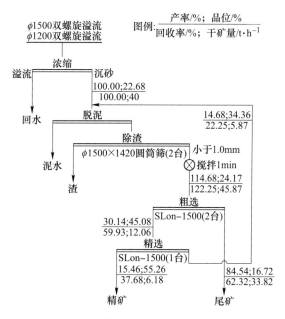

图 3　上厂铁矿细粒尾矿回收生产流程

4　攀钢选钛厂回收细粒钛铁矿的应用

攀枝花地区储存有大量的钒钛磁铁矿。攀钢现有 1 个磁选厂和 1 个钛铁矿选矿厂。采出的矿石首先进入磁选厂碎矿、磨矿并用弱磁选机选出含钒的磁铁矿，磁选厂的尾矿进入选钛厂回收钛铁矿。

攀钢选钛厂是我国最大的钛精矿生产基地，其原有的生产流程为首先将入选物料分级为大于 0.1mm 粗粒级、0.1～0.045mm 中粒级和小于 0.045mm 细粒级。粗粒级和中粒级用螺旋溜槽和电选机等设备分选，而细粒级因无合适的选矿工艺和设备一直作为尾矿抛弃。

导致其全厂钛回收率仅有17%~20%。

其细粒级产率占全厂总给料的35%左右，TiO_2品位与总给料基本一致。随着采出矿石中难选矿石的增加，磨矿细度变细，其细粒级的产率有进一步上升的趋势。因此，有效地回收这部分细粒级钛铁矿对提高全厂钛回收率具有重要的意义。

经过多年的小试、中试和工业试验，攀钢选钛厂于1997年底建成了第一条以SLon-1500磁选机为主体选矿设备的磁选—浮选流程生产线回收小于0.045mm细粒级，其数质量流程如图4所示。

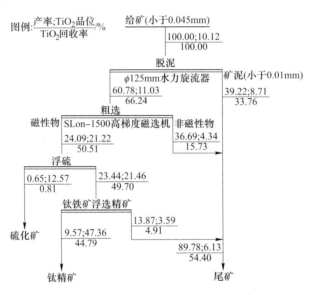

图4 攀钢选钛厂细粒选矿数质量流程

该流程小于0.045mm细粒级首先用ϕ125mm水力旋流器脱泥浓缩，该作业可将固体浓度为5%左右的矿浆浓缩到28%左右，水力旋流器的沉砂（每小时25.25t干矿）用SLon-1500磁选机粗选。如图4所示，该机将含TiO_2 11.03%的给矿富集到TiO_2 2.22%，其作业回收率为76.24%。水力旋流器和SLon-1500磁选机为浮选精选作业创造了非常好的条件，这两个作业丢弃了产率为75.91%的尾矿和绝大部分小于0.01mm的矿泥，仅有占产率24.09%的磁选精矿进入浮选精选作业。使最终浮选精矿品位含TiO_2达到47.36%，作业回收率88.68%，全流程TiO_2回收率44.79%的良好指标。该流程于1997年投产，1998年和1999年分别产出TiO_2>47%的优质钛精矿10000t和15000t。在此基础上，攀钢选钛厂在2000年将进一步增加生产线扩大产量。

5 开发安徽来安县长石资源的应用

安徽省来安县境内储藏有丰富的长石资源。矿石主要有钾长石、斜长石，少量石英、黑云母、角闪石和磁铁矿组成。矿石中除Fe_2O_3含量较高外，其他物理、化学指标均能达到玻璃、陶瓷工业的要求。因此，当地长石矿资源是否可变成商品矿，关键在于含铁杂质能否有效地除去。

1998年来安县成立了皖东长石有限公司，在屯仓水库旁边建成了一座年产3万吨长石精矿的选矿厂。其原矿来自民采丢弃在屯仓水库内的长石废石。该矿的选矿流程如图 5 所示。含 Fe_2O_3 的原矿经棒磨机和高频振动细筛组成的闭路磨矿筛分至小于 0.8mm，进入筒式中磁机除去强磁性铁物质；然后进入一台 SLon-1250 磁选机除去弱磁性铁矿物。磁选作业的非磁性产品含铁 0.26%。完全满足玻璃工业和陶瓷工业的要求。SLon-1250 磁选机在该流程中有效地除去了绝大部分弱磁性铁矿物，为来安县长石矿产资源的开发利用起到了关键作用。

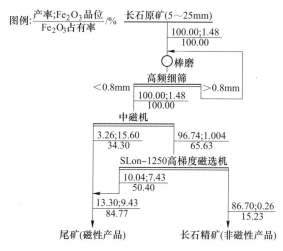

图 5 来安长石矿选矿流程

6 结论

（1）SLon 型立环脉动高梯度磁选机在昆明罗次铁矿用于分选赤铁矿，在保证综合铁精矿品位达到 59% 以上的前提下，使磁选作业回收提高了 15.9%，折算成全流程铁回收率提高了 7.99%，而且降低了入选原矿品位，每年降低原矿采购费用 600 万元。

（2）SLon 型磁选机在昆钢上厂铁矿的尾矿中回收细粒铁精矿应用成功，该矿每年可多产 3 万吨铁精矿，使该尾矿库中 1000 万吨的尾矿重新获得开采价值。预计可延长该矿的服务年限 10 年左右。

（3）SLon 型磁选机在攀钢选钛厂的应用，解决了该厂小于 0.045mm 粒级钛铁矿长期不能回收的技术难题，建成了一条年产 20000t 细粒优质钛精矿的生产线。经 2 年多的运转，该生产流程稳定，钛精矿品位稳定在 TiO_2 47% 以上。

（4）SLon 型磁选机在安徽来安县长石矿应用成功。解决了来安县长石矿因含铁偏高（Fe_2O_3 1.48%左右）而不能作为商品矿的问题，现已建成年产 3 万吨长石精矿的选矿厂。经两年的生产表明，该生产线稳定可靠，长石精矿含 Fe_2O_3 稳定在 0.26% 的水平，是优质的玻璃、陶瓷工业原料。

写作背景 本论文介绍了 SLon 立环脉动高梯度磁选机安装在昆钢罗次铁矿分选氧化铁矿、在昆钢上厂铁矿从尾矿中回收赤铁矿、在攀钢选钛厂回收细粒级钛铁矿以及在安徽来安县皖东长石厂用于长石提纯，大幅度地提高了这些矿物的资源利用率。

提高姑山赤铁矿生产指标的工业试验研究

熊大和[1] 杨庆林[2] 汤桂生[2] 钱士湖[2]

(1. 赣州有色冶金研究所；2. 马钢姑山矿业公司)

摘 要 姑山铁矿为进一步提高生产指标，近年来，采用新型、高效的SLon立环脉动高梯度磁选机改造流程，经过72h工业试验，实现了$\beta=60.17\%$、$\varepsilon=76.59\%$的良好指标，创姑山矿选矿指标历史新高。

关键词 SLon立环脉动高梯度磁选机 赤铁矿 红矿选矿

Experimental Research of Improving the Hematite Production Indexes of Gushan Iron Mine

Xiong Dahe[1] Yang Qinglin[2] Tang Guisheng[2] Qian Shihu[2]

(1. Ganzhou Nonferrous Metallurgy Research Institute;
2. Gushan Mining Company, Magang Iron and Steel Co.)

Abstract To further improve the production indexes, Gushan Iron Mine applied new and efficient SLon vertical ring pulsating high gradient magnetic separators in its flowsheet transformation. Seventy-two hour industry test achieved good indexes of $\beta = 60.17\%$ and $\varepsilon = 76.59\%$, which created historically new metallurgical performances.

Keywords SLon vertical ring pulsating high gradient magnetic separator; Hematite; Mineral processing of red iron ore

马钢姑山选矿厂所处理的是宁芜地区较典型的难选红矿，选矿工艺虽经多次改进，生产指标也有较大提高，但仍存在一些问题，主要表现为以下几个方面：

(1) 钢铁行业竞争激烈，冶炼部门为降低焦比和提高高炉利用系数，对铁精矿品位要求日益提高。姑山铁精矿品位低于60%，缺乏市场竞争能力。

(2) 随着矿山服务年限的延长，矿床深部矿石开采量日益增加，矿石中含磷量增高，而冶炼部门对铁精矿质量要求日益提高，因此生产与需求之间的矛盾越来越突出。实验研究结果表明，铁精矿品位越高，含磷就越低，二者呈反比关系。

(3) 选矿生产流程中用于粗选和精选作业的平环式强磁选机已使用多年，设备老化，齿板堵塞频繁，导致选矿指标偏低且不稳定，设备维修工作量大。

(4) 近年来中矿（球磨给矿）品位提高，粗选作业一次抛尾的尾矿品位偏高，现有

---

① 原载于《金属矿山》，2000 (12)：31~33。

的生产流程和设备已不适应矿石品位提高的变化。

为了进一步提高姑山赤铁矿的选矿指标，马钢姑山矿业公司与赣州有色冶金研究所合作，通过小型试验、闭路流程工业试验、开路流程工业试验等工作，获得了成功。

1 探索试验

为了进一步提高姑山铁精矿品位，降低含磷量，解决平环强磁选机齿板堵塞等问题，姑山铁矿于1998年初提供矿样委托赣州有色冶金研究所进行了广泛深入的探索试验。探索试验设备采用新型高效的SLon脉动高梯度磁选机，选矿流程模拟现有生产流程，获得的结果如表1所示。由结果可以看出，铁精矿品位不小于60%，回收率78%~81%，铁精矿含P<0.25%，且试验的重现性较好，这说明生产指标仍有相当大的潜力可挖。

表1 姑山铁矿全流程探索试验指标 （%）

现场取样日期	磨矿细度（小于74μm（200目））	给矿品位 TFe	给矿品位 P	精矿产率	精矿品位 TFe	精矿品位 P	尾矿品位	铁回收率
1998年6月	80	36.76	0.43	49.63	60.32	0.23	13.55	81.44
1999年4月	80	40.50	0.45	52.67	60.05	0.242	18.74	78.10

2 工业试验的准备工作

1999年初，笔者针对姑山铁矿粗选平环强磁选机齿板堵塞频繁、尾矿品位偏高等问题，专门研制了用于粗选作业的SLon-1750立环脉动高梯度磁选机。该机在姑山选厂取代1台SQC-6-2770A型湿式强磁选机（平环），并于1999年4月1日~6月30日进行了3个月的平行对比试验，试验指标见表2。

表2 SLon-1750与SQC-6-2770A磁选机粗选作业工业对比指标 （%）

时间（1999年）	设备	给矿品位	精矿 品位	精矿 产率	精矿 回收率	尾矿品位
4月	SLon	40.45	51.55	69.71	88.83	14.91
4月	SQC	42.42	48.48	73.85	84.39	25.31
5月	SLon	41.94	51.11	73.39	89.44	16.65
5月	SQC	41.84	52.52	62.01	77.83	24.41
6月	SLon	41.55	50.96	73.85	90.57	14.97
6月	SQC	未取样	—	—	—	—
4~5月累计平均	SLon	41.20	51.33	71.50	89.09	15.78
4~5月累计平均	SQC	42.13	50.50	67.36	80.74	24.86
4~5月累计平均	差值	-0.93	+0.83	+4.14	+8.35	-9.08

由表2可见，SLon-1750立环脉动高梯度磁选机取代SQC-6-2770A湿式强磁选机用于粗选作业，在给矿品位和精矿品位相近的条件下，SLon-1750磁选机的尾矿品位低仅为9.08%左右，作业回收率高8.35%左右。该机的选矿指标与实验室的试验指标相吻合，该机在生产上运转至今已有15个月，选矿指标一直保持稳定，设备运转平稳，磁介质从不堵塞。该机研制成功为提高姑山铁矿粗选作业回收率起到了重要的作用。

3 闭路流程工业试验

为了使姑山铁矿粉精矿品位尽早达到 60% 以上和铁回收率 75%~78% 的生产水平，1999 年 11 月 24~27 日姑山铁矿组织了闭路流程工业试验。该流程中，粗选、精选、扫选作业全部采用了 SLon-1750 立环脉动高梯度磁选机。试验流程和选矿指标如图 1 所示。

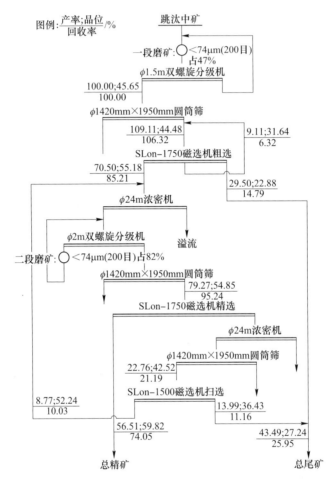

图 1 闭路流程工业试验数质量流程

分析图 1 流程，发现其存在以下问题：

（1）粗选作业回收率偏低，用于粗选作业的 SLon-1750 立环脉动高梯度磁选机自身产生的中矿返回给矿，虽然精矿品位提高至 55% 左右，但是降低了该机的给矿浓度，增大了给矿量和给矿体积，导致其作业回收率仅 85.21%，尾矿品位高达 22.88%。

（2）用于精选作业的 SLon-1750 磁选机磁介质偏粗（ϕ4mm 棒介质），经筛分水析发现其对微细粒铁矿物的回收率偏低。

（3）扫选作业的尾矿品位偏高，达 36.43%，原因是扫选精矿返回二段球磨分级机造成一部分微细粒铁矿物的回收率偏低。

4 改进

在上述试验的基础上,用于精选作业的 SLon-1750 立环脉动高梯度磁选机又进行了如下改进:

(1) 用 φ2mm 棒介质取代 φ4mm 棒介质,加强微细矿粒的回收。
(2) 修改矿浆通道和排水通道,加大了脉动作用、漂洗作用和排泥水作用。
(3) 加大脉动冲程冲次,进一步提高设备的富集比。
(4) 对转环转速、液位高度、整流参数等进行优化试验,并根据优化试验的最佳参数进行调整。

通过努力,笔者于 2000 年 5 月研制出了适应于精选作业的 SLon-1750K(精选型)立环脉动高梯度磁选机,并于 6 月初成功地安装在姑山选厂精选作业位置,取代了原有的 2 号 SQC-6-2770 湿式强磁选机。经半个月的现场调试,该机达到了最佳工作状态。

5 实现铁精矿品位60%和回收率76%的工业试验

姑山铁矿于 2000 年 6 月 27~30 日组织了 72h 的工业试验。本次试验采用了与生产上完全一致的开路流程(如图 2 所示),其中粗选作业采用 1 台 SLon-1750 立环脉动高梯度磁选

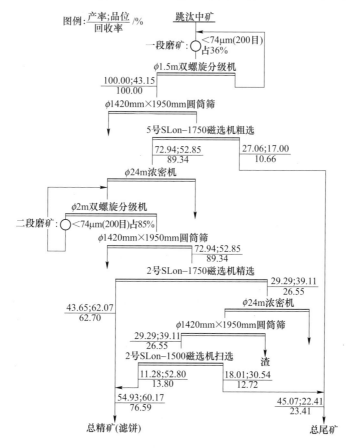

图 2 工业试验数质量流程

机，精选作业采用 1 台 SLon-1750K 立环脉动高梯度磁选机，扫选作业采用生产上原有的 SLon-1500 立环脉动高梯度磁选机。这 3 台磁选机与 1 个球磨系统构成了 1 条完整的工业试验流程。试验期间，该系统每小时处理干矿量 33t 左右，与目前生产上的处理量一致。

工业试验数质量流程分析如下：

（1）与图 1 所示的闭路流程比较，开路流程取消了粗选作业的中矿循环，粗选中矿直接排入粗精矿中去，虽然粗精矿品位从 55% 左右降至 52% 左右，但因给矿浓度加大，给矿量和给矿体积减小，粗选尾矿从 22% 左右降至 17%，保证了粗选作业回收率达 89% 左右，与实验室小型试验指标基本吻合。

（2）新研制的 SLon-1750 立环脉动高梯度磁选机用于精选作业具有很大富集比，其铁精矿品位高达 62.07%（理论最高品位 64.05%），综合回收率达 62.79%（理论综合回收率 65.23%），为开路流程的总精矿品位实现 $\beta = 60\%$ 以上起到了关键作用。

（3）扫选精矿不再循环而直接进入总精矿中去，虽然扫选精矿品位较低，仅 52.08%，但其产率仅为精选精矿的 1/4 左右，总精矿品位仍保持在 60% 以上，扫选精矿的开路避免了微细矿粒的不良循环，使扫选尾矿品位大幅度下降，即从 36% 左右降至 30% 左右，对全流程回收率的提高起到了重要作用。

（4）通过引进 SLon-1750 立环脉动高梯度磁选机应用于精选作业及对流程的改进，全流程的 72h 选矿指标为：给矿（跳汰中矿）品位 43.15%，总精矿品位 60.17%，总尾矿品位 20.41%，总精矿回收率 76.59%。

6 结语

（1）为进一步提高姑山赤铁矿选矿指标，解决生产难题，通过小型试验、闭路流程工业试验、开路流程工业试验及多次的改进工作，终于在最近的开路流程工业试验中获得了成功，实现全流程 72h 连续运转的工业试验指标为：给矿（跳汰中矿）品位 43.15%，总精矿品位 60.17%，总尾矿品位 22.41%，总精矿铁回收率 76.59%。该指标创姑山铁矿历史新高，初步实现了姑山矿业公司提出的 $\beta = 60\%$ 和 $\varepsilon = 75\% \sim 78\%$ 的奋斗目标。

（2）本工业试验全部采用新型高效的 SLon 型立环脉动高梯度磁选机。该机具有富集比大、选矿效率高、磁介质不堵塞、设备处理量大、运转稳定、作业率高的优点。SLon 型立环脉动高梯度磁选机的全面应用，为姑山赤铁矿选矿生产指标创历史新高，大幅度降低选矿生产成本和提高市场竞争能力起到了重要的作用。尤其是 SLon-1750K 立环脉动高梯度磁选机在精选作业中的应用，其优异的选矿性能和较大的富集比使精选作业精矿品位达到 62.07%，为总精矿品位达到 60.17% 起到了关键作用。工业试验完成后，所有参试的 SLon 立环脉动高梯度磁选机直接转化成为生产设备。

（3）存在的问题和改进建议。与 20 世纪 90 年代初期比较，姑山铁矿的跳汰中矿品位从 36% 左右上升至目前的 44% 左右，目前原设计的强磁粗选一次抛尾的作业已不太适合于目前的矿石，即粗选作业的尾矿品位偏高，因此今后生产上有必要增加一道扫选作业，以利于进一步提高回收率。

（4）扫选作业的精矿品位偏低。如图 2 所示，扫选精矿品位仅 52.80% 左右，而实验室小型试验扫精品位可达 55% 左右。因此，加强扫选作业的研究，是进一步提高姑山赤铁矿选矿指标的途径之一。

参 考 文 献

［1］ 熊大和．SLon-1500立环脉动高梯度磁选机的研制［J］．金属矿山，1990（7）：43~46．
［2］ 王重渝，等．应用SLon-1500立环脉动高梯度磁选机回收姑山细粒赤铁矿尾矿的工业试验［J］．金属矿山，1991（5）：38~43．
［3］ 熊大和．SLon型立环脉动高梯度磁选机的改进及其在红矿选矿的应用［J］．金属矿山，1994（6）：30~34．
［4］ 王重渝，等．强磁—脉动高梯度磁选、阶段磨选工艺研究及生产实践［J］．金属矿山，1994（9）：36~38．
［5］ 熊大和，等．SLon-1750立环脉动高梯度磁选机的研制与应用［J］．金属矿山，1999（10）：23~26．

写作背景 本文介绍了SLon立环脉动高梯度磁选机在马钢姑山铁矿分选赤铁矿的工业试验应用，SLon-1750和SLon-1500磁选机组成的优化选矿流程，显著地提高了姑山铁矿细粒氧化铁矿的选矿指标。

新一代强磁选设备——SLon 立环脉动高梯度磁选机

熊大和

（赣州有色冶金研究所）

摘　要　SLon 立环脉动高梯度磁选机结合磁力、脉动流体力和重力选矿，具有富集比大、分选效率高、磁介质不易堵塞、适应性强的优点。现已在我国多个大中型厂矿广泛应用，是当今国内外发展最快、性能最好的强磁选设备。

关键词　磁选　SLon 立环脉动高梯度磁选机　赤铁矿　菱铁矿　长石　霞石

某些平环强磁选机和平环高梯度磁选机因给矿方向与排磁性产品的方向一致，粗粒和强磁性物质必须穿过磁介质堆才能排出来，存在着富集比较低和磁介质易堵塞、设备检修工作量大的问题，且选矿指标随着磁介质的堵塞程度波动，因而影响了它们在工业上的应用。

为了提高选矿效率和解决磁介质堵塞的问题，我们从 1986 年开始研制 SLon 立环脉动高梯度磁选机，通过 15 年的不断改进与发展，迄今已研制出 SLon-750、SLon-1000、SLon-1250、SLon-1500、SLon-1750、SLon-2000 等型号的立环脉动高梯度磁选机。该系列磁选机分选粒度下限可达 $10\mu m$，并且具有富集比大，分选效率高，磁介质不易堵塞，对给矿粒度、浓度和品位波动适应性强，工作可靠，操作维护方便等优点。

1　设备构造和工作原理

SLon 型立环脉动高梯度磁选机结构如图 1 所示，它主要由脉动机构、激磁线圈、铁轭、转环和各种矿斗、供水装置等组成。用导磁不锈钢制成的钢板网或圆棒作磁介质。该机工作原理如下：激磁线圈通以直流电，在分选区产生感应磁场，位于分选区的磁介质表面产生非均匀磁场即高梯度磁场；转环作顺时针旋转，将磁介质不断送入和运出分选区；矿浆从给矿斗给入，沿上铁轭缝隙流经转环；矿浆中的磁性颗粒吸附在磁介质表面上，被转环带至顶部无磁场区，被冲洗水冲入精矿斗；非磁性颗粒在重力、脉动流体力的作用下穿过磁介质堆，与磁性颗粒分离，然后沿下铁轭缝隙流入尾矿斗排走。

该机转环采用立式旋转方式，对于每一组磁介质而言，冲洗磁性精矿的方向与给矿方向相反，粗颗粒不必穿过磁介质堆便可冲洗出来。该机的脉动机构驱动矿浆产生脉动。可使位于分选区磁介质堆中的矿粒群保持松散状态，使磁性矿粒更容易被捕获，使非磁性矿粒尽快穿过磁介质堆进入到尾矿中去。

显然，反冲精矿和矿浆脉动可防止磁介质堵塞，脉动分选可提高磁性精矿的质量。这

---

❶ 原载于《有色金属》2000，52（04）：185~187。

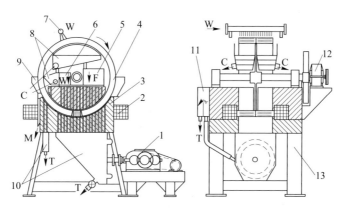

图 1 SLon 立环脉动高梯度磁选机结构图

1—脉动机构；2—激磁线圈；3—铁轭；4—转环；5—给矿斗；6—漂洗水斗；7—精矿冲洗装置；
8—精矿斗；9—中矿斗；10—尾矿斗；11—液位斗；12—转环驱动机构；13—机架；
F—给矿；W—清水；C—精矿；M—中矿；T—尾矿

些措施保证了该机具有较大的富集比、较高的分选效率和较强的适应能力。

SLon 系列立环脉动高梯度磁选机主要技术参数见表 1。

表 1 SLon 磁选机主要技术参数

机型	转环外径 /mm	背景场强 /T	激磁功率 /kW	传动电动机 /kW	脉动冲程 /mm	脉动冲次 /次·min^{-1}	给矿粒度 /mm	给矿浓度 /%	矿浆通过能力 /m^3·h^{-1}	干矿处理量 /t·h^{-1}	主机质量 /t
SLon-2000	2000	0~1.0	0~82	5.5+7.5	0~30	0~300	0~2.0	10~45	100~200	50~80	50
SLon-1750	1750	0~1.0	0~62	4+4	0~26	0~300	0~1.3	10~45	75~150	30~50	35
SLon-1500	1500	0~1.0	0~44	3+4	0~20	0~300	0~1.3	10~45	50~100	20~30	20
SLon-1250	1250	0~1.0	0~35	1.5+2.2	0~20	0~300	0~1.3	10~45	20~50	10~18	14
SLon-1000	1000	0~1.2	0~28.6	1.1+2.2	0~20	0~300	0~1.3	10~45	10~20	4~7	6
SLon-750	750	0~1.0	0~20.4	0.55+0.75	0~50	0~400	0~1.0	10~45	1.0~2.0	0.1~0.5	3

2 在工业生产中的应用

2.1 在梅山铁矿降磷工程中的应用

梅山铁矿矿石类型为磁铁矿、菱铁矿、赤铁矿和黄铁矿的混合矿，因硫和磷含量偏高，铁精矿不能作为球团原料，只能作为配料，直接影响了该公司产品销量和经济效益。为解决上述问题，梅山铁矿联合多家科研院所进行了大量选矿试验，经过分析比较，该矿于 1995 年采用如图 2 所示的磁选流程进行工业试验。该工艺采用弱磁选机回收磁铁矿，2 台 SLon-1500 强磁选机分别作粗选和扫选，以回收赤铁矿和菱铁矿。通过强磁作业，有效地回收了细粒赤铁矿和菱铁矿，降低了尾矿品位。全流程选矿指标见表 2。

试验结果表明，铁精矿质量显著提高，完全满足冶炼要求。该矿自 1996 年至 2000 年又订购了 16 台 SLon-1500 磁选机。经过数年的应用，SLon-1500 磁选机运转稳定，综合铁精矿含磷量稳定到 0.25% 以下，为解决梅山铁精矿含磷偏高的问题作出了较大的贡献。

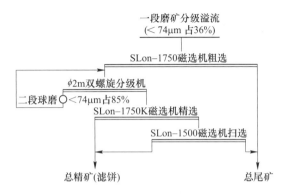

图 2 梅山铁矿降磷硫工业试验流程

表 2 梅山铁矿降磷硫工业试验指标（276h） （%）

产品名称	产率	品位			回收率		
		Fe	S	P	Fe	S	P
给矿	100.0	52.77	0.44	0.399	100.0	100.0	100.0
精矿	88.90	56.08	0.29	0.246	94.51	59.10	54.76
冶炼要求		>55.0	<0.35	<0.25	>94		
尾矿	11.10	26.13	1.61	1.626	5.49	40.90	45.24

2.2 在马钢姑山铁矿的全面应用

姑山铁矿属中温热液矿床，主要金属矿物为赤铁矿，矿石结构以致密块状和角砾状为主，其次为斑点状和条带状，铁矿物与脉石共生紧密，矿石硬度大，嵌布粒度细，且不均匀，理论上铁精矿品位最高仅能达到 64% 和铁回收率 84.11% 左右，该矿属宁芜地区较典型的难选红矿。

姑山选厂 1978 年投产，其球磨主厂房采用浮选流程生产铁精矿，仅能达到铁精矿品位 49% 和铁回收率 49% 左右。20 世纪 80 年代改为重选流程，生产指标提高到品位为 55% 和回收率为 61% 左右。90 年代改为磁选流程，生产指标进一步提高到品位为 58% 和回收率为 73% ~ 75%。

为了进一步提高铁精矿品位，降低含磷量，解决平环强磁选机齿板堵塞等问题，姑山铁矿从 1999 ~ 2000 年开展了全面采用 SLon 立环脉动高梯度磁选机取代平环强磁选机的探索试验和工业试验并获得成功。如图 3 所示，原矿（跳汰中矿）磨至小于 74μm 占 36% 左右，用 SLon-1750 磁选机粗选抛尾，粗精矿进行二段磨矿至小于 74μm 占 85% 左右，用 SLon-1750K 磁选机精选和 SLon-1500 磁选机扫选，工业试验 72h 所达到平均选矿指标为：给矿品位 43.15%，铁精矿品位 60.17%，铁回收率 76.59%，尾矿品位 22.41%。选矿指标创历史新高。目前，姑山铁矿已完成一个系统全面采用 SLon 磁选机的工业改造（共 3 个系统），改造后系统生产指标稳定，设备作业率高，不存在磁介质堵塞问题。与原生产流程比较铁精矿品位提高 2%，铁回收率提高 3%，该系统每年可多产优质铁精矿 2 万吨左右，经济效益显著。

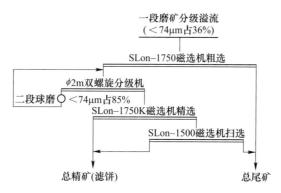

图 3 SLon 磁选机分选姑山赤铁矿工业试验原则流程

2.3 SLon 磁选机在其他矿山的应用简介

SLon 立环脉动高梯度磁选机已在国内外广泛应用于赤铁矿、褐铁矿、菱铁矿、钛铁矿等金属矿物的选矿和长石、霞石等非金属矿物的提纯。表 3 列举了该机近年在金属矿山和非金属矿山的应用实例。

表 3 SLon 磁选机在一些厂矿应用简介

厂矿名称	矿石类型	选矿流程	SLon 磁选机型号	综合选矿指标/%			
				给矿品位	精矿品位	尾矿品位	回收率
攀钢选钛厂	钛铁矿（<45μm）	磁—浮流程	SLon-1500（1台）	10.12TiO_2	47.36TiO_2	6.13TiO_2	44.79TiO_2
重钢太和铁矿	钛铁矿	高梯度粗选	SLon-1500（3台）	12.06TiO_2	21.23TiO_2	3.56TiO_2	84.68TiO_2
昆钢罗次铁矿	赤铁矿	弱磁—高梯度磁选	SLon-1500（3台）	45.86TFe	62.51TFe	24.74TFe	76.22TFe
昆钢上厂铁矿	赤铁矿、褐铁矿（尾矿再回收）	高梯度一粗一精	SLon-1500（3台）	22.68TFe	55.26TFe	16.72TFe	37.68TFe
宁夏中卫铁厂	褐铁矿	高梯度一粗一扫	SLon-1500（2台）	46.36TFe	53.06TFe	28.43TFe	83.32TFe
安徽来安长石厂	长石	中磁—高梯度磁选	SLon-1250（1台）	1.48Fe_2O_3	0.26Fe_2O_3	9.43Fe_2O_3	除铁率 84.77
四川南江	霞石	高梯度一次磁选	SLon-1000（工业试验）	1.61Fe_2O_3	0.179Fe_2O_3	6.23Fe_2O_3	除铁率 91.52
唐山陶瓷原料厂	长石	高梯度磁选	SLon-1000	0.21Fe_2O_3	0.1Fe_2O_3	0.89Fe_2O_3	除铁率 59.00

3 结论

（1）在总结国内外强磁选机和高梯度磁选机优缺点的基础上，通过 15 年的设备研制、工业试验、生产应用的不断创新和改进，SLon 立环脉动高梯度磁选机已初步形成了一个型号较齐全的系列，其处理能力从实验室型的每周期给矿 100g 发展到工业生产型处理能力 80t/h 的大型设备。

（2）经过长期的生产实践证明，SLon 立环脉动高梯度磁选机具有富集比大、选矿效率高、磁介质不易堵塞、适应性强、工作可靠的优点。该机以其优异的选矿性能和机电性能赢得了用户的信任，1999~2000年，SLon立环脉动高梯度磁选机已跃升为国内销量最大的强磁选设备。

（3）SLon立环脉动高梯度磁选机在我国多个大中型厂矿的成功应用，为充分利用我国的矿产资源作出了一定的贡献。随着设备的不断改进和推广应用，该机作为新一代强磁选设备将为我国的矿产资源综合利用作出更大的贡献。

参 考 文 献

[1] 陈志新，任祥君，艾光华. 海南某小型铁矿选矿厂生产流程的研究及应用 [J]. 矿业快报，2004（10）.
[2] 陈剑，李晓波. 高梯度磁选机的发展现状及应用 [J]. 矿业工程，2005（4）.
[3] 陈剑，李晓波. 高梯度磁选机的发展及应用 [J]. 矿业快报，2005（9）.

写作背景　经过15年的研制与发展，SLon立环脉动高梯度磁选机已研制出多种型号，具有优异的设备性能，在多个矿山推广应用。本文预测到该机已成为国内外新一代强磁选设备。

上厂铁矿尾矿复选赤铁矿新技术

熊大和¹ 苏 卫²

(1. 赣州有色冶金研究所；2. 昆明钢铁公司上厂铁矿)

摘 要 昆钢上厂铁矿尾矿库堆存含铁20%~28%的尾矿近千万吨，采用细粒跳汰和SLon立环脉动高梯度磁选机分选的新技术，每年从尾矿中回收铁品位55%~58%的优质赤铁矿精矿7万吨。

关键词 尾矿 赤铁矿 跳汰机 SLon立环脉动高梯度磁选机

上厂铁矿是昆明钢铁公司主要原料基地之一，其铁矿物以赤铁矿和褐铁矿为主，具有含磷、硫等杂质元素低的优点，是优质的钢铁原料。该矿设计的年采选规模为100万吨，于1972年投产，经28年的开采，矿山资源已趋于枯竭，目前只能靠民工回采残存矿石，选厂收购民采的矿石每年仅能选出5万吨铁精矿。然而上厂铁矿的尾矿已堆存含铁20%~28%的尾矿近千万吨，折合成铁金属量200余万吨。为了充分利用尾矿资源，延长矿山服务年限，该矿从前几年开始回采尾矿，并与赣州有色冶金研究所合作，采用该所研制的具有国际领先水平的SLon立环脉动高梯度磁选机回收细粒铁矿，至今已建立了以跳汰机和SLon立环脉动高梯度磁选机为主体选矿设备的尾矿复选系统，已形成每年从尾矿中回收7万吨铁精矿的生产能力。

1 原矿性质和原有选矿流程

上厂铁矿是露天开采的铁帽型风化矿，采出的原矿氧化程度较深，磁铁矿、假象磁铁矿含量极少，含铁矿物主要是赤铁矿和褐铁矿，主要脉石矿物是石英和铝土矿。铁矿物的赋存状态以块状富集体为主，铁矿物的粒度从数百毫米至几微米不等。矿石粒度越粗，则铁品位越高且可选性越好；矿石粒度越细，则铁品位越低，且因含泥大而可选性越差。因此，上厂铁矿原设计的选矿流程是以感应辊强磁选机和跳汰选矿机回收粗粒铁矿物为主，细粒部分未经任何选别就作尾矿排放。上厂铁矿原有选矿生产原则流程如图1所示。

由图1可见，采场原矿经高压水冲洗、筛分、破碎、槽洗等工序分成大于2mm和小于2mm两个级别，大于2mm部分经筛分后分别用大粒跳汰、CS1和CS2感应辊强磁选机、粗粒跳汰分别选出大粒和粗粒铁精矿，而小于2mm部分未经选别即排入尾矿库。这部分尾矿含铁品位在20%~28%，而且排放年代越早，品位越高。

---

❶ 原载于《矿山环保》，2001（03）：3~5。

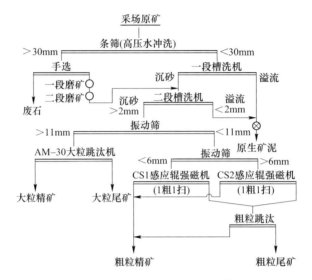

图 1 上厂铁矿原有选矿生产原则流程
（原生细泥排入尾矿库）

2 细粒尾矿性质和回收工艺流程

2.1 细粒尾矿性质

上厂铁矿细粒尾矿包括两部分，一部分是采场原矿选矿系统槽洗机的溢流，另一部分是从现有尾矿库采出的老尾矿。这两部分尾矿性质基本相同，因此可采用同一流程进行回收，以节省设备和投资。

上厂铁矿细粒尾矿中含铁矿物以赤铁矿为主，褐铁矿次之，另有少量的磁铁矿、碳酸铁、硫化铁和硅酸铁，具有含泥量大、有害杂质元素低的特点。

细粒尾矿多元素化学分析、铁化学物相分析、矿物组成及解离度测定见表1~表3。

表 1 细粒尾矿多元素化学分析结果 （%）

元素	TFe	SiO_2	Al_2O_3	S	P	Na_2O	K_2O	CaO
含量	22.60	37.05	12.84	<0.05	0.044	0.075	2.12	0.78

表 2 细粒尾矿铁的化学物相分析结果 （%）

铁物相	赤褐铁矿	硅酸矿	磁铁矿	碳酸铁	硫化铁	合 计
铁含量	15.98	4.48	0.87	0.58	0.87	22.78
铁分布率	73.86	17.23	3.34	2.23	3.34	100.00

表 3 细粒尾矿的矿物组成及解离度测定结果 （%）

矿物名称	赤铁矿	褐铁矿	磁铁矿	脉石
矿物含量	31.28	2.80	微量	65.92
解离度	41.93	47.72		

2.2 细粒尾矿回收工艺流程

上厂铁矿目前每年从细粒尾矿中回收 7 万吨铁精矿,其中约 2 万吨是从采场原矿选矿流程所产生的细粒级中回收的,约 5 万吨是尾矿库回采的尾矿中回收的。按铁精矿产率 18% 计,每年需要从尾矿库中采出 28 万吨细粒尾矿入选。目前采法为:用排水沟排水和水泵排水等方法疏干尾矿库,用两台斗式挖掘机从上往下挖起尾矿,用汽车运至建在尾矿库旁边的堆场堆存备用,或运至预选系统处理后用泵扬送至选厂分选。采矿部分采用白天一班 8 小时工作制,选厂采用二班 16 小时工作制,采矿堆存的部分供小夜班选矿或雨天选矿用。

上厂铁矿细粒尾矿回收的选矿流程如图 2 所示。

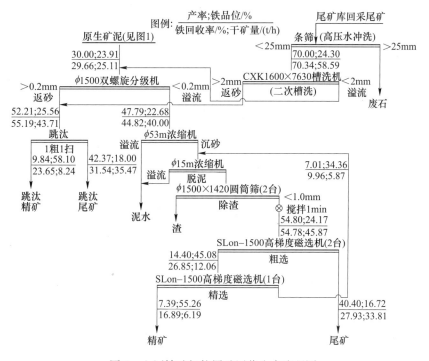

图 2 上厂铁矿细粒尾矿回收生产流程图

上厂铁矿细粒尾矿回收工艺流程的特点如下:

(1) 原生细泥与回采尾矿混合入选。原生细泥为采场原矿进入槽洗机的溢流,该部分细泥已高度分散,因此直接进入 φ1500mm 双螺旋分级机分成大于 0.2mm 和小于 0.2mm 两部分;从尾矿库回采的尾矿因已堆存多年,矿泥已形成团块,采用高压水枪将矿泥打散,用条筛隔除粗粒石块后用 CXK1600mm×7630mm 槽洗机分成大于 2mm 和小于 2mm。小于 2mm 粒级与原生细泥合并后进入 φ1500mm 螺旋分级机。φ1500mm 螺旋分级机的返砂即大于 0.2mm 部分和槽洗机大于 2mm 部分合并后,用细粒跳汰分选,溢流即小于 0.2mm 部分进高梯度磁选系统分选。

(2) φ1500mm 螺旋分级机的溢流即小于 0.2mm 部分具有浓度小、含泥量大的特点,首先用 φ53m 和 φ15m 浓缩机浓缩并脱去部分微泥。φ53m 和 φ15m 浓缩机既具有浓缩脱洗作用又具有储存一定矿量和均匀给矿的作用,使高梯度磁选机的给矿浓度控制在 25%~

30%，给矿体积和给矿量控制在140m³/h和40~50t/h（干矿）左右。

（3）φ15m浓缩机的底流用2台φ1500×1420圆筒筛除去大于1.0mm粗粒和杂物，筛下矿浆用搅拌桶搅拌1min左右，使微细泥团进一步分散，然后给入2台SLon-1500立环脉动高梯度磁选机粗选。其粗选精矿合并后直接给入1台SLon-1500立环脉动高梯度磁选机精选，其最终精矿用沉淀池沉淀，到一定量后更换沉淀池，用抓斗行车装入汽车外运。精选尾矿返回φ15m浓缩机，形成闭路流程，粗选尾矿排入另一个尾矿库堆存。

（4）采用重磁选流程回收细粒铁精矿，具有投资少、成本低、见效快、无污染的优点。

3 细粒尾矿回收的成本分析

上厂铁矿细粒尾矿回收系统工程投资500万元，目前每年回收细粒铁精矿7万吨。从尾矿中回收铁精矿，采矿费用低，不需要交纳矿产资源税，省去了破碎、磨矿等作业，具有选矿成本低的优点。

上厂铁矿的细粒尾矿选矿成本分析见表4。由表4可见，该矿建成尾矿回收系统后，每年新增产值1340万元，新增利税539万元，获得了显著的经济效益。

表4 上厂铁矿细粒尾矿选矿技术经济指标分析

项目	尾矿处理量/万吨·a⁻¹	细粒跳汰精矿产量/万吨·a⁻¹	高梯度磁选精矿产量/万吨·a⁻¹	选矿比/倍	给矿品位TFe/%	细粒跳汰精矿品位TFe/%	高梯度磁选精矿品位TFe/%	细粒跳汰尾矿品位TFe/%	高梯度磁选尾矿品位TFe/%	铁精矿产率/%
技术经济指标	40.6	4	3	5.80	24.18	58.10	55.26	18.00	16.72	17.234
备注	其中回采尾矿28.4万吨				平均值			尾矿产率60%	尾矿产率40%	

项目	精矿铁回收率/%	新增投资/万元	回采尾矿成本/万元·a⁻¹	尾矿加工费/万元·a⁻¹	设备折旧/万元·a⁻¹	跳汰精矿售价/万元·a⁻¹	高梯度精矿售价/万元·a⁻¹	总成本/万元·a⁻¹	总售价/万元·a⁻¹	经济效益（利税总额）/万元·a⁻¹
技术经济指标	40.54	500	142	609	50	800	540	801	1340	539
备注			5元/t×28.4万吨	15元/t×40.64万吨	年折旧率10%	200元/t×4万吨	180元/t×3万吨			

4 结论

（1）上厂铁矿是昆钢的主要原料基地之一，其铁精矿含杂质元素很低，是优质的钢铁原料，但是经过28年开采后，至今已面临资源枯竭的问题。该矿积极地采用最新的选矿技术和设备，从尾矿中二次回收有用矿物，已建成年产7万吨铁精矿的尾矿回收系统。年新增产值1340万元，年新增利税539万元，不仅获得了显著的经济效益，而且可延长矿山服务年限 10 年以上。

（2）上厂铁矿尾矿回收系统采用了跳汰—高梯度磁选流程，其中采用的 3 台 SLon-1500 立环脉动高梯度磁选机是回收微细粒铁精矿的关键设备。该机转环立式旋转，反冲精矿，配有矿浆脉动机构，具有富集比大、选矿效率高、磁介质不堵塞、适应性强的优点。该机的选矿性能和机电性能具有国内外领先水平，在国内外弱磁性矿物选矿和非金属矿除铁提纯中广泛应用。SLon-1500 立环脉动高梯度磁选机在上厂铁矿的应用，解决了该矿微细粒（<0.2mm）的铁矿物长期因无合适设备而不能回收的技术难题。

（3）上厂铁矿采用重—磁选流程回收细粒铁精矿，减少尾矿的排放量，有利于充分利用国家的矿产资源和生态环境保护。

写作背景　昆钢上厂铁矿的尾矿库堆存有约 1000 万吨品位为 20%~28% 的尾矿，本文较详细地介绍采用 3 台 SLon-1500 立环脉动高梯度磁选机从尾矿中回收赤铁矿的选矿流程和选矿指标。

SLon 立环脉动高梯度磁选机发展与应用现状

熊大和

（赣州有色冶金研究所）

摘　要　本文详细介绍了 SLon 立环脉动高梯度磁选机近年的发展及其在冶金矿山的应用现状。
关键词　SLon 立环脉动高梯度磁选机　选矿　细粒氧化铁矿　钛铁矿

1　设备特点

SLon 立环脉动高梯度磁选机（以下简称 SLon 磁选机）利用磁力、脉动流体力和重力的综合力场连续分选细粒弱磁性矿物。其分选环立式旋转、分选环内安装导磁不锈钢棒或菱形网作为磁介质；采用水内冷大电流低电压的激磁系统；在磁系下部配有脉动机构。该机具有富集比大、对弱磁性矿物回收率高、磁介质不易堵塞、分选效果优良和设备作业率高的优点。

至今已经研制出了 SLon-750、SLon-1000、SLon-1250、SLon-1500、SLon-1750、SLon-2000 等磁选机。它们的背景磁感强度可达到 1.0~1.2T，处理量涵盖 0.1~80t/h，可满足多数弱磁性金属矿和非金属矿的选矿要求。它们的主要技术参数见表 1。

表 1　SLon 立环脉动高梯度磁选机主要技术参数

机型	转环直径/mm	背景磁感强度/T	激磁功率/kW	驱动功率/kW	脉动冲程/mm	脉动冲次/次·min^{-1}	给矿粒度/mm	给矿浓度/%	矿浆流量/m^3·h^{-1}	干矿处理量/t·h^{-1}	机重/t
SLon-2000	2000	0~1.0	0~82	5.5+7.5	0~30	0~300	0~2.0	10~45	100~200	50~80	50
SLon-1750	1750	0~1.0	0~62	4+4	0~30	0~300	0~1.3	10~45	75~150	30~50	35
SLon-1500	1500	0~1.0	0~44	3+4	0~30	0~300	0~1.3	10~45	50~100	20~35	20
SLon-1250	1250	0~1.0	0~35	1.5+2.2	0~30	0~300	0~1.3	10~45	20~50	10~20	14
SLon-1000	1000	0~1.0	0~26	1.1+2.2	0~30	0~300	0~1.3	10~45	10~20	4~7	6
SLon-750	750	0~1.2	0~20.4	0.55+0.75	0~30	0~300	0~1.3	10~45	1~2	0.1~0.5	3

经过 20 余年的不断开发、技术创新、工业试验、生产检验和多次改进，SLon 磁选机以其优异的产品质量和技术性能迅速地在全国的众多厂矿得到应用。

2　SLon 磁选机近年在铁矿山的应用

2.1　在梅山铁矿的应用

宝钢梅山铁矿是位于江苏南京市附近的一座较大的铁矿山，其矿石由磁铁矿、赤铁

---

[1] 原载于《第四届全国矿山采选技术进展报告会论文集》，矿业快报，2001：517~520。

矿、菱铁矿和黄铁矿组成。由于其早期生产的铁精矿含磷偏高而缺乏市场竞争力。为了解决这一严重困扰梅山铁矿发展的问题,梅山铁矿于1995年组织了两个工业试验,一个为浮选流程工业试验,而另一个为磁选流程工业试验。通过数百个小时的对比,磁选流程显示了选矿指标较好、成本较低、有利于环保、铁精矿易于脱水和生产流程稳定等优点。该矿最终采用了磁选流程改造和扩建生产流程,其中采用了16台SLon-1500磁选机用于降磷生产和2台SLon-1500磁选机用于其重选车间的细粒铁矿选别作业。梅山铁矿现行的降磷流程和生产指标见图1和表2。

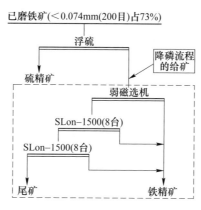

图1 梅山铁矿降磷生产流程

表2 梅山铁矿降磷指标 （％）

产品	产率	品位			回收率		
		Fe	S	P	Fe	S	P
给矿	100.00	52.77	0.44	0.40	100.00	100.00	100.00
精矿	88.90	56.08	0.29	0.25	94.51	59.10	54.76
要求	—	>55.0	<0.35	<0.25	—	—	—
尾矿	11.10	26.13	1.61	1.63	5.49	40.90	45.24

2.2 SLon磁选机在马钢姑山铁矿的应用

马钢姑山铁矿是一个较大的难选氧化铁矿生产基地。自从1978年建立选厂以来,其球磨主厂房的选矿工艺流程和生产指标经历了多次改进,最初采用浮选流程所获得的生产指标为:铁精矿品位49%和铁回收率49%;20世纪80年代采用螺旋溜槽和离心选矿机的生产指标为:铁精矿品位55%和铁回收率61%;1992年改为以SQC-6-2770湿式强磁选机为主及SLon-1500磁选机为辅的全磁选流程,SQC磁选机存在齿板容易堵塞问题及铁精矿品位一直为57%~58%。姑山铁矿近年改为以SLon-1750磁选机和SLon-1500磁选机的全磁选流程,新的生产流程已达到铁精矿品位59.5%和铁回收率76%的良好指标,在3个球磨系统未增容的条件下,铁粉精矿产量由1990年的15万吨增至2000年的31万吨。2001年可望达到38万吨左右。技术进步给姑山铁矿带来了显著的经济效益。姑山铁矿球磨主厂房现行的选矿流程和指标见图2和表3。

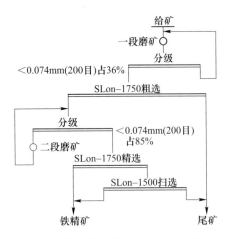

图 2　分选难选赤铁矿的生产流程

表 3　姑山铁矿磁选流程与重选流程指标比较　　　　　　　　　　　（%）

流程	年份	品位			回收率
		给矿	精矿	尾矿	
重选	1988 年	37.16	55.22	24.47	61.32
SLon 磁选	2000 年	42.65	59.50	22.41	76.13

2.3　从尾矿中回收细粒赤铁矿

昆钢上厂铁矿的尾矿库中储存有含铁量 20%~28%Fe 的尾矿约 1000 万吨，近年该矿采用了一种以细粒跳汰机和 3 台 SLon-1500 磁选机为主体的选矿新工艺流程，从尾矿中回收赤铁矿，每年从尾矿中回收 7 万吨品位为 55%~58%Fe 的铁精矿，这些铁精矿含有害杂质元素很低，是优质的炼铁原料。其选矿流程如图 3 所示。

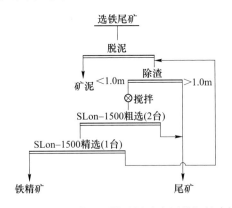

图 3　SLon-1500 磁选机从上厂铁矿尾矿中回收细粒赤铁矿流程

2.4　齐大山选厂的技术改造

鞍钢矿业公司齐大山选矿厂原采用煤气焙烧法处理 25~75mm 粒级的鞍山式氧化矿，

矿石还原成磁铁矿后再磨矿并用弱磁选机分选。但该生产工艺流程存在能耗大、生产成本高、污染严重且铁精矿石品位在63%左右难以提高的缺点。为了提高技术经济指标，解决污染问题，鞍钢矿业公司决定采用阶段磨矿、重选强磁反浮选的流程对原焙烧磁选流程进行全面的技术改造。2001年1月，该公司与赣州有色冶金研究所合作，将一台SLon-1500磁选机安装于齐大山选厂粉矿车间进行强磁选工业试验。通过3个月的对比试验，证明了SLon-1500磁选机选矿指标和设备作业率明显高于其他强磁选设备。工业试验原则流程和选矿指标见图4和表4。

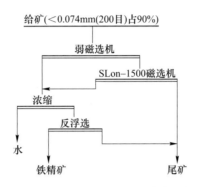

图4　SLon-1500磁选机应用于分选齐大山氧化铁矿

表4　氧化铁矿工业试验全流程选矿指标　　　　　　　　　　　　　　（%）

产品	产率	品位	回收率
给矿	100	29.50	100.00
铁精矿	34.76	65.13	76.73
尾矿	65.24	10.52	23.27

3　SLon磁选机应用于钛铁矿选矿

3.1　攀钢回收细粒钛铁矿

攀枝花是我国最大的钛铁矿产地，攀枝花选钛厂小于0.045mm粒级曾因无高效的选矿设备和工艺而作为尾矿排弃，其全厂钛铁矿的回收率仅有17%~20%。为了提高选钛回收率，1997年该厂建成一条磁浮流程分选小于0.045mm粒级的钛铁矿。一台SLon-1500磁选机用于粗选作业，然后用浮选精选。磁浮流程的综合选矿指标为：给矿品位11% TiO_2，钛精矿品位不小于47% TiO_2，TiO_2回收率44%。该流程每年从细粒尾矿中回收1.5万~2万吨细粒钛精矿。2001年攀钢钛业公司又购买了4台SLon-1500磁选机用于完善其细粒钛铁矿的选矿流程，该工程结束后，TiO_2的综合回收率将提高10%~12%，钛精矿产量每年将增加6万~8万吨。该微细粒选钛流程如图5所示。

3.2　太和铁矿选钛流程改造

重庆钢铁公司太和铁矿的矿石为磁铁矿与钛铁矿的混合矿。1994年以前，该矿建立了弱磁选流程分选磁铁矿。1994~1995年，该矿又建立了以螺旋溜槽、摇床和浮选为主的钛

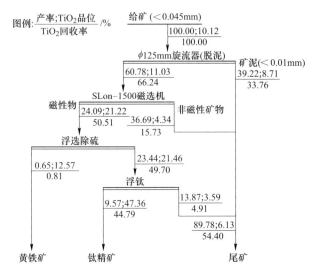

图 5 攀枝花选钛厂小于 0.045mm 钛铁矿选矿流程

铁矿选矿流程,从弱磁选尾矿中回收钛铁矿。但是,由于螺旋溜槽和摇床都不能回收细粒钛铁矿,而浮选又难以回收粗粒钛铁矿,TiO_2 从细粒和粗粒中大量流失,因此选钛回收率很低,仅占弱磁选尾矿的 10%。

2000 年,该矿安装了 3 台 SLon-1500 磁选机,其中两台取代螺旋溜槽,一台取代摇床。这 3 台磁选机有效地回收了各粒级的钛铁矿。新流程在保证钛精矿品位不小于 47% 前提下,使 TiO_2 的综合回收率提高至 40%,该流程如图 6 所示。

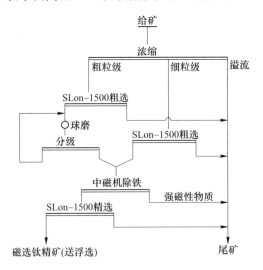

图 6 SLon-1500 磁选机在太和铁矿应用于提高 TiO_2 回收率

4 结论

SLon 立环脉动高梯度磁选机越来越多地应用于我国的赤铁矿、菱铁矿、钛铁矿和非金属矿的选矿,并逐步在我国的大、中型选矿厂的主体选矿流程中应用,促进了细粒弱磁性矿物选矿技术的发展。

写作背景 本文介绍了多种型号的 SLon 立环脉动高梯度磁选机，SLon-1500 磁选机在鞍钢齐大山选矿厂的工业试验和在四川西昌太和铁矿用于分选钛铁矿的工业应用，均获得了良好的选矿指标。证明了 SLon 磁选机在这些厂矿有很大的应用潜力。

SLon型磁选机在齐大山选矿厂的应用

熊大和

（赣州立环磁电设备高技术有限责任公司）

摘　要　SLon立环脉动高梯度磁选机是新一代高效强磁选设备，具有富集比大、选矿效率高、磁介质不易堵塞、设备工作稳定的优点。2001年鞍钢齐大山选矿厂在重选—强磁—反浮选技改中采用该机控制细粒赤铁矿尾矿品位获得成功，全流程的选矿指标为：给矿品位30.15%，铁精矿品位67.00%，尾矿品位11.05%，铁回收率75.86%，选矿指标创厂历史新高。

关键词　赤铁矿选矿　SLon立环脉动高梯度磁选机　螺旋溜槽　反浮选

Application of SLon Magnetic Separator at Qidashan Concentrator

Xiong Dahe

(Ganzhou Vertical Ring Magnetic-electrical Equipment Hi-tech Co., Ltd.)

Abstract　SLon verticalring pulsating high gradient magnetic separator is a new generation of highly efficient high intensity magnetic separation equipment. It has advantages of great enrichment ratio, high separation efficiency, unblocking magnetic medium and stable operation. In the technical reform of gravity separation-high intensity magnetic separation-reverse flotation flowsheet made at Qidashan Concentrator in 2001, it was successfully used to control the tailings grade of fine hematite ore. The concentration results of the whole flowsheet were: feed grade, 30.15%; iron concentrate grade, 67.00%; tailing sgrade, 11.05%; and iron recovery 75.86%, which was a new historical record of the concentrater.

Keywords　Hematite ore dressing; SLon vertical ring pulsating high gradient magnetic separator; Spiral sluice; Reverse flotation

　　SLon立环脉动高梯度磁选机利用磁力、脉动流体力和重力的综合力场连续分选细粒弱磁性矿物。其分选环立式旋转、分选环内安装导磁不锈钢棒或菱形网作为磁介质；采用水内冷大电流、低电压的激磁系统；在磁系下部配有脉动机构。该机具有富集比大，对弱磁性矿物回收率高，磁介质不易堵塞，分选效果优良和设备作业率高的优点。至今已经研制出了SLon-750、SLon-1000、SLon-1250、SLon-1500、SLon-1750、SLon-2000等磁选机。它们背景磁感应强度可达到1.0~1.2T，处理量为0.1~80t/h，可满足多数弱磁性金属矿

---

❶ 原载于《金属矿山》，2002（04）：42~44。

和非金属矿的选矿要求。

1 矿石性质

鞍钢矿业公司齐大山铁矿石属于高硅、微量磷硫锰的鞍山式贫赤铁矿石，矿石主要的铁矿物有赤铁矿、磁铁矿和褐铁矿，主要脉石矿物有石英、透闪石、阳起石、绢云母、绿泥石等，其中石英占脉石矿物总量的80%~85%。矿石硬度系数为12~16。

齐大山铁矿石以典型的条带状构造，其次为揉皱状和致密块状构造，属于矿物嵌布粒度不均匀的细粒浸染贫铁矿石。铁矿物最大粒度可超过1mm，最小粒度在0.005mm以下，平均粒度为0.05mm，石英平均粒度0.085mm。铁矿物单体解离度为：在磨矿细度小于0.074mm（200目）分别为34%、45%、55%、73%、80%和90%时，对应的单体解离度为67%、75%、78%、85%和98%。由此可知，当磨矿细度达到小于0.074mm（200目）为45%~55%的条件下，已有75%~78%的铁矿物单体解离，这一特性适合于阶段磨矿，在粗磨条件下可以选别出量较大的粗粒精矿。

齐大山选矿厂是鞍钢主要铁精矿供应基地，年处理铁品位为30%左右的鞍山式氧化铁矿800万吨。该厂原矿经粗破和中破后筛分成0~20mm和20~75mm两个粒级。2001年以前的选矿工艺为：0~20mm粒级进一选车间，采用阶段磨矿—重选—强磁—浮选流程分选[1,2]；20~75mm粒级进二选车间用煤气焙烧，竖炉进行还原焙烧成磁铁矿，然后进磨矿和弱磁选流程[3]。

多年来，煤气还原焙烧—弱磁选工艺流程存在着能耗高、污染大、工人工作环境差、生产成本高、煤气管道腐蚀大的问题。此外，该工艺所得到的最终铁精矿品位仅63%左右，越来越满足不了冶炼的精料方针。

在2000年下半年，齐大山选厂处理0~20mm粒级的一选车间成功地由阶段磨矿—重选—强磁—正浮选流程改造成为阶段磨矿—重选—强磁—反浮选流程，使铁精矿品位由64%提高到66%，为二选车间的技术改造打下了良好的基础，鞍钢矿业公司决定将二选车间焙烧—弱磁选的工艺流程也改造成为阶段磨矿—重选—强磁—反浮选流程。其中强磁设备的选型是决定该流程改造成功与否的关键之一。

2 SLon立环脉动高梯度磁选机探索试验

鞍钢矿业公司于2000年12月将齐大山一选车间的强磁给矿寄至赣州有色冶金研究所，经该所用SLon-100周期式脉动高梯度磁选机试验，获得选矿指标见表1。

表1 SLon磁选机探索试验结果 （%）

项目	给矿品位	精矿			尾矿	
		品位	产率	回收率	品位	产率
SLon磁选机探索试验平均指标	17.52	31.52	36.13	65.00	9.60	63.87
现场强磁选机指标	17.45	27.12	32.42	50.39	12.81	67.58

通过探索试验表明，SLon磁选机在给矿品位17.52%时，可获得磁选精矿品位31.52%、尾矿品位9.60%、作业回收率65%的良好指标，其尾矿产率占给矿的63.87%。与现场强磁选机比较，强磁选作业的精矿品位提高4.4%，作业回收率提高14.61%。

3 SLon-1500 立环脉动高梯度磁选机现场工业试验

为了证实 SLon 立环脉动高梯度磁选机在工业生产中的适应能力，2001 年 1 月鞍钢矿业公司及齐大山选矿厂与赣州有色冶金研究所合作，在齐大山选矿厂一选车间安装了 1 台 SLon-1500 立环脉动高梯度磁选机，分选现场强磁选机的给矿，连续运转 3 个月所获得的工业试验选别指标见表 2。

表 2 SLon-1500 磁选机齐选工业试验指标　　　　　　　　（%）

给矿品位	精矿			尾矿		备 注
	品位	产率	回收率	品位	产率	
15.78	30.06	31.77	60.61	9.10	68.23	SLon-1500 磁选机平均处理量为 32.56t/(台·时)

工业试验证明 SLon 立环脉动高梯度磁选机选矿指标明显优于其他强磁选设备，其设备运转率（即设备工作时间与日历时间之比）高达 99%，且具有磁介质不堵塞、操作维护工作量小的优点，在现场受到欢迎。SLon-1500 立环脉动高梯度磁选机工业试验成功，为齐选二选车间技术改造提供了可靠的技术依据。

4 SLon-1750 磁选机在二选车间技改中的应用

4.1 改造后的原则流程

SLon-1750 立环脉动高梯度磁选机是近年采用高技术研制的新一代高场强磁选机，该机最初在马钢姑山铁矿获得应用[4,5]，为齐选厂二选车间技改提供了设备可靠性保证。

2001 年 5 月，鞍钢矿业公司订购了 6 台 SLon-1750 立环脉动高梯度磁选机用于齐选二选车间的技术改造。该车间年处理矿石 420 万吨，新的选矿流程采用阶段磨矿、螺旋溜槽重选、SLon-1750 立环脉动高梯度磁选机强磁选及反浮选的工艺流程，新流程于 2001 年 12 月投产并调试成功。新流程的数质量原则流程见图 1。

4.2 新流程特点

一段磨矿的分级溢流（小于 0.074mm（200 目）占 60% 左右）经旋流器分成粗粒级和细粒级（小于 0.074mm（200 目）占 91% 左右）。粗粒级采用螺旋溜槽重选选出一部分铁精矿，螺旋溜槽尾矿经中磁机扫选后抛弃，螺旋溜槽中矿及中磁机磁性产品进入二段球磨，再磨后返回流程。旋流器溢流用弱磁选机选出磁铁矿，弱磁选尾矿经圆筒筛隔除 1.3mm 以上的渣后进入 6 台 SLon-1750 立环脉动高梯度磁选机回收细粒赤铁矿和褐铁矿；弱磁精矿和强磁精矿合并进入反浮选精选，强磁尾矿和浮选尾矿均作为最终尾矿抛弃。螺旋溜槽精矿和浮选精矿合并为最终精矿。

该流程的优点为：重选可在一段磨矿后拿出大部分粗粒铁精矿，并通过中磁机抛弃部分尾矿，显著减少了二段磨矿量（一段磨矿和二段磨矿球磨机台数之比为 5∶1），6 台 SLon-1750 立环脉动高梯度磁选机尾矿品位仅 8.91%（含圆筒筛筛上产物），其抛尾产率占本作业给矿的 73.28%，占全流程原矿的 31.30%，占全流程尾矿量的 47.53%，很好地控制了

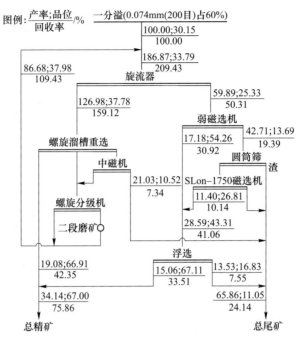

图 1 齐大山选矿厂新建二选车间数质量原则流程

细粒级尾矿品位,保证了全流程获得较高的回收率;反浮选保证了获得较高的细粒级最终精矿品位。该流程与焙烧—还原—磁选流程比较,选矿指标大幅度提高,生产成本显著下降。

4.3 新流程稳定运行指标

新流程调试正常后,鞍钢矿山公司研究所对新流程的生产指标进行了连续 7 天共 168h 生产指标统计,结果见表 3。

表 3 生产统计指标

原矿品位/%		总精矿品位/%	总尾矿品位/%	铁回收率/%	重精品位/%	浮精品位/%	强磁尾品位/%	中磁尾品位/%	浮尾品位/%	球磨处理量/t·h^{-1}
TFe	FeO									
30.51	6.85	67.56	10.15	78.53	66.84	67.11	7.18	9.56	16.34	89.53

统计结果表明,给矿品位 30.51%,总精矿的品位高达 67.56%,铁回收率高达 78.53%。强磁尾(SLon-1750 磁选机尾矿)品位仅 7.18%。全流程运行顺畅,生产指标稳定。

5 结语

6 台 SLon-1750 立环脉动高梯度磁选机在鞍钢矿业公司齐大山选厂二选车间技改工程中,用于控制细粒氧化铁矿的尾矿品位,其尾矿品位控制在 7%~9%,抛尾量占全流程尾矿量的 47.53%。其磁介质不堵塞,操作简单,维护工作量小,设备运转率高达 99% 以上。该设备为保证全流程获得较高的回收率发挥了重要的作用。

齐大山选厂二选车间选矿工艺流程由煤气还原焙烧—磁选流程改为阶段磨矿—重选—强磁—反浮选流程后,生产成本大幅度下降,铁精矿品位由63%提高到67%,铁回收率达到75%~78%,创我国红矿工业选矿的历史最高水平。新流程具有选矿指标好,生产稳定且成本低,污染小的优点。SLon-1750立环脉动高梯度磁选机在该厂的成功应用,证明了该机是居国内领先水平的新一代强磁选工业生产设备。

参 考 文 献

[1] 彭润泽. 齐大山选矿厂一选车间改造设计与生产实践 [J]. 金属矿山,1995 (12):46~49.
[2] 徐冬林. 简化齐大山选矿厂浮选车间工艺流程的试验研究 [J]. 金属矿山,2001 (6):34~36.
[3] 宋乃斌. 齐大山选矿厂优化二选生产工艺的研究 [J]. 金属矿山,2000 (10):21~33.
[4] 熊大和. SLon-1750立环脉动高梯度磁选机的研制与应用 [J]. 金属矿山,1999 (10):23~26.
[5] 熊大和,杨庆林,等. 提高姑山铁矿生产指标的工业试验研究 [J]. 金属矿山,2000 (12):31~33.

写作背景 2001年上半年,一台SLon-1500立环脉动高梯度磁选机在鞍钢齐大山选矿厂进行了3个月分选细粒赤铁矿的工业对比试验,选矿指标和设备性能明显优于其他强磁选设备。鞍钢矿业公司决定对齐大山选矿厂进行技术改造,采用SLon-1750立环脉动高梯度磁选机作为强磁选设备,全流程的选矿指标有重大突破。之后的两年中,齐大山选矿厂将原有的其他强磁选设备拆除,全面采用SLon立环脉动高梯度磁选机,选矿指标创历史新高。SLon磁选机在齐大山选矿厂应用成功,在全国各地取到了良好的示范作用,从此SLon磁选机在全国得到了迅速的推广应用。

SLon 立环脉动高梯度磁选机的应用研究

熊大和

(赣州有色冶金研究所)

摘　要　SLon 立环脉动高梯度磁选机是新型高效的强磁选设备。该机在姑山铁矿、梅山铁矿、齐大山选矿厂、攀枝花选钛厂等地应用于赤铁矿、菱铁矿、钛铁矿等金属矿的选矿，显著地提高了铁精矿或钛精矿的品位和回收率；该机还在江西乐平、安徽来安、四川南江、福建南平应用于石英、长石、霞石等非金属矿的提纯，为这些厂矿生产优质非金属矿成品发挥了重要作用。

关键词　SLon 立环脉动高梯度磁选机　红铁矿　钛铁矿　石英　长石　霞石　强磁选

1　SLon 立环脉动高梯度磁选机简介

SLon 立环脉动高梯度磁选机利用磁力、脉动流体力和重力的综合力场连续分选细粒弱磁性矿物。其分选环立式旋转，分选环内安装导磁不锈钢棒或菱形网作为磁介质；采用水内冷大电流低电压的激磁系统；在磁系下部配有脉动机构。该机具有富集比大、对弱磁性矿物回收率高、磁介质不易堵塞、分选效果优良和设备作业率高的优点。

SLon 立环脉动高梯度磁选机主要由脉动机构、激磁线圈、铁轭、转环和各种矿斗、供水装置等组成，其结构如图 1 所示，其主要技术参数见表 1。该机转环内装有导磁不锈钢棒或钢板网磁介质。选矿时，转环作顺时针旋转，矿浆从给矿斗给入，沿上铁轭缝隙流

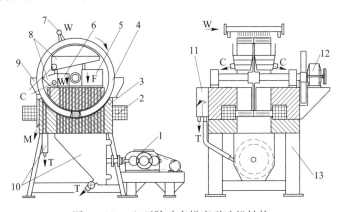

图 1　SLon 立环脉动高梯度磁选机结构

1—脉动机构；2—激磁线圈；3—铁轭；4—转环；5—给矿斗；6—漂洗水斗；7—精矿冲洗装置；
8—精矿斗；9—中矿斗；10—尾矿斗；11—液位斗；12—转环驱动机构；13—机架
F—给矿；W—清水；C—精矿；M—中矿；T—尾矿

---

❶ 原载于《西部矿产资源开发利用与保护学术会议论文集》，2002：137～141。

经转环,转环内的磁介质在磁场中被磁化,磁介质表面形成高梯度磁场,矿浆中磁性颗粒被吸着在磁介质表面,随转环转动被带至顶部无磁场区,用冲洗水冲入精矿斗中,非磁性颗粒沿下铁轭缝隙流入尾矿斗中排走。

表 1　SLon 立环脉动高梯度磁选机主要技术参数

机型	转环直径/mm	背景磁感强度/T	激磁功率/kW	驱动功率/kW	脉动冲程/mm	脉动冲次/次·min⁻¹	给矿粒度/mm	给矿浓度/%	矿浆流量/m³·h⁻¹	干矿处理量/t·h⁻¹	机重/t
SLon-2000	2000	0~1.0	0~82	5.5+7.5	0~30	0~300	0~2.0	10~45	100~200	50~80	50
SLon-1750	1750	0~1.0	0~62	4+4	0~30	0~300	0~1.3	10~45	75~150	30~50	35
SLon-1500	1500	0~1.0	0~44	3+4	0~30	0~300	0~1.3	10~45	50~100	20~35	20
SLon-1250	1250	0~1.0	0~35	1.5+2.2	0~30	0~300	0~1.3	10~45	20~50	10~20	14
SLon-1000	1000	0~1.0	0~26	1.1+2.2	0~30	0~300	0~1.3	10~45	10~20	4~7	6
SLon-750	750	0~1.2	0~20.4	0.55+0.75	0~30	0~300	0~1.3	10~45	1~2	0.1~0.5	3

2　SLon 磁选机在姑山铁矿选矿流程中的应用

马钢姑山铁矿是一个较大的难选氧化铁矿生产基地。自从 1978 年建立选厂以来,其球磨主厂房的选矿工艺流程和生产指标经历了多次改进,最初采用浮选流程所获得的生产指标为:铁精矿品位 49% 和铁回收率 49%;20 世纪 80 年代采用螺旋溜槽和离心选矿机的生产指标为:铁精矿品位 55% 和铁回收率 61%;1992 年改为以 SQC-6-2770 湿式强磁选机为主及 SLon-1500 磁选机为辅的全磁选流程,由于 SQC 磁选机存在齿板容易堵塞问题及铁精矿品位徘徊在 57%~58% 的水平,姑山铁矿近年改为以 SLon-1750 磁选机和 SLon-1500 磁选机的全磁选流程,新的生产流程已达到铁精矿品位 59.5% 和铁回收率 76% 的良好指标,在 3 个球磨系统未增容的条件下,铁粉精矿产量由 1990 年的 15 万吨增至 2001 年的 36 万吨。2002 年可望达到 40 万吨左右。技术进步给姑山铁矿带来了显著的经济效益。姑山铁矿球磨主厂房现行的选矿流程和指标见图 2 和表 2。

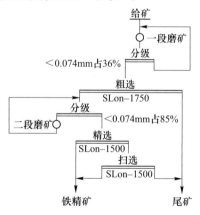

图 2　SLon 磁选机在姑山铁矿分选难选赤铁矿的生产流程

表2 姑山铁矿磁选流程与重选流程指标比较

流程	年份	铁品位/%			铁回收率/%
		给矿	精矿	尾矿	
重选	1988	37.16	55.22	24.47	61.32
SLon磁选	2000	42.65	59.50	22.41	76.13

3 SLon 磁选机在梅山铁矿降磷工程中的应用

宝钢梅山铁矿是位于江苏南京市附近的一座较大的铁矿山,其矿石由磁铁矿、赤铁矿、菱铁矿和黄铁矿组成。由于其早期生产的铁精矿含磷偏高而缺乏市场竞争力。为了解决这一个严重困扰梅山铁矿发展的问题,梅山铁矿于1995年组织了两个工业试验,一个为浮选流程工业试验,而另一个为磁选流程工业试验。通过数百个小时的对比试验,磁选流程显示了选矿指标较好、成本较低、有利于环保、铁精矿易于脱水和生产流程稳定等优点。该矿最终采用了磁选流程改造和扩建生产流程,其中采用了16台SLon-1500磁选机用于降磷生产和2台SLon-1500磁选机用于其重选车间的细粒铁矿选别作业。梅山铁矿现行的降磷流程和生产指标见图3和表3。

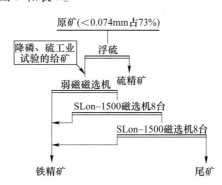

图3 梅山铁矿降磷磁选流程

表3 梅山铁矿降磷指标 (%)

产品	产率	品位			回收率		
		Fe	S	P	Fe	S	P
给矿	100.00	52.77	0.44	0.399	100.00	100.00	100.00
精矿	88.90	56.08	0.29	0.246	94.51	59.10	54.76
冶炼要求		>55.0	<0.35	<0.25			
尾矿	11.10	26.13	1.61	1.626	5.49	40.90	45.24

4 从昆钢上厂铁矿的尾矿中回收细粒赤铁矿

昆钢上厂铁矿的尾矿库中储存有含铁量20%~28%Fe的尾矿约1000万吨,近年该矿采用了一种以细粒跳汰机和3台SLon-1500磁选机为主体的选矿新工艺流程从尾矿中回收赤铁矿,每年从尾矿中回收7万吨品位为55%~58%Fe的铁精矿,这些铁精矿含有害杂质元素很低,是优质的炼铁原料。其选矿流程如图4所示。

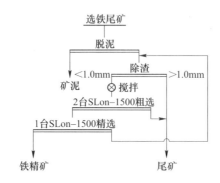

图 4 磁选机从尾矿中回收细粒赤铁矿流程

5 SLon 磁选机应用于鞍山齐大山选厂技术改造

鞍钢鞍山矿业公司齐大山选矿厂原采用煤气焙烧法处理 25~75mm 粒级的鞍山式氧化矿，矿石还原成磁铁矿后再磨矿并用弱磁选机分选。但该生产工艺流程存在能耗大、生产成本高、污染严重且铁精矿品位在 63% 左右难以提高的缺点。为了提高技术经济指标，解决污染问题，鞍山矿业公司决定采用阶段磨矿、重选强磁反浮选的流程对原焙烧磁选流程进行全面的技术改造。鞍山矿业公司把 6 台 SLon-1750 磁选机用于齐大山的技术改造，该工程已于 2001 年 11 月底竣工。新流程中，SLon 磁选机有效地将细粒级尾矿品位控制在 7%~9%Fe。全流程铁精矿品位由 63% 提高到 66%~67%，铁回收率达到 75%~78%，创我国红矿选矿的历史最高水平。新流程具有选矿指标好、生产稳定、成本低、污染小的优点。其数质量流程如图 5 所示。

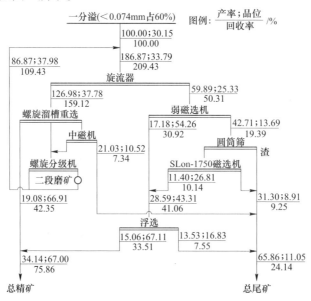

图 5 齐大山选矿厂新建二选车间数质量原则流程图

6 SLon 磁选机在攀钢用于回收细粒钛铁矿

攀枝花是我国最大的钛铁矿产地，攀枝花选钛厂小于 0.045mm 粒级曾因无高效的选矿设备和工艺而作为尾矿排弃，其全厂钛铁矿的回收率曾仅有 17%~20%。为了提高选钛回收率，2001 年该厂建成了磁浮流程分选小于 0.045mm 粒级的钛铁矿。6 台 SLon-1500 磁选机用于粗选作业，然后用浮选精选。磁浮流程的综合选矿指标为：给矿品位 11%TiO_2，钛精矿品位不小于 47%TiO_2，TiO_2 回收率 44%。该流程每年从细粒尾矿中回收 6 万吨细粒钛精矿。TiO_2 的综合回收率提高了 10%~12%。该微细粒选钛流程如图 6 所示。

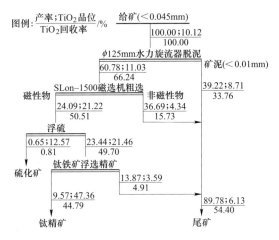

图 6　攀枝花选钛厂小于 0.045mm 钛铁矿选矿流程

7　SLon 磁选机应用于太和铁矿选钛流程改造

重庆钢铁公司太和铁矿的矿石为磁铁矿与钛铁矿的混合矿。1994 年以前，该矿建立了弱磁选流程分选磁铁矿。1994~1995 年，该矿又建立了以螺旋溜槽、摇床和浮选为主的钛铁矿选矿流程，从弱磁选尾矿中回收钛铁矿。但是，由于螺旋溜槽和摇床都不能回收细粒钛铁矿，而浮选又难以回收粗粒钛铁矿，TiO_2 从细粒和粗粒中大量流失，因此选钛回收率很低，仅占弱磁选尾矿的 10%。

2000 年，该矿安装了 3 台 SLon-1500 磁选机，其中 2 台取代螺旋溜槽，1 台取代摇床。这 3 台磁选机有效地回收了各粒级的钛铁矿。新流程在保证钛精矿品位不小于 47%前提下，使 TiO_2 的综合回收率提高至 40%，该流程如图 7 所示。

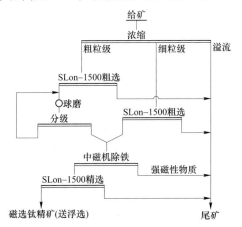

图 7　SLon-1500 磁选机在太和铁矿应用于提高 TiO_2 回收率

8 SLon 磁选机在稀土矿选矿中的应用

四川德昌大陆槽稀土矿，为残坡积角砾状萤石重晶石型及英碱正长岩型脉状氟碳铈矿床，REO 地质品位 4.16%~6.19%，已建成日处理矿石 21000t 的选矿厂。原生产流程为单一重选流程，精矿品位为 50%REO 时，总回收率约 15%。为寻求合理的选矿工艺流程，该矿曾先后委托国内科研院所开展重—浮及磁—重流程的探索试验研究。研究结果表明，原矿中重晶石、天青石比重与氟碳铈矿相近，含量约占 26%，是造成单一重选流程指标低的根本原因。若采用 SLon 立环脉动高梯度磁选机预先抛弃大量重晶石、天青石等脉石矿物，可大幅度提高摇床的分选效率，获得较高的磁—重流程指标：采用 2 台 SLon-1500 磁选机进行一粗一扫选别（如图 8 所示），磁选精矿再用摇床一次选别的磁重流程，在原矿品位 REO 为 5.72% 时，获得了稀土精矿品位 REO 53.11%、回收率为 55.63% 的良好指标。

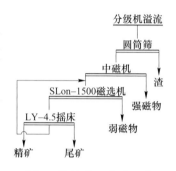

图 8 德昌稀土矿磁—重工业生产流程

9 闽南钽铌矿提高长石精矿质量的应用

南平钽铌矿属大型钽铌伟晶岩矿床，矿石经重选回收钽铌精矿后，尾矿中含有大量的钠长石、钾长石。尾矿经脱泥、隔粗、采用磁选技术去除含铁矿物后，即成为长石产品，可作为陶瓷原料使用。

随着市场要求的提高，原有长石综合回收工艺流程的长石产品质量难以达到要求，为进一步提高产品质量，闽宁钽铌矿于 2001 年 8 月安装了一台 SLon-1500 立环脉动高梯度磁选机组成二段除铁工艺流程。重选尾矿入浓密箱浓缩脱水后，经高频细筛隔除大于 0.4mm 粗粒部分，小于 0.4mm 物料经强磁一次除铁后的非磁性产品入 SLon-1500 立环脉动高梯度磁选机进一步除去暗色矿物，非磁性产品经 $\phi1500mm$ 螺旋分级机脱水、脱泥后返砂即为最终长石产品，长石生产调试工艺流程如图 9 所示。

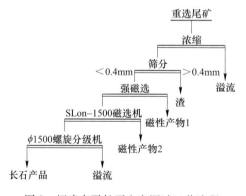

图 9 福建南平长石生产调试工艺流程

工业调试结果表明：重选尾矿，经浓密箱脱除部分细泥，入高频筛隔粗后，含 Fe_2O_3 0.49%~0.64%，SLon-1500 磁选机非磁性产品含 Fe_2O_3 0.19%~0.24%，经 $\phi1500mm$ 螺旋分级脱泥、脱水后，长石产品含 Fe_2O_3 0.15%~0.18%，平均为 Fe_2O_3 0.167%，成为优质的玻璃和陶瓷原料。

10 开发安徽来安县长石资源的应用

安徽省来安县境内储藏有丰富的长石资源，矿石主要由钾长石、斜长石、少量石英、黑云母、角闪石和磁铁矿组成。矿石中除 Fe_2O_3 含量较高外，其他物理、化学指标均能达到玻璃、陶瓷工业的要求。1998 年来安县皖东长石有限公司建成了一座年产 3 万吨长石精矿的选矿厂，其选矿流程如图 10 所示。含 Fe_2O_3 1.48% 的原矿经棒磨和高频细筛组成的闭路磨矿筛分至小于 0.8mm，进入滚筒式中磁机除去强磁性铁物质，然后进入一台 SLon-1250 磁选机除去弱磁性矿物。磁选作业的非磁性产品含 Fe_2O_3 0.26%，满足玻璃工业和陶瓷工业 Fe_2O_3 含量小于 0.3% 的要求。

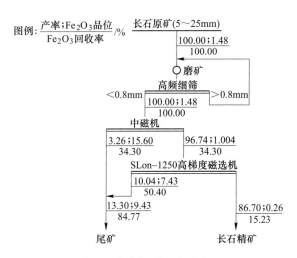

图 10 来安长石厂选矿流程

11 四川南江霞石矿除铁的应用

四川南江霞石矿是我国第一个具有较高工业开采价值的霞石矿床。矿石中钠、钾含量较高，具有熔点低和助熔性能强等优点，故在建材、化工等工业中有着广泛的用途。但该霞石矿铁含量高，不能直接开采利用。通过对矿物性质考查及可选性试验研究，确定采用 SLon 立环脉动高梯度磁选机作为除铁主体设备开展一粗一精全磁流程的提纯工艺，工业试验流程如图 11 所示。工业试验表明，当给矿含 Fe_2O_3 1.69% 时，经一粗一精选别可获得含 Fe_2O_3 0.179% 的霞石精矿，回收率 76.36%，除铁率达 91.52%，产品质量达到出口类一级霞石精矿指标。目前该矿采用 SLon-1000 网介质磁选机分选霞石精矿，经济效益可观。

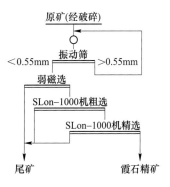

图 11 南江霞石矿一粗一精流程

12 江西乐平石英砂除铁提纯的应用

江西乐平县境内具有丰富的沉积石英砂矿，由于含铁和含泥量偏高而不能直接作为工业原料。2000年10月，乐平华源硅砂厂建起了一条年产5万吨石英精矿的生产线，其中用永磁、中磁选矿机和SLon-1500立环脉动高梯度磁选机为主体选矿设备除铁，使石英精矿含铁量从原矿的 0.1% Fe_2O_3 左右降至 0.02% Fe_2O_3，使石英精矿成为玻璃、建材的优质原料。其选矿流程如图12所示。

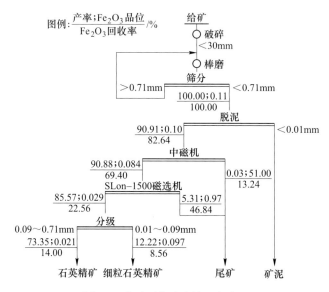

图 12　乐平石英矿除铁生产流程

13 结语

SLon立环脉动高梯度磁选机是新一代强磁选设备，具有优异的选矿性能和机电性能，在我国选矿工业中已广泛应用于赤铁矿、菱铁矿、钛铁矿等弱磁性金属矿的选矿和石英、长石、霞石、红柱石等非金属矿的提纯，对提高我国矿产资源的选矿工业技术水平起到了日益重要的作用。

参 考 文 献

[1] 熊大和，杨庆林，汤桂生，等．提高姑山赤铁矿生产指标的工业试验研究［J］．金属矿山，2000（12）：31～33．

[2] 熊大和，苏卫．上厂铁矿尾矿复选赤铁矿新技术［J］．矿山环保，2001（3）：3～6．

[3] 杨君勤．SLon立环脉动高梯度磁选机在梅山铁矿降磷工程中的应用［J］．矿山机械，1997，25（10）：46～47．

[4] 熊大和．SLon型磁选机在齐大山选矿厂的应用［J］．金属矿山，2002（4）：42～44．

[5] 许新邦．SLon磁选机分选攀钢微细粒钛铁矿的工业试验［J］．金属矿山，1997（21）：17～20．

[6] 张云红．SLon立环脉动高梯度磁选机提高某稀土精矿质量的研究［J］．江西有色金属，1998，12（3）：34～36．

[7] 王兆元.SLon 立环脉动高梯度磁选机选别霞石矿的试验研究［J］.非金属矿，1997（6）：40~42.
[8] 王兆元.SLon 立环脉动高梯度磁选机在长石及霞石非金属矿除铁提纯的工业试验应用［J］.江西有色金属，1998，12（2）：24~27.

写作背景 本文介绍了多种型号的 SLon 立环脉动高梯度磁选机和在多个厂矿的应用，其中德昌稀土矿、福建南平钽铌矿、四川南江霞石矿和江西乐平石英砂都是 SLon 磁选机新开辟的应用矿点。

SLon 立环脉动高梯度磁选机在提高红矿质量的应用

熊大和

(赣州立环磁电设备高技术有限责任公司)

摘　要　SLon立环脉动高梯度磁选机具有富集比大、选矿效率高、磁介质不易堵塞、设备作业率高的优点。该机在梅山铁矿的降磷工程中应用于回收细粒赤铁矿和菱铁矿，梅山铁精矿的铁品位从52.77%提高到56.08%，铁精矿的含磷量从0.399%降低至0.246%；用于姑山铁矿细粒赤铁矿全磁选流程中，使综合铁精矿从55%提高到59.5%，铁回收率从61%提高到75%~76%；应用于齐大山的选矿厂二选车间的阶段磨矿、重选—强磁—反浮选流程中，综合铁精矿品位从63%提高至66%~67%，铁回收率达到75%~78%；应用于东鞍山烧结厂一选车间阶段磨矿、重选—强磁—反浮选工业试验中，综合铁精矿品位从60%提高至64%。铁回收率达到70%以上。该机为提高我国氧化铁矿的选矿水平做出了重要的贡献。

关键词　SLon立环脉动高梯度磁选机　红铁矿　提质降杂

Application of SLon Vertical Ring Pulsating High Gradient Magnetic Separators in Raising Oxidized Iron Ore's Quality

Xiong Dahe

(*Ganzhou Vertical Ring Magnetic-electric Equipment Hi-tech Co.,Ltd.*)

Abstract　SLon vertical ring pulsating high gradient magnetic separators have the advantages of great concentration ratio, high separation efficiency, non-blockage of magnetic medium and high equipment availability. When used in the phosphorus project of Meishan Iron Mine for recovering fine hematite and siderite, they raised the iron concentrate grade from 52.77% to 56.08% and reduced the phosphorus content from 0.399% to 0.246%. When used in the whole magnetic separation flowsheet of Gushan Iron Mine's fine hematite, they raised the iron grade of comprehensive concentrate from 55% to 59.5% and the iron recovery from 61% to 75%~76%. When used in the stage grinding, gravity separatoin-high intensity magnetic separation reverse flotation flowsheet of No. 2 workshop of Qidashan Concentrator, they raised the grade of comprehensive iron concentrate from 63% to 66%~67% and the iron recovery to 75%~78%. When used in the industrial test of stage grinding, gravity separation-high intensity magnetic separatoin-reverse flotation at No. 1 workshop of Donganshan Sintering Plant, they raised the iron grade of comprehensive iron concentrate from 60% to 64% and the iron recovery to over 70%. As a result,

---

[1]　原载于《金属矿山》2002 (09): 238~241。

they have made important contribution to the improvement of China's beneficiation level of oxidized iron ore.

Keywords SLon vertical ring pulsating high gradient magnetic separator; Oxidized iron ore; Quality improvement and impurity reduction

1 SLon 立环脉动高梯度磁选机简介

SLon 立环脉动高梯度磁选机是一种适合于分选细粒弱磁性矿物的高效磁选设备,其转环立式旋转、反冲精矿、并配有脉动机构. 具有富集比大、分选效率高、磁介质不易堵塞、对给矿粒度、浓度和品位波动适应性强、工作可靠、操作维护方便等优点。通过近20年的研究。已研制出了 SLon-750、SLon-1000、SLon-1250、SLon-1500、SLon-1750、SLon-2000 等型号的立环脉动高梯度磁选机,它们已在工业生产上广泛应用于赤铁矿、菱铁矿、褐铁矿、钛铁矿等金属矿的选矿及长石、石英、霞石等非金属矿的提纯。SLon 立环脉动高梯度磁选机主要技术参数见表1。

表1 SLon 立环脉动高梯度磁选机主要技术参数

机型	转环直径/mm	背景磁感强度/T	激磁功率/kW	驱动功率/kW	脉动冲程/mm	脉动冲次/r·min^{-1}	给矿粒度/mm	给矿浓度/%	矿浆流量/m^3·h^{-1}	干矿处理量/t·h^{-1}	机重/t
SLon-2000	2000	0~1.0	0~82	5.5+7.5	0~30	0~300	0~2.0	10~45	100~200	50~80	50
SLon-1750	1750	0~1.0	0~62	4+4	0~30	0~300	0~1.3	10~45	75~150	30~50	35
SLon-1500	1500	0~1.0	0~44	3+4	0~30	0~300	0~1.3	10~45	50~100	20~35	20
SLon-1250	1250	0~1.0	0~35	1.5+2.2	0~30	0~300	0~1.3	10~45	20~50	10~20	14
SLon-1000	1000	0~1.0	0~26	1.1+2.2	0~30	0~300	0~1.3	10~45	10~20	4~7	6
SLon-750	750	0~1.2	0~20.4	0.55+0.75	0~30	0~300	0~1.3	10~45	1~2	0.1~0.5	3

2 SLon 磁选机在梅山铁矿降磷提质中的应用

宝钢梅山铁矿是位于南京市附近的一座大型铁矿山,其矿石由磁铁矿、赤铁矿、菱铁矿和黄铁矿组成。该矿于20世纪70年代建成年采选200万吨、年产铁精矿130万吨的生产能力,其铁精矿用于球团原料。然而,由于其早期生产的铁精矿含磷和硫偏高,其铁精矿不能作为单独的球团原料而只能作为辅料,因缺乏市场竞争力使产量受到限制。

20世纪80年代,梅山铁矿计划将采选规模由200万吨扩大至400万吨,面临的最大问题是如何降低铁精矿的磷含量。为此,该矿组织了多个科研院所进行了广泛的研究和试验。最终于1995年选定了1个磁选流程和1个浮选流程分别进行工业试验。两个流程获得了近似的结果。由于磁选流程具有生产成本低,精矿易过滤和环境污染小的优点,自1996~2000年,该矿改造了其第一期200万吨生产线并新建了第二期200万吨生产线,共有16台 SLon-1500 立环脉动高梯度磁选机应用于该流程中。梅山铁矿的矿物组成及元素分析见表2和表3,其降磷提质的数质量流程如图1所示。

表 2　梅山矿石的矿物组成　　　　　　　　　　　　（％）

铁矿物	磁铁矿	菱铁矿	赤铁矿	黄铁矿	硅酸盐	合计
原矿	21.84	10.32	8.58	2.76	3.00	46.50
精矿	26.92	10.34	14.46	0.441	1.60	53.76

表 3　梅山矿石的化学成分　　　　　　　　　　　　（％）

元素	Fe	S	P	SiO_2	Al_2O_3	CaO	MgO
原矿	46.5	1.91	0.366	9.98	1.87	4.87	1.85
降磷前铁精矿	52.77	0.44	0.399	5.07	0.86	4.10	1.81
降磷后铁精矿	56.08	0.29	0.246	3.47	0.651	3.43	1.79
冶炼要求	>55.0	<0.35	<0.25				

由表 3 可见，整个降磷作业使铁精矿品位从 52.77% 提高到 56.08%，含磷和硫分别从 0.399% 和 0.44% 降低至 0.246% 和 0.29%，使铁精矿质量大幅度提高，满足了冶炼要求。目前，梅山铁矿年处理铁矿石 400 万吨，年产铁精矿 260 万吨，铁精矿销售状况良好。根据梅山铁矿的性质，如图 1 所示的降磷流程中浮选后的铁精矿用弱磁选机选出磁铁矿，用 SLon-1500 立环脉动高梯度磁选机组成一次粗选、一次扫选作业回收赤铁矿和菱铁矿。

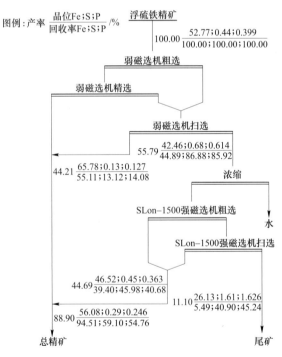

图 1　梅山铁矿提质降杂的数质量流程

3　SLon 磁选机对提高姑山铁矿精矿粉质量的作用

姑山铁矿是马钢主要原料基地之一，主要金属矿物为赤铁矿，矿石硬度大、嵌布粒度细且不均匀。属难选红铁矿。自从1978年建成选厂以来，其球磨机主厂房的选矿工艺流程和生产指标经历了多次改进。最初采用浮选流程所获得的生产指标为铁精矿品位49%和铁回收率49%；20世纪80年代采用螺旋溜槽和离心选矿机的生产指标为铁精矿品位55%和铁回收率61%；1992年改为以SQC-6-2770湿式强磁选机为主及SLon-1500磁选机为辅的全磁选流程。由于SQC磁选机存在齿板容易堵塞及铁精矿品位徘徊在57%~58%的水平，姑山铁矿近年改为以SLon-1750和SLon-1500磁选机的全磁选流程。新的生产流程已达到铁精矿品位59.5%和铁回收率75%~76%的良好指标。在3个球磨系统未增容的条件下，铁精矿粉产量由1990年的15万吨增至2001年的36万吨，2002年可望达到40万吨左右。SLon磁选机的应用给姑山铁矿带来了显著的经济效益。

4 SLon磁选机在齐大山选矿厂的应用

齐大山选矿厂是鞍钢主要铁精矿供应基地，年处理铁品位为30%左右的鞍山式氧化铁矿800万吨。该厂原矿经粗碎和中碎后筛分成0~20mm和20~75mm两个粒级。2001年以前的选矿工艺为0~20mm粒级进一选车间，采用阶段磨矿—重选—强磁—浮选流程分选；20~75mm粒级进二选车间用煤气焙烧，竖炉进行还原焙烧成磁铁矿。然后进磨矿和弱磁选流程。

多年来，煤气还原焙烧—弱磁选工艺流程存在着能耗高、污染大、工人工作环境差、生产成本高、煤气管道腐蚀大的问题。此外，该工艺所得到的最终铁精矿品位仅63%左右，越来越满足不了冶炼的精料方针。

2001年5月，鞍钢矿业公司订购了6台SLon-1750立环脉动高梯度磁选机用于齐选二选车间的技术改造。该车间年处理矿石420万吨，新的选矿流程采用阶段磨矿、螺旋溜槽重选、SLon-1750立环脉动高梯度磁选机强磁选及反浮选的工艺流程，新流程于2001年12月投产并调试成功，新流程的数质量原则流程如图2所示。

齐大山选厂二选车间选矿工艺流程由煤气还原焙烧—弱磁选流程改为阶段磨矿—重选—强磁—反浮选流程后，生产成本大幅度下降，铁精矿品位由63%提高到66%~67%，铁回收率达到75%~78%，创我国红矿工业选矿的历史最高水平，铁精矿的生产成本由改造前的每吨300元降低到改造后的每吨240元。新流程具有选矿指标好，生产稳定且成本低、污染小的优点，6台SLon-1750磁选机在该流程中发挥了重要的作用。

5 SLon磁选机在东鞍山选矿厂的工业试验

东鞍山烧结厂是鞍钢铁精矿原料基地之一，其铁矿石结晶粒度细，氧化程度较深，是鞍山地区最难选的红矿。东鞍山烧结厂一选车间多年来一直采用两段连续磨矿，用氧化石蜡皂和塔尔油混合药剂作捕收剂的单一碱性正浮选工艺流程，铁精矿品位一直徘徊在60%左右。由于铁精矿品位较低，随着鞍钢对精矿质量要求不断提高，东鞍山烧结厂一选车间面临停产的严峻形势。为此鞍钢集团鞍山矿业公司决定对东烧一选车间进行全面的技术改造，改造目的要使铁精矿品位达到64%以上，铁回收率达到70%以上。为此，2001年鞍山矿业公司研究所与东鞍山烧结厂在东烧一选车间开展了两个新流程的工业试验，其一为重选—强磁—反浮选流程（如图3所示），其二为强磁—重选—反浮选流程（如图4所示）。

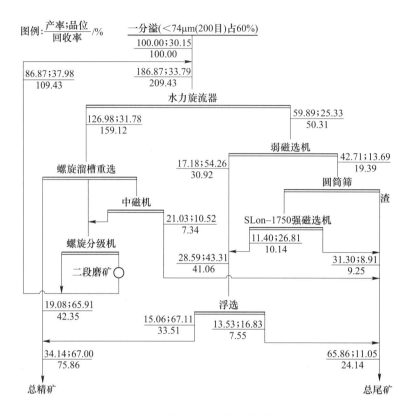

图 2　齐大山选矿新建二选车间数质量原则流程

如图 3 所示的流程中，二段磨矿的分级溢流采用二次旋流器分级，二次旋流器的沉砂用螺旋溜槽选出一部分粗粒铁精矿、溢流经浓缩后用 SLon-1750 立环脉动高梯度磁选机作为强磁选抛尾作业。强磁精矿用反浮选精选，浮选精矿和螺旋溜槽重选精矿混合为最终精矿。该流程的综合选矿指标为原矿品位 31.38%，精矿品位 64.08%，尾矿品位 13.77%，铁回收率 71.47%。

如图 4 所示的流程中，二段磨矿分级溢流采用一次旋流器分级，旋流器溢流用弱磁选机和 SLon-1750 立环脉动高梯度磁选机分别选出磁铁矿和氧化铁矿，并预先抛去产率为 40% 左右的尾矿；磁选精矿经旋流器分级，沉砂用螺旋溜槽选出部分粗粒铁精矿，溢流采用反浮选精选，浮选精矿和重选精矿混合为最终精矿。该流程的综合选矿指标为原矿品位 32.94%，精矿品位 64.74%，尾矿品位 14.68%，铁回收率 71.69%。在上述两个工业试验流程中，SLon-1750 立环脉动高梯度磁选机承担了脱泥抛尾的重要作业。该机一次选别将尾矿品位控制在 12%～13%。抛尾量占原矿的 40%～43%，占总尾矿的 63%～66%，保证了全流程获得较高的回收率，并为浮选作业获得较高的精矿品位和减少浮选药剂用量创造了良好的条件。

东烧厂一选车间的工业试验获得了良好的技术经济指标。目前该技改项目已全面展开，预计 2002 年底完成，届时东烧厂一选车间的铁精矿品位可望从目前的 60% 左右提高到 64% 以上。

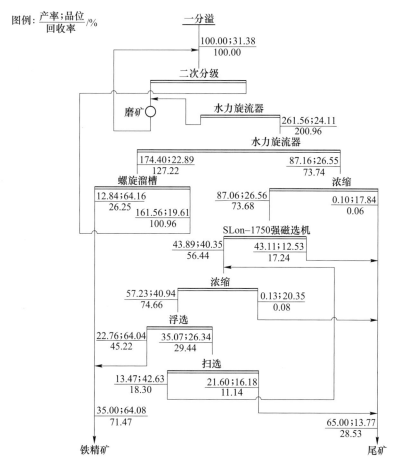

图 3　东烧一选车间重选—强磁—反浮选数质量原则流程

6　结论

（1）该机用于梅山铁矿降磷工程中，使铁精矿铁品位从 52.77% 提高到 56.08%，铁精矿含磷和硫分别从 0.399% 和 0.44% 降低至 0.246% 和 0.29%，铁精矿质量大幅度提高并满足了冶炼要求。

（2）该机用于姑山铁矿细粒赤铁矿全磁选流程中，使铁精矿品位从 55% 提高到 59.5%。在球磨机基本未增容的条件下，细粒铁精矿产量由 1990 年的 15 万吨增至 2001 年的 36 万吨，选矿成本大幅度下降。

（3）该机用于齐大山二选车间的技术改造，全流程的铁精矿品位从 63% 提高到 66%~67%，铁回收率达到 75%~78%，创我国红矿工业选矿的历史最高水平，铁精矿的生产成本从改造前的 300 元/t 降低至 240 元/t。

（4）该机用于东鞍山烧结厂的工业试验中，全流程的铁精矿品位可望从原生产的 60% 左右提高到 64% 以上，铁回收率保持在 70% 以上，目前该技改工程正在施工之中。

以上应用实践证明，SLon 立环脉动高梯度磁选机是可广泛应用于红矿提质降杂的新一代高效强磁选设备，可显著提高铁精矿质量和降低生产成本。该机将为提高我国红矿选矿技术水平作出日益重要的贡献。

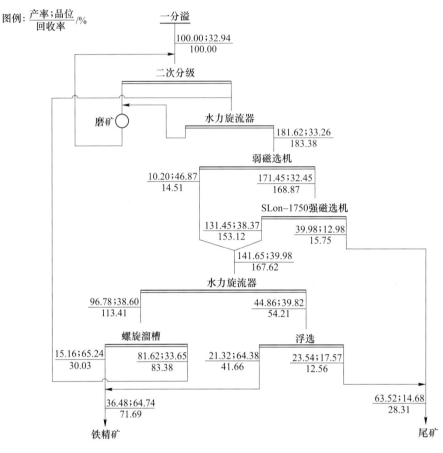

图 4　东烧一选车间强磁—重选—反浮选数质量原则流程

参 考 文 献

[1] 衣德强. 立环脉动高梯度磁选机在梅山铁矿的应用 [J]. 冶金矿山设计与建设, 1999 (2): 35~39.
[2] 熊大和, 杨庆林, 汤桂生, 等. 提高姑山赤铁矿生产指标的工业试验研究 [J]. 金属矿山, 2000 (12): 31~33.
[3] 熊大和. SLon 型磁选机在齐大山选矿厂的应用 [J]. 金属矿山, 2001 (4): 42~44.
[4] 任长彬. 浅谈东鞍山烧结厂一选红矿选别指标低的原因和对策 [J]. 金属矿山, 1999 (1): 25~27.

写作背景　本文介绍了 SLon 立环脉动高梯度磁选机在几个大中型氧化铁矿选矿厂提高铁精矿质量所发挥的作用, 如梅山铁矿的铁精矿降磷提质、马钢姑山铁矿、鞍钢齐大山选矿厂、鞍钢东鞍山烧结厂提高铁精矿品位。其中东鞍山烧结厂应用 SLon 磁选机的工业试验是第一次公布。

The Development of A New Type of Industrial Vertical Ring and Pulsating HGMS Separator[①]

Xiong Dahe   Liu Shuyi   Liu Yongzhi   Chen jin

(Department of Mineral Engineering, Central South University of Technology
Changsha Hunan, The People's Republic of China)

Abstract   This new type industrial vertical ring and pulsating high gradient magnetic separator utilizes the combined force field of magnetic force, pulsating fluid and gravity force to treat fine weakly magnetic particles. It flushes concentrate in the opposite direction of feeding and possesses a slurry pulsating mechanism. Commercial test was carried out in Gushan Iron Mine of MaAnshan Iron and Steel Company, Anhui, China, to treat fine and refractory hematite ore. It had been successfully running for 3000 hours by the end of march, 1988, and keeps continuously running up to now in the production line. Compared with centrifugal separator to be substituted, it increases the iron concentrate grade over 4% and the recovery by 10%-20%. Moreover, under a new contract another larger separator of the same type is under construction. In this paper, the structure, working principles and commercial test results of the separator will be described.

Productivity and technology in the metallurgical industries edited by P. D. Koch and J. C. Taylor the Minerals, Metals & Materials Society, 1989.

1 Introduction

High gradient magnetic separation (HGMS) features the advantage of high recovery of fine weakly magnetic particles[1-4]. Cyclic HGMS separators have been successfully applied to kaoline purification and wastewater treatment since 1970s. However continuous HGMS separators, designed to treat ores containing large portion of weakly magnetic minerals such as hematite ore, are not so competitive in metal mineral processing industry because of their relatively low grade of magnetics and frequent blockage.

We began to study pulsating high gradient magnetic separation (PHGMS) in 1981[5]. Great efforts have been done on preventing magnetic matrix from blockage and improving the quality of magnetic product. The first SLon-1000 vertical ring and pulsating high gradient magnetic separator was worked out under the cooperation of Central South University of Technology and Ganzhou Research Institute of Non-ferrous metallurgy in September, 1987. Then it was installed into the hematite processing flowsheet of Gushan Iron Mine to carry out commercial test. After a running period of 3000 hours test, it still keeps running to meet the industrial needs and up to now (November,

---

① 原载于《世界冶金工业生产力与技术会议文集》,1989。

1988) it has been running over 8000 hours. Commercial test verified that this kind of PHGMS separator possesses a series of advantages for industrial application.

2 Structure and working principles of SLon-1000 separator

SLon-1000 separator consists of mainly thirteen parts, as shown in Fig. 1. The pulsating mechanism, magnetic yoke, energizing coil and working ring are the key parts. Pulsating frequency and ring rotating speed are controlled by speed variable motors. Although pulsating stroke is adjustable, we usually take 13mm as applied to the slurry in the working zone. The energizing coil is made of hollow copper tube and cooled internally by water. Along the periphery of the ring there are a number of rectangular rooms in which expanded nets sized 0.3:3.2:8mm made of magnetic conductive stainless steel are filled as matrix. The height of matrix pile is 7cm. Other kind of matrix may also be used depending on the requirement of different ores. While the separator is working, the ring rotates clockwise. Slurry fed from the feeding box enters the ring from the gaps in the upper yoke. The average speed of slurry in the working zone is about 6cm per second. The matrix in the working zone is magnetized. Magnetic particles are attracted from slurry onto matrix surface, brought to the top by the ring, where magnetic field is negligible, and flushed out to the concentrate box. Non-magnetic particles enter the tailing box along the gaps in lower yoke under the combined action of slurry pulsation, gravity and drag force. The pulsating mechanism drives the rubber diaphragm on the tailing box to move back and forth. As long as slurry level is adjusted above the level line, the kinetic energy due to pulsation can be effectively transmitted to the working zone.

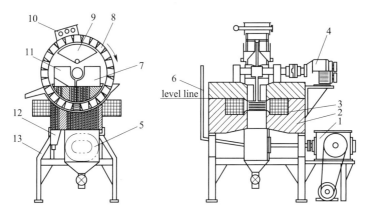

Fig. 1 SLon-1000 separator

1—pulsating mechanism; 2—magnetic yoke; 3—energizing coil; 4—ring driver; 5—tailing box;
6—slurry level tube; 7—feeding box; 8—working ring; 9—concentrate box; 10—concentrate flusher;
11—washing water box; 12—draining box; 13—support frame

As the flushing direction of magnetic concentrate is opposite to that of feeding relative to each matrix pile, coarse particles can be washed out without the need of passing through matrix pile. Slurry pulsation can keep the particles in matrix pile in a loose state all the time. Evidently, discharging concentrate in an opposite way and slurry pulsating up and down help to prevent matrix from blockage, and magnetic separation can enhance the quality of magnetic concentrate. Moreover, these measures not only ensure the effective recovery of size down to about 10 microns, but also extend the feeding size limit up to 800 microns, thus enlarging particle size range to be treated and simplifying classification unit.

The major parameters and electromagnetic features are shown in Table 1 and Table 2 respectively. The highest background magnetic induction in the working zone is 1.2T, though 1.0T is usually enough for most oxidized iron ores.

Table 1 Parameters of SLon-1000 separator

Ring diameters(outer, inner)/mm	1000;800
Ring width/mm	300
Ring rotating speed/r · min^{-1}	0.5-5
Feed size/mm	0.8-0
Throughput/t · h^{-1}	4-7
Driving power/kW	1.1;1.5
Pulsating stroke/mm	0-20
Pulsating frequency/min^{-1}	0-400
Separator weight/t	5.34

Table 2 Electromagnetic features of SLon-1000 separator

D.C. current/A	300	390	510	720	930	1100	1200	1480
D.C. voltage/V	4.5	6.0	8.0	11.5	15.0	18.0	20.0	24.5
Energizing power/kW	1.35	2.34	4.08	8.28	14.0	19.8	24.0	36.3
Magnetic field/T	0.30	0.38	0.50	0.70	0.90	1.00	1.08	1.20

3 Commercial test

Gushan Iron Mine is one of the major raw material bases of Ma Anshan Iron and Steel Company. The ore is of finely disseminated and colloidal hematite quartzite. The size range of most hematite rain is 30-5 microns. This kind of ore is hard to grind and difficult to beneficiate. The existing flowsheet is that the ore is ground to 50%<200mesh, then classified by hydrocyclones of D500 and D350mm. Their underflows are treated by spiral concentrator and the overflow of D350mm hydrocyclone is treated by centrifugal separator. The particle size of the overflow of D350mm hydrocyclone is about 90%<200mesh. It contains a large amount of slime and light particles. Hence it is very difficult for various separators to get good results. For example, the results of flotation in 1982 were: feed 34.14%Fe, concentrate 49.23%Fe, tailings 24.26%Fe, recovery 49.09%; and the results of centrifuge in 1987 were: feed 34.12%Fe, concentrate 52.36%Fe, tailing 26.79% Fe, recovery 43.99%. Because the concentrate grade could not meet metallurgical requirement (Fe= 55% or more), the overflow of D350mm hydrocyclone was mostly discarded as tailing and the mine suffered a heavy loss.

SLon-1000 separator was installed in line with the existing flowsheet to treat the overflow of D350mm hydrocyclone in parallel with centrifuger. The following are some of the results it achieved during the 3000hr site test.

4 Effect of background magnetic induction

The results of background magnetic induction test are listed in Table 3. When magnetic induction is increased from 0.53T to 0.92T, iron concentrate grade slightly decreased from 56.66% to 56.16%, while recovery increases from 56.57% to 69.44%. When magnetic induction is higher than 0.92T, the results keep almost the same. The suitable magnetic induction is 0.76-0.92T.

Table 3 Results of magnetic induction test (%)

Induction/T	0.53	0.76	0.92	1.10
Feed grade	28.70	28.95	30.20	28.45
Concentrate Grade	56.66	56.28	56.16	55.79
Tailing grade	17.47	15.23	14.73	13.98
Recovery	56.57	64.98	69.44	67.87

5 Effect of pulsating frequency

The results of pulsating frequency test are listed in Table 4. When no pulsation ($N=0$), the concentrate grade is only 48.80% due to serious mechanical trap of non-magnetic particles. As frequency is increased, mechanical trap is greatly reduced and concentrate grade is increased rapidly. When $N=350/\min$, concentrate grade reaches 58.41% gained by about 10% compared with that in the case of no pulsation. Moreover, because pulsation improves operating condition in the working zone, in a certain range, tailing grade slightly decreases and recovery remains almost unchanged. When $N=400/\min$, concentrate grade continues rising, but tailing grade also rises and recovery drops due to rapid increase of competing force. The proper frequency is 300-400/min at 13mm slurry stroke.

Table 4 Results of pulsating frequency test (%)

$N/\min$	0	200	300	350	400
Feed grade	34.45	33.22	34.45	33.95	33.70
Concentrate grade	48.80	53.42	56.78	58.41	59.53
Tailing grade	19.97	16.72	16.47	15.97	16.97
Recovery	71.15	72.52	73.52	72.89	69.44

Note: Magnetic induction is fixed at 0.9T and matrix volume package at 6%.

6 Effect of feed size

The statistic results according to feed size are listed in Table 5. Although feed size fluctuates in the range of 60%-98%<200 mesh, concentrate grade keeps at about 57% and tailing grade remains at about 20%. These results demonstrated that SLon-1000 separator possesses strong adaptability to wide range of particle size.

Table 5 Statistic results related to feed size				(%)
<200mesh in feed/%	60-70	70-80	80-90	90-98
Feed grade	34.57	37.92	36.49	35.18
Concentrate grade	58.88	57.60	57.54	57.09
Tailing grade	19.87	21.90	20.01	19.16
Recovery	64.18	68.16	69.24	68.54

7 Effect of feeding slurry density

The statistic results related to feeding slurry density are shown in Table 6. When density (solid weight percentage in feeding slurry) is 25%-30%, the concentrate grade and recovery are fairly good, 58.54% and 67.74% respectively. However, the effect of feeding slurry density seems small. In practice, higher feeding slurry density is always preferred so that the capacity of the separator can be fully utilized.

Table 6 Statistic results related to feed density				(%)
Feed density/%	10-20	20-25	25-30	30-42
Feed grade	35.07	35.75	37.22	36.75
Concentrate grade	56.76	57.38	58.54	57.04
Tailing grade	19.21	19.43	21.09	22.41
Recovery	68.36	69.02	67.74	64.27

8 Effect of feed grade

The statistic results according to feed grade are shown in Table 7. Although feed grade changes from 23% to 40%, concentrate grade is kept stable at 56.07%-58.19%. Recovery increases as feed grade increases. It is very important for the mine to maintain concentrate quality. The results show that SLon-1000 separator possesses the ability to adapt large fluctuations of feed grade.

Table 7 Statistic results related to feed grade					(%)
Feed grade	23-27	27-30	30-35	35-38	38-40
Concentrate Grade	56.07	56.45	56.81	57.40	58.19
Tailing grade	15.16	16.53	18.32	20.72	21.22
Recovery	52.35	60.78	68.46	67.90	71.56

9 One roughing and one cleaning

In order to further improve the quality of iron concentrate, the overflow of D350mm hydrocyclone was tested by one roughing and one cleaning with SLon-1000 separator. The results are shown in Table 8. The grade and recovery of final concentrate are 61.90% and 62.49% respectively. The test shows a practical way for the mine to improve the quality of iron concentrate.

Table 8 Results of one roughing and one cleaning   (%)

Product	Feed	Rough Concentrate	Clean Concentrate	Middling	Tailing
Grade	38.28	57.61	61.90	42.65	19.16
Weight	100.00	49.73	38.65	11.08	50.27
Recovery	100.00	74.84	62.49	12.35	25.16

10 Conclusion

Commercial test of 3000 hours demonstrated that SLon-1000 vertical ring and pulsating high gradient magnetic separator possesses the advantages of high beneficiation ratio, not easy to be blocked, and strong adaptability to large fluctuations of feeding size, feed grade and slurry density. This type of separator can be widely applied to process weakly magnetic minerals.

References

[1] Oberteuffer J A. High gradient magnetic separation[J]. IEEE Trans. On Mag., 1973(3):303-306.

[2] Kelland D R. HGMS applied to mineral beneficiation[J]. IEEE Trans. On Mag., 1973(3):307-309.

[3] Oberteuffer J A. HGMS of steel mill process and waste water[J]. IEEE Trans. on Mag., 1975(5):1591-1593.

[4] Ianicelli J. Development of high extraction magnetic filtration by the koaline industry of Georgian[J]. IEEE Trans. on Mag., 1976(5):489.

[5] Liu Shuyi, et al. Pulsating HGMS of the mixed materials of fine grain tungsten and cassiterite from shanhu Tin Mine[J]. Mining and Metallurgical Engineering, P. R. China, 1983(2):30-34.

写作背景　这是作者发表的第一篇英文论文，作者与中南工业大学傅崇悦教授参加了1989年在联邦德国科隆召开的世界冶金工业生产力与技术会议。此文向国外同行介绍了第一台SLon-1000立环脉动高梯度磁选机的研制及其在马钢姑山铁矿的试验。

New Development of the SLon Vertical Ring and Pulsation HGMS Separator[0]

Xiong Dahe

(Ganzhou Nonferrous Metallurgy Research Institute
341000 Ganzhou, Jiangxi Province, P. R. China)
(Received August 27, 1993)

Abstract SLon-1000 and SLon 1500 vertical ring and pulsation high-gradient magnetic separators have been developed in recent years. Because of their ability to treat fine weakly magnetic minerals they are being rapidly applied in several mineral processing plants to treat oxidized iron ores. In this paper, their structure, working principles and new applications will be described.

Keywords High-gradient magnetic separation; Slurry pulsation; Combined force field; Hematite beneficiation; Desulphurisation

1 Introduction

The first SLon-1000 vertical ring and pulsation high-gradient magnetic separator (SLon-1000 VPHGMS) was successfully developed in 1988 at Ganzhou Non-Ferrous Metallurgy Research Institute. It attracted a considerable interest of numerous domestic and foreign specialists owing to its unique design and its excellent mineral beneficiation capability. In recent years an effort was made to scale up the separator to production size. The designers of the separator not only successfully developed the SLon-1500 VPHGMS, improved considerably its design, increased its reliability and beneficiation efficiency, but also successfully applied the machine on the production or pilot-plant scales at the Gushan Iron Mine of Ma Anshan Iron and Steel Company, Meishan Iron Ore Mine of Shanghai Iron and Steel Company and Gong Changling Mineral Processing Plant of Anshan Iron and Steel Company and promoted the progress of technology at these mines and plants.

2 Design and working principles of the SLon VPHGMS

The SLon VPHGMS separator utilises the combined force field of magnetic force, pulsating fluid and gravity to beneficiate fine weakly magnetic particles. Its unique feature is the vertically rotating ring whereby the magnetic fraction is flushed in the direction opposite to that of the feed, and the mechanism of slurry pulsation.

When treating fine weakly magnetic minerals, the separator has a higher beneficiation ratio, the matrix cannot be easily blocked and the separator is flexible and adaptable compared to other high-gradient magnetic separators.

---

❶ 原载于《Magnetic and Electrial Separation》,1994,8。

The SLon VPHGMS consists of thirteen main parts, as shown in Fig. 1. The pulsating mechanism, energising coil, magnetic yoke and working ring are the key parts. The frequency of pulsation and the stroke are adjustable. The energising coil is made from hollow copper tube and cooled internally with water. Along the periphery of the ring there is a number of rectangular chambers in which the matrix of the expanded metal sheets made of magnetic stainless steel are placed. When the separator is in operation, the ring rotates clockwise.

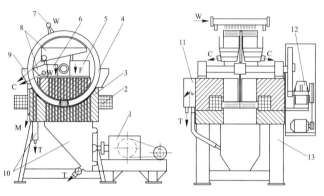

Fig. 1 SLon VPHGMS

1—pulsating mechanism; 2—energising coil; 3—magnetic yoke; 4—working ring; 5—feeding box; 6—wash water box; 7—concentrate flush; 8—concentrate box; 9—middlings chute; 10—tailings box; 11—slurry level box; 12—ring drive; 13—support frame; F—feed; W—water; C—concentrate; M—middlings; T—tailings

The slurry fed from the feed box enters the ring through slots in the upper yoke. The matrix in the working zone is magnetised. The magnetic particles are attracted from the slurry onto the surface of the matrix, then brought to the top of the ring where the magnetic field strength is negligible, and are then flushed out into the concentrate box.

The non-magnetic particles pass through the matrix and enter the tailings box through slots in the lower yoke under the combined action of the slurry pulsation, gravity and the hydrodynamic drag. The pulsating mechanism drives the rubber diaphragm on the tailings box so that it moves back and forth. As long as the slurry level is adjusted above the fixed level in the slurry level box, kinetic energy due to the pulsation can be effectively transmitted to the working zone.

As the direction of flush of the magnetic fraction is opposite to that of the feed, relative to each matrix pile, coarse particles can be flushed without having to pass through the entire depth of the matrix. The slurry pulsation can keep particles within the matrix in a loose suspended state all the time. It is clear that the reverse flush and the slurry pulsation allow to prevent the matrix clogging while the pulsation improves the quality of the concentrate.

Moreover, these measures not only ensure the effective recovery of weakly magnetic particles as small as $10\mu m$, but also extend the size range of the feed material up to about 1 mm, thus increasing the upper limit of the particle size to be treated and simplifying a classification operation.

The major parameters of SLon VPHGMS are summarised in Table 1 while the overall view of the SLon-1500 separator is shown in Fig. 2. The average magnetic field strength in the working zone is up to 1.0 Tesla which is usually sufficient for most oxidised iron ores.

Table 1 Parameters of the SLon-1000 and SLon-1500 VPHGMS

Parameter	SLon-1000	SLon-1500
Ring diameter/mm	1000	1500
Background magnetic induction/T	0-1.0	0-1.0
Energising power/kW	25.5	38
Driving power/kW	1.1+1.1	3+4
Pulsating stroke/mm	0-30	0-30
Pulsating frequency/r · min^{-1}	0-400	0-400
Feed size/mm	−1.0	−1.0
Throughput/t · h^{-1}	4-7	20-35
Mass of machine/t	6	20

Fig. 2 Vertical-ring and pulsation high-gradient magnetic separator SLon-1500

3 New applications of SLon VPHGMS

Beneficiation of Hematite Ore at Gushan Iron Ore Mine

Gushan Iron Ore Mine is one of the major bases of raw materials of Ma Anshan Iron and Steel Company. The ore is finely disseminated with colloidal hematite and quartzite. The size range of most hematite particles is 5-30μm. The ore is difficult to grind and beneficiate. The beneficiation plant was established in 1978; the gravity-based flowsheet used jigs, spirals and centrifugal separators as major beneficiation equipment.

After the ore was crushed down to <12mm, jigs removed part of coarse iron concentrate and part of coarse tailings. The jig middlings amounting to approximately 50 percent of the feed ore were ground in a ball mill, classified into two classes and then treated by a spiral concentrator and a centrifugal separator, respectively. The beneficiation results of the milled fraction were quite poor for many years. For instance, the production results of the milled fraction were, in 1988, as follows:

Feed grade	37.16%Fe
Concentrate grade	55.22%Fe
Tailings grade	24.47%Fe
Iron recovery	61.32%

In order to improve the production results, the flowsheet of the milled fraction was modified in,

1989-1992, into a stage grinding-high-intensity and high-gradient magnetic separation. Two SLon-1500 VPHGMS separators are being used as scavengers and the SQC high-intensity magnetic separators are installed as roughers or cleaners. The flowsheet of the plant is shown in Fig. 3.

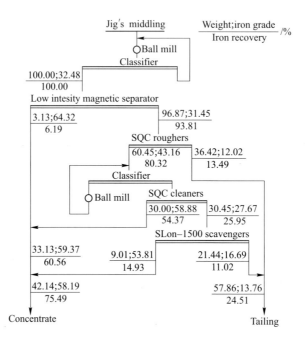

Fig. 3 The application of the SLon-1500 VPHGMS separator in the magnetic separation flowsheet of the Gushan Iron Ore Mine

The overall results of this flowsheet are as follows:

Feed grade	32.48%Fe
Concentrate grade	58.19%Fe
Tailings grade	13.76%Fe
Iron recovery	75.49%

Compared with the previous gravity flowsheet, the grade of the iron concentrate is 2.87% higher and the recovery is 14.17% higher. The increase in the recovery is mainly due to the SLon-1500 VPHGMS separators which recover part of the iron from the tailings of the SQC cleaner, which otherwise would be difficult to recover.

3.1 The application in desulphurisation and dephosphorisation

There is a large iron ore deposit at Meishan Iron Ore Mine. The ore contains a mixture of magnetite, siderite, hematite and pyrite. The iron concentrate is supplied mainly to the Shanghai Metallurgical Company as a ball raw material. However, because of a high content of sulphur and phosphorus, the iron concentrate alone cannot be used as a ball raw material but rather as an auxiliary material. Production of the iron concentrate is thus limited.

In order to reduce the concentration of sulphur and phosphorus, the Meishan Iron Ore Mine cooperated with several research institutes and performed site trials in 1990 with a magnetic separation flowsheet for desulphurisation and dephosporisation of the iron concentrate. Principal flowsheet is shown in Fig. 4. The SLon-1000 VPHGMS was applied as scavenger which effectively controlled the grade of the tailings and made important contribution to guarantee the total recovery of iron.

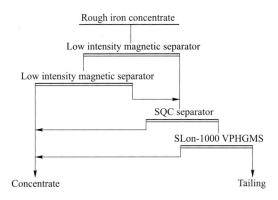

Fig. 4 The application of SLon-1000 to desulphurisation and dephosporisation of the Meishan iron concentrate

The test results summarised in Table 2 show that 49.99% of sulphur and 45.55% of phosphorus reported into the tailings, the loss of iron being 4.37% only. The grade of the iron concentrate was increased from 53.69% to 57.50% Fe while the concentrations of S and P were greatly reduced. The quality of the iron concentrate was thus considerably improved. Presently, this flowsheet is being implemented in the second extension stage.

Table 2 The test results of desulphurisation and dephospohorisation of the Meishan iron concentrate

Product	Mass /%	Grade/%			Recovery/%		
		TFe	S	P	TFe	S	P
Concentrate	89.3	57.5	0.14	0.23	95.6	50.0	54.5
Tailing	10.7	21.9	1.20	1.62	4.4	50.0	45.5
Rougher Concentrate	100.0	53.7	0.26	0.38	100.0	100.0	100.0

3.2 Processing of the Gongchangling low-grade hematite ore

Gongchangling Mineral Processing Plant is one of the major raw material bases of Anshan Iron and Steel company. Its hematite-processing plant treats 2.8 million tonnes of ore annually, with stage grinding and beneficiation flowsheet. The fine fraction (80%<200 mesh) representing approximately 50% of the feed ore is treated by $\phi1600\times900$ centrifugal separators with one roughing stage to eliminate the tailings, and with two cleaning stages to produce the iron concentrate.

The main problem in the production is that the concentration of iron in the tailings is too high which seriously affects the recovery of iron into the concentrate. In September 1989 the Ganzhou Non-Ferrous Metallurgy Research Institute initiated a joint project with the Anshan Mine Company and its Gong Changling Mineral Processing Plant. Pilot-plant tests using SLon-1000 magnetic separators were carried out, continuously treating, for a period of one month, the fine fraction of the ore. The SLon-1000 VPHGMS was installed at the position of the roughing centrifugal separator and treated the same feed. During the trials the separator worked for 713 hours and its availability was as high as 99%. Comparison of the metallurgical results achieved using the SLon separator and the centrifugal separator are shown in Table 3.

Table 3 Results of SLon-1000 and of the centrifugal separator

Separator	Grade/%			Mass of Concentrate/%	Recovery of Concentrate/%
	Feed	Concentrate	Tails		
SLon-1000	31.7	50.8	9.1	54.3	86.9
Centrifuger	31.8	46.7	19.2	45.9	67.3

Compared to the centrifugal separator, the grade of the concentrate produced by the SLon separator was 4.12% higher. The grade of the tailings 10.16% lower and the recovery of iron 19.65% higher. It can thus be seen that the results obtained using the SLon separator were considerably better. The pilot-scale tests laid firm foundations for the production-scale application of the SLon VGHGMS in the plant.

In order to demonstrate that the SLon VPHGMS can fully replace the roughing centrifugal separators on the production scale, the Anshan Mine Company and Gong Changling Mineral Processing plant, in cooperation with the Ganzhou Non-Ferrous Metallurgy Research Institute and Ganzhou Non-Ferrous Metallurgical Machinery Plant installed, in the period from March 1991 to July 1992, five SLon-1500 VPHGMS separators in the 7-8 lines of the hematite processing plant. These separators replaced 24 units of the $\phi 1600mm \times 900mm$ roughing separators and the production-scale tests were carried out for 16 months. Results of the total test flowsheet, for the last three months (from May to July 1992) and of the original flowsheet (lines 5-6) are shown in Table 4.

It can be seen that after the installation of the SLon-1500 VPHGMS in lines 7-8 the grade of the tailings drop sharply, the recovery of iron increased by 15.99% and the grade of the concentrate increased by 1.53% Fe. These results established an efficient way of increasing the recovery of the Anshan-type hematite ore. Presently, five SLon-1500 VGHGMS separators are in operation and they are shown in Fig. 5.

Table 4 The overall results as obtained using either the original flowsheet or the modified circuit which employed the SLon-1500 VPHGMS

System	Grade of feed ore	Concentrate(Fe)/%			Grade of tails
		Grade	Mass	Recovery	
7-8(test)	28.5	64.3	31.8	71.9	11.7
5-6(unmodified)	28.5	62.8	25.4	55.9	16.8
Difference	0	+1.5	+6.4	+16.0	-5.1

Fig. 5 SLon-1500 magnetic separators at the Gong Changling Mineral Processing Plant of the Anshan Iron and Steel Company

4 Conclusions

The application of the SLon vertical ring and pulsation high-gradient magnetic separators at the Gushan Iron Ore Mine, Meishan Iron Ore Mine and Gong Changling Mineral Processing plant demonstrated that the separator can efficiently beneficiate fine weakly magnetic minerals.

The SLon-1500 VPHGMS was applied at the Gushan Iron Ore Mine and Gong Changling Mineral Processing plant to treat fine hematite ore and considerable increase in the recovery of iron was achieved. SLon-1000 VPHGMS, as applied to desulphurisation and dephospohorisation of the rough iron concentrate at the Meishan Iron Ore Mine substantially reduced the concentration of sulphur and phosphorus in the iron concentrate. Successful application of magnetic separation technology at these plants and mines promoted advances of their technology and production.

In general, SLon VPHGMS separator which combines theories of gravity concentration and of magnetic separation became the first industrial continuous mineral processing separator which utilises the combined force of a pulsating fluid, gravity and high-gradient magnetic force. It not only possesses advantages of conventional HGMS separators in recovering fine weakly magnetic minerals, but it also considerably increases the grade of the magnetic concentrate, eliminates the matrix clogging, extends the range of particle sizes of the feed material and improves mechanical stability.

It represent a new type of an efficient magnetic separator for fine particle treatment and may contribute to a technological reform of old plants and to a construction of new mines. It is expected that the separator will be further developed and applied in a near future.

写作背景 这是作者第一次在国际性杂志《磁电选矿》上发表的英文论文。此文介绍了 SLon-1000 和 SLon-1500 立环脉动高梯度磁选机的工作原理,在马钢姑山铁矿、南京梅山铁矿和鞍钢弓长岭选矿厂的工业应用。

Development and Commercial Test of SLon-2000 Vertical Ring and Pulsating High-gradient Magnetic Separator

Xiong Dahe

(*Ganzhou Nonferrous Metallurgy Research Institute*
44 *Qingnian Road*, *Ganzhou* 341000, *Jiangxi Province*, *P. R. China*)
(*received March* 22, 1996, *accepted May* 15, 1996)

Abstract SLon-2000 vertical ring and pulsating high-gradient magnetic separator is an efficient industrial equipment for processing weakly magnetic minerals. It has been recently developed at the Ganzhou Nonferrrous Metallurgy Research Institute. A six-month commercial testwork to process low-grade hematite ore was completed in Chong Changeling Mineral Processing Plant in 1995. Compared with WHIMS-2000 wet high-intensity magnetic separator, the grade of the concentrate is by 7.21% higher the grade of the tailings lower by 5.41% and the iron recovery higher by 7.36%. Particularly, the matrix is always clean so that the matrix clogging usual in the WHIMS-2000 machine has been overcome.

Keywords High-gradient magnetic separator; Slurry pulsation; Combined force field; Hematite beneficiation; Desulphurisation

1 Introduction

As the mining scale of oxidised iron ore increases and reserves of rich iron ores decreases, the minerals processing industry is facing problems with treating low-grade and finely disseminated oxidised iron ores. High-intensity magnetic separation has a very important position in the processing of oxidised iron ore. Unfortunately, most high-intensity magnetic separators face a problem of matrix clogging.

Such a problem is, for instance, serious in processing in beneficiation of the hematite ore in Gong Changeling Mineral Processing Plant of Anshan Iron and Steel Company. The grade of the mined ore was about 30% Fe in 1980s and has dropped to about 28% Fe in 1990s. with annual rate of decrease of about 0.2% Fe.

There are five wet high-intensity magnetic separator installed at the plant. Their ring diameter is 2000mm (WHIMS-2000), with grooved plates as matrix. The separators are used as roughers for −2mm hematite. Because the feed contains many coarse particles and some strongly magnetic particles such as magnetite and maghemite, the matrix can be very easily clogged. This happens even

---

❶ 原载于《Magnetic and Electrical Separation》,1997,2。

when the gaps between the grooved plates is set to 3.5mm which results in unstable beneficiation results and in demanding maintenance.

In order to solve these problems, a contract was signed between Gong Changeling Mining Company and Ganzhou Nonferrous Metallurgy Research Institute (GNMRI). It was agreed that GNMRI designs and builds SLon-2000 Vertical Ring and Pulsating High-Gradient Magnetic Separator. It was also agreed that GNMRI delivers the separator to Gong Changeling Mineral Processing Plant in order to carry out commercial tests for a period of six months. The aim of the tests was to compare the results with those obtained with WHIMS-2000. The target of the testwork was that SLon-2000 must eliminate the matrix clogging and that the grade of the tailings must be lower by 2% than the tailings produced by WHTMS-2000.

2 Structure and principles of operation of SLon-2000

SLon-2000 magnetic separator consists of the pulsating mechanism, the energising coil, magnetic yoke separating ring, the feed and product boxes, as is shown in Fig. 1. Its operating principles are as follows: while the direct electric current flows through the energising coil, a magnetic field is built in the separating zone. Stainless steel 5mm rods were used as a matrix in these commercial tests.

Fig. 1 SLon-2000 vertical-ring and pulsating high-gradient magnetic separator

The ring with the magnetic matrix rotates around its horizontal axis. A slurry fed from the feed box enters the matrix located in the separating zone. Magnetic particles are attracted from the slurry onto the surface of the matrix and are then brought to the top of the ring where the magnetic field is negligible. Magnetic particles are then flushed into the concentrate box. The non-magnetic particles pass through the matrix and enter the tailings box under the combined force of the slurry pulsation, gravity and hydrodynamic drag.

Since the ring rotates in the vertical plane, direction of the flush is opposite to that of the feed. Relative to each segment of the matrix. The pulsating mechanism drives the slurry in the separating zone up and down keeping particles in the matrix section in a loose state at the time. Magnetic particles can thus be more easily captured by the matrix and the non-magnetic particles can be more easily dragged to the tailings box through the segment of the matrix.

Therefore, the opposite flushing and pulsation help to prevent the matrix clogging, and the pulsation help to purify the magnetic product. These measures guarantee that the SLon-2000 separator possesses an advantage of higher ratio of beneficiation, higher efficiency and considerable flexibility. The main specifications of the SLon-2000 separator and the plant operational data in this commercial testwork are summarised in Table 1. The electrical and magnetic specifications are listed in Table 2.

Table 1 Specifications of SLon-2000 separator

Parameters	Designed data	Applied data
Ring diameter×width/mm×mm	2000×900	
Background field/T	0~1.0	0.85
Energizing current/A	0~1080	800
Energizing voltage/V	0~76	52
Energizing power/kW	0~82	41.6
Driving motor/kW	5.5+7.5	
Pulsating stroke/mm	0~30	14
Pulsating frequency/min^{-1}	0~300	300
Pressure of water/MPa	0.3~0.5	0.3
Water consumption/m$^3 \cdot$h^{-1}	100~200	180
Feed size/mm	0~2.0	0~2.0
Feed solid density/%	10~45	35~45
Slurry throughput/m$^3 \cdot$h^{-1}	100~200	100~170
Ore throughput/t$\cdot$h^{-1}	50~80	60~70
Mass of machine/t	50	
Dimensions($L \times B \times H$)/mm×mm×mm	4200×3500×4300	

Table 2 Electrical and magnetic data of SLon-2000 separator

Energizing current/A	Energizing voltage/V	Energizing power/kW	Background field/T
100	8.0	0.80	0.109
200	15.2	3.04	0.230
300	20.0	6.00	0.342
400	26.0	10.4	0.453
500	33.8	16.9	0.580
600	39.6	23.8	0.694
700	46.0	32.2	0.781
800	52.0	41.6	0.851
900	58.0	52.2	0.918
1000	64.0	64.0	0.975
1080	71.0	76.7	1.026

3 Commercial tests

3.1 The ore and the flowsheet

The Ore is the Anshan-type iron ore containing mainly magnetite, maghemite and hematite intergrown with quartz and other gangue minerals. It contains very little sulphur, phosphorus and other harmful gangue minerals. Crystal size of the iron minerals is mainly from 0.037mm to 0.125mm. and the average size of quartz is about 0.13mm. The grade of the mined ore is about 28.4%Fe.

As the mined ore contains low percentage of iron it is very important to discard as much gangue as possible, and as early in the process as possible. Since it contains very low concentrations of harmful elements, it is relatively easy to use magnetic and gravity methods to concentrate the iron minerals. Schematic diagram of the hematite beneficiation is shown in Fig. 2 in which flowsheet the SLon-2000 separator was installed in parallel with WHIMS-2000 separator to carry out the commercial tests.

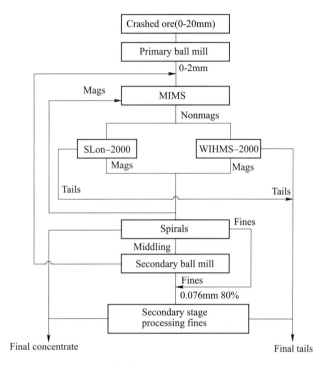

Fig. 2 A simplified flowsheet use in the commercial tests

The ore is ground by two primary D2700×3600 ball mills to about −2mm (about 40%<200 mesh). four D1050×L2100 medium intensity drum permanent magnetic separators (MIMS, the average magnetic field on the drum surface is 0.18 to 0.20 Tesla) are used to remove magnetite and a portion of maghemite. The non-magnetic product from MIMS containing mainly hematite and a portion of maghemite is fed in parallel into the SLon-2000 and WHIMS-2000 separators. These separators discard immediately most of the final tailings.

The magnetic fractions from MIMS, SLon-2000 and WHIMS-2000 are further cleaned to the grade of 63.5% Fe by spirals and other separators. The middlings from the spirals are further ground by a secondary D2700×3600 ball mill and returned to join the MIMS feed.

An advantage of this flowsheet is that the SLon-2000 separator and the WHIMS-2000 separator can discard approximately half (by mass to the primary ball mill feed) of the final tailings in the coarse stage, so that all of the following equipment and the processing costs can be reduced. The feed characteristics of SLon-2000 and WHIMS-2000 are shown in Table 3.

Table 3 The feed into the SLon-2000 and WHIMS-2000 separators

Grade (Fe)/%	FeO /%	Solid density /%	Size (0.076mm)/%
25.34	2.25	39.2	50.5

3.2 Determination of the SLon-2000 operating parameters

In order to optimise the operating parameters of SLon-2000, conditional tests of the background magnetic field, pulsation and the ore throughput were done, as is shown in Tables 4-6.

Table 4 Results of SLon-2000 background magnetic field tests

Background field/T	Grade(Fe)/%			Mass of mags/%	Recovery of iron/%
	Feed	Mags	Tails		
0.78	22.48	36.53	8.32	50.19	81.57
0.85	22.48	35.88	7.96	52.01	83.01
0.92	22.48	34.72	7.70	54.70	84.48
1.00	22.48	34.60	7.41	55.42	85.31

As the background magnetic field increases, the recovery of iron increases while the grades of the mags and the tailings decrease. Reasonable background magnetic field is 0.85 T for this type of an ore, and the corresponding energising current is 800A, voltage 52V and power 41.6kW.

Table 5 Results of the SLon-2000 pulsation tests

Pulsation		Grade(Fe)/%			Mass of mags/%	Recovery of iron/%
Frequency /mm	Stroke /mm	Feed	Mags	Tails		
0	0	23.32	34.71	8.27	56.92	84.72
300	14	23.32	39.55	7.91	48.70	82.60
Difference		0	+4.84	-0.36	-8.22	-2.12

Table 5 shows that the slurry pulsation increases the grade of the mags by 4.84% Fe, and reduces both the grade of the tailings by 0.36% and the mass of the mags by 8.22%, compared to a situation when no pulsation was used in the same SLon-2000 separator. The fact the recovery with pulsation is slightly lower is a results of the fact that the grade of the mags is higher.

The throughput tests demonstrated that the SLon-2000 separator can treat up to about 100 tons per hour of such a hematite ore, as can be seen in Table 6. To get better tailings and consider its tonnage, it is more advantageous to set the throughput to 60 to 70 tons per hour.

Table 6 Results of the SLon-2000 throughput tests

Feed			Grade(Fe)/%			Recovery of iron/%
Volume /m^3·h^{-1}	Throughput /t·h^{-1}	Solid density/%	Feed	Mags	Tails	
105.0	55.2	38.3	25.76	40.40	9.50	82.53
117.2	61.2	38.7	27.33	41.85	10.39	82.45
130.6	76.9	41.5	28.18	44.98	12.70	76.54
166.5	102.4	43.7	29.76	45.62	13.33	78.00

3.3 Comparison tests with WHIMS-2000

During the commercial tests the SLon-2000 separator ran for six months in parallel with the WHIMS-2000 separator, from September 1, 1994 till February 28, 1995. Its installation on the plant is shown in Fig. 3. The efficiency of beneficiation, operating hours, water and electricity consumption were compared with WHIMS-2000. The employed operating parameters of SLon-2000 are shown in Table 1 while the test results are summarised in Table 7.

Fig. 3   SLon-2000 in operation in Gong Changeling Mineral Processing Plant

Table 7   The average comparative results of 6-month tests

Separator	Grade(Fe)/%			Mass of mags/%	Recovery of iron/%
	Feed	Mags	Tails		
SLon-2000	25.94	41.21	10.74	49.89	79.25
WHIMS-2000	25.94	34.00	16.15	54.85	71.89
Difference	0	7.21	-5.41	-4.96	7.36

Table 7 shows that the SLon-2000 separator can achieve a much higher grade and the recovery into the mags, much lower grade of the tailings and lower mass yield into the mags than WHIMS-2000. A higher grade of the mags is mainly due to the contribution of its pulsating mechanism. a lower grade of the tailings and a higher recovery of iron is mainly due to the fact that the matrix is always kept clean and that the magnetic force acting on the iron mineral particles is stronger. The lower mass yield into the mags is favourable for the subsequent processing procedure and the secondary ball mill.

The operating time and the availability of SLon-2000 are 4220 hours and 98.8%, respectively, in the 6-month commercial test, as is shown in Table 8. It is 869 hours and 20.3% higher than with WHIMS-2000 separator. The matrix clogging and the ring problem are the major problems affecting the operation of the WHIMS-2000 separator. It has been demonstrated that the magnetic matrix of the SLon-2000 separator is always kept clean which is very important for maintaining high availability and for reducing the maintenance work.

Table 8   Comparison of the operating time in six months

Separator	Calendar time /h	Ball mill work time /h	Operat time /h	Operat ratio /%	Matrix and ring problem /h	Other problem /h
SLon-2000	4344	4270	4220	98.8	0	50
WHIMS-2000	4344	4270	3351	78.5	793	126
Difference	0	0	+869	20.3	-793	-76

Table 9 shows that SLon-2000 and WHIMS-2000 consume the same amount of water ($3m^3/t$) for each ton of ore treated. Table 10 indicates that SLon-2000 consumes only 0.86kW · h per ton of feed, 0.26kW · h less than WHIMS-2000.

Table 9 Comparison of water consumption

Separator	Ore throughput /t·h^{-1}	Water consumption /m^3·h^{-1}	Unit water consumption /m^3·t^{-1}
SLon-2000	60	180	3
WHIMS-2000	40	120	3

Table 10 Comparison of electricity consumption

Separator		Installed power /kW	Measured power/kW	Unit power consumption /kW·h·t^{-1}
SLon-2000	Energizing	82	41.6	
	Driving	5.5+7.5	4.1+5.6	0.86
	Total	95	51.3	
WHIMS-2000	Energizing	52	26	
	Driving	25	18.8	1.12
	Total	77	44.8	
Difference		+18	+6.5	−0.26

4 Conclusions

The six-month commercial testwork in processing low-grade hematite ore in Gong Changeling Mineral Processing Plant demonstrated that the SLon-2000 vertical ring and pulsating magnetic separator is a reliable and efficient equipment for processing weakly magnetic minerals. The separator increased the grade of the magnetic concentrate by 7.21%Fe and the recovery of iron by 7.36%. Its throughput is 50% higher compared to the WHIMS-2000 wet high-intensity magnetic separator.

Because the ring of the separator rotates vertically, allowing the magnetic fraction to be flushed in a direction opposite the that of the feed, and because the pulsating mechanism keeps the slurry in the separation zone pulsating, the matrix is always kept clean. Onerous problem of matrix blockage that exists in wet high-intensity magnetic separators with a horizontal ring is thus eliminated. The maintenance work is thus considerably reduced and the availability of the separator increased.

The commercial tests fully satisfied the conditions of the contract signed between Gong Changeling Mining Company and Ganzhou Nonferrous Metallurgy Research Institute.

References

[1] Xiong Dahe, et al. The development of a new type of industrial vertical ring and pulsating HGMS separator [C]. Proc. Conf. Productivity and Technology in the Metallurgical Industries, Cologne, Germany, 1989: 947-952.

[2] Liu Shuyi, et al. A new type of commercial pulsating high-gradient magnetic separator and its separating principles[C]. Proc. XVIIth Int. Min. Proc. Congress Dresden, Germany, 1991: 152-159.

[3] Yang Peng, et al. The separation performance of the pulsating high-gradient magnetic separator [J]. Magn. Electr. Sep., 1993(4): 211-221.

[4] Xiong Dahe. New development of the SLon vertical ring and pulsating high-gradient HGMS separator [J]. Magn. Electr. Sep. 1994(5): 211-222.

写作背景 本文发表在国际性杂志《磁电选矿》上，介绍了首台 SLon-2000 立环脉动高梯度磁选机的技术性能，在鞍钢弓长岭选矿厂用于分选氧化铁矿的工业试验。该机与现场原有的 ϕ2000 平环强磁选机对比的工业试验流程和选矿指标。

Development and Applications of SLon Vertical Ring and Pulsating High Gradient Magnetic Separators

Xiong Dahe

(*Ganzhou Nonferrous Metallurgy Research Institute, 44, Qingnian Road, Ganzhou, Jiangxi Province, P. R. China*)

Abstract Pulsating High Gradient Magnetic Separation(PHGMS) is efficient to treat weakly magnetic minerals. It is essential to develop commercial PHGMS separators to apply them in industry. During the past decade, the SLon-1000, SLon-1500, SLon-2000 Vertical ring and Pulsating High Gradient Magnetic Separators have been developed. With the advantages of better beneficial results, matrix not easy to be blocked and strong adaptability, they are now applied to beneficiate hematite, limonite and siderite in several Chinese mines. This paper sums up the development and applications of SLon separators.

1 Introduction

The Central South university of Technology began to study vibrating and pulsating high gradient magnetic separation in 1981[1]. A variety of laboratory tests were done on cyclic high gradient magnetic separators with mechanical vibrating system or slurry pulsating system. Compared with non vibrating and non pulsating HGMS, the following facts were demonstrated.

Separating a mixture of wolframite and cassiterite, the wolframite concentrate grade was 4%-20% units WO_3 higher while recovery was about the same; crassiterite recovery was 8%-33% units higer while SnO_2 grade remined almost the same.

For low grade wolframite slime contaning 0.4%-0.5% WO_3, concentrate grade was 5%-7% units WO_3 higher while WO_3 recovery remained almost the same.

For low grade hematite ore, hematite concentrate grade was 8%-12% units Fe higher while Fe recovery remained about the same.

The above facts showed that pulsating high gradient magnetic separation (PHGMS) is an efficient method to treat weakly magnetic minerals. However, laboratory results could not be transferred to industry without commercial PHGMS equipment. In this respect, the Ganzhou Nonferrous Metallurgy Research Institute (GNMRI) developed the SLon Vertical ring and Pulsating High Gradient Magnetic Separators(SLon VPHGMS or SLon).

2 Structure and working principle

SLon mainly consists of the pulsating mechanism, energizing coil, magnetic yoke, separating ring,

---

❶ 原载于《第二十届国际选矿会论文集》,1997。

feed and product boxes as shown in Fig. 1. Expanded nets or round bars made of magnetic conductive stainless steel are used as matrix. While a direct electric current follows through the energizing coil, a magnetic field is built up in the separating zone. The ring with matrix rotates clockwise. Slurry from the feed box enters into matrix piles located in the separating zone. Magnetic particles are attracted from slurry onto the surface of the matrix, brought to the top of the ring where magnetic field is negligible, then flushed out into the concentrate box. Non-magnetic particles pass through the matrix and enter the tailing box under the combined force action of slurry pulsation, gravity, and hydrodynamic drag.

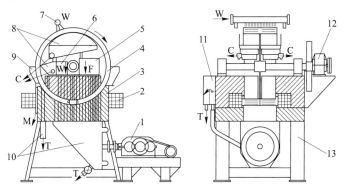

Fig. 1 SLon vertical ring and pulsating high gradient magnetic separator

1—Pulsating mechanism; 2—energizing coil; 3—magnetic yoke; 4—working ring; 5—feeding box; 6—wash water box; 7—concentrate flush; 8—concentrate box; 9—middling chute; 10—tailings box; 11—slurry level box; 12—ring driver; 13—support frame; F—feed; W—water; C—concentrate; M—middlings; T—tailings

As the ring rotates vertically, flushing direction of mags is opposite to that of feed relative to each matrix pile, coarse particles can be flushed out without having to pass through the entire depth of matrix pile. The pulsating mechanism drives slurry in the separating zone up and down, keeping the particles in the matrix pile in loose state, so that magnetic particles can be more easily captured by the matrix and nonmagnetic particles can be more easily dragged to the tailing box through the matrix pile. Obviously, opposite flushing and pulsation help to prevent matrix clogging, and pulsating helps to purity magnetic product.

The major parameters of SLon VPHGMS are listed in Table 1.

Table 1 Parameters of SLon VPHGMS

Parameters	SLon-2000	SLon-1500	SLon-1000
Ring diameter/mm	2000	1500	1000
Background induction/T	0-1.0	0-1.0	0-1.0
Energizing power/kW	0-82	0-38	0-26
Pulsating stoke/mm	0-20	0-20	0-20
Pulsating frequency/min^{-1}	0-300	0-300	0-300
Feed size/mm	0-2.0	0-1.3	0-1.0
Feed solid density/%	10-45	10-45	10-45
Slurry throughput/$m^3 \cdot h^{-1}$	100-200	50-100	10-20
Ore throughput/$t \cdot h^{-1}$	50-80	20-35	4-7
Mass of machine/t	50	20	6

3 The development and application of SLon-1000

The first SLon-1000 was designed and built in 1985 to 1988[2,3]. It was installed at the hematite processing flowsheet in Gushan Iron Mine to carry out 3000h commercial test. Gushan iron ore is a finely disseminated and colloidal hematite quartzite. According to the original flowsheet the ore was ground to 50%<200 mesh. Then classified by hydrocyclones of 500 and 350mm diameter. Their underflow was treated by spirals and the overflow of 350mm hydrocyclone was treated by centrifugal separators. The particle size of the overflow was about 90%<200 mesh. It contained a large amount of slimes and light product. So it was very difficult for various separators to get good results. The SLon-1000 was installed in parallel with the $D1600 \times B900$ centrifugal separators and fed with the same hydrocyclone overflow. After 3000h commercial testing, the results of SLon-1000, together with the results of the centrifugal separator, the historical results of flotation, wet high intensity magnetic separators and ordinary (horizontal ring) high gradient magnetic separator, are listed in Table 2. The results of SLon-1000 are the best for processing such a fine hematite. The higher concentrate grade is due to its efficient pulsating mechanism.

Table 2 Results of SLon-1000 compared with other separators

Separator	Year	Scale	Fe/%			Recov./%
			Feed	Concen.	Tail	
Floatation	1982	Production	32.14	49.23	24.08	49.09
Centrifug. sep.	1983	Production	31.37	49.70	24.26	44.28
Centrifug. sep	1987	Production	34.12	52.36	26.79	43.99
WHIMS	1983	Pilot test	30.36	53.90	19.52	55.98
Ordin. HGMS	1982-1983	Commer. test	32.73	52.94	20.56	60.80
SLon-1000	1987-1988	Commer. test	33.26	56.78	19.31	63.56

4 The development and application of SLon-1500

As the capacity of SLon-1000 is too small for iron ore processing, in 1989 Gushan Iron Mine supported GNMRI to design and build the first SLon-1500 and it was brought to the mine for a 3000h-test on production scale before GNMRI transferred it to the mine.

In 1989-1992, Gushan Iron Mine rebuilt its hematite processing flow sheet from gravity separation to magnetic separation. The present flowsheet is [4] as follows:

The ore is ground to about 50%<200 mesh in primary ball mill, four SQC-6-2770 WHIMS separators are used an rougher to eliminate parts of the tailings(nonmags), the mags are reground to about 80%<200 mesh, then two SQC-6-2770 WHIMS are used as cleaners and two SLon-1500 as scavengers. The total results of both, the new and the old flowsheets are shown in Table 3. Compared with the gravity flowsheet, the grade of iron concentrate is 2.95% units higher and the recovery is 13.80% units higher. The recovery increase is mainly due to the two SLon-1500 which recover the fine hematite particles from the tailing of SQC WHIMS cleaner, which otherwise would be difficult to recover.

Table 3 Gushan production results of gravity and magnetic flowsheets

Flowsheet	Year	Grade(Fe)/%			Recovery of concen. (Fe)/%
		Feed	Concern.	Tail	
Gravity	1988	37.16	55.22	24.47	61.32
Magnetic	1993	35.51	58.17	16.32	75.12
Difference		-1.65	2.95	-8.15	13.80

5 SLon-1500 applied in desulphurisation and dephosphorisation

There is a large iron ore deposit at Meishan Iron Mine. The ore contains a mixture of magnetite hematite, siderite, and pyrite. The iron concentrate is used as pellet feed. However, because of a high content of sulphur and phosphorus, the iron concentrate cannot be used alone as pellet feed but as an auxiliary material. Thus, production of the iron concentrate is limited by the market.

In order to reduce S and P, Meishan Iron Mine cooperated with several research institutes and did various tests. Finally a magnetic flowsheet was chosen and a commercial test was carried out in 1995. Low intensity magnetic separators were used to recover magnetite. A SLon-1500 as rougher and another SLon-1500 as scavenger were applied to recover hematite and siderite. The two SLon-1500 effectively recovered the fine hematite and siderite, controlled the tailing grade, and guaranteed the total results to meet the metallurgic requirement. The commercial test was run for 276h and 15393t ores were treated. The total results are shown in Table 4, which shows that the quality of concentrate was considerably improved and met the requirements of metallurgy. In 1996, Meishan Iron Mine ordered another eight SLon-1500 separators to built its desulphurisation and dephosphorisation flowsheet.

Table 4 Results of 276h commercial test on Meishan iron ore   (%)

Product	Mass	Grade			Recovery		
		Fe	S	P	Fe	S	P
Feed	100.00	52.77	0.44	0.399	100	100	100
Concen.	88.90	56.08	0.29	0.246	94.51	59.10	54.76
Metallurgy Require		>55.0	<0.35	<0.25	>94		
Tailing	11.10	26.13	1.61	1.626	5.49	40.90	45.24

6 SLon-1000, SLon-1500 applied to Gongchangling hematite ore

Gongchangling Mineral Processing Plant treats low grade hematite ore by a multi-stage grinding and sorting flowsheet. The fine fraction (80%<200 mesh) weighing about 50% of mined ore is treated by $D1600 \times B900$ centrifugal separators with one roughing stage to eliminate tailings and two cleaning stages to produce iron concentrate.

The main problem in the production is that iron content of the tailing is too high which seriously affects iron recovery. In 1989 a SLon 1000 was brought to the plant and installed in parallel with the roughing centrifugal separators to treat the same feed. Comparative results of 713h showed that the SLon-1000 can get much better results compared with the centrifugal separator[3,4]. The recovery of iron concentrate was raised by 19.65% units and the grade of concentrate was also raised by 4.12% units. This test laid the foundation for the application of SLon-1500 in the plant.

In 1991-1992, the plant, in cooperation with the GNMRI, installed five SLon-1500 separators in line 7 and 8 of the hematite processing plant. The five SLon-1500 replaced 24 units of $D1600mm \times B900mm$ roughing centrifugal separators and the productional scale tests were carried out for 16 months. Results of the total test flowsheet and of the original flowsheet (line 5-6) are shown in Table 5. The five SLon-1500 reduced the grade of total tails by 5.71% units Fe and increased the recovery of iron by 18.56% units.

Table 5 Results of SLon-1500 applied in Gong Changling

Flowsheet	Grade(Fe)/%			Mass of concen./%	Recovery of concen.(Fe)/%
	Feed	Concen.	Tails		
7-8(test)	28.47	64.31	11.73	31.84	71.92
5-6(unmodified)	28.47	63.65	17.44	23.87	53.36
Difference	0	+0.66	-5.71	+7.97	+18.56

7 SLon-1500 applied to beneficiate limonite

In 1993-1994, two SLon-1500 were installed in Zhongwei Iron Mine, a small limonite processing plant in Ningxia Province. The flowsheet is that the limonite ore is ground in one stage mill to about 60%<200 mesh, a SLon-1500 is used as rougher and other SLon-1500 is used as scavenger. The total results are shown in Table 6.

Table 6 Results of SLon-1500 applied to Zhongwei limonite　　　　(%)

Operation	Grade(Fe)			Mass of concen.	Recovery of concen.(Fe)
	Feed	concen.	Tails		
Roughing	46.36	54.17	38.60	49.84	58.24
scavenging	38.60	50.65	28.43	22.96	25.08
Total	46.36	53.06	28.43	72.80	83.32

Table 6 shows that the total concentrate grade is about 53% and recovery is about 83%. The results are good due to the efficient processing ability of SLon-1500. This flowsheet is very simple and the operating cost is very low. It is suitable for various mines to exploit limonite, hematite, siderite and other minerals, which are simply associated with gangues.

8 SLon-1500 applied in Luoci Iron Mine

Luoci Iron Mine is one of the iron ore suppliers of Yunnan Province. The major iron minerals are magnetite, hematite and limonite. The previous mineral processing flowsheet consisted of a single stage grinding. Then low intensity magnetic separators were used to recover magnetite, several wet high intensity magnetic separators (with 5mm balls as matrix) were used as a rougher and a cleaner. The existing problem is that the tailing grade is too high and iron recovery in the final concentrate too low. To improve the productional efficiency, three SLon-1500 were applied in 1994-1996 to replace part of the previous WHIMS, two SLon-1500 as rougher and the other as cleaner. After the application of the three SLon-1500, the total iron recovery was increased from 66.25% to 75.46%. The comparative results are listed in Table 7.

Table 7 Comparative results of SLon-1500 with WHIMS in Luoci   (%)

Separators	Grade(Fe)			Mass of concen.	Recovery of concen. (Fe)
	Feed	Concen.	Tails		
WHIMS(previous)	45.88	56.90	31.60	51.00	66.25
SLon-1500(now)	45.51	60.41	25.88	56.85	75.46
Difference	-0.37	+0.81	-5.72	+5.85	+9.21

9 The development and commercial test of SLon-2000 VPHGMS

Because of the increase of oxidized iron ore and the decrease of rich iron ore in Gong Changling Mineral Processing Plant, wet high intensity magnetic separation (WHIMS) takes a important position in processing the low grade hematite ore. But matrix clogging is the biggest problem with WHIMS in the plant. There are five Wet High Intensity Magnetic Separators in the plant. Their ring diameter is 2000mm (WHIMS-2000). They are applied for roughing 0-2mm hematite. Because the feed contains a lot of coarse particles and some strong magnetic particles such as magnetite and maghemite, their matrixes are very easy to be clogged even if the gap between the grooved plate is set at 3.5mm, causing the processing results unstable and the equipment maintenance work hard.

To solve this problem, a contract was signed between Gong Changling Mining Company and GNMRI 1993. The main clause is that GNMRI designs and builds a SLon-2000 VPHGMS and shift it to Gong Changling Mineral Processing Plant to carry out a six-month commercial test, to compare with one of the WHIMS-2000. The SLon-2000 must solve the problem of matrix clogging and its tailing grade must 2% units Fe lower than the WHIMS-2000.

In 1994, the first SLon-2000 was installed in parallel with the five WHIMS-2000 to treat 0~2mm (about 40%<200 mesh) low grade hematite in the primary grinding stage. During the commercial test the SLon-2000 ran six months in parallel with the WHIMS-2000 from September 1, 1994 to February 28, 1995. Beneficiating efficiency and operating hours were compared with the WHIMS-2000. The test results are shown in Table 8 and Table 9.

Table 8 Average results of 6-month beneficial comparison   (%)

Separators	Grade(Fe)			Mass of mags	Recovery of iron
	Feed	Mags	Tails		
SLon-2000	25.94	41.21	10.74	49.89	79.25
WHIMS-2000	25.94	34.00	16.15	54.85	71.89
Difference	0	7.21	-5.41	-4.96	7.36

Table 8 shows that the SLon-2000 can get much higher grade and recovery of mags, much lower grade of tails and lower mass of percentage of mags than the WHIMS-2000. Higher grade of mags is mainly due to the contribution of its pulsating mechanism. Lower grade of tails and higher iron recovery is mainly due to the facts that its magnetic matrix always keeps clean and its magnetic force is relatively stronger. The operating time and ratio of SLon-2000 are 4220 hours and 98.8% respectively in the 6-month commercial test, 869 hours and 20.3% units more than the WHIMS-2000. Matrix clogging and ring problems are the major causes affecting the working ratio of WHIMS-2000. It demonstrated that the magnetic matrix of SLon-2000 always keeps clean. This is very important to keep its operating ratio and to reduce its maintenance work.

Table 9 Comparison of operating hours in 6 months

Separator	Calendar time/h	Ball mill work time/h	Operating time/h	Operating ratio/%	Matrix problems /h	Other problems /h
SLon-2000	4344	4270	4220	98.8	0	50
WHIMS-2000	4344	4270	3351	78.5	793	126
Difference	0	0	+869	20.3	−793	−76

10 Conclusions

The application of SLon-1500, together with SQC WHIMS, in the Gushan Iron Mine increases the iron recovery from 61.32% to 75.12% and the concentrate grade from 55.22%Fe to 58.17%Fe.

The application of SLon-1500 in the Gong Changling mineral processing plant increased the iron recovery from 53.36% to 71.92%.

The commercial test of SLon-1500 in the Meishan Iron Mine demonstrate that it is capable to reduce S and P in the iron concentrate to meet metallurgy requirements. It may solve the biggest problem of this Mine in the coming years.

The application of SLon-1500 in Zhongwei Iron Mine to treat limonite demonstrated that about 83% iron recovery and 53%Fe concentrate can be recovered with one SLon-1500 as rougher and one SLon-15000 as scavenger.

The application of SLon-1500 in Luoci Iron Mine for processing hematite and limonite increases the iron recovery from 66.25% to 75.46%.

The commercial test processing low grade hematite ore in Gong Changling Mineral processing plant demonstrate that the SLon-2000 can raise the mags grade 7.21% units Fe and iron recovery 7.36% units compared with the WHIMS-2000.

All of the applications demonstrated that SLon VPHGMS is a reliable and efficient equipment. It is a new generation of continuously working HGMS with higher concentration ratio and without matrix clogging. It is becoming a major industrial equipment for processing hematite, siderite, limonite and other weakly magnetic minerals.

References

[1] Yang Peng, et al. The separation performance of the PHGMS [J]. Magnetic and Electrical Separation, 1993(4):211-221.

[2] Xiong Dahe, et al. The development for a new type of industrial VPHGMS[C] // Proc. Productivity and Technology in Metallurgical Industries, 1989:947-952.

[3] Liu Shuyi, et al. A new type of commercial PHGMS and its separating principles[C]// Proc. XVⅡth Int. Min. Proc. Congress,1991:152-159.
[4] Xiong Dahe. New development of the SLon VPHGMS[J]. Magnetic and Electrical Separation,1994(5):211-222.

写作背景 这是作者第一次在国际选矿会议上发表的论文,在德国亚琛召开的第二十届国际选矿会议上介绍了 SLon-1000、SLon-1500 和 SLon-2000 立环脉动高梯度磁选机的研制和工业应用。SLon 磁选机从此在国际上有了一定的知名度。

New Technology of Pulsating High Gradient Magnetic Separation[1]

Xiong Dahe[1] Liu Shuyi[2] Chen Jin[2]

(1. Ganzhou Nonferrous Metallurgy Research Institute, Ganzhou, Jiangxi Province, P. R. China;
2. Central South University of Technology, Changsha, Hunan Province, P. R. China)

Abstract Extensive laboratory and commercial test works have been done on Pulsating High Gradient Magnetic Separation (PHGMS) and it is demonstrated that it is an efficient method to treat weakly magnetic minerals. To investigate the PHGMS principle, a cyclic pulsating high gradient magnetic separator was built for laboratory test work. The relationships between mineral processing results and pulsating stroke and frequency are discussed. To apply the new technology of PHGMS in industry, the SLon-1000, SLon-1500, SLon-2000 vertical ring and pulsating high gradient magnetic separators have been developed, which possess the advantages of better beneficiation results, a matrix that is not easily clogged and a strong adaptability. Now they have been applied to beneficiate hematite, limonite and siderite in mineral processing industry. The working principle and commercial applications of SLon separators are also described in this paper. © 1998 Elsevier Science B. V. All rights reserved.

Keywords High gradient magnetic separator; Pulsating; Hematite; Siderite; Limonite

1 Introduction

High gradient magnetic separation (HGMS) has the advantage of high recovery of fine weakly magnetic particles (Oberteuffer, 1973, 1974; Svoboda, 1987). Cyclic HGMS separators have been successfully applied to kaolin purification since 1970s. However, when conventional HGMS separators are applied in industry to treat metallic ores such as hematite, matrix clogging and mechanical entraining of non-magnetic particles become the biggest problems, because such ores contain large portions of weakly magnetic minerals and relatively coarser particles. The mechanical entrainment of non-magnetic particles in the matrix lowers the grade of the magnetic product and matrix clogging makes the maintenance of the equipment a hard task.

To solve such problems, a pulsating high gradient magnetic separation (PHGMS) technology has been developing fast since 1981 (Xiong et al., 1989; Liu et al., 1991; Yang et al., 1993; Xiong, 1994, 1997). PHGMS can greatly reduce the entrainment effect and can obviously increase the grade of the magnetic product. It is also very helpful for reducing the clogging of the matrix.

2 Cyclic PHGMS separator and its working principle

The cyclic PHGMS separator used for laboratory test work is shown in Fig. 1. A pulsating mecha-

---

[1] 原载于《International Journal of Mineral Processing》, 1998(07): 111~127。

nism is arranged under the lower magnetic pole, which drives a rubber forth and back. Expanded metals or round bars made of magnetic stainless steel are used as matrix. The working procedure is as follows. While a direct electric current flows through the energizing coils, a magnetic field is built up in the separating zone. At first, the separating zone is filled up with water so that the pulsating energy can be transmitted to the separating zone. Then slurry is fed from the feeding box and enters the matrix pile located in the separating zone. The level of water or slurry and their flowrates can be adjusted by the bottom valve. Magnetic particles are attracted from slurry onto the surface of the matrix. Non-magnetic particles pass through matrix and go out through the product box under the combined force actions of slurry pulsation, gravity and hydrodynamic drag. The

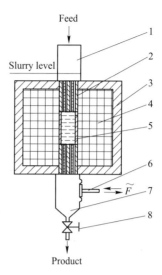

Fig. 1 Cyclic pulsating high-gradient magnetic separator

1—feeding box; 2—magnetic pole; 3—magnetic yoke; 4—energizing coils; 5—magnetic matrix; 6—pulsating mechanism; 7—product box; 8—valve

pulsating mechanism drives the slurry in the separating zone up and down, keeping the particles in the matrix pile in a loose state, so that magnetic particles can be more easily captured by the matrix and nonmagnetic particles can be more easily dragged out through the matrix pile. The cyclic PHGMS separator can only be fed periodically. When a batch of feed is finished, the energizing electric current is switched off and the magnetic product (mags) is washed out with water.

3 Principles of PHGMS

3.1 The behaviour of slurry and particle in the separating zone

While the separating zone is filled up with slurry and the pulsating mechanism is working, the pulsating energy is transmitted to the separating zone and drives the slurry up and down. The velocity curve is shown in Fig. 2. The pulsating velocity, maximum pulsating velocity and average pulsating velocity are respectively:

$$\tilde{v} = \frac{1}{2} S\omega \sin(\omega t) \tag{1}$$

$$\tilde{v}_{\max} = \frac{1}{2} S\omega \tag{2}$$

$$\bar{v} = \frac{1}{2\pi} \int_0^{2\pi} \left| \frac{1}{2} S\omega \sin(\omega t) \right| d(\omega t) = \frac{S\omega}{\pi} \tag{3}$$

The velocity of the slurry is the sum of feeding velocity and pulsating velocity:

$$v = v_0 + \tilde{v} = v_0 + \frac{1}{2} S\omega \sin(\omega t) = v_0 + \tilde{v}_{\max} \sin(\omega t) \tag{4}$$

where $\tilde{v}, \tilde{v}_{max}, \bar{\tilde{v}}$ are pulsating velocity, maximum pulsating velocity and average pulsating velocity, respectively; v, v_0 are slurry velocity and feeding velocity, respectively.

The pulsating acceleration, maximum pulsating acceleration and average pulsating acceleration are respectively:

$$\tilde{a} = \frac{1}{2}S\omega^2 \cos(\omega t) \quad (5)$$

$$\tilde{a}_{max} = \frac{1}{2}S\omega^2 \quad (6)$$

$$\bar{a} = \frac{1}{2\pi}\int_0^{2\pi}\left|\omega \frac{1}{2}S\omega^2\cos(\omega t)\right|d(\omega t) = \frac{S\omega^2}{\pi} \quad (7)$$

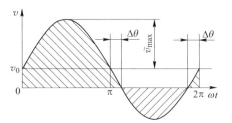

Fig. 2 Velocity curve of PHGMS

where $\tilde{a}, \tilde{a}_{max}, \bar{a}$ are pulsating acceleration, maximum pulsating acceleration and average pulsating acceleration, respectively; S, ω, t are pulsating stroke, pulsating angle speed and time variable, respectively.

When $\tilde{v}_{max} > v_0$, slurry velocity is reversed periodically as shown in Fig. 2. The shadowed area below the ωt axis means that the slurry moves upwards, while that above the ωt axis means downwards corresponding to Fig. 1.

It is very important that $\tilde{v}_{max} > v_0$ can effectively reduce the mechanical trap of nonmagnetic particles in the matrix as shown in Fig. 3. Non-magnetic particles may be trapped on the matrix caused by a bridge effect or other reasons if slurry is fed in one direction and $\tilde{v}_{max} \leq v_0$. When $\tilde{v}_{max} > v_0$, the flowing direction of the slurry is reversed periodically and bridge effects can be damaged or destroyed.

Fig. 4 shows the trajectories of a particle passing through the matrix pile with non-pulsation (Fig. 4a) and with pulsation (Fig. 4b). Suppose a particle enters the matrix pile from point C_1, then comes out from point C_2. For visibility, we artificially set point C_2 to point C'_2; the particle may actually comes out from point C_2.

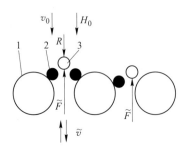

Fig. 3 Mechanical trap of non-magnetic particles
1—magnetic matrix wire; 2—magnetic particles;
3—non-magnetic particles

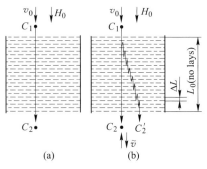

Fig. 4 Trajectories of a particle passing through the matrix pile
(a) non-pulsation; (b) with pulsation

When the pulsating velocity is zero or small, that is $\tilde{v}_{max} \leq v_0$, the velocity direction of the slurry will not be reversed, and the particle passes matrix layers n times, n being equal to the number of matrix layers:

$$n = n_0 = \frac{L_0}{\Delta L} \tag{8}$$

where n is the number of times that the particle passed matrix layers, n_0 is the number of matrix layers, L_0 is the height of matrix pile, and ΔL is the distance between two neighbouring layers.

When $\tilde{v}_{max} > v_0$, the actual velocity direction of the slurry reverses, and the slurry moves down and up. There are more chances for the particle to pass through the matrix layers.

The reversed velocity in each pulsating cycle is located between the angles of $\pi + \Delta\theta$ and $2\pi - \Delta\theta$ as shown in Fig. 2. The distance L_T that the particle travelled in a pulsating cycle in the vertical direction in the working zone can be calculated by integration:

$$L_T = \int_0^T |v_0 + \tilde{v}_{max}\sin(\omega t)|dt = \begin{cases} v_0 T & \tilde{v}_{max} \leqslant v_0 \\ \frac{2}{\pi}v_0 T\left[\arcsin\frac{v_0}{\tilde{v}_{max}} + \sqrt{\left(\frac{\tilde{v}_{max}}{v_0}\right)^2 - 1}\right] & \tilde{v}_{max} > v_0 \end{cases} \tag{9}$$

If the capacity of a separator and the feeding volume of slurry are constant, then the feeding velocity v_0 is constant, no matter whether the slurry pulsates or not, and the time that the particle passes the matrix pile is constant:

$$\Delta t = \frac{L_0}{v_0} \tag{10}$$

When the stroke number per minute is N, the stroke number in Δt is:

$$\Delta N = \frac{N}{60}\Delta t = \frac{\Delta t}{T} = \frac{L_0}{Tv_0} \tag{11}$$

From Eq. (9) to Eq. (11) we get the vertical distance that the particle travelled in the matrix pile:

$$L = \Delta N L_T = \begin{cases} L_0 & \tilde{v}_{max} \leqslant v_0 \\ \frac{2}{\pi}L_0\left[\arcsin\frac{v_0}{\tilde{v}_{max}} + \sqrt{\left(\frac{\tilde{v}_{max}}{v_0}\right)^2 - 1}\right] & \tilde{v}_{max} > v_0 \end{cases} \tag{12}$$

When L is divided by ΔL, we get the passing times:

$$n = \begin{cases} n_0 & \tilde{v}_{max} \leqslant v_0 \\ \frac{2}{\pi}n_0\left[\arcsin\frac{v_0}{\tilde{v}_{max}} + \sqrt{\left(\frac{\tilde{v}_{max}}{v_0}\right)^2 - 1}\right] & \tilde{v}_{max} > v_0 \end{cases} \tag{13}$$

According to Eq. (13), we can calculate the chances that the particle passes matrix layers as shown in Table 1, which shows that the passing times increase as the pulsating velocity increases.

Table 1 Particle passing times against pulsating velocity

$\tilde{v}_{max} > v_0$	0	≤1	2	3	4	5	6	7	8
n/n_0	1	1	1.4	2.0	2.6	3.3	3.9	4.5	5.1

3.2 The effect of passing times on recovery of magnetic particles

Suppose the capture possibility for a particle passing through a matrix layer is p; then when it passes n matrix layers the capture possibility is:

$$P = 1 - (1 - p)^n \qquad (14)$$

If n is big enough, $(1-p)^n \approx \exp(-pn)$, then:

$$P = 1 - \exp(-p^n) \qquad (15)$$

Suppose the maximum recovery for a group of magnetic particles is R_{max}; then the recovery R is:

$$R = R_{max}P = R_{max}[1 - \exp(-p^n)] \qquad (16)$$

3.3 Pulsating effect to the capture possibility of a single matrix layer

When a weakly magnetic particle passes through a matrix layer, the possibility that it can be captured is proportional to the magnetic force and inversely proportional to the competing force:

$$p = K\frac{F_m}{F_c} \qquad (17)$$

where K is a constant, F_m is the magnetic force, F_c is the competing force,

$$F_c = F_d + \overline{F_p} + F_g + F_a \qquad (18)$$

F_d = hydrodynamic drag force caused by feeding velocity, $F_d = 6\pi\eta bv_0$; $\overline{F_p}$ = hydrodynamic drag force caused by pulsating velocity, $\overline{F_p} = 6\pi\eta b\tilde{v} = 6\eta bS\omega$; F_g = gravitational force, $F_g = mg$; F_a = acceleration force caused by pulsating, $F_a = m\overline{a} = (m/\pi)S\omega^2$.

Because $\omega = 2\pi N/60$, so:

$$F_c = (6\pi\eta bv_0 + mg) + \frac{\pi}{5}\eta bSN + \frac{\pi}{900}mSN^2 = K_1 + K_2SN + K_3SN^2 \qquad (19)$$

where b, m, η are particle radius, particle mass and slurry viscosity, respectively; K_1; K_2; K_3 are constants.

From Eq. (19) we can see that F_c is a linear function of stroke S and a square function of frequency N. So when N increases to a certain extent, F_c increases very fast.

From Eqs. (16) and (17) we get the recovery formula related to pulsating parameters which can be calculated with Eqs. (18) and (19):

$$R = R_{max}\left[1 - \exp\left(-KF_m\frac{n}{F_c}\right)\right] \qquad (20)$$

3.4 Pulsating effect to grade of magnetic product

The pulsating effect on the grade of the magnetic product can be qualitatively described with a formula given by Oberteuffer (1974):

$$G_m = \frac{G_{max}}{1 + A_{nm}K'\frac{F_i}{F_c}} \quad (21)$$

where G_m, G_{max} are grade and maximum grade of the magnetic product, A_{nm} is the mass ratio of non-mags to mags in feed, K' is a constant, F_i is the interaction force between magnetic and non-magnetic particles, F_c is the competing force as expressed by Eq. (19).

From Eq. (21) it is easy to understand that G_m increases as F_c increases. Because F_c increases with pulsating stroke and frequency, G_m generally increases with pulsating stroke and frequency.

4 Laboratory test results and discussion

Table 2 and Fig. 5 show a group of test results of hematite ore from the Anshan Mining Company with the cyclic PHGMS separator.

Table 2 Test result of the cyclic PHGMS separator with hematite ore

Pulsating frequency /min^{-1}	Mass of mags /%	Grade of mags (Fe)/%	Recov. in mags (Fe)/%	Grade of tails (Fe)/%	Efficiency /%
0	58.50	44.91	89.70	7.27	28.85
100	54.24	48.71	90.21	6.27	37.50
150	50.59	52.18	90.13	5.83	45.00
200	50.31	53.00	91.04	5.28	47.74
250	47.34	55.21	89.23	5.99	50.65
300	45.59	56.42	87.81	6.56	51.71
350	40.87	57.63	82.03	9.70	48.46
400	38.43	58.50	76.76	11.06	44.67
450	37.56	59.34	73.59	11.21	42.59

Grade of feed is 29.27% Fe; feed size 80% <0.074mm; feeding velocity v_0 = 7cm/s; background magnetic induction 0.6T; stroke 10mm; efficiency = (recov. of mags-mass of mags)(grade of mags-grade of feed)/[(100-mass of mags)(maximum grade of mags-grade of feed)]×100%.

From Fig. 5 we can see the flowing phenomena.

(1) The iron grade of mags goes up as pulsating frequency increases, at first faster, then slower and slower as G_m approaches the maximum iron grade G_{max} of hematite.

(2) The tails grade slightly goes down to its lowest point as the pulsating frequency increases, at first because pulsation increases the passing times of particles though the matrix layers, then it goes up as the competing force increases faster and faster with increases of the pulsating frequency.

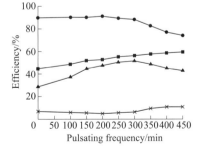

Fig. 5 Curves of test results against pulsating frequency
■—Grade of mags(Fe%);
●—Recovery in mags(Fe%);
×—Grade of tails(Fe%); ▲—Efficiency(%)

(3) Iron recovery slightly increases, reaching its highest point as the tails grade goes down; then it goes down as the competing force increases faster and faster with increases of the pulsating force.

(4) Efficiency of benefit goes up rapidly at first because the tails grade goes down and the mags grade goes up at the same time. After the tails grade has reached its lowest point, the efficiency still goes up slightly and reaches its highest point as the mags grade still goes up. Then the efficiency goes down as the tails grade increases faster and the mags grade goes up slower and slower.

5 Particle build-up model in PHGMS

Nesset (1981) studied the static build-up model of magnetic particles on a single wire, from which we can draw the following two major conclusions.

(1) The amount of magnetic particles captured on a single wire decreases as competing force increases.

(2) If the flowing direction of the slurry is constant, magnetic particles accumulate almost only on one side of the wire against a flowing direction v_0 even of up to 23.3 cm/s.

Combining the analysis of this paper and the above conclusions we may qualitatively draw a build-up model of particles on a single wire in PHGMS as shown in Fig. 6.

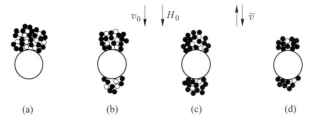

Fig. 6 Particle build-up on a single wire in PHGMS
(a) no pulsation; (b) pulsation is weak; (c) pulsation isfair; (d) pulsation is strong

Fig. 6(a). When $v_{max} = 0$, or there is no pulsation, particles accumulate almost only on the upper side of the matrix wire against the feeding direction and a lot of non-magnetic particles are trapped in or on the magnetic particles, obviously reducing the purity or grade of the mags.

Fig. 6(b). As pulsating velocity starts up and when $\tilde{v}_{max} > v_0$, the slurry direction is reversed frequently. Particles begin to accumulate on the lower side of the matrix wire. Non-magnetic particles are trapped less and less as v_{max} increases.

Fig. 6(c). As $\tilde{v}_{max}$ increases further, the chance for particles to accumulate on both sides of the wire becomes more and more even. Lesser non-magnetic particles get trapped.

Fig. 6(d). When $\tilde{v}_{max}$ is very large, particles are almost equally accumulated on both sides of the wire. Very few non-magnetic particles are trapped in the mags. However, the ability of attracting magnetic particles of the wire drops dramatically as the competing force becomes too large. Under this condition, although the purity of mags still goes up as $\tilde{v}_{max}$ increases, it usually can not compensate the decrease of mags recovery.

6 SLon VPHGMS separators for industrial application

In order to apply PHGMS in industry, we have developed several types of SLon Vertical ring and Pulsating High Gradient Magnetic Separators (SLon VPHGMS or SLon, Fig. 7) since 1986 and applied them in twelve mines for beneficiation of hematite, limonite, siderite, ilmenite and other minerals.

6.1 Structure and working principles of SLon

The SLon mainly consists of a pulsating mechanism, energizing coils, magnetic yoke, working ring, feed and product boxes as shown in Fig. 7. Expanded metals or round bars made of magnetic stainless steel are used as matrix. Its working principles are as follows.

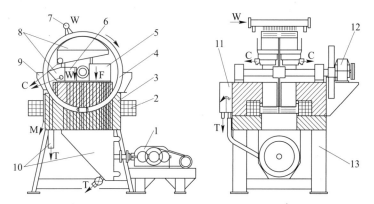

Fig. 7 SLon vertical ring and pulsating high-gradient magnetic separator

1—pulsating mechanism; 2—energizing coils; 3—magnetic yoke; 4—working ring; 5—feeding box; 6—wash water box; 7—concentrate flush; 8—concentrate box; 9—middling chute; 10—tailings box; 11—slurry level box; 12—ring driver; 13—support frame; F—feed; W—water; C—concentrate; M—middling; T—tailings

While a direct electric current flows through the energizing coil, a magnetic field is build up in the separating zone. The working ring with magnetic matrix rotates around its horizontal axis. Slurry fed from the feeding boxes enters into the matrix located in the separating zone. Magnetic particles are attracted from the slurry onto the surface of the matrix, brought to the top of the ring where the magnetic field is negligible, then flushed into the concentrate boxes. Non-magnetic particles pass through the matrix and enter the tailings box under the combined force of pulsating slurry, gravity and hydrodynamic drag.

Since the working ring rotates vertically, the flushing direction of the mags is opposite to that of the feed relative to each matrix pile; coarse particles can be flushed out without having to pass through the entire depth of the matrix pile. The pulsating mechanism drives the slurry in the separating zone up and down, keeping particles in the separating zone in a loose state all the time. Magnetic particles can thus be more easily captured by the matrix and non-magnetic particles can be more easily dragged to the tailings box through the matrix pile.

Therefore, opposite flushing and pulsation help to prevent matrix clogging, and pulsation helps to purify the magnetic product. These give SLon VPHGMS the advantages of a higher ratio of beneficiation, higher efficiency and considerable flexibility.

The major parameters of SLon VPHGMS are listed in Table 3.

Table 3 Parameters of SLon VPHGMS

Parameters	SLon-2000	SLon-1500	SLon-1000
Ring diameter/mm	2000	1500	1000
Background magnetic induction/T	0-1.0	0-1.0	0-1.0
Energizing power/kW	0-82	0-44	0-26
Driving power/kW	5.5+7.5	3+4	1.1+2.2
Energy consumption/kW·h·t^{-1}	0.2-1.9	0.3-2.6	0.5-7.3
Pulsating stroke/mm	0-20	0-20	0-20
Pulsating frequency/min^{-1}	0-300	0-300	0-300
Feed size/mm	0-2.0	0-1.3	0-1.0
Feed solid density/%	10-45	10-45	10-45
Slurry throughput/m^3·h^{-1}	100-200	50-100	10-20
Ore throughput/t·h^{-1}	50-80	20-35	4-7
Mass of machine/t	50	20	6

7 Examples of applications of SLon

In the following a few examples are given of SLon separators applied in industry.

7.1 The application of SLon-1500 in the Gushan Iron Mine of the Maanshan Iron and Steel Company

Gushan iron ore is a finely disseminated and colloidal hematite quartzite. The assay and mineralogical composition of the ore are shown in Tables 4 and 5, respectively.

Table 4 Assay of Gushan iron ore (%)

Total Fe	SiO$_2$	Al$_2$O$_3$	CaO	MgO	S	P
35.92	37.50	5.97	0.39	0.26	0.049	0.25

Table 5 Mineralogical composition of Gushan iron ore (%)

Hematite	Limonite	Magnetite	Siderite	Pyrite	Quartz	Kaolin	Other	Total
51.38	1.40	0.73	0.23	0.30	39.43	5.03	1.50	100.00

The old flowsheet was that the ore was ground to 50%<0.074mm, and then classified by hydrocyclones of D500mm and D350mm. Their underflow was treated by spirals and the overflow of the D350mm hydrocyclone was treated by a centrifugal separator.

In 1989-1992, the Gushan Iron Mine reformed its hematite processing flowsheet from gravity to magnetic separation. The present flowsheet is shown in Fig. 8. The ore is ground to about 48% < 0.074mm in a primary ball mill. Four SQC-6-2770 wet high-intensity magnetic separators (SQC WHIMS) are used as rougher to throw away part of the tailing (non-mags), the mags are reground to about 87% < 0.074mm, then two SQC-6-2770 WHIMS are used as cleaner and two SLon-1500 as scavenger. The total results of both new and old flowsheets are shown in Table 6. Compared with the gravity flowsheet, the grade of iron concentrate is 2.95% higher and the recovery is 13.80% high-

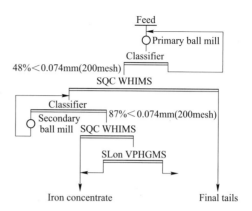

Fig. 8 Flowsheet of the Gushan iron mine

er. The recovery increase is mainly due to the two SLon-1500 which recover the fine hematite particles from the tailing of SQC WHIMS cleaner, which otherwise would be difficult to recover.

Table 6 Gushan production results of gravity and magnetic flowsheets

Flowsheet	Year	Grade (Fe)/%			Recovery of concentr. (Fe)/%
		Feed	Concentr.	Tails	
Gravity	1988	37.16	55.22	24.47	61.32
Magnetic	1993	35.51	58.17	16.32	75.12
Difference		-1.65	2.95	-8.15	13.80

7.2 SLon-1500 applied in desulphurization and dephosphorization

There is a large iron ore deposit at Meishan Iron Mine in Jiangsu Province. The ore consists of magnetite, hematite, siderite and pyrite. The iron ore concentrate is used as raw material of pellets. However, because of relatively high contents of sulphur and phosphorus, the iron concentrate cannot be used alone as raw material of pellets but as an auxiliary material. Thus, production of the iron concentrate is limited by the market. The assay and mineralogical composition of the milled iron ore are shown in Tables 7 and 8, respectively.

Table 7 Assay of Meishan iron ore (%)

Total Fe	S	P	CaO	MgO	SiO_2	Al_2O_3
52.53	2.46	0.399	3.87	1.78	5.69	0.94

Table 8 Mineralogical composition of Meishan iron ore (%)

Iron mineral	Magnetite	Siderite	Hematite	Pyrite	Silicate iron	Total
Fe	30.57	10.35	7.58	2.43	1.60	52.53
Fe distr.	58.20	19.70	14.43	4.62	3.05	100.00

In order to reduce S and P, the Meishan Iron Mine cooperated with several research institutes and did variety tests. Finally a magnetic flowsheet was chosen to carry out a commercial test in 1995 as shown in Fig. 9. Low-intensity magnetic separators (LIMS) were used to clean out magnetite. A SLon-1500 as rougher and another SLon-1500 as scavenger were applied to recover hematite and siderite. The two SLon-1500 effectively recovered the fine hematite and siderite, controlled the tailing grade, and guaranteed the total results to meet the metallurgical requirements. The commercial test was run for 276h, and 15393t of ores were treated. The results are shown in Table 9, and show that the quality of concentrate was considerably improved and met the metallurgical requirements. In 1996, the mine bought another eight SLon-1500's to reform its flowsheet according to the commercial test.

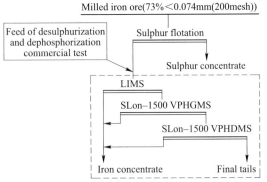

Fig. 9 Desulphurization and dephosphorization flowsheet of Meishan iron ore

Table 9 Results of a 276 h commercial test on Meishan iron ore (%)

Product	Mass	Grade			Recovery		
		Fe	S	P	Fe	S	P
Feed	100.00	52.77	0.44	0.399	100.00	100.00	100.00
Concentrate	88.90	56.08	0.29	0.246	94.51	59.10	54.76
Metallurgical requirement		>55.0	<0.35	<0.25	>94		
Tailing	11.10	26.13	1.61	1.626	5.49	40.90	45.24

7.3 SLon-1500 applied to beneficiate limonite

In 1993-1994, two SLon-1500 were installed in the Zhongwei Iron Mine, a small limonite processing plant in Ningxia Province. The order of processing is that the ore is ground by one stage mill to about 60% <0.074mm, one SLon-1500 is used as rougher and the other SLon-1500 as scavenger as shown in Fig. 10. The results are shown in Table 10.

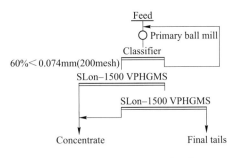

Fig. 10 Limonite processing flowsheet of the Zhongwei iron mine

Table 10 Results of SLon-1500 applied to Zhongwei limonite (%)

Operation	Grade (Fe)			Mass of concentrate	Recovery of concentrate (Fe)
	Feed	Concentrate	Tails		
Roughing	46.36	54.17	38.60	49.84	58.24
Scavenging	38.60	50.65	28.43	22.96	25.08
Total	46.36	53.06	28.43	72.80	83.32

Table 10 shows that the total concentrate grade is about 53% and the recovery is about 83%. The results are satisfactory due to the efficient processing ability of SLon-1500. The flowsheet is very simple and the operating cost is very low. It is suitable for a variety of mines to exploit limonite, hematite, siderite and other minerals which are simply associated with gangues

7.4 The applications of SLon-2000 in the Gongchangling mineral processing plant

As the mining scale of oxidized iron ore increases and rich ore decreases in the Gongchangling mineral processing plant, wet high-intensity magnetic separation (WHIMS) takes a very important position in processing low-grade hematite ore because it can throw away about half of the low grade tails immediately after the primary ball mill. But matrix clogging is the biggest problem with WHIMS in the plant. There are five wet high-intensity magnetic separators in the plant. Their ring diameter is 2000mm (WHIMS-2000). They are applied for roughing 0-2.2mm hematite. Because the feed contains a lot of coarse particles and some strong magnetic particles such as magnetite and maghemite, their matrixes are very easily clogged even the gap between the grooved plate is put to 3.5mm, causing unstable processing results and making equipment maintenance a difficult task. To solve this problem, in 1993-1994 the first SLon-2000 was designed, manufactured and installed in parallel with the five WHIMS-2000's to treat the 0-2.2mm (about 50% <0.074mm) low-grade hematite in the primary

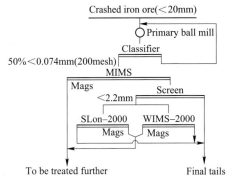

Fig. 11 Flowsheet of the commercial comparison test of SLon-2000 with WHIMS-2000

grinding stage as shown in Fig. 11. The assay and mineralogical composition of the iron ore are shown in Tables 11 and 12. In the flowsheet, permanent middle-intensity (0.22T) magnetic separators (MIMS) were used to move the magnetite and part of the maghemite. The non-mags of MIMS were treated with the SLon-2000 and WHIMS-2000 to throw away final tails. During the commercial test, the SLon-2000 ran six months in parallel with the WHIMS-2000 from September 1, 1994 to February 28, 1995. The compared results are shown in Table 13.

Table 11 Assay of Gongchangling iron ore (%)

Total Fe	SiO$_2$	Al$_2$O$_3$	CaO	MgO	P	S
28.59	52.15	0.78	0.21	0.29	0.043	0.018

Table 12 Mineralogical composition of Gongchangling iron ore (%)

Iron mineral	Hematite	Magnetite	Maghemite	Silicate iron	Siderite	Total
Fe	17.07	7.30	2.70	1.30	0.40	28.77
Fe distr.	59.33	25.37	9.38	4.52	1.40	100.00

Table 13 Average results of a 6-month comparison of beneficiation (%)

Separator	Grade (Fe)			Mass of mags	Recovery of iron
	Feed	Mags	Tails		
SLon-2000	25.94	41.21	10.74	49.89	79.25
WHIMS-2000	25.94	34.00	16.15	54.85	71.89
Difference	0	7.21	5.41	-4.96	7.36

Table 13 shows that the SLon-2000 can get a much higher grade and recovery of mags, a much lower grade of tails and a lower mass percentage of mags than the WHIMS-2000. The higher grade of mags is mainly due to the contribution of the pulsating mechanism. The lower grade of tails and higher iron recovery are mainly due to the fact that the magnetic matrix is kept clean and the magnetic force is relatively stronger.

8 Conclusion

(1) Slurry pulsating increases the number of particle collisions with matrix layers. It also increases the possibility for magnetic particles to be captured on both sides of the matrix wire and is favourable for recovering magnetic particles. On the other hand, it increases the competing force acting on all particles and is unfavourable for recovering magnetic particles. With simultaneous increases of collision times and competing force in a certain region the recovery of mags remains almost the same or slightly goes up.

(2) Pulsating increases the competing force acting on non-magnetic particles. It is favourable for eliminating the mechanical trap of the non-magnetic particles. So the grade of mags generally goes up with the increase of pulsating velocity.

(3) For each certain mineral, there is a certain pulsating frequency and a certain stroke at which the benefit efficiency reaches a maximum.

(4) The SLon vertical ring and pulsating high-gradient magnetic separator is a new generation of continuously working HGMS with a higher beneficiation ratio and without matrix clogging. It is becoming a major industrial equipment for processing hematite, siderite, limonite and other weakly magnetic minerals.

References

[1] Liu Shuyi, Chen Jin, Xiong Dahe. A new type of commercial PHGMS and its separating principles[C] // Proc. XVIIth Int. Miner. Process. Congr., 1991:152-159.

[2] Nesset, J E. The static (buildup) model of particle accumulation separation: experimental confirmation[J]. IEEE Trans. 1981,4:1506-1508.

[3] Oberteuffer, J A. High gradient magnetic separation[J]. 1973, 3:303-306.

[4] Oberteuffer, J A. Magnetic separation: A review of principles, devices, and applications[J]. 1974, 2: 223-238.

[5] Svoboda, J. Magnetic methods for the treatment of minerals[J]. Developments in Mineral Processing 8, Elsevier, Amsterdam.

[6] Xiong Dahe. New development of the SLon vertical ring and pulsating high gradient magnetic separator[J]. Magn. Electr. Sep. 1994, 5:211-222.

[7] Xiong Dahe. Development and commercial test of SLon-2000 vertical ring and pulsating high gradient magnetic separator[J]. Magn. Electr. Sep. 1997,8:89-100.

[8] Xiong Dahe, Liu Shuyi, Liu Yongzhi, et al. The development of a new type of industrial vertical ring and pulsating high gradient magnetic separator[M]. Koch, M., Taylor, J. C. (Eds.), Productivity and Technology in Metallurgical Industries. The Minerals, Metals and Materials Society of Germany, 1989:947-952.

[9] Yang Peng, Liu Shuyi, Chen Jin. The separation performance of the pulsating high gradient magnetic separator[J]. Magn. Electr. Sep. ,1993,4:211-221.

写作背景 本文是作者在《国际矿物加工杂志》上发表的第一篇论文,文中介绍了脉动高梯度磁选的理论基础,SLon 立环脉动高梯度磁选机的工作原理,SLon-1500 磁选机在南京梅山铁矿用于氧化铁矿提质降磷,在宁夏中卫铁厂分选褐铁矿,SLon-2000 磁选机在鞍钢弓长岭选矿厂工业试验。

Research and Commercialisation of Treatment of Fine Ilmenite with SLon Magnetic Separators[1]

Xiong Dahe

(Ganzhou Nonferrous Metallurgy Research Institute, 36 Qingnian Road, Ganzhou 341000, Jiangxi Province, P. R. China)
(Received 16 September 1999; Accepted 27 September 1999)

Abstract The fine ilmenite material finer than 0.045mm at the Pan Zhi Hua Ilmenite Processing Plant was discarded for the lack of efficient processing equipment and method. The total recovery of ilmenite in the plant had been only about 17%-20%. To increase the ilmenite recovery, numerous laboratory, pilot and commercial tests were conducted in 1995-1996. Finally, a magnetic separation-flotation flowsheet was chosen for the commercial test, in which the fine fraction was fed into 125mm diameter cyclones for desliming. The underflow of the cyclones was then fed into the magnetic separation-flotation flowsheet. A SLon-1500 vertical ring and pulsating high gradient magnetic separator was applied as a rougher and was followed by flotation as a cleaner. The overall results of the magnetic separation-flotation flowsheet in the commercial test were: feed 11.03% TiO_2, concentrate 47.36% TiO_2, recovery 44.79% TiO_2. The flowsheet of the commercial test has been used as a production process since 1997. The recovery accounted for about 12% TiO_2 of the feed into the ilmenite processing plant.

Keywords Ilmenite; Magnetic separation; SLon pulsating magnetic separator

1 Introduction

The Panzhihua Ilmenite Processing Plant is the biggest producer of ilmenite in China. The deposit consists of vanadium, titanium and iron minerals and other gangues. Magnetite is the major iron mineral and vanadium exists in magnetite. Ilmenite is the major titanium mineral. An iron processing plant recovers magnetite and vanadium before the ilmenite processing plant. The tails of the iron processing plant represent the feed into the ilmenite processing plant.

The previous flowsheet of the ilmenite processing plant was based on classification into >0.1mm (coarse), 0.1-0.045mm (middle) and <0.045mm (fine) fractions. The coarse and the middle particle fractions were treated with spirals and electrical separators. The fine fraction was discarded as the final tails for the lack of efficient beneficiation technology. The overall recovery of ilmenite had been only 17%-20% of the feed of the ilmenite processing plant.

The mass of the fine fraction is about 35% and the TiO_2 grade is almost the same as that of the feed to the ilmenite plant. The recovery of ilmenite from the fine particle fraction is thus of great

---

[1] 原载于《Magnetic and Electrical Separation》,1999(09):121~130。

significance for increasing the overall recovery of the ilmenite plant.

2 Mineralogical analysis

Major minerals and their physical properties in the feed into the ilmenite plant are shown in Table 1.

Table 1 Minerals and their properties in the feed of the ilmenite plant

Mineral	Molecule	Mass /%	Grade (TiO_2) /%	Specific density /g·cm^{-3}	Magnetic susceptibility /cm^3·g^{-1}	Specific resistance /Ω·cm
Vanadium-titanium magnetite	$(V,Ti,Fe)_3O_4$	2~5	13.4	4.6~4.8	>1000×10^6	1.38×10^6
Ilmenite	$FeTiO_3$	15.3~17.5	51.89	4.6~4.7	240×10^6	1.75×10^5
Titaniferous augite	$Ca(Mg,Fe,Ti)[(Si,Al)_2O_6]$	46~46.5	1.85	3.2~3.3	100×10^6	3.13×10^{13}
Pyrite	FeS_2	2~2.5		4.4~4.9	4100×10^6	1.25×10^4
Plagioclase	$(100-n)Na[AlSi_3O_8] \cdot nCa[Al_2Si_2O_8]$	31~34	0.097	2.7	14×10^6	>10^{14}

3 The laboratory testwork on fine ilmenite

3.1 Laboratory tests of magnetic separation

Ilmenite is our target mineral to be recovered and the other minerals are considered to be tailings. It can be seen from Table I that magnetic susceptibilities of vanadium-titanium magnetite, ilmenite and pyrite are relatively stronger than the others. But magnetic susceptibility of titaniferous augite is close to that of ilmenite. It is possible to remove plagioclase and part of titaniferous augite by magnetic separation.

The fine particle fraction (<0.045mm) of the ilmenite plant feed was used as the feed in our laboratory test. SLon-100 cyclic pulsating high gradient magnetic separator (SLon-100) was used as a rougher. In the test, the matrix type, the background magnetic induction, pulsating stroke and pulsating frequency were optimized. Typical results of the magnetic separation test are shown in Table 2.

Table 2 Test results of SLon-100 magnetic separation of fine ilmenite

Product	Mass /%	Grade (TiO_2) /%	Recovery (TiO_2) /%	Separating conditions			
				Matrix type	Background induction /T	Pulsating stroke /mm	Pulsating frequency /min^{-1}
Concentrate	27.71	21.77	65.22	2mm dia. rod	0.46	12	200
Tails	72.29	4.45	34.78				
Feed	100.00	9.25	100.00				
Concentrate	26.95	22.29	64.93	4mm dia. rod	0.58	12	225
Tails	73.05	4.44	35.07				
Feed	100.00	9.25	100.00				

Continued Table 2

Product	Mass /%	Grade (TiO_2) /%	Recovery (TiO_2) /%	Separating conditions			
				Matrix type	Background induction /T	Pulsating stroke /mm	Pulsating frequency /min^{-1}
Concentrate	28.21	20.55	62.67	Expanded mesh	0.46	12	225
Tails	71.79	4.81	37.33				
Feed	100.00	9.25	100.00				

Table 2 shows that with 4mm dia. rod matrix in SLon-100 magnetic separation better results can be obtained than with 2mm dia. rod matrix or expanded mesh. When the feed is 9.25% TiO_2 and with 4mm dia. rod matrix, the concentrate grade is 22.29% TiO_2, the tails grade is 4.44% TiO_2, recovery in the concentrate is 64.93% TiO_2, the mass of the concentrate is only 26.95% and the mass of the tails is 73.05%. The importance of magnetic separation is that it can discard about 73% by mass of the low-grade tails and only about 27% by mass of the concentrate enters into next cleaning stage. For example, if flotation is used for the cleaning, a lot of flotation reagents can be saved.

3.2 Laboratory tests with flotation as a cleaning stage

In the laboratory flotation tests, several reagents as collectors were tested and the flowsheet was optimized. Typical results are shown in Table 3.

Table 3 Results of the laboratory test of flotation in the cleaning stage

Product	Mass /%	Grade (TiO_2)/%	Recovery (TiO_2)/%	Flotation flowsheet
Concentrate	40.43	47.10	85.44	One roughing, one cleaning, one scavenging (closed circuit)
Tails	59.57	5.45	14.56	
Feed	100.00	22.29	100.00	
Concentrate	28.65	46.87	60.24	One roughing, one cleaning, one scavenging (open circuit)
Tails	71.35	12.42	39.76	
Feed	100.00	22.29	100.00	
Concentrate	27.21	47.54	58.04	One roughing, one cleaning, one scavenging (open circuit)
Tails	72.79	12.85	41.96	
Feed	100.00	22.29	100.00	

The Table shows that the grade of the ilmenite concentrate can be upgraded to about 47% TiO_2 by flotation. Recovery of TiO_2 in the flotation stage can reach about 60% in the open circuit or about 85% in the closed circuit.

3.3 Summary of the laboratory tests

Table 4 shows the overall results of magnetic and flotation separation in laboratory tests. With SLon 100 magnetic separation as a rougher and flotation as a cleaner, we can get the final concentrate with 47.10% TiO_2 grade and 55.47% recovery in the closed flotation circuit or 46.87% TiO_2 grade and 39.12% recovery in the open flotation circuit.

Table 4 Summary of the results of the laboratory tests of magnetic and flotation separation

Product	Mass /%	Grade (TiO_2)/%	Recovery (TiO_2)/%	Flotation flowsheet
Final concentrate	10.90	47.10	55.47	SLon-100 roughing, flotation cleaning (closed circuit)
Total tails	89.10	4.62	44.52	
SLon-100 feed	100.00	9.25	100.00	
Final concentrate	7.22	46.87	39.12	SLon-100 roughing, flotation cleaning (open circuit)
Total tails	92.28	6.10	60.88	
SLon-100 feed	100.00	9.25	100.00	

The laboratory tests-scale showed the direction for subsequent pilot and commercial tests.

3.4 Pilot-plant tests on fine ilmenite

In 1995, SLon-1000 vertical ring and pulsating high gradient magnetic separator[1] was moved to Pan Zhi Hua Ilmenite Procesing Plant for pilot-plant tests to recover ilmenite from the discarded <0.045mm ilmenite slime. SLon-1000 was installed in the production flowsheet and a slurry stream of <0.045mm ilmenite slime was used as its feed. The magnetic product (the rougher concentrate) was used as the flotation feed in the cleaning stage. The flowsheet of the pilot test is shown in Fig. 1.

In the pilot-plant tests technical parameters were first optimized. Then SLon-1000 ran continuously for 72h to get more reliable results. As is shown in Fig. 1, SLon-1000 removes about 75% by mass of the low-grade tails, and most of the slimes (<10μm fraction) which are harmful to the flotation reaction, reported into the tails. Only about 25% of the mass enters the flotation cleaning stage.

The overall results showed that a good ilmenite concentrate containing 47.60% TiO_2 with the total recovery of 44.20% TiO_2 can be obtained.

4 The commercial tests

Based on the results of the pilot test a commercial test was carried out in 1996. SLon-1500 vertical ring and pulsating high gradient magnetic separator was installed in the production flowsheet as shown in Fig. 2. In order to increase the throughput of SLon-1500, 125mm diameter cyclones were used to dewater and deslime the <0.01mm slime fraction. The operational results of the cyclones, of SLon-1500 and of flotation are shown in Tables 5-7 respectively.

Table 5 shows that the cyclone operation is important because the solids concentration in the feed is only 5.42%. The solids concentration of the cyclone underflow is 28.32%. Most water and <0.01mm slimes report into the cyclone overflow. This creates a much better condition for the SLon-

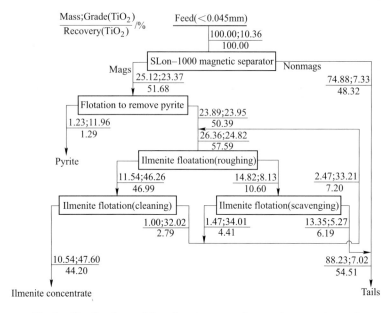

Fig. 1　The flowsheet of the pilot-plant test of magnetic separation and flotation of the <0.045mm ilmenite slime

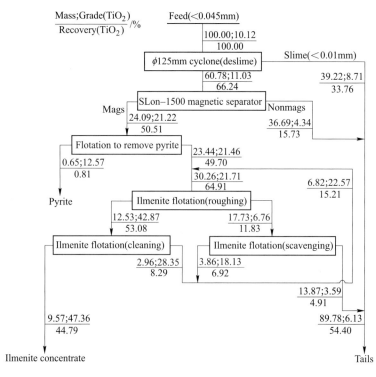

Fig. 2　The flowsheet of the commercial test of magnetic separation and flotation of the <0.045mm ilmenite fraction

1500 operation. As shown in Table 6, the capacity of SLon-1500 reaches 25.25t/h with 76.24%

TiO$_2$ operational recovery. Fig. 2 shows that the combined TiO$_2$ recovery of the cyclone and SLon-1500 is 50.51%, almost the same compared with the pilot test shown in Fig. 1.

Table 5 The operational results of 125mm diameter cyclones (%)

Feed			Under flow product			Overflow product		
Solid density	TiO$_2$	TiO$_2$ distribution	Solid density	TiO$_2$	TiO$_2$ distribution	Solid density	TiO$_2$	TiO$_2$ distribution
5.42	10.12	100.00	28.32	11.03	66.24	2.94	8.71	33.76

Table 6 The average operational results of SLon-1500 (%)

Feed			Concentrate			Tails		
t/h	<0.045mm	TiO$_2$ grade	Mass	TiO$_2$ grade	TiO$_2$ recovery	Mass	TiO$_2$ grade	TiO$_2$ distribution
25.25	75	11.03	39.63	21.22	76.24	60.37	4.34	23.76

The operation of the cyclone and SLon-1500 created very good conditions for the flotation stage. The first two operations discarded 75.91% of mass and most of the <0.01mm slimes into the tails. Only 24.09% of mass of the magnetic concentrate enters into the flotation cleaning stage. As shown in Table 7, the flotation results are very good. The final grade of ilmenite concentrate reaches 47.36% TiO$_2$ and the flotation recovery is 88.68% TiO$_2$. The overall ilmenite recovery of the full flowsheet is 44.79% TiO$_2$.

Table 7 The average operational results of flotation (%)

Product	Mass	TiO$_2$ grade	TiO$_2$ distribution
Pyrite	2.70	12.57	1.60
Ilmenite concentrate	39.73	47.36	88.68
Tails	57.57	3.59	9.72
Feed(SLon-1500 Mags)	100.00	21.22	100.00

5 The production-scale application

The commercial test shown in Fig. 2 was completed at the end of 1997, and the flowsheet was implemented on production scale. In 1998, about 10000t of > 47% TiO$_2$ ilmenite concentrate was recovered from the <0.045mm fine fraction. After more than a one year-long production practice, it was demonstrated that the SLon-1500 VPHGMS magnetic separator works very reliably. In 1999, the production scale of the <0.045mm fine fraction is expected to reach 15000-20000t of the ilmenite concentrate as the flowsheet becomes more optimized.

6 Conclusions

A magnetic separation-flotation flowsheet is applied to the recovery of the <0.045mm fine ilmenite minerals from the previous tails in the Panzhihua Ilmenite Processing Plant. SLon-1500 vertical ring and pulsating high gradient magnetic separator is used in the roughing stage. It is a key equipment for the flotation cleaning stage. The overall flowsheet is relatively simple and reliable. It has been put into production by the end of 1997. The overall results of the flowsheet are as follows: feed grade 11% TiO_2, ilmenite concentrate grade $\geqslant$ 47% TiO_2 and TiO_2 recovery 44%. Approximately 10000-20000t of such ilmenite concentrate can be recovered from the previously abandoned tails. It is a significant technological innovation in the history of the Panzhihua Ilmenite Processing Plant.

References

[1] Xiong Dahe. New development of the SLon vertical ring pulsation HGMS separator [J]. Magnetic and Electrical Separation, 1994(5):211.

写作背景 攀钢选钛厂过去不能回收 45μm 以下的钛铁矿，全部当尾矿排放。应用 SLon 立环脉动高梯度磁选机后，这部分钛铁矿得到有效的回收。此文介绍 SLon 磁选机回收细粒钛铁矿的小型试验、半工业试验和工业试验。

A Large Scale Application of SLon Magnetic Separator in Meishan Iron Ore Mine[①]

Xiong Dahe

(Ganzhou Nonferrous Metallurgy Research Institute, Ganzhou, 341000, Jiangxi, China)
(Received 18 September 2000; In final form 12 October 2000)

Abstract Meishan Iron Mine processes 4 million tons of iron ore and produces about 2.6 million tons of iron concentrate annually. Because of a relatively high content of phosphorus, the iron concentrate does not meet the market requirements. In recent years, 18 SLon-1500 magnetic separators have been applied to reduce phosphorus to upgrade the concentrate quality. This paper describes research, test work and industrial application of SLon magnetic separator in the mine.

Keywords Hematite; Siderite; Dephosphorisation; SLon pulsating magnetic separator

1 Introduction

There is a large iron ore deposit at Meishan Iron Mine in Jiangsu Province of China. In 1970s, a iron ore mining and processing line was built up, producing about 1.3 million tons of iron concentrate a year. The iron concentrate was used as raw material for pellets. However, because of a relatively high content of sulfur and phosphorus (phosphorus is very diffult to remove by metallurgical method). the iron concentrate could not be used alone as a raw material for pellets but as an auxilliary material. Thus, the production of the iron concentrate was limited by the market.

In 1980s, the mine planned to expand its mining and processing scale to 4 million tons a year. The biggest problem was how to reduce phosphorus in the iron concentrate. The mine cooperated with several research institutes and conducted various research work and tests. Finally a magnetic flowsheet and a flotation flowsheet were chosen to carry out separate commercial tests in 1995. Both flowsheets obtained very close results but the magnetic one at much lower cost and much more environmentally friendly. From 1996 till 2000, the mine innovated its first processing line and built a second processing line with the magnetic flowsheet and 18 SLon-1500 Vertical Ring and Pulsating High Gradient Magnetic Separators (SLon-1500 VP HGMS) were applied in the 4 million ton production line.

2 The previous processing flowsheet

In 1970s, Meishan Iron Mine built its first iron ore processing flowsheet as shown in Fig. 1. The mined ore was crushed and screened to 30-2mm, then treated by a gravity vibrating chute.

The enriched iron ore was milled down to about 70% <200 mesh with ball mills and treated by

---

[①] 原载于《Magnetic and Electrical Separation》,2002,02。

flotation to remove sulfur minerals such as pyrite. The concentrate was the final product but it contained too much phosphorus.

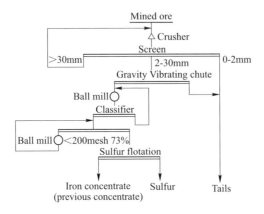

Fig. 1 The previous processing flowsheet of Meishan iron ore

3 Mineralagical studies of the Meishan iron ore

The mineralogical compositions of the Meishan mined ore and the concentrate are shown in Table 1.

Table 1 Mineralogical composition of the Meishan mined ore (Fe) (%)

Iron mineral	Magnetite	Siderite	Hematite	Pyrite	Silicate	Total
Mined ore	21.84	10.32	8.58	2.76	3.00	46.50
concentrate	26.92	10.34	14.46	0.441	1.60	53.76

The assay of the Meishan mined ore and the assay of the iron concentrate (see Fig. 1) are shown in Table 2.

Table 2 The assay of the Meishan iron ore and concentrate (%)

Element	Fe	S	P	SiO_2	Al_2O_3	CaO	MgO
Mined ore	46.5	1.91	0.366	9.98	1.87	4.87	1.85
Concentrate	53.96	0.45	0.393	5.07	0.86	4.10	1.81
Metallurgical requirement		<0.35	<0.25				

Table 1 shows that the iron minerals in the mined ore and the concentrate consist mainly of magnetite, siderite and hematite. Table 2 shows that the concentrate contains 0.393% P, which is much higher than the metallurgical requirement of <0.25%. Sulfur is also higher than the metallurgical requirement of <0.35%.

4 A commercial test of the flotation flowsheet

In order to compare the results and reliability of the magnetic flowsheet, a commercial test of the flotation flowsheet was carried out in 1995 before magnetic flowsheet test. The flotation flowsheet is shown in Fig. 2. In the flowsheet low intensity magnetic separators (LIMS) were used to remove magnetite, the nonmagnetic fraction from LIMS was dewatered and then fed into the flotation stage to remove phosphorus. The 176h commercial test results of flotation flowsheet are shown in Table 3.

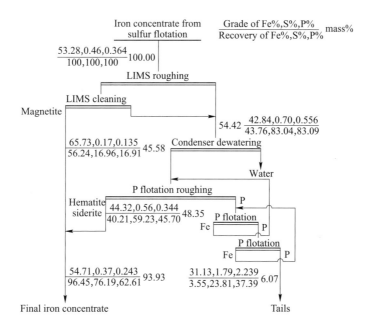

Fig. 2 A flowsheet and results of the flotation commercial test

Table 3 176h commercial test results of flotation flowsheet (%)

Product	Mass	Grade			Recovery		
		Fe	S	P	Fe	S	P
Feed	100.00	53.28	0.46	0.364	100.00	100.00	100.00
Concentrate	93.93	54.71	0.37	0.243	96.45	76.19	62.61
Tailing	6.07	31.13	1.79	2.239	3.55	23.81	37.39

5 A commercial test of the magnetic flowsheet

In 1995, a commercial test of the magnetic flowsheet was carried out following the flotation commercial test. The magnetic flowsheet is shown in Fig. 3. In the flowsheet, low-intensity magnetic separators (LIMS) were used to remove magnetite. A SLon-1500 VPHGMS as rougher and another SLon-1500 VPHGMS as scavenger were applied to recover hematite and siderite. These two SLon-1500s shown in Fig. 4 effectively recovered fine hematite and siderite, controlled the tailings grade, and guaranteed the total results to the metallurgical requirements. The commercial test was run for 276h, and 15393t of the ore were treated. The results are shown in Table 4, which shows that the quality of the concentrate was considerably improved and met the metallurgical requirements.

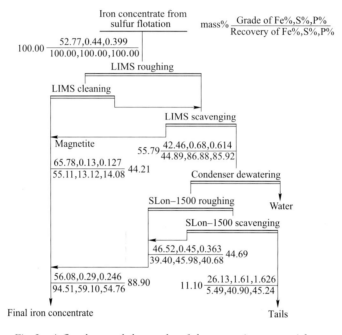

Fig. 3 A flowsheet and the results of the magnetic commercial test

Fig. 4 SLon-1500 VPHGMS magnetic separators applied in the Meishan Iron Mine

Table 4 276h commercial test results of magnetic flowsheet (%)

Product	Mass	Grade			Recovery		
		Fe	S	P	Fe	S	P
Feed	100.00	52.77	0.44	0.399	100.00	100.00	100.00
Concentrate	88.90	56.08	0.29	0.246	94.51	59.10	54.76
Metallurgical requirement		>55.0	<0.35	<0.25			
Tailing	11.10	26.13	1.61	1.626	5.49	40.90	45.24

6 A comparison of the magnetic flowsheet with the flotation flowsheet

Comparing the results of the magnetic flowsheet and the flotation flowsheet, one can see that the magnetic flowsheet possesses the following advantages:

(1) The Fe grade of the iron concentrate is higher (56.08%-54.71%).
(2) The S grade of iron concentrate is lower (0.29%-0.37%).
(3) The P recovery in the iron concentrate is lower (54.76%-62.61%).
(4) Magnetic flowsheet does not need any reagent. Its running cost is much lower.
(5) Magnetic flowsheet does not cause any environment problem.
(6) Magnetic concentrate is much easier to dewater.

7 Industrial application of magnetic flowsheet

Because the magnetic flowsheet possesses a number of advantages compared to the flotation flowsheet, Meishan Iron Mine finally adopted the magnetic flowsheet to reduce phosphorus in the iron concentrate. Since 1996, the mine has innovated its previous flowsheet with 8 SLon-1500 vertical ring and pulsating high gradient magnetic separators. After several years of running, SLon-1500 showed good performance and reliability. The mine acquired confidence and bought further eight SLon-1500 separators to build additional 2 million ton production line from 1998 to 2000. Moreover, in order to increase the iron recovery, the mine bought another two SLon-1500 separators to recover iron minerals from the <2mm screening products (see Fig. 1) in 1999. Up to now, Meishan Iron Mine installed eighteen SLon-1500 VPHGMS magnetic separators in its mineral processing line. This is the largest application of SLon VPHGMS separators in China.

8 Conclusions

In order to solve the biggest problem of a high conclusions of phosphorus and sulphur in the iron concentrate the Meishan Iron Mine arranged for two commercial tests in the 1995. One test was the flotation flowsheet and the other was the magnetic flowsheet. Through comparison of hundreds of hours, the magnetic flowsheet showed a number of advantages, namely better processing results, much lower cost, environment friendliness and easier dewatering of the iron concentrate. The mine finally adopted the magnetic flowsheet for innovation and expansion of its production line. In the course of several years of operation SLon-1500 vertical ring and pulsating high gradient magnetic separators demonstrated their high efficiency and reliability. Up to now eighteen SLon-1500 VPHGMS magnetic separators have been installed in the 4 million ton iron ore processing line of the Meishan Iron mine.

References

[1] Xiong Dahe. New development of SLon vertical ring and pulsating HGMS separator[J]. Magnetic and Electrical Separation, 1994(5):211-222.

[2] Xiong Dahe. Development and commercial test of SLon-2000 vertical ring and pulsating high gradient magnetic separator[J]. Magnetic and Electrical Separation,1997(8):89-100.

[3] Xiong Dahe. Development and applications of SLon vertical ring and pulsating HGMS separators[C]//Proceedings of the XX IMPC-Aachen,1997:21-26.

[4] Xiong Dahe, Liu Shuyi, Chen Jin. New technology of pulsating high gradient magnetic separation[J]. Miner Process,1998(54):111-127.

[5] Xiong Dahe. Research and commercialization of treatment of fine ilmenite with SLon magnetic separators[J]. Magnetic and Electrical Separation,2000(10):121-130.

写作背景 本文详细介绍了SLon立环脉动高梯度磁选机在南京梅山铁矿应用于氧化铁矿提质降磷的工业试验。工业试验共有两个选矿流程，即弱磁选—浮选流程和弱磁选—SLon强磁选流程，通过试验表明：弱磁选—SLon强磁选流程具有明显的优越性，最终在大规模生产中采用了这一流程。

SLon 磁选机分选东鞍山氧化铁矿石的应用

熊大和

(赣州立环磁电高技术有限责任公司)

摘 要 在东鞍山烧结厂选矿一车间的重选—强磁—反浮选新流程中，10 台 SLon-1750 立环脉动高梯度磁选机用于控制细粒级尾矿品位，另 10 台 SLon-1750 立环脉动中磁机用于控制螺旋溜槽尾矿品位。全流程的铁精矿品位从改造前的 60%左右提高到 64%~65%，铁回收率保持在 70%左右。SLon 磁选机为提高东鞍山难选氧化铁矿的选矿指标作出了重要的贡献。

关键词 鞍山式氧化铁矿 SLon 立环脉动高梯度磁选机 反浮选 技术改造

Applications of SLon Magnetic Separators in Processing Donganshan Oxidized Iron Ore

Xiong Dahe

(*Ganzhou Vertical Ring Magnetic-electrical Equipment Hi-tech Co., Ltd.*)

Abstract In the new flowsheet of gravity separation-high intensity magnetic separation-reverse flotation of No. 1 Workshop of Donganshan Sintering Plant, ten SLon-1750 vertical ring pulsating high gradient magnetic separators were used to control the grade of fine tailings while the other ten SLon-1750 medium intensity ones to control the grade of spiral tailings. As a result, the grade of iron concentrate of the new flowsheet has been raised from about 60% of the previous one to 64%-65%, with the iron recovery kept at about 70%. SLon magnetic separators have made significant contribution to the improvement of the beneficiation performances of Donganshan refractory oxidized iron ore.

Keywords Anshan type oxidized iron ore; SLon vertical ring pulsating high gradient magnetic separator; Reverse flotation; Technical transformation

东鞍山烧结厂是鞍钢铁精矿原料基地之一，其铁矿石结晶粒度细，氧化程度较深，是鞍山地区最难选的氧化铁矿。东烧一选车间年处理氧化铁矿 470 万吨，长期采用 2 段连续磨矿、用氧化石蜡皂和塔尔油混合药剂作捕收剂的单一碱性正浮选工艺流程，铁精矿品位一直徘徊在 60%左右。由于铁精矿品位较低，随着鞍钢对精矿质量要求不断提高，东烧一选车间曾一度面临停产的严峻形势。

为了提高鞍山式氧化铁矿选矿指标，多年来国内外许多科研院所进行了广泛的研究和试验。前几年鞍钢调军台选矿厂强磁反浮选工艺流程的应用及齐大山选矿厂阶段磨矿—重

---

① 原载于《金属矿山》，2003 (6)：21~24。

选—强磁—反浮选工艺流程的技改获得了巨大的成功,目前这2个厂的选矿指标稳定在铁精矿品位67.2%、铁回收率75%左右(给矿品位30%左右),创我国氧化铁矿选矿工业生产的历史最高水平。

借鉴调军台、齐大山的成功经验,鞍钢集团鞍山矿业公司组织多个科研院所对东鞍山氧化铁矿进行了一系列研究。在此基础上,制定了重选—强磁—反浮选和强磁—重选—反浮选的试验方案,并按这2种方案进行了详细全面的连选试验和工业试验。根据试验结果,东鞍山烧结厂一选车间按重选—强磁—反浮选流程进行了全面技术改造。改造后的初步生产实践表明,东鞍山氧化铁矿的选矿指标得到了显著提高,而SLon立环脉动高梯度磁选机在其中发挥了重要作用。

1 实验室连选试验

连选试验由鞍山矿业公司研究所完成,进行了多种工艺流程的试验。其中强磁选作业采用SLon-750立环脉动高梯度磁选机。

1.1　2段连续磨矿、中矿再磨、重选—强磁—反浮选流程连选试验

该连选试验数质量流程如图1所示。

由于东鞍山矿石嵌布粒度较细,因此采用2段连续磨矿至小于74μm(200目)占75%左右,用旋流器分级。粗粒级进螺旋溜槽分选得出一部分合格铁精矿;细粒级先用弱磁机选出磁铁矿,弱磁尾矿用强磁机(SLon立环脉动高梯度磁选机)选出赤铁矿、褐铁矿等弱磁性铁矿。弱磁选精矿和强磁选精矿合并、浓缩后进入反浮选作业精选。浮选精矿和重选精矿合并后为最终精矿,浮选尾矿和强磁选尾矿合并为最终尾矿。该流程选矿指标为:给矿品位32.02%,铁精矿品位64.30%,尾矿品位12.90%,铁回收率74.70%。

该流程的优点为:螺旋溜槽可拿出部分粒度较粗的铁精矿,强磁机可抛出产率为44.87%(占尾矿量的71.45%)的低品位尾矿,为控制全流程的尾矿品位及为浮选作业创造良好条件起到了重要的作用。

该流程的缺点是螺旋溜槽中矿循环量较大,返回旋流器的中矿循环产率达到100.35%。

1.2　2段连磨、中矿返2次磨矿、重选—强磁—反浮选流程连选试验

为了减少螺旋溜槽中矿循环量,该流程增加了中磁机抛尾作业,中磁机抛去一部分粗粒尾矿,使中矿的循环产率下降至86.73%,详见图2。该流程选矿指标为:给矿品位32.87%,精矿品位63.80%,尾矿品位13.24%,铁回收率75.36%。

1.3　其他流程试验

鞍山矿业公司研究所还进行了先进行强磁选的强磁—重选—反浮选等几种流程试验,其中有中矿进第3段磨矿或中矿返回至2段磨矿等措施。均获得了全流程铁精矿品位64%左右、铁回收率75%左右的良好指标。

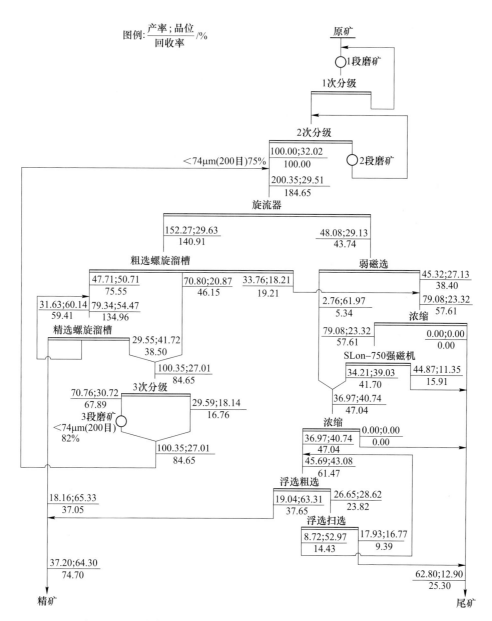

图 1 2 段连续磨矿、中矿再磨、重选—强磁—反浮选连选试验流程

2 工业试验

为了给工业改造提供可靠的技术依据，鞍山矿业公司于 2001 年投资了 600 余万元在东烧一选车间 12 号系统进行了几个工艺流程的工业试验。工业试验中强磁机采用 SLon-1750 立环脉动高梯度磁选机，中磁机最初采用滚筒式永磁中磁机（滚筒表面磁感应强度 0.6T），但因其不能正常工作而中止。

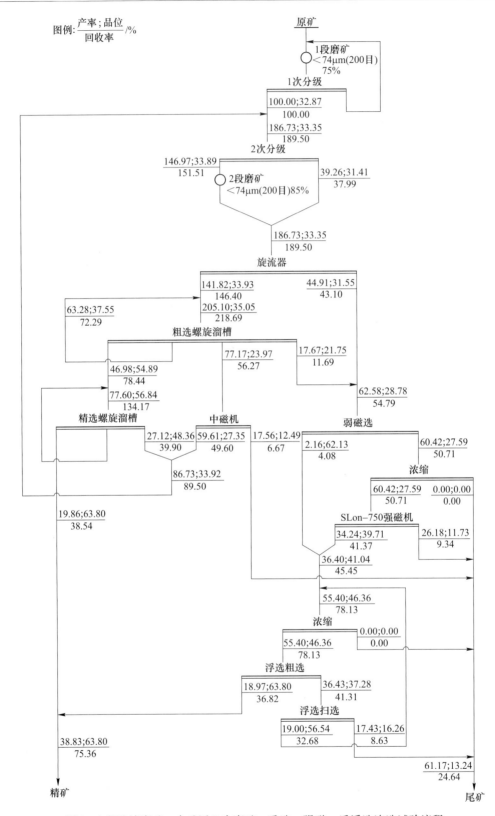

图2 2段连续磨矿、中矿返2次磨矿、重选—强磁—反浮选连选试验流程

2.1 2段连续磨矿、重选—强磁—反浮选流程工业试验

由于生产现场不具备第3段的磨矿条件，且因滚筒式中磁机存在着精矿卸矿困难和尾矿品位偏高的问题而中止使用，本工业试验流程中的螺旋溜槽中矿返回至2次分级的旋流器，试验流程见图3。

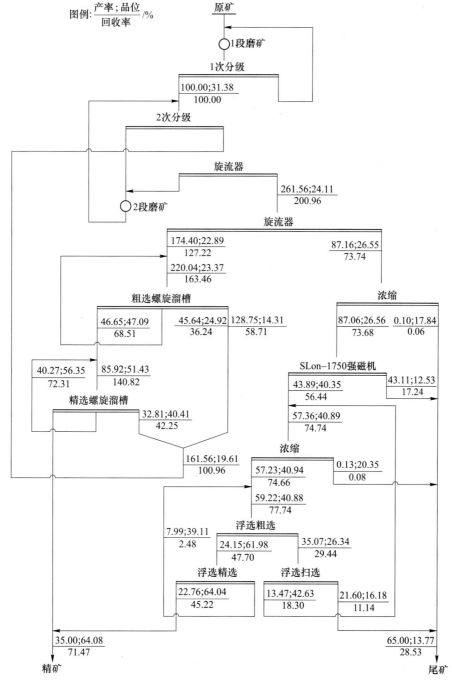

图 3 2段连续磨矿、重选—强磁—反浮选工业试验流程

该流程获得的选矿指标为:给矿品位 31.38%,精矿品位 64.08%,尾矿品位 13.77%,铁回收率 71.47%,与如图 1 所示的连选试验指标相近。

但是,由于该流程螺旋溜槽中矿没有用中磁机抛尾,造成中矿循环产率高达 161.56%,2 段磨矿负荷很重,整个系统处理能力仅达到 40t/h 左右(原系统处理能力为 55t/h)。

2.2 2段连磨、中矿返2次分级、强磁—重选—反浮选流程工业试验

试验流程如图 4 所示。该流程中采用了 2 台 SLon-1750 立环脉动高梯度磁选机,系统

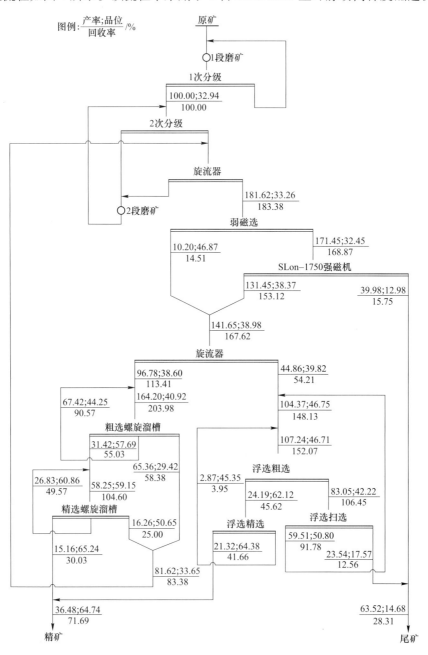

图 4 2 段连续磨矿、重选—强磁—反浮选工业试验流程

处理能力为42t/h。2台SLon-1750磁选机的给矿量合计为72t/h。

该流程的特点为：用2段磨矿将矿石磨至小于74μm（200目）占70%左右，紧接着用弱磁机和强磁机抛去40%左右的尾矿，弱磁选精矿和强磁选精矿合并后用旋流器分级。较粗的粒级用螺旋溜槽选出部分粒度较粗的铁精矿，细粒级用反浮选选出细粒铁精矿。该流程达到的选矿指标为：给矿品位32.94%，精矿品位64.74%，尾矿品位14.68%，铁回收率71.69%。

3 SLon磁选机在工业生产中的应用

在上述连选试验和工业试验的基础上，鞍山矿业公司于2002年对东烧一选车间12个生产系统进行了全面的技术改造，新建设的选矿流程和设计指标如图5所示。

新流程中采用了10台SLon-1750立环脉动高梯度磁选机（额定背景磁场1.0T）作为强磁机控制细粒级尾矿品位；另外采用了10台SLon-1750立环脉动高梯度中磁机（额定背景磁场0.6T）分选螺旋溜槽尾矿；中磁机的精矿和精选螺旋溜槽的尾矿合并分级后，粗粒部分进第3段球磨机，磨矿后返回流程中分选。

由于SLon-1750中磁机的应用，可及时抛出产率为22.84%的粗粒尾矿，使中矿循环量大幅度减少。据测算，粗粒尾矿抛出量与循环量减少的量的比例为1:3的关系。比较图5和图3可知：中矿循环产率由161.56%降低至90.00%。2003年1~3月份的生产运行表明，中矿循环量显著减少，生产系统的矿石处理能力与工业试验时的40t/h比较有大幅度提高，每个生产系统的矿石处理能力恢复到改造前的水平，即55t/h左右。

新生产流程所达到的月平均指标为：给矿品位32%~33%，铁精矿品位64%~65%，尾矿品位15%~16%，铁回收率70%左右。目前生产流程还在进一步调试中。

4 结语

（1）通过多年的探索试验、连选试验和工业试验，东鞍山难选氧化铁矿的选矿工艺逐渐发展为目前的2段连续磨矿、中矿再磨、重选—强磁—反浮选生产流程。通过大规模的技术改造，东鞍山氧化铁矿选矿指标显著提高，铁精矿品位从60%左右提高到64%~65%，铁回收率达到70%左右，目前生产流程还在进一步调试之中，选矿指标有望进一步提高。

（2）10台SLon-1750立环脉动高梯度磁选机用于控制细粒级尾矿品位，为提高全流程的回收率和为浮选创造良好的作业条件起到了重要作用。

（3）10台SLon-1750立环脉动高梯度中磁机用于控制螺旋溜槽尾矿品位，提前抛出部分粗粒尾矿，为大幅度减少中矿循环量和提高全系统的生产能力发挥了重要作用，全流程的中矿循环量由161.56%降低至90%以下，每个系统的生产能力由工业试验的40t/h提高到55t/h左右。

（4）SLon-1750立环脉动高梯度磁选机及SLon-1750立环脉动高梯度中磁机在东烧一选车间的成功应用，为提高我国难选氧化铁矿的选矿技术水平又作出了新的贡献。

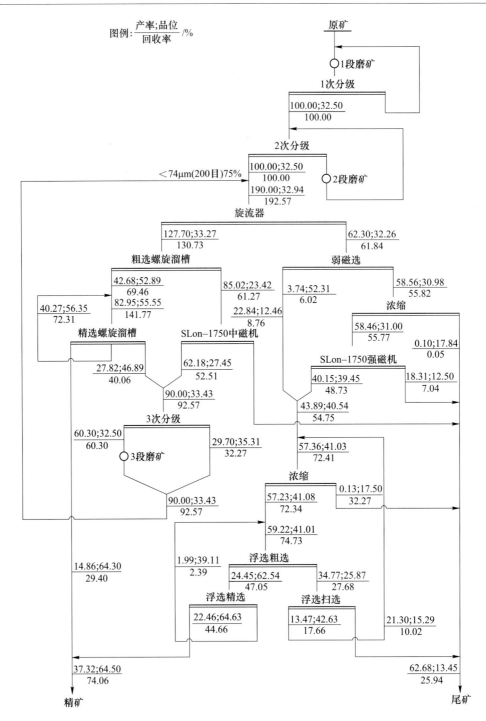

图 5 2 段连续磨矿、中矿再磨、重选—强磁—反浮选生产流程

参 考 文 献

[1] 刘动. 反浮选应用于铁精矿提铁降硅的现状及展望 [J]. 金属矿山, 2003（2）: 38~42.
[2] 熊大和. SLon 型磁选机在齐大山选矿厂的应用 [J]. 金属矿山, 2002（4）: 42~44.

写作背景 东鞍山氧化铁矿是鞍山地区最难选的氧化铁矿,鞍钢东鞍山烧结厂过去采用连续磨矿—浮选流程,铁精矿品位仅能达到60%左右。此文详细介绍了SLon磁选机分选东鞍山氧化铁矿的小型试验、连选试验、工业试验及工业应用。SLon-1750强磁机和SLon-1750中磁机及新流程的应用。使鞍钢东鞍山烧结厂的铁精矿品位达到了64.5%以上。

SLon 立环脉动高梯度磁选机选别鞍山红矿的研究与应用[❶]

熊大和

(赣州立环磁电设备高技术有限责任公司)

摘 要 SLon 立环脉动高梯度磁选机具有富集比大、选矿效率高、适应性强、磁介质不易堵塞、设备作业率高的优点。近年对 SLon-1500、SLon-1750、SLon-2000 等机型做了多方面的改进,并新研制出处理量较大的 SLon-1750 和 SLon-1500 中磁机。这些磁选机在鞍山式红矿的选矿方面取得了成功,促进了我国细粒弱磁性矿石的选矿技术水平的提高。

关键词 高梯度磁选机 红铁矿 弱磁性矿石

1 SLon 立环脉动高梯度磁选机的改进

SLon 立环脉动高梯度磁选机(以下简称 SLon 强磁机)是一种新型高效的强磁选设备,目前已广泛应用于我国的红铁矿、钛铁矿的选矿和非金属矿的提纯。为了进一步提高 SLon 强磁机选矿指标和设备工作可靠性,近年对该设备又进行了多方面的改进。

(1) 在分析磁力、脉动流体力、重力在分选过程中的相互作用机理的基础上优化磁系结构,合理布置分选空间,可获得优良的选矿指标。

(2) 根据选矿原理采用计算机精确定位使磁介质做到了最佳有序排列组合,有利于获得优良的选矿指标,并做到磁介质长期不堵塞。

(3) 采用优质导线和优质的绝缘材料绕制激磁线圈,其耗电低、寿命长。

(4) 采用耐磨材料制造脉动冲程箱,数值式调节冲程的偏心连杆机构,使脉动冲程箱具有高效耐磨、使用寿命长、易于调节的优点。

(5) 研制了数值式整流控制电路,提高了整机的控制精度,并且可根据矿石性质的波动而方便调节激磁电流。

2 SLon-1500 和 SLon-1750 立环脉动高梯度中磁机的研制

在东鞍山和齐大山的阶段磨矿—重选—强磁—反浮选的分选鞍山式红铁矿的流程中,螺旋溜槽尾矿需要用中磁机扫选。中磁机尾矿作为最终尾矿抛弃,中磁机精矿返回流程再磨再选。东鞍山的工业试验和齐大山的生产实践证明,用钕铁硼永磁块制成的滚筒式中磁机(表面场强 0.4~0.6T)抛尾存在着尾矿品位偏高的问题。为此根据生产的需要新研制了 SLon-1500 和 SLon-1750 立环脉动高梯度中磁机(以下简称 SLon 中磁机)。这种中磁机作为鞍山式红铁矿粗粒级的抛尾设备具有处理量大、尾矿品位低、可调性好、工作稳定、性能价格比较高的优点。SLon 中磁机、SLon 强磁机和钕铁硼滚筒永磁机性能比较见表 1。

[❶] 原载于《中国矿山可持续发展与矿业创新科技大会论文集》,2003。

表1 SLon中磁机、SLon强磁机和钕铁硼滚筒永磁机性能比较（处理鞍山式红矿）

机型	SLon-1500		SLon-1750		$\phi1050mm\times2400mm$ 钕铁硼永磁机
	中磁机	强磁机	中磁机	强磁机	
额定磁感应强度/T	0.4	1.0	0.6	1.0	0.4~0.6（滚筒表面）
额定激磁功率/kW	16	44	38	62	0
干矿处理量/台·h^{-1}	30~50	20~35	40~70	30~50	30~50
给矿粒度（<74μm（200目））/%	20~50	50~95	20~50	50~95	20~50

由表1可见，SLon中磁机与SLon强磁机比较，前者处理量大（高40%左右），但因其场强较低而只能处理粒度较粗的给矿。SLon中磁机与钕铁硼滚筒永磁机比较，前者具有磁场梯度大、尾矿品位低的优点，而处理每吨矿石的激磁功耗仅0.2kW·h电（齐大山）和0.7 kW·h电（东鞍山）。因此，SLon中磁机用于处理粒度粗的氧化铁矿、钛铁矿等中等磁性的矿石具有一定的技术经济优势。

3 SLon磁选机在分选鞍山式红矿中应用

通过多年的选矿试验研究、现场的工业试验及设备改进，SLon磁选机显示出了明显的技术经济优势，近年在鞍山式红矿的选矿工业中取得了成功应用。

3.1 SLon-1750强磁机应用于齐大山选厂二选车间的技术改造

齐大山选矿厂二选车间原采用煤气焙烧法处理25~75mm粒级的鞍山式氧化矿，矿石还原成磁铁矿后再磨矿并用弱磁选机分选。但该生产工艺流程存在能耗大、生产成本高、污染严重且铁精矿品位在63%左右难以提高的缺点。为了提高技术经济指标，解决污染问题，鞍山矿业公司决定采用阶段磨矿、重选、强磁反浮选的流程对原焙烧磁选流程进行全面的技术改造。鞍山矿业公司采用6台SLon-1750强磁机用于齐大山选厂二选车间的技术改造，该工程已于2001年11月底竣工。新流程中，SLon磁选机有效地将细粒级尾矿品位控制在7%~9%，全流程铁精矿品位由63%提高到67.2%，铁回收率达到75%~78%，创我国红矿选矿的历史最高水平。新流程具有选矿指标好、生产稳定、成本低、污染小的优点。其流程如图1所示。

3.2 SLon-1500中磁机及SLon-1750强磁机在齐大山选厂一车间的工业试验与应用

齐大山选厂一选车间采用阶段磨矿、重选、强磁反浮选的流程处理0~25mm的氧化铁矿，原来采用5台φ2000mm平环式强磁选机作业为细粒铁矿的强磁抛尾设备。由于平环强磁选机的齿板易堵塞、尾矿品位偏高、设备作业率较低，该车间于2002年采用了4台SLon-1750强磁机取代了原有5台φ2000mm平环式强磁选机。4台SLon-1750强磁机有效地将细粒级尾矿品位控制在7%~9%。在作业铁精矿品位提高3.83%的基础上，其尾矿品位降低3.15%，作业回收率提高19.26%，使全流程的尾矿降低1%以上，全流程铁回收率提高2.74%。齐大山选厂一选车间生产流程如图2所示。

如图2（a）所示流程中，用于扫选螺旋溜槽尾矿的钕铁硼中磁机存在着尾矿品位较高的缺点。为了进一步降低尾矿品位，齐大山选矿厂已采用两台SLon-1500中磁机进行工业试验，如图2（b）所示。工业试验表明，SLon-1500中磁机取代钕硼中磁机可将作业尾矿品位降低1.7%。

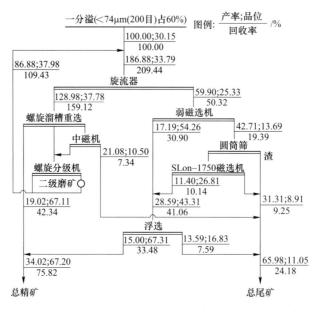

图 1　齐大山选厂新建二选车间流程图

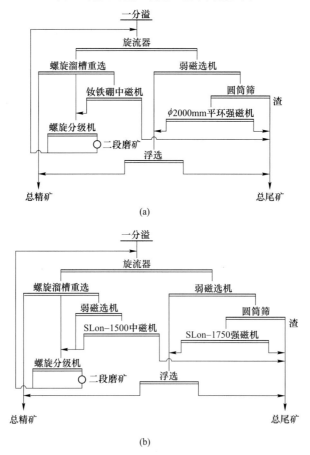

图 2　齐大山选厂一选车间流程图
（a）原流程图；（b）现流程图

齐大山选矿厂一车间目前的生产指标已达到：给矿品位 30% 左右，铁精矿品位 67.2%，尾矿品位 11.5%，铁回收率 74%~75% 的水平。若全面推广 SLon-1500 中磁机，其全流程回收率可望进一步提高。

3.3 SLon 磁选机在东鞍山选矿厂的应用

东鞍山烧结厂是鞍钢铁精矿原料基地之一，其铁矿石结晶粒度细，氧化程度较深，是鞍山地区最难选的红矿。东鞍山烧结厂一选车间多年来一直采用两段连续磨矿，用氧化石蜡皂和塔尔油混合药剂作捕收剂的单一碱性正浮选工艺流程，铁精矿品位一直徘徊在 60% 左右。由于铁精矿品位较低，鞍钢鞍山矿业公司决定采用阶段磨矿、重选、强磁反浮选工艺流程对东鞍山烧结厂一选车间进行全面的技术改造，新建设的选矿流程和设计指标如图 3 所示。

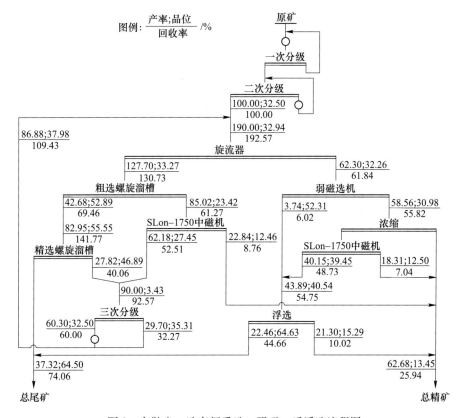

图 3　东鞍山一选车间重选—强磁—反浮选流程图

该流程中，采用 10 台 SLon-1750 强磁机承担脱泥抛尾的重要作业，该设备一次选别将尾矿品位控制在 12.5% 左右，保证了全流程获得较高的回收率，并为浮选作业获得较高的精矿品位和减少浮选药剂用量创造了良好的条件。

该流程中还采用 10 台 SLon-1750 中磁机扫选螺旋溜槽尾矿，使粗粒尾矿品位控制在 12.5% 左右，提前抛出部分粗粒尾矿，为大幅度减少中矿循环量和提高全系统的生产能力发挥了重要作用。

东鞍山一选车间的技改流程已于 2002 年底完成，综合铁精矿品位已从过去的 60% 左右提高到目前的 64.5%，铁回收率达到 70% 左右。目前生产流程尚在调试之中。

4 结论

（1）近年采用最新的先进技术对 SLon-1500、SLon-1750 和 SLon-2000 立环脉动高梯度磁选机进行了多方面的改进，使它们具有更优良的选矿性能和工作可靠性，目前已成为我国工业生产中广泛应用的新一代强磁选设备。

（2）针对鞍山式红矿阶段磨矿、重选、强磁反浮选流程中粗粒抛尾的生产需要，研制了 SLon-1500 和 SLon-1750 立环脉动高梯度中磁机，它们的额定背景磁感强度为 0.4～0.6T，设备处理量增加 40% 左右，具有良好的性能价格比。

（3）10 台 SLon-1750 立环脉动高梯度磁选机在齐大山选矿厂一车间和二选车间的应用，有效地将细粒级红矿的尾矿品位控制在 7%～9%，为齐大山选厂全厂实现精矿品位平均达到 67.2% 和回收率达到 75% 的历史最高水平提供了一种先进的生产设备。

（4）10 台 SLon-1750 强磁机和 10 台 SLon-1750 中磁机在东鞍山一选车间技改流程中应用成功，为该车间铁精矿品位 60% 提高到 64.5% 创造了良好的条件。

参 考 文 献

[1] 熊大和．SLon 型磁选机在齐大山选矿厂的应用口［J］．金属矿山，2002（4）：42~44．
[2] 徐冬林，等．SLon 立环脉动高梯度强磁选机在齐大山选矿厂试验与应用［J］．金属矿山（增刊），2002（9）：242~244．

写作背景　本文介绍了 SLon 立环脉动高梯度磁选机在鞍钢齐大山选矿厂和东鞍山烧结厂的应用。其中 SLon-1500 中磁机和 SLon-1750 中磁机是专门为取代滚筒永磁中磁机而研制的新设备，在齐大山选矿厂一选车间和东烧结厂得到良好的应用。

SLon 磁选机在钛铁矿选矿工业中的研究与应用

熊大和

（赣州立环磁电设备高技术有限责任公司）

摘　要　SLon 立环脉动高梯度磁选机具有富集比大、选矿效率高、适应性强、磁介质不易堵塞、设备作业率高的优点。近年该磁选机在攀钢选钛厂、攀钢综合回收厂和重钢太和铁矿的钛铁矿选矿工业中取得了成功的应用，促进了我国微细粒级钛铁矿选矿工业的快速发展。

关键词　SLon 立环脉动高梯度磁选机　钛铁矿　微细粒级

Research and Application of SLon Vertical Ring Pulsating High Gradient Magnetic Separators in Ilmenite Processing Industry

Xiong Dahe

(Ganzhou Vertical Ring Magnetic-electric Equipment Hi-tech Co., Ltd.)

Abstract　SLon vertical ring pulsating high gradient magnetic separators possess the advantages of large concentration ratio, high separation efficiency, strong adaptability, magnetic matrix not easy to be blocked and high equipment availability. In recent years, these magnetic separators have been successfully applied in beneficiating the ilmenite ore at the Panzhihua Ilmemte concentrator, Panzhihua Comprehensive Recovering Plant and Taihe Iron Mine of Chong Steel, promoting the rapid development of fine ilmenite processing industry in China.

Keywords　SLon vertical ring pulsating high gradient magnetic separator; Ilmenite; Fine fraction

1　简介

SLon 立环脉动高梯度磁选机（简称 SLon 磁选机，下同）利用磁力、脉动流体力和重力的综合力场选矿，适用于黑色、有色金属弱磁性矿的选矿和非金属矿的提纯。它采用半封闭式铁轭和水内冷空心矩形线圈组成电磁磁系；导磁不锈钢板网或圆棒为磁介质；分选环立式旋转，转环下部选矿，上部冲洗磁性产品，冲洗磁性产品的方向与给矿方向相反，粗颗粒不必穿过磁介质便可容易地冲洗出来；选矿区下部配有脉动机构，驱动分选区矿浆脉动，使磁介质堆中的矿粒群始终保持松散状态。该机具有适应性强、分选粒度范围宽、选矿效率高、

---

❶ 原载于《全国破碎、磨矿及选别设备学术研讨与技术交流会论文集》，2003。

磁系工作可靠寿命长、脉动冲程箱摩擦阻力小和易于调节冲程、磁介质不堵塞、耗电量低、设备运转率高达99%的优点。该机可有效地回收细粒红铁矿、钛铁矿及用于非金属矿提纯工业，具有对细粒弱磁性矿物回收率高、精矿品位高的优异选别性能。

2 SLon磁选机在钛铁矿选矿工业中的应用

在我国攀西地区蕴藏着占世界储量35%，占我国储量91.78%的钛资源。合理开发利用攀西地区钛资源，对发展我国钛工业具有重大的意义。

由于SLon立环脉动高梯度磁选机具有选矿效率高、富集比较大、脱泥效果好、工作性能稳定的优点，近年该机在选钛工业中的应用得到了较快的发展。

2.1 SLon磁选机在攀枝花选钛厂用于回收细粒钛铁矿

攀枝花是我国最大的钛铁矿产地，攀钢（集团）矿业公司选钛厂是一座大型原生钛铁矿选厂，入选原料（原矿）为选铁厂所排出的磁选尾矿。原来生产上采用重选—浮选—电选工艺流程，回收粗粒级（>0.1mm）、细粒级（0.1~0.045mm）两级别钛铁矿，而对产率和TiO_2金属量均占入选原料45%的小于0.045mm微细粒级钛铁矿难于回收而直接丢弃，导致其全厂钛回收率仅有17%~20%。

为了提高选钛回收率，1997~2001年该厂建成了磁选浮选流程分选小于0.045mm粒级的钛铁矿。6台SLon-1500磁选机用于粗选作业，然后用浮选精选。磁浮流程的综合选矿指标为给矿品位10.12%TiO_2，钛精矿品位不小于47%TiO_2，TiO_2回收率44%。该流程2002年从细粒尾矿中回收了7.6万吨品位为47.5%TiO_2的细粒钛精矿。TiO_2的综合回收率提高了10%~12%。第一条工业试验生产线的数质量流程见图1（注：该厂第二条生产线即4万吨/a钛精矿生产车间取消了ϕ125mm水力旋流器脱泥作业）。

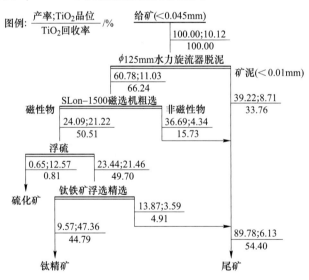

图1 攀枝花选钛厂小于0.045mm钛铁矿选矿流程

由于经济效益显著，2003年攀枝花钛业公司又订购了5台SLon-1500磁选机建设1座设计规模为年产6万吨微细粒级钛精矿的新车间。

2.2 SLon 磁选机用于从攀钢总尾矿中回收钛铁矿

攀钢综合回收厂为攀钢附属集体企业，主要从攀钢排放的总尾矿中回收铁精矿和钛精矿。原选钛流程采用螺旋溜槽粗选，摇床精选的单一重选流程。由于螺旋溜槽和摇床都不能回收细粒级，因此选钛回收率很低。2001~2002 年，该厂相继采用了 3 台 SLon-1500 磁选机建成了如图 2 所示的选钛流程。

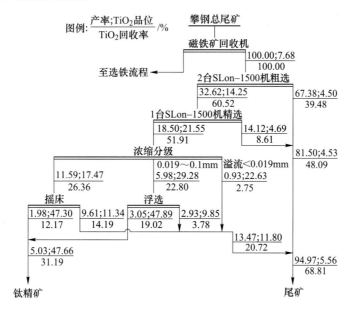

图 2 攀枝花综合厂选钛流程

该流程的特点为用 2 台 SLon-1500 磁选机尽可能多地处理攀钢总尾矿，每台 SLon-1500 磁选机处理量为 45t/(台·h)，2 台共处理 90t/(台·h)；用 1 台 SLon-1500 磁选机精选。这 3 台 SLon-1500 磁选机使 TiO_2 品位从 7.68% 提高到 21.55%，而且抛弃了产率为 81.50% 的最终尾矿，它们为后续的浮选作业和重选作业创造了非常好的精选条件。该流程具有处理量大、钛精矿品位高、生产成本低的优点。目前该流程已调试成功，具备了年产约 3 万吨优质钛精矿的生产能力，TiO_2 回收率达到原来单一重选流程的 4 倍以上。

2.3 SLon 磁选机应用于太和铁矿选钛流程

重庆钢铁公司太和铁矿的矿石为磁铁矿与钛铁矿的混合矿。1994 年以前，该矿建立了弱磁选流程分选磁铁矿。1994~1995 年，该矿又建立了以螺旋溜槽、摇床为粗选和浮选为精选的钛铁矿选矿流程从弱磁选尾矿中回收钛铁矿。但是，由于螺旋溜槽和摇床都不能回收细粒钛铁矿，而浮选又难以回收粗粒钛铁矿，TiO_2 从细粒和粗粒中大量流失。因此选钛回收率很低，仅占弱磁选尾矿中 TiO_2 的 10%。

2000 年，该矿安装了 3 台 SLon-1500 磁选机，其中 2 台取代螺旋溜槽，1 台取代摇床，这 3 台磁选机与原有的浮选作业组成磁浮流程分选钛铁矿。该流程首先采用浓缩分级将给矿大致分为大于 0.1mm（粗粒级）、0.1~0.02mm（细粒级）和小于 0.02mm（溢流），粗粒级和细粒级分别各用 1 台 SLon-1500 磁选机粗选，粗粒级磁选精矿再磨后与细粒级精矿合并，用永磁中磁机除去磁铁矿等强磁性物质后再用 1 台 SLon-1500 磁选机精选,其磁选精矿再进入浮选精选。该流程如图 3 所示,其特点主要包括:

(1) 粗、细分级磁选可采用不同的磁选参数以获得最佳的选矿指标。
(2) 浮选给矿品位较高,有利于节约浮选药剂降低生产成本。
(3) 浮选尾矿返回流程中形成闭路,有利于提高全流程的 TiO_2 回收率。

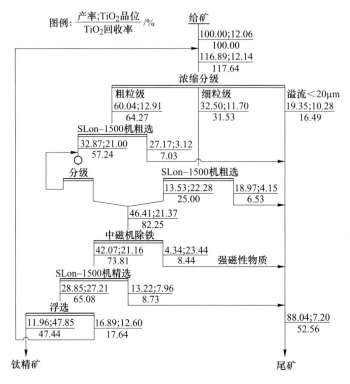

图 3　太和铁矿选钛流程

该流程有效地回收了各粒级的钛铁矿,新流程在保证钛精矿品位不小于 47% 前提下,使 TiO_2 的综合回收率提高至 47% 左右。目前,太和铁矿的钛精矿产量已由改造前的约 2000t/a 提高到 2 万吨/a 的能力(注:该矿每年正常开机时间为 6~7 个月)。

3　结论

(1) 新型高效 SLon 型磁选机具有富集比大、选矿效率高、磁介质不堵塞、设备处理量大、运转稳定、作业率高的优点,近年在我国的选钛工业中得到了成功应用。

(2) 6 台 SLon-1500 磁选机在攀枝花选钛厂用于小于 0.045mm 微细粒级钛铁矿的磁浮选矿流程中,该机有效地回收了这部分微细粒级钛铁矿,为浮选作业创造了良好的条件。全流程在给矿品位 10% 左右的条件下,实现钛精矿品位大于 47% TiO_2,TiO_2 回收率 44% 的良好指标。2002 年该厂生产品位为 47.5% TiO_2 的微细粒级钛精矿 7.6 万吨,该厂微细粒级钛精矿产量可望在 1~2 年内实现 15 万吨/a 左右。

(3) 3 台 SLon-1500 磁选机在攀枝花综合回收厂用于从攀钢总尾矿中回收钛精矿,其中 2 台粗选和 1 台精选,磁选精矿分级后细粒级用浮选精选,粗粒级用摇床精选,可从全粒级中回收钛精矿。该流程的选钛指标为给矿品位 7.68%TiO_2,钛精矿品位 47.66%TiO_2,TiO_2 回收率 31.19%,已具备年产 3 万吨左右优质钛精矿的能力。

(4) 3 台 SLon-1500 磁选机在重钢太和铁矿回收钛铁矿,采用分级磁选、粗粒再磨、磁选和浮选精选的流程,该流程的选钛指标为给矿品位 12.06%TiO_2,钛精矿品位 47.85%TiO_2,TiO_2 回收率 47.44%,已具备半年生产 2 万吨左右优质钛精矿的能力。

(5) SLon 磁选机在钛铁矿选矿工业中的成功应用,为解决我国微细粒级钛铁矿选矿技术难题,大幅度提高钛铁矿的回收率和降低选钛生产成本发挥了重要的作用。

写作背景 过去我国钛铁矿选矿回收率很低。SLon 立环脉动高梯度磁选机在钛铁矿选矿中应用后,钛铁矿选矿回收率大幅度提高。此文介绍了 SLon 磁选机在攀钢选钛厂、攀钢综合回收厂、重钢太和铁矿分选钛铁矿的工业试验、生产流程和选钛指标。

SLon Magnetic Separator Applied to Upgrading the Iron Concentrate[①]

Xiong Dahe

(*Ganzhou Nonferrous Metallurgy Research Institute*, 36 *Qingnian Rd.*, 34100, *Ganzhou*, *Jiangxi province*, *P. R. China*)

(*Received* 21 *December* 2002; *In final form* 8 *January* 2003; *Accepted* 13 *January* 2003)

Abstract SLon vertical ring and pulsating high gradient magnetic separator is a new generation of a highly efficient equipment for processing weakly magnetic minerals[1-3]. It possesses advantages of a large beneficiation ratio, high recovery, a matrix that cannot easily be blocked and excellent performance. In the technical reform of upgrading the iron concentrate in Qidashan Mineral Processing Plant of Anshan Iron and Steel Company in 2001 to 2002, ten SLon-1750 magnetic separators were successfully applied to process oxidized iron ores. The iron concentrate of the plant was upgraded from 63.22%Fe to 67.11%Fe. The overall results of the reformed flowsheet are: the feed grade 29.84%Fe, the iron concentrate grade 67.11%Fe, the tailings grade 11.27%Fe, and the iron recovery 74.79%, which set up a new historical record of the plant.

1 Introduction

Qidashan Mineral Processing Plant is a large oxidized iron ore processing base that belongs to the Anshan Iron and Steel Company, in Liaoning Province, Northeast China. It processes eight million tons of oxidized iron ore and produces 2.7 million tons of the iron concentrate annually. In previous years, the grade of the iron concentrate could reach only about 63%Fe. In order to increase the concentrate grade, extensive test work had been done during the past several years. Finally a gravity separation-high intensity magnetic separation-reverse flotation flowsheet was chosen, in which high-intensity magnetic separation is a very important process. Through comparison of several types of high-intensity magnetic separators, ten SLon-1750 Vertical ring and Pulsating High Gradient Magnetic Separators were chosen for the high-intensity magnetic separation process.

2 Analysis of the Qidashan oxidized iron ore

Qidashan iron ore is a poor iron deposit, which consists mainly of hematite, magnetite and quartz. The crystal particle sizes of iron minerals are between 0.005mm and 1.0mm, on average 0.05mm. The Quartz particle sizes are 0.085mm on average. It contains very low concentrations of harmful elements such as S and P. It is thus expected that very good iron concentrate could be obtained using an efficient mineral process. Tables 1 and 2 summarise the assay and mineralogical composition of the ore.

[①] 原载于《Physical Separation in Science and Engineering》,2003,12(2):63~69。

Table 1 Assay of Qidashan iron ore   (%)

Total Fe	SiO$_2$	Al$_2$O$_3$	CaO	MgO	P	S	K$_2$O	Na$_2$O
29.10	56.10	0.79	0.19	0.75	0.04	0.084	0.18	0.29

Table 2 Mineralogical composition of Qidashan iron ore   (%)

Mineral	Hematite	Magnetite	Mag-hematite	Silica iron	Total
Mass	17.13	4.85	4.46	2.15	28.00
Distribution	59.90	16.99	15.59	7.52	100.00

3 The previous iron ore processing flowsheet

The previous iron ore processing flowsheet is shown in Fig. 1.

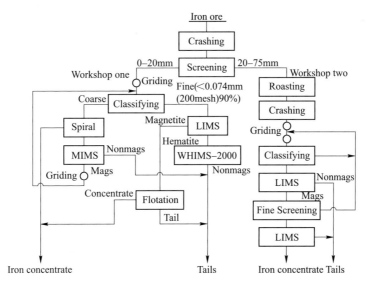

Fig. 1 The previous flowsheet of Qidashan Mineral Processing Plant
LIMS—low intensity magnetic separation; MIMS—mid intensity magnetic separation;
WHIMS—wet high intensity magnetic separation

The crushed iron ore was classified into two fractions of <20mm and 75-20mm. The <20mm fraction was treated with a spiral-high intensity magnetic-flotation flowsheet. The 75-20mm fraction was treated with a roast-low intensity magnetic separation (LIMS) flowsheet. The overall results were: the feed grade 29.86%Fe, the iron concentrate grade 63.22%Fe, the tails grade 11.92% Fe, the iron recovery 74.04%.

Problems of the previous flowsheet were that the grade of the iron concentrate could reach only about 63% and the roast process was costly, caused heavy pollution and the working conditions were difficult.

In 2000, reverse flotation was tested at the Workshop One. It demonstrated that the iron concentrate grade could reach 65%Fe.

4　The comparative test of Slon VPHGMS

In early 2001, a SLon-1500 vertical ring and pulsating high-gradient magnetic separator (SLon-1500 VPHGMS) was installed at the Workshop One. A commercial comparative test was carried out between the SLon-1500 VPHGMS, a WHIMS-2000 horizontal ring high-intensity magnetic separator and a DGYC-1860 vertical ring permanent high gradient magnetic separator. This three months long comparative test, demonstrated that SLon-1500 VPHGMS was the most efficient and most reliable magnetic separator. Comparison of the results is shown in Table 3.

Table 3　Comparison results of three types of magnetic separators

Magnetic separator	Ore throughput /t·h^{-1}	Feed grade (Fe)/%	Mags grade (Fe)/%	Tails grade (Fe)/%	Iron recovery /%	Operating ratio /%
SLon-1500	35.53	14.25	27.45	7.80	63.23	99.03
DGYC-1860	31.19	15.19	24.56	9.98	57.78	98.66
WHIMS-2000	36.01	14.33	23.62	10.95	43.97	89.50
SLon-1500 vs. DGYC-1860	+4.34	-0.94	+2.89	-2.18	+5.45	+0.37
SLon-1500 vs. WHIMS-2000	-0.48	-0.08	+3.83	-3.15	+19.26	+9.53

Table 3 shows that with SLon-1500 as compared to DGYC-1860, grade of the magnetics and the iron recovery are 2.89 and 5.45 percent higher. Comparison of SLon-1500 with WHIMS-2000, shows that its mags grade and the iron recovery are 3.83 and 19.26 percent higher. SLon-1500 VPHGMS not only achieved much better beneficiation results, but also it possesses advantages of a higher operating ratio, higher ore throughput capacity, no matrix blockage problem (matrix is always kept clean) and it is much easier to maintain than the other types of magnetic separators.

5　The flowsheet innovation

In 2001 to 2002, a technical reform was carried out in Qidashan Mineral Processing Plant. The roast LIMS flowsheet in the Workshop Two had been changed into a spiral-SLon VPHGMS-flotation flowsheet, in which six SLon-1750 VPHGMS were installed. Additional four SLon-1750 VPHGMS were installed in the Workshop One to replace the previous five WHIMS-2000 high-intensity magnetic separators. The entire direct flotation was changed into reverse flotation. The new principle flowsheet is shown in Fig. 2 and the installation of SLon separators is shown in Fig. 3.

The results of the new flowsheet as compared to the previous flowsheet are shown in Table 4.

Table 4 shows that the new flowsheet achieves very good quality iron concentrate, the grade of which reaches 67.11%, which is much higher than 63.22%, achieved by the previous flowsheet.

As the novel flowsheets of the Workshop One and the Workshop Two are similar, let us have a look at the details of the new flowsheet of the Workshop Two.

As shown in Fig. 4, the 75-20mm iron ore is crushed and ground to 60% <200 mesh by a primary ball mill and classified by a cyclone to two fractions. The coarse fraction is treated by spirals and MIMS (medium-intensity magnetic separator). The spirals remove part of the coarse iron

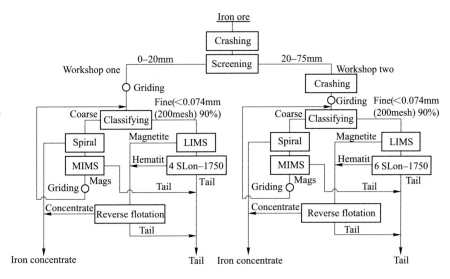

Fig. 2 The novel flowsheet of Qidashan Mineral Processing Plant

LIMS—low intensity magnetic separation; MIMS—mid intensity magnetic separation;
SLon-1750—SLon-1750 VPHGMS

Fig. 3 Ten SLon-1750 vertical ring and pulsating high-gradient magnetic
separators as applied in the new flowsheet of Qidashan Mineral Processing Plant

Table 4 Results of a new flowsheet vs previous flowsheet

Flowsheet		Feed grade (Fe)/%	Concentrate grade (Fe)/%	Tails grade (Fe)/%	Concentrate mass ratio /%	Iron recovery /%	Ore throughput /Mt·a^{-1}
New flowsheet	Workshop One	29.50	67.00	11.50	32.43	73.66	3.8
	Workshop Two	30.15	67.20	11.05	34.02	75.82	4.2
	Total	29.84	67.11	11.27	33.26	74.79	8.0
Previous flowsheet	Workshop One	29.50	63.47	12.50	33.35	71.76	3.8
	Workshop Two	30.19	63.02	11.36	36.45	76.09	4.2
	Total	29.86	63.22	11.92	34.97	74.04	8.0
New flowsheet vs previous flowsheet		-0.04	+3.89	-0.65	-1.71	+0.75	

concentrate and the MIMS discharge part of the coarse tailings. The spiral middlings and the mags of MIMS are sent to the secondary ball mill and then returned to the cyclone feed. The fine fraction (90% <200mesh) is treated by LIMS (low-intensity magnetic separator) to concentrate magnetite, then by six SLon-1750 VPHGMS concentrate hematite and discharge part of the fine tailings. The mags of LIMS and SLon-1750 are mixed and further cleaned by reverse flotation. The spiral concentrate and flotation concentrate are joined as the final concentrate. The MIMS tailings, SLon-1750 tailings and flotation tailings are joined as the final tailings.

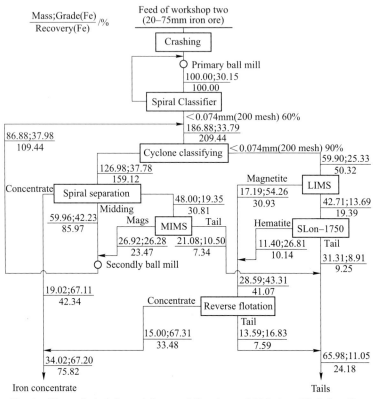

Fig. 4 The technical data of the novel flowsheet of Qidashan Workshop Two

Advantages of this flowsheet are the following: the spiral takes out most of the liberated coarse iron concentrate and the MIMS discharges part of the coarse tailings, which greatly reduces the feed into the secondary ball mill (10 primary ball mills to 3 secondly ball mills). The tailings grade of 6 SLon-1750 VPHGMS only is 8.91%, the mass ratio of the tailings to the feed of SLon-1750 is 73.31% and 31.31% to the feed of the flowsheet, 47.45% to the total tailings of the flowsheet. Because the tailings grade of SLon-1750 is very low and the proportion is relatively large, SLon-1750 not only effectively controlled the fine fraction tailings, but also guaranteed the high iron recovery of the total flowsheet. Reverse flotation can achieve good quality of the fine concentrate.

6 Conclusions

Most Chinese iron ores are very poor. To get good quality iron concentrate, it is necessary to apply efficient equipment and advanced technology to treat iron ores of low grade.

Qidashan Mineral Processing Plant is a major iron ore concentrator of the Anshan Iron and Steel Company. The grade of its mined iron ore is only about 30%Fe.

Its previous oxidized iron processing flowsheet can only obtain iron concentrate with the grade of 63.22%Fe. In recent years, the plant developed a novel flowsheet consisting of spiral-SLon VPH-GMS-reverse flotation, which replaced the previous flowsheet of coal gas roasting-low-intensity magnetic separation. Its iron concentrate grade reached 67.11%Fe. The iron concentrate quality has been greatly upgraded and the processing cost has also been reduced.

Ten SLon-1750 vertical ring and pulsating high gradient magnetic separators (SLon-1750 VPH-GMS) are installed at the plant. Their successful application in the 8 million oxidized iron ore processing plant demonstrated that they are very efficient and reliable. They successfully control the tailings of the fine fraction. The iron recovery of the plant reaches 74.79%. The average iron concentrate grade and recovery are expected to reach 67.20%Fe and 76.80% respectively as the flowsheet is being improved in the coming years.

References

[1] Xiong Dahe. Development and commercial test of SLon-2000 vertical ring and pulsating high gradient magneticseparator[J]. Magn. Electr., 1997(8):89-100.

[2] Xiong Dahe, et al. New technology of pulsating high gradient magnetic separation[J]. Int. J. Miner. Process, 1998(54):111-127.

[3] Xiong Dahe. A large scale application of SLon magnetic separator in Meishan Iron Ore Mine. Magn. Electr., 2002(11):1-8.

写作背景 本文向国外读者介绍了SLon立环脉动高梯度磁选机在鞍钢齐大山选矿厂分选氧化铁矿的工业试验和生产应用。其中列出了SLon-1500磁选机与DGYC永磁立环高梯度磁选机和φ2000平环强磁选机的选矿对比指标。SLon磁选机选矿指标明显高于其他强磁选机。SLon磁选机和新流程的应用，齐大山选矿厂的选矿指标有了大幅度的提高。

SLon-2000 磁选机在调军台选矿厂的工业试验与应用[1]

熊大和[1] 张国庆[2]

(1. 赣州立环磁电设备高技术有限责任公司; 2. 鞍钢集团新钢铁齐大山铁矿)

摘 要 SLon-2000 立环脉动高梯度磁选机分选细粒弱磁性矿物,具有富集比大、回收率高、磁介质不易堵塞、设备作业率高的优点。采用 1 台 SLon-2000 磁选机在调军台选矿厂与 Shp-3200 平环强磁选机进行分选鞍山式细粒氧化铁矿的工业对比试验,当脉动冲程为 20mm 时,在给矿量和给矿性质一致的条件下,SLon-2000 磁选机铁精矿品位高 1.19%、尾矿品位低 1.56%、作业回收率高 8.19%,取得了良好的技术经济指标。

关键词 SLon-2000 立环脉动高梯度磁选机 鞍山式氧化铁矿 强磁选 工业试验

Commercial Test and Application of SLon-2000 Magnetic Separator in Diaojuntai Concentrator

Xiong Dahe[1] Zhang Guoqing[2]

(1. Ganzhou Vertical Ring Magnetic-electric Equipment Hi-tech Co., Ltd.;
2. Qidashan Iron Mine of New Iron & Steel Co., Ltd., An Steel (Group))

Abstract SLon-2000 vertical ring pulsating high gradient magnetic separators have the advantages of large enrichment ratio, high recovery, uneasy-blocked matrix and high availability when used in processing fine weakly magnetic minerals. In a comparative commercial test with Shp-3200 horizontal ring high intensity magnetic separator in processing Anshan type fine oxidized iron ore, SLon-2000 magnetic separator achieved good technical and economical performances. At a pulsating stroke of 20mm, under the conditions of the same feed rate and feed properties, SLon-2000 separator obtained an iron concentrate grade 1.19 percentage points higher, tailings grade 1.56 percentage points lower and an operation recovery 8.19 percentage points higher than those of Shp-3200 separator.

Keywords SLon-2000 vertical ring pulsating high gradient magnetic separator; Anshan type oxidized iron ore; High intensity magnetic separation; Commercial test

调军台选矿厂是鞍钢于 20 世纪 90 年代新建的一座具有现代化生产水平的大型选矿厂,现由鞍钢集团新钢铁齐大山铁矿管理,其设计规模为年处理鞍山式氧化铁矿 900 万

---

[1] 原载于《金属矿山》,2003,12:37~39。

吨。该厂于1998年3月投产，采用连续磨矿—弱磁—中磁—强磁—反浮选的选矿流程，其中强磁选采用15台Shp-3200平环强磁选机。这些平环强磁选机存在着齿板易堵塞，设备故障率较高，检修和维护较困难的问题，其设备作业率只能达到80%~90%。

SLon立环脉动高梯度磁选机是新一代高效强磁选设备，分选细粒弱磁性矿物，具有富集比大、回收率高、磁介质不易堵塞、设备作业率高的优点。鞍钢集团鞍山矿业公司齐大山选矿厂在其技改流程中已应用10台SLon-1750立环脉动高梯度磁选机[1]、1台SLon-1500立环脉动高梯度磁选机和10台SLon-1500立环脉动中磁机；该公司东鞍山烧结厂一选车间在其技改流程中已应用10台SLon-1750立环脉动高梯度磁选机和10台SLon-1750立环脉动中磁机[2]。这些设备在鞍山式氧化铁矿选矿生产中发挥了重要的作用。实践证明，SLon立环脉动高梯度磁选机各项性能均超过了平环强磁选机[3~5]。

为了提高调军台选矿厂选矿指标和解决Shp-3200平环强磁选机存在的问题，鞍钢集团新钢铁齐大山铁矿与赣州立环磁电设备高技术有限责任公司合作，开展了SLon-2000立环脉动高梯度磁选机在调军台选矿厂的工业试验。

1 选矿流程

调军台选矿厂和齐大山选矿厂的原矿均来自齐大山铁矿，为鞍山式贫氧化铁矿。原矿铁品位30%左右；主要的铁矿物有赤铁矿、磁铁矿和褐铁矿，主要脉石有石英、透闪石、阳起石、绢云母、绿泥石等，其中石英占脉石总量的80%~85%；矿石硬度系数为12~16；矿石中的硫、磷等有害元素很低。调军台选矿厂采用连续磨矿—弱磁—中磁—强磁—反浮选的选矿流程，如图1所示。

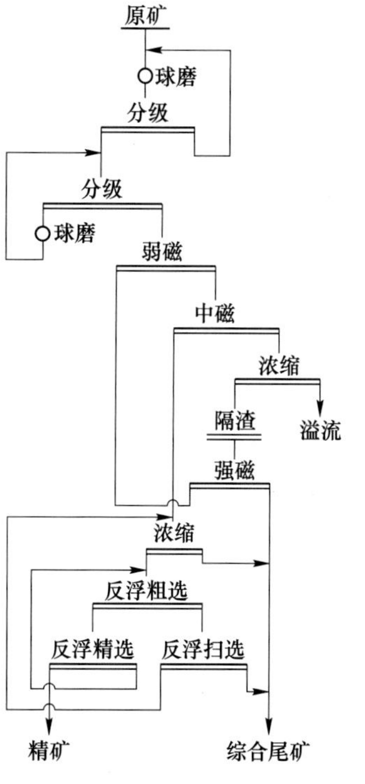

原矿经连续磨矿至小于74μm（200目）占90%以上，首先采用滚筒式弱磁选机和中磁选机选出磁铁矿和假象磁铁矿等强磁性矿物，然后采用强磁选机选出赤铁矿和褐铁矿。这3种磁选机的精矿进入反浮流程精选，由浮选获得最终精矿。

从该流程可见，强磁选作业是主要的抛尾设备，其抛弃的尾矿量占原矿量的42%左右，占全流程尾矿量的62%左右。因此强磁选机尾矿品位的高低对全流程的铁回收率具有很大的影响。

图1 调军台选矿厂原则流程

2 SLon-2000磁选机简介

SLon-2000立环脉动高梯度磁选机是由赣州有色冶金研究所设备研究室（现改制为赣州立环磁电设备高技术有限责任公司）自20世纪90年代初开始研制的高效大型强磁选设备。研制该机采用了一系列的新技术。

（1）采用120°的最佳磁系包角，扩大了磁场的有效分选空间，并有利于矿浆液位的控制。

（2）采用水内冷空心铜管绕制激磁线圈，采用多重绝缘和装备不锈钢防护外罩等措施，使其具有更可靠的防水性能和绝缘效果。

（3）采用导磁不锈钢棒为磁介质，利用计算机进行优化组合排列并加工成盒装形式，便于磁介质的安装和提高选矿指标，以及防止磁介质堵塞。

（4）采用耐磨材料制造脉动冲程箱，数值式调节冲程的偏心连杆机构，使脉动冲程箱具有高效耐磨、使用寿命长、易于调节的优点。

（5）给矿斗、精矿斗及各矿浆通道均采用优化设计，易受矿浆冲刷的地方均采用耐磨橡胶衬里或增加缓冲装置，显著延长了其使用寿命。

首台SLon-2000立环脉动高梯度磁选机于1994年4月研制出来，在鞍钢弓长岭选矿厂与Shp-2000平环强磁选机进行了6个月的工业对比试验，获得了作业铁精矿品位高7.21个百分点、尾矿品位低5.41个百分点、作业回收率高7.36个百分点、设备处理量大50%以上的优良指标[6,7]。该机在弓长岭选矿厂安全运转2年余，后因弓长岭选矿厂改氧化铁矿生产为磁铁矿生产而停止。

2001~2002年，在总结首台SLon-2000立环脉动高梯度磁选机研制成功的基础上，运用研制SLon-1750、SLon-1500、SLon-1250、SLon-1000等机型所采用的新技术，对SLon-2000磁选机又做了较大的改进，使该机的选矿性能和机电性能有了进一步提高。经改进后的SLon-2000立环脉动高梯度磁选机主要技术参数见表1。

表1 SLon-2000立环脉动高梯度磁选机主要技术参数

转环外径/mm	转环宽度/mm	转环转速/r·min^{-1}	介质盒尺寸（长×宽×高）/mm×mm×mm	给矿粒度/mm	给矿浓度/%	矿浆通过能力/m^3·h^{-1}	干矿处理量/t·h^{-1}	额定背景磁感应强度/T	额定激磁电流/A	额定激磁电压/V
2000	900	3~4	426×154×144	0~2.0	10~40	75~150	50~100	1.0	1400	53

额定激磁功率/kW	转环电动机功率/kW	脉动电动机功率/kW	脉动冲程/mm	脉动冲次/次·min^{-1}	供水压力/MPa	耗水量/m^3·h^{-1}	主机重量/t	最大部件重量/t	外形尺寸（长×宽×高）/mm×mm×mm
74	5.5	7.5	6~26	0~300	0.2~0.3	100~200	50	15	4200×3500×4300

3 小型探索试验

2000年12月，调军台选矿厂取了强磁给矿样到赣州有色冶金研究所进行小型探索试验，采用SLon-100周期式脉动高梯度磁选机，模拟工业生产条件，所获得的选矿指标，与现场Shp-3200平环强磁选机生产指标对比见表2。

通过探索试验可知，SLon-100周期式脉动高梯度磁选机与现场Shp-3200平环强磁选机比较，平均铁精矿品位高2.20%，铁回收率高9.03%。但是由于Shp-3200机的给矿品位偏低，因此，试验指标的可比性要在工业试验中验证。

4 工业试验

2002年10月，赣州立环磁电设备高技术有限责任公司运送1台SLon-2000立环脉动高梯度磁选机到调军台选矿厂，与该厂共同开展工业试验。该厂拆去1台Shp-3200平环强磁选机，将SLon-2000磁选机安装在该Shp-3200磁选机的基础上，其给矿系统、精矿和尾矿排放系统均采用原有的设施，因此具有安装费用低、给料性质和给矿量一致、工业试验可比性好的优点。

表 2 调军台强磁给矿采用 SLon 磁选机探索试验结果

编号	产品	产率/%	品位/%	铁回收率/%	背景磁感应强度/T
1	精矿	32.65	36.45	66.23	0.7
	尾矿	67.35	9.01	33.77	
	给矿	100.00	17.97	100.00	
2	精矿	34.54	35.37	67.85	0.8
	尾矿	65.46	8.84	32.15	
	给矿	100.00	18.00	100.00	
3	精矿	35.62	35.59	69.41	0.9
	尾矿	64.38	8.68	30.59	
	给矿	100.00	18.27	100.00	
4	精矿	36.91	34.95	71.33	0.9
	尾矿	63.09	8.22	28.67	
	给矿	100.00	18.09	100.00	
5	精矿	34.17	34.76	66.60	0.7
	尾矿	65.83	9.46	33.40	
	给矿	100.00	18.11	100.00	
6	精矿	34.40	35.44	67.25	0.9
	尾矿	65.60	9.05	32.75	
	给矿	100.00	18.13	100.00	
平均	精矿	35.33	35.43	68.55	
	尾矿	64.67	8.88	31.45	
	给矿	100.00	18.26	100.00	
现场 Shp-3200 强磁机	精矿	30.50	33.23	59.52	
	尾矿	69.50	9.92	40.48	
	给矿	100.00	17.03	100.00	

工业试验于 2002 年 11 月 17 日开始至 2002 年 12 月 26 日结束。工业试验期间 SLon-2000 磁选机与现场现存 14 台 Shp-3200 平环强磁机中工业状况最好的 1 台进行对比试验。

工业试验期间，着重考察了 SLon-2000 立环脉动高梯度磁选机的脉动冲程对精矿品位的影响。11 月 17~18 日，脉动冲程为 16mm；11 月 19~25 日，脉动冲程为 20mm；11 月 26 日~12 月 19 日，脉动冲程为 24mm。不同脉动冲程的试验结果见表 3。

表 3 不同脉动冲程时的试验结果

设备名称	给矿品位/%	精矿品位/%	尾矿品位/%	精矿产率/%	精矿回收率/%	脉动冲程/mm
SLon-2000	16.82	33.50	7.40	36.09	71.88	
Shp-3200	16.82	32.60	8.59	34.28	66.44	16
差值	0.00	0.90	-1.19	1.81	5.44	
SLon-2000	15.30	29.75	6.42	38.06	74.01	
Shp-3200	15.18	28.56	7.98	34.99	65.82	20
差值	0.12	1.19	-1.56	3.07	8.19	
SLon-2000	18.42	34.50	8.32	38.58	72.26	
Shp-3200	18.45	32.12	9.55	39.43	68.65	24
差值	-0.03	2.38	-1.23	-0.85	3.61	

由表3可见，脉动冲程越大，SLon-2000立环脉动高梯度磁选机的精矿品位越高，但回收率在冲程过高时有所下降。强磁选机是抛尾设备，提高回收率是最主要的目标，当脉动冲程为20mm时，在给矿量、给矿品位和给矿性质一致的条件下，SLon-2000磁选机获得铁精矿品位高1.19%、尾矿品位低1.56%、铁回收率高8.19%的良好指标。

从2002年11月18日~12月26日1个多月对比试验期间全部累计的对比指标（包括冲程为16mm、20mm、24mm时）见表4。

表4 工业试验累计对比指标 （%）

设备名称	给矿品位	精矿品位	尾矿品位	精矿产率	精矿回收率
SLon-2000	17.75	33.80	7.90	38.03	72.42
Shp-3200	17.91	31.92	9.19	38.36	68.37
差　值	-0.16	1.88	-1.29	-0.33	4.05

由表4可见，通过1个多月的工业对比试验，SLon-2000立环脉动高梯度磁选机累计给矿品位低0.16%、精矿品位高1.88%、尾矿品位低1.29%、精矿产率低0.33%、精矿回收率高4.05%，各项技术指标全面超过Shp-3200平环强磁选机。

5　生产应用

经过上述1个多月的工业试验，确定SLon-2000立环脉动高梯度磁选机在调军台选矿厂的较好分选条件见表5。

表5　SLon-2000立环脉动高梯度磁选机工业生产条件

背景磁感应强度/T	激磁电流/A	激磁功率/kW	传动功率/kW	脉动冲程/mm	脉动冲次/r·min^{-1}	处理量/t·h^{-1}
0.75~0.85	900~1050	24~36	13	20	300	60~100

工业试验完成后，SLon-2000立环脉动高梯度磁选机已转化为生产设备，在表5所示条件下，已安全运转近1年时间，运转平稳，磁介质长期保持清洁无堵塞，设备作业率高达99%以上。该机处理量达到70~100t/h，处理每吨矿石的电耗仅为0.5kW·h左右。目前该机仍在运转之中。

6　结语

（1）采用1台SLon-2000立环脉动高梯度磁选机代替调军台选矿厂Shp-3200平环强磁选机进行为期1个多月的工业对比试验，SLon-2000磁选机累计铁精矿品位高1.88%、尾矿品位低1.29%、铁回收率高4.05%。

（2）当脉动冲程为20mm时，SLon-2000立环脉动高梯度磁选机与Shp-3200平环强磁选机比较，铁精矿品位高1.19%、尾矿品位低1.56%、铁回收率高8.19%。由于强磁机在调军台选矿厂主要是控制尾矿品位以提高铁回收率，因此生产上该机已采用脉动冲程20mm的技术参数。

（3）工业对比试验期间，由于矿量的波动及同系统强磁设备运转的台数不同，SLon-2000立环脉动高梯度磁选机的处理量为60~80t/h，最大处理量达100t/h，其处理量与同系统的Shp-3200平环强磁选机相同。

（4）SLon-2000立环脉动高梯度磁选机已作为生产设备在调军台选矿厂运转近1年时间，与Shp-3200平环强磁选机相比，该机具有运行可靠、设备作业率高达99%、磁介质不堵塞、选矿指标好、操作维护方便、省水省电等优点，获得了优良的技术经济指标。

参 考 文 献

[1] 熊大和. SLon型磁选机在齐大山选矿厂的应用 [J]. 金属矿山，2002（4）：42~44.
[2] 熊大和. SLon磁选机分选东鞍山氧化铁矿石的应用 [J]. 金属矿山，2003（6）：21~24.
[3] 徐冬林，陈志华，崔玉环，等. SLon立环脉动高梯度磁选机在齐大山选矿厂试验与应用 [J]. 金属矿山，2002（S1）：242~244.
[4] 宋乃斌，徐冬林. 依靠技术进步缩小齐大山选矿厂一选和二选技术指标差距 [J]. 金属矿山，2003（S1）：118~122.
[5] 赫荣安，陈平，熊大和. SLon强磁机选别鞍山式贫赤铁矿的试验及应用 [J]. 金属矿山，2003（9）：5~10.
[6] 熊大和. SLon-2000立环脉动高梯度磁选机的研制 [J]. 金属矿山，1995（6）：32~34.
[7] 王棣华. SLon-2000立环脉动高梯度磁选机分选弓长岭贫赤铁矿的工业试验 [J]. 金属矿山，1995（9）：29~32.

写作背景 鞍钢调军台选矿厂（现名为：齐大山铁矿选矿分厂）当时是我国最大的氧化铁矿选矿厂，其中采用了15台Shp-3200平环强磁选机。由于存在齿板易堵塞问题，2003年以后被SLon-2000立环脉动高梯度磁选机取代。此文介绍了当时该厂的选矿流程和这两种设备的工业试验和选矿对比数据。

SLon 磁选机研究与应用新进展

熊大和

(赣州立环磁电设备高技术有限责任公司)

摘 要 SLon 立环脉动高梯度磁选机具有优异的选矿性能。介绍了该机的最新发展。该机在分选氧化铁矿、钛铁矿和多种非金属矿的工业生产中得到广泛应用,其成功应用为提高我国细粒弱磁性矿物的选矿水平作出了重要的贡献。

关键词 SLon 立环脉动高梯度磁选机 氧化铁矿 钛铁矿 非金属矿

1 SLon 高梯度磁选机研制新进展

1.1 SLon 高梯度磁选机的改进

SLon 立环脉动高梯度磁选机(以下简称 SLon 强磁选机)是为满足我国量大面广的细粒弱磁性矿石的选矿需要而设计的一种新型高效强磁选设备,是国内外第一代成功应用于大规模工业生产的连续式高梯度磁选机。目前已研制成功 SLon-500、SLon-750、SLon-1000、SLon-1250、SLon-1500、SLon-1750、SLon-2000 等型号 SLon 强磁选机,其背景磁感强度可在 0~1.0 T 范围调节,它们的干矿处理量为 0.05~80t/h,可满足实验室、连选试验和工业生产的需要。

为了进一步提高 SLon 立环脉动高梯度磁选机的选矿指标和设备工作可靠性,近年对该设备又进行了多方面的改进。

(1)在分析磁力、脉动流体力、重力在分选过程中的相互作用机理的基础上优化磁系结构,合理布置分选空间,可获得优良的选矿指标。

(2)根据选矿原理采用计算机精确定位实现了磁介质最佳有序排列组合,有利于获得优良的选矿指标,并做到磁介质长期不堵塞。

(3)采用优质导线和优质的绝缘材料绕制激磁线圈,其耗电低、寿命长。

(4)采用耐磨材料制造脉动冲程箱,数值式调节冲程的偏心连杆机构,使脉动冲程箱具有高效耐磨、使用寿命长、易于调节的优点。

(5)研制了数值式整流控制电路,提高了整机的控制精度,并且可根据矿石性质的波动而方便调节激磁电流。

1.2 SLon-1500 和 SLon-1750 中磁机的研制

在东鞍山和齐大山的阶段磨矿—重选—强磁—反浮选工艺分选鞍山式红铁矿的流程中,螺旋溜槽尾矿需要用中磁机扫选。中磁机尾矿作为最终尾矿抛弃,中磁机的精矿返回流程再磨再选。东鞍山的工业试验和齐大山的生产实践证明,用钕铁硼永磁块制成的滚筒

---

❶ 原载于《矿冶工程》,2004,24(z1):12~14。

式中磁机（表面场强 0.4~0.6 T）抛尾存在着尾矿品位偏高的问题。为此根据生产的需要新研制了 SLon-1500 和 SLon-1750 立环脉动高梯度中磁机（以下简称 SLon 中磁机）。这种中磁机作为鞍山式红铁矿粗粒级的抛尾设备具有处理量大、尾矿品位低、可调性好、工作稳定、性能价格比高的优点。

SLon 强磁机和 SLon 中磁机主要技术参数见表 1。

表 1 SLon 磁选机主要技术参数

机型	转环外径/mm	给矿粒度/mm	给矿浓度/%	矿浆通过能力/m³·h⁻¹	干矿处理量/t·h⁻¹	额定背景磁感强度/T	额定激磁功率/kW	传动功率/kW	脉动冲程/mm	脉动冲次/r·min⁻¹	主机质量/t
SLon-500 强磁机	500	<1.0	10~40	0.5~1.0	0.05~0.25	1.0	13.5	0.18+0.37	0~50	0~400	1.5
SLon-750 强磁机	750	<1.0	10~40	1.0~2.0	0.1~0.5	1.0	22	0.55+0.75	0~50	0~400	3
SLon-1000 强磁机	1000	<1.3	10~40	12.5~20	4~7	1.0	28.6	1.1+2.2	0~30	0~300	6
SLon-1250 强磁机	1250	<1.3	10~40	20~50	10~18	1.0	35	1.5+2.2	0~20	0~300	14
SLon-1500 强磁机	1500	<1.3	10~40	50~100	20~30	1.0	44	3+4	0~30	0~300	20
SLon-1750 强磁机	1750	<1.3	10~40	75~150	30~50	1.0	62	4+4	0~30	0~300	35
SLon-2000 强磁机	2000	<1.3	10~40	100~200	50~80	1.0	74	5.5+7.5	0~30	0~300	50
SLon-1500 中磁机	1500	<1.3	10~40	75~150	30~50	0.4	16	3+4	0~30	0~300	15
SLon-1750 中磁机	1750	<1.3	10~40	75~150	30~50	0.6	38	4+4	0~30	0~300	35

2 SLon 磁选机在氧化铁矿选矿中的应用

近年我国的钢铁工业得到飞速发展，我国的铁矿资源绝大多数为贫矿，钢铁工业对优质铁精矿的巨大需求促进了以 SLon 立环脉动高梯度磁选机、反浮选等新技术在我国贫、细、杂氧化铁矿选矿工业中的高速发展和应用，使我国氧化铁矿的选矿水平达到国际先进水平。表 2 列举了 SLon 磁选机在细粒氧化铁矿选矿工业中的应用状况。

表 2 SLon 磁选机在细粒氧化铁矿工业中的应用

厂家名称	设备型号	使用台数	分选矿种	选矿流程	分选指标/%	备注
鞍钢齐大山选矿厂	SLon-1750 强磁机 SLon-1500 中磁机	11 11	鞍山式赤铁矿	阶段磨矿—重选—强磁—反浮选	α=30.50, β=67.50, θ=10.15, ε=78.53	全流程选矿指标
鞍钢东鞍山烧结厂一选车间	SLon-1750 强磁机 SLon-2000 强磁机 SLon-1750 中磁机	8 2 10	鞍山式赤铁矿	阶段磨矿—重选—强磁—反浮选	α=32.50, β=64.50, θ=14.50, ε=71.45	全流程选矿指标
马钢姑山铁矿	SLon-1750 强磁机 SLon-1500 强磁机	7 4	宁芜式赤铁矿	阶段磨矿—强磁—粗—精—扫	α=43.15, β=60.17, θ=22.41, ε=76.59	磨选主厂房全流程指标
宝钢南京梅山铁矿	SLon-1500 强磁机	18	赤铁矿、黄铁矿、菱铁矿	弱磁—强磁—粗—扫联合降磷流程	α(Fe)=52.77, α(S)=0.44, α(P)=0.399, β(Fe)=56.08, β(S)=0.29, β(P)=0.246, ε(Fe)=94.51	降磷流程综合指标
昆钢大红山铁矿	SLon-1500 强磁机	4	磁铁矿和赤铁矿混合矿	阶段磨矿—弱磁—强磁流程	α=37.69, β=64.24, θ=12.14, ε=83.59	全流程选矿指标
海南钢铁公司选矿厂	SLon-2000 强磁机	4	磁铁矿少量赤铁矿为主	阶段磨矿—弱磁—强磁—粗—精—扫	α=51.09, β=64.20, θ=24.50, ε=84.16	细粒选矿流程综合流程
首钢秘鲁铁矿	SLon-1750 强磁机	2	含硫磁铁矿、赤铁矿	阶段磨矿—弱磁—强磁—浮硫	α=55.20, β=65.09, θ=28.10, ε=86.39	全流程指标

注：α 为原矿 Fe 品位，%；β 为铁精矿 Fe 品位，%；θ 为尾矿 Fe 品位，%；ε 为铁回收率，%。

3 SLon 磁选机在钛铁矿选矿中的应用

随着国民经济的发展，我国钛白粉、钛金属的产量与日俱增，钛工业的发展对钛精矿的需求量迅速增加。四川攀西地区是我国最大的钛精矿产地，近年来，微细粒级选钛技术的发展和 SLon 立环脉动高梯度磁选机的广泛应用使我国的选钛水平有了大幅度提高，以攀钢钛业公司选钛厂为例，10 年前采用重选—电选工艺流程只能回收大于 0.045mm 的钛精矿，钛精矿年产量为 5 万吨，钛精矿 TiO_2 品位为 47%~48%，钛的综合回收率仅 17%~20%。1997~2004 年该厂先后建起了 3 条微细粒级选钛生产线，共采用 SLon-1500 立环脉动高梯度磁选机和浮选工艺流程分选小于 0.045mm 钛精矿，至今微细粒级钛精矿的年产量达到 10 万吨以上，微细粒级钛精矿 TiO_2 品位为 48%，微细粒级钛精矿回收率为 44% 左右。目前攀钢钛业公司钛精矿年产量已达到 20 万吨，TiO_2 的综合回收率达到 35% 左右。

攀钢钛业公司选钛技术的发展，带动了攀西地区众多选钛厂的选钛技术发展和 SLon 立环脉动高梯度磁选机在选钛工业中的广泛应用。表 3 列举了 SLon 磁选机在攀西地区选钛工业中的应用状况。

表 3 SLon 磁选机在攀西地区选钛工业中的应用

厂家名称	设备型号	使用台数	分选矿种	选钛流程	分选指标 /%	钛精矿产量 /万吨·a^{-1}
攀钢钛业公司选钛厂	SLon-1500 强磁机	12	<0.045mm 微细粒级钛铁矿	SLon 强磁粗选—浮选精选	$\alpha=10.50$，$\beta=48.00$，$\theta=6.50$，$\varepsilon=44.06$	10（微细粒级）
攀钢兴矿公司综合厂	SLon-1500 强磁机	3	从攀钢总尾矿回收钛精矿	SLon 强磁磁选—分级—粗粒重选精选—细粒浮选精选	$\alpha=7.68$，$\beta=47.66$，$\theta=5.56$，$\varepsilon=31.19$	3
重钢太和铁矿	SLon-1500 强磁机 SLon-1750 强磁机	3 1	从钒钛磁铁矿选铁尾矿中回收钛精矿	SLon 强磁粗选—粗粒再磨—粗细合并 SLon 再选—浮选精选	$\alpha=12.06$，$\beta=47.82$，$\theta=7.20$，$\varepsilon=47.44$	4
攀枝花永琦商贸公司	SLon-1750 强磁机	1	从钒钛磁铁矿选钛尾矿中回收钛精矿	SLon 强磁粗选—摇床精选	$\alpha=14.50$，$\beta=47.60$，$\theta=8.30$，$\varepsilon=51.80$	3.8
攀枝花红发有限公司	SLon-1750 强磁机	1	从钒钛磁铁矿选钛尾矿中回收钛精矿	SLon 强磁粗选—摇床精选	$\alpha=14.00$，$\beta=47.50$，$\theta=8.10$，$\varepsilon=50.80$	3.5
盐边述伦矿业有限公司	SLon-1750 强磁机	1	从钒钛磁铁矿选钛尾矿中回收钛精矿	SLon 强磁粗选—摇床精选	$\alpha=11.00$，$\beta=47.20$，$\theta=7.20$，$\varepsilon=40.76$	3.3

注：α 为原矿 TiO_2 品位，%；β 为钛精矿 TiO_2 品位，%；θ 为尾矿 TiO_2 品位，%；ε 为 TiO_2 回收率，%。

4 SLon 磁选机在非金属矿加工中的应用

随着人民生活水平提高，我国对优质建材的需求量迅速增加，我国非金属矿加工和提纯工业也在迅速发展。近年 SLon 立环脉动高梯度磁选机在长石、霞石、石英、煤系高岭土等加工工业中广泛应用于除铁作业。SLon 磁选机在非金属矿的应用实例见表 4。

5 结语

（1）采用最新的先进技术对 SLon 立环脉动高梯度磁选机进行了多方面的改进，并研制了 SLon-1750 和 SLon-1500 立环脉动中磁机，使它们具有更优良的选矿性能和工作可靠

表4 SLon磁选机在非金属矿加工工业中的应用

厂矿名称	设备型号	使用台数	分选矿种	选钛流程	分选指标/%	精粉生产规模/万吨·a^{-1}
安徽明光泰达长石厂	SLon-1500强磁机	2	长石	SLon磁选机—粗—精	$\alpha=1.45$, $\beta=0.26$, $\theta=9.35$, $\varepsilon=86.91$	6
安徽来安长石厂	SLon-1250强磁机	1	长石	滚筒中磁机粗选—SLon磁选机精选	$\alpha=1.48$, $\beta=0.26$, $\theta=9.43$, $\varepsilon=86.70$	3
安徽皖东长石厂	SLon-1500强磁机 SLon-1000强磁机	1 1	长石	滚筒中磁机粗选—SLon磁选机精选	$\alpha=1.47$, $\beta=0.26$, $\theta=9.43$, $\varepsilon=86.80$	7
安徽新华长石厂	SLon-1200强磁机	1	长石	滚筒中磁机粗选—SLon磁选机精选	$\alpha=1.46$, $\beta=0.25$, $\theta=8.06$, $\varepsilon=84.50$	2
四川南江霞石厂	SLon-1250强磁机	2	霞石	弱磁选—SLon磁选机—粗—精	$\alpha=1.69$, $\beta=0.179$, $\theta=6.57$, $\varepsilon=94.93$	3
广东英德中信公司	SLon-1000强磁机	6	长石	SLon磁选机—粗—精	$\alpha=0.053$, $\beta=0.025$, $\theta=0.57$, $\varepsilon=94.93$	5
江西乐平华源硅砂厂	SLon-1500强磁机	1	石英	滚筒中磁机粗选—SLon磁选机精选	$\alpha=0.11$, $\beta=0.021$, $\theta=0.35$, $\varepsilon=73.35$	5
陕西洋县石英砂厂	SLon-1000强磁机	1	石英	弱磁选—SLon磁选机强磁选	$\alpha=0.018$, $\beta=0.012$, $\theta=0.098$, $\varepsilon=93.00$	2
四川乐山石英砂厂	SLon-1000强磁机	1	石英	弱磁选—SLon磁选机强磁选	$\alpha=0.06$, $\beta=0.03$, $\theta=0.37$, $\varepsilon=91.20$	2
淮北金岩高岭土公司	SLon-1500强磁机	1	煤系高岭土	SLon磁选机一次性除铁	$\alpha=0.65$, $\beta=0.44$, $\theta=3.21$, $\varepsilon=92.42$	1.2

注：α为原矿Fe_2O_3品位；β为精矿Fe_2O_3品位，%；θ为尾矿Fe_2O_3品位，%；ε为非磁性产品（精矿）产率，%。

性，目前已成为我国工业生产中广泛应用的新一代强磁选设备。

（2）SLon立环脉动高梯度磁选机在氧化铁矿的广泛应用，使我国氧化铁矿选矿技术水平有了大幅度的提高，以鞍钢齐大山选矿厂为代表的阶段磨矿—重选—强磁—反浮选流程，使氧化铁矿的选矿指标达到：原矿品位30%左右，铁精矿品位67.50%，尾矿品位10%左右，铁回收率78%左右的历史最高水平。

（3）SLon立环脉动高梯度磁选机在钛铁矿选矿工业中的成功应用，为解决我国微细粒级钛铁矿选矿技术难题，大幅度提高钛铁矿的回收率和降低选钛生产成本发挥了重要作用。但是，我国钛铁矿总体回收率还是较低，在保证钛精矿品位的前提下，如何进一步提高钛铁矿的回收率是今后研究的方向。

（4）SLon立环脉动高梯度磁选机在我国非金属矿加工工业中的应用，使许多低等级的非金属矿如长石、石英、霞石等变成较高等级的建筑材料，使煤系高岭土变为造纸涂料级的优质产品。随着国民经济的发展，SLon磁选机在非金属矿加工工业中将得到越来越广泛的应用。

写作背景 本文是作者在《矿冶工程》组织的学术会上发表的论文，文章中介绍了多种型号的SLon立环脉动高梯度磁选机（SLon强磁机和SLon中磁机）以及它们在全国各地应用于氧化铁矿、钛铁矿的选矿和非金属矿的提纯。由此可见，SLon磁选机已得到较为广泛的应用。

SLon 磁选机在淮北煤系高岭土除铁中的应用[1]

熊大和

(赣州立环磁电设备高技术有限责任公司)

摘 要 安徽淮北金岩高岭土公司采用 SLon-1500 立环脉动高梯度磁选机作为煤系高岭土除铁设备,建成一条年生产1万吨煅烧高岭土的生产线,煅烧高岭土最终产品白度达到93%,粒度小于 $2\mu m$ 为90%。

关键词 SLon 立环脉动高梯度磁选机 煤系高岭土 除铁 提纯

Application of SLon Magnetic Separator for Iron Removal Huaibei Coal-Series Kaolin Mine

Xiong Dahe

(*Ganzhou Vertical Ring Magnetic-electric Equipment Hi-tech Co., Ltd.*)

Abstract In the last two years, Anhui Huaibei Jinyan Kaolin Company built up a calcined kaolin production line with the capacity of 10 kt/a of final product, in which a SLon-1500 vertical-ring pulsating high-gradient magnetic separator was applied as iron removing equipment. The final kaolin product reached the whiteness of 93% and the grain size of <$2\mu m$ 90%.

Keywords SLon vertical-ring pulsating high-gradient magnetic separator; Coal-series kaolin; Iron-removal; Purification

自20世纪80年代以来,随着高岭土开发技术的发展和我国工业对高岭土需求量的增大,煤系高岭土的开发利用越来越受到重视。实践证明,经过深加工后的煤系高岭土白度高、质量好,其开发和应用前景将越来越广阔。

2001年初,我们与安徽淮北金岩高岭土公司开始共同研究煤系高岭土的除铁提纯问题。至2003年底,该公司建成了一座年产1万吨优质煤系高岭土的生产线,其中除铁工序采用 SLon-1500 立环脉动高梯度磁选机,使 SLon 立环脉动高梯度磁选机在高岭土除铁工业应用中首次获得成功。

1 脉动高梯度磁选探索试验

1.1 试料性质

试样为淮北金岩高岭土公司提供的煤系高岭土粉料,其有害杂质铁的来源主要是赋存

---

[1] 原载于《非金属矿》,2004 (05):44~46。

在原矿中的微粒磁铁矿、褐铁矿以及加工过程中混入的机械铁。共 3 个试样，其粒度组成和 Fe_2O_3、TiO_2 含量，见表 1。

表 1　淮北煤系高岭土粉料粒度和杂质含量

试样编号	粒度/μm(目)	Fe_2O_3/%	TiO_2/%
1	<140（100）	1.76	0.57
2	<74（200）	1.07	0.59
3	<46（300）	0.69	0.63

1.2　试验设备

探索试验在 SLon-100 周期式脉动高梯度磁选机上进行。该机的主要结构，如图 1 所示。

SLon-100 周期式脉动高梯度磁选机的下部装有一脉动机构，由偏心连杆机构产生的交变力 $\widetilde{F}$ 推动橡胶鼓膜往复运动，使分选腔内的矿浆产生脉动。用导磁不锈钢制成的钢板网或圆棒作磁介质。激磁线圈通以直流电，在分选区产生感应磁场。工作时，先在分选腔内灌满水，使脉动能量传递到分选腔，然后从给矿盒给入矿浆，调节下部阀门，可控制矿浆的流速和液位高度。分选腔内的磁性矿物和非磁性矿物，在磁力、脉动流体力、重力的综合力场作用下得到分离，磁性矿粒被吸附在磁介质表面上，非磁性矿粒随矿浆从下部排走。当橡胶鼓膜在脉动机构的驱动下作往复运动时，脉动流体力使矿粒群在分选过程中始终保持松散状态，从而可有效地消除非磁性颗粒的机械夹杂，明显地提高磁性产品品位或非磁性产品的产率。周期式脉动高梯度磁选机为阶段性给矿，给矿完毕后，切断激磁电流，然后用清水将磁性物冲洗出来，即完成一个周期的磁选。

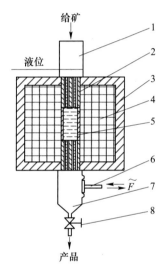

图 1　周期式脉动高梯度磁选机结构图
1—给矿盒；2—磁极头；3—磁轭；4—激磁线圈；5—磁介质；6—脉动机构；7—接矿斗；8—阀门

1.3　探索试验

采用 SLon-100 周期式脉动高梯度磁选机（背景磁感强度分别为 1.05T 和 1.53T）对表 1 所列 3 种试料进行除铁试验，结果见表 2。

由表 2 可见，淮北煤系高岭土粉碎至小于 140μm（100 目）、小于 74μm（200 目）或小于 46μm（300 目）后，用 SLon-100 周期式脉动高梯度磁选机一次选别，均可获得 Fe_2O_3 低于 0.5% 的指标，二次选别精矿中 Fe_2O_3 的含量与一次选别基本未变，说明第二次选别作用不大。从该表还可见，TiO_2 的去除率较低，这可能是试样所含的 TiO_2 以金红石为主所致，但金红石的存在基本上不影响高岭土煅烧后的白度。

表2 煤系高岭土高梯度磁选除铁指标

背景磁感强度/T	选别流程	试料编号	精矿产率/%	精矿品位/% Fe_2O_3	精矿品位/% TiO_2	除铁率/%	除钛率/%
1.05	一次选别	1	91.00	0.43	0.55	77.77	12.19
1.05	一次选别	2	93.67	0.41	0.58	64.11	7.92
1.05	一次选别	3	95.00	0.40	0.57	44.93	14.05
1.05	一粗一精	1	83.33	0.40	0.53	81.06	22.52
1.05	一粗一精	2	90.00	0.46	0.54	61.31	17.63
1.05	一粗一精	3	91.00	0.40	0.52	47.25	24.89
1.53	一次选别	1	86.00	0.43	0.54	78.99	18.53
1.53	一次选别	2	93.33	0.46	0.58	59.88	8.25
1.53	一次选别	3	94.33	0.36	0.55	50.78	17.65
1.53	一粗一精	1	76.00	0.43	0.57	81.43	24.00
1.53	一粗一精	2	86.17	0.41	0.58	66.98	15.29
1.53	一粗一精	3	91.33	0.42	0.59	44.41	14.45

注：精矿，即非磁性产品。

试验所获精矿经煅烧后，可获得白度为93%的优质高岭土产品。该产品经进一步磨细后，可作为优质造纸涂料。

2 半工业试验

2001年10月，我们在淮北金岩高岭土公司的生产现场，采用了1台SLon-1000立环脉动高梯度磁选机进行高岭土除铁的半工业试验，处理小于74μm（200目）的煤系高岭土粉料，所获选矿指标见表3。

表3 SLon-1000磁选机半工业试验结果

给矿 处理量/t·h⁻¹	给矿 浓度/%	给矿 Fe_2O_3/%	精矿 产率/%	精矿 Fe_2O_3/%	尾矿 Fe_2O_3/%	除铁率/%
1.6	30	0.70	93.05	0.48	3.65	36.19
			91.49	0.46	3.28	39.88
			93.15	0.52	3.15	30.80
			91.34	0.49	2.91	36.06
			93.27	0.52	3.19	30.71
			93.12	0.56	2.59	25.50
			90.23	0.47	2.82	39.42
			90.76	0.47	2.96	39.06
平均			92.05	0.49	3.07	35.57
1.2	30	0.72	88.37	0.40	3.15	50.91
			89.29	0.41	3.30	49.15
			90.02	0.41	3.52	48.74
			93.16	0.49	3.85	36.60
			92.48	0.44	4.16	43.48
			91.53	0.43	3.85	45.34
平均			90.81	0.43	3.64	45.77

注：背景磁感强度，1.0T；尾矿，即磁性产品。

从表3可见，SLon-1000立环脉动高梯度磁选机的处理量对高岭土除铁效果有一定的影响，处理量较低时，除铁效果较好。该机在处理量为1.2t/h时，高岭土精矿Fe_2O_3含量为0.43%，达到了Fe_2O_3%含量小于0.5%的要求，与小型探索试验指标基本一致。

3 SLon-1500磁选机在工业中的应用

在小型试验和半工业试验成功之后，年产1万吨精粉的第一期工程已于2002~2003年建成。淮北金岩高岭土公司决定投资建设一条年产2万吨煅烧高岭土精粉的工厂，该生产线中采用了1台SLon-1500立环脉动高梯度磁选机作为除铁设备。

3.1 用于高岭土除铁的SLon-1500立环脉动高梯度磁选机简介

SLon立环脉动高梯度磁选机是一种连续的高效强磁选设备，其转环立式旋转，反冲精矿，并配有脉动机构，具有富集比大、分选效率高、磁介质不堵塞，对给矿粒度、浓度和品位的波动适应性强、工作可靠、操作维护方便的优点。该机在冶金矿山已广泛应用于氧化铁矿、钛铁矿的选矿中，在非金属矿如石英、长石、霞石矿等的除铁方面也得到较为广泛的应用。但是，SLon立环脉动高梯度磁选机在工业生产中应用于高岭土除铁还是第一次。高岭土中的矿物嵌布粒度极细，高岭土除铁是各种非金属矿除铁最难的一种，因此，对除铁设备有非常高的要求。

针对淮北煤系高岭土的特性，我们对SLon-1500立环脉动高梯度磁选机做了如下专项改进：（1）分选转环采用变频调速，以便在生产上根据不同的给矿量和矿浆体积，将转环转速控制在最佳状态；（2）采用较细的导磁不锈钢网作为磁介质，使微细粒级铁矿物能有效地从非磁性产品中分离；（3）非磁性产品的排料部分及阀门全部采用不锈钢制造，尽可能避免二次铁锈污染；（4）磁性产品的冲洗水采用双管路，其中一路为循环水，以节约生产用水；另一路为清洁水，使磁介质冲洗干净后进入下一个作业循环。改进后的SLon-1500立环脉动高梯度磁选机的机械结构和工作原理，见图2。

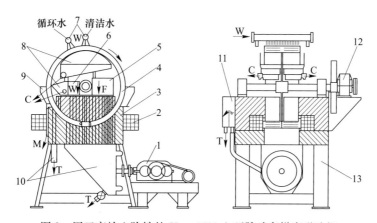

图2 用于高岭土除铁的SLon-1500立环脉动高梯度磁选机

1—脉动机构；2—激磁线圈；3—铁轭；4—转环；5—给矿斗；6—漂洗水斗；7—精矿冲洗水装置；
8—精矿斗；9—中矿斗；10—尾矿斗；11—液位计；12—转环驱动机构；13—机架；
F—给矿；W—清水；C—精矿；M—中矿；T—尾矿

SLon 立环脉动高梯度磁选机的工作原理：转环内装有导磁不锈钢棒或钢板网磁介质（也可根据需要充填导磁不锈钢毛等磁介质）。选矿时，转环作顺时针旋转，矿浆从给矿斗给入，沿上铁轭缝隙流经转环，转环内的磁介质在磁场中被磁化，磁介质表面形成高梯度磁场，矿浆中磁性颗粒被吸附在磁介质表面，随转环转动被带至顶部无磁场区，用冲洗水冲入精矿斗中，非磁性颗粒沿下铁轭缝隙流入尾矿斗中排走。

脉动机构驱动橡胶鼓膜做往复运动，使分选室的矿浆做上下往复运动，脉动流体力使矿粒群在分选过程中始终保持松散状态，从而有效地消除非磁性颗粒的机械夹杂，提高非磁性产品的产率。此外，矿浆脉动对防止磁介质的堵塞也大有益处。

3.2 煤系高岭土加工生产线简介

2002～2003年，淮北金岩高岭土公司建成了一条年产1万吨煅烧煤系高岭土精粉的生产线，其工艺流程如图3所示。

该煤系高岭土加工流程为：原矿经破碎、磨矿至小于74μm（200目）占100%，经调浆至浓度30%进入一台SLon-1500立环脉动高梯度磁选机除铁，其背景磁感强度为1.0T，使非磁性产品 Fe_2O_3% 含量小于0.5%。非磁性产品用旋流器脱水，脱水后用立式搅拌磨研磨至小于2μm占90%，然后用煅烧炉煅烧，高岭土中的有机物质经煅烧后被剔除。煅烧后的高岭土白

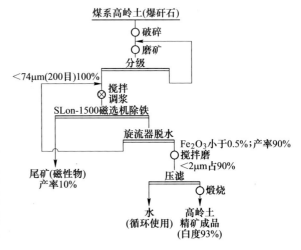

图3 淮北煤系高岭土加工流程

度达到93%，粒度小于2μm占90%，可作为造纸用的优质涂料。经一年的生产运转，该生产线产品质量稳定，各项指标达到设计要求。

4 结论

（1）安徽淮北金岩高岭土公司采用SLon-1500立环脉动高梯度磁选机作为煤系高岭土的除铁设备，在原料含 Fe_2O_3 为0.7%左右时，一次性除铁达到 Fe_2O_3<0.5%。最终煅烧产品的白度达到93%，可作为优质的造纸涂料。

（2）高岭土除铁技术难度较大，SLon立环脉动高梯度磁选机在高岭土除铁提纯工业应用中首次获得成功，标志着我国高岭土除铁技术达到一个新的高度。

写作背景 这是SLon立环脉动高梯度磁选机第一次在煤系高岭土中的应用，文章中介绍了SLon磁选机应用于煤系高岭土除铁的小型试验、半工业试验和一台SLon-1500立环脉动高梯度磁选机成功应用于生产。

脉动高梯度磁选垂直磁场与水平磁场对比研究

熊大和

（赣州立环磁电设备高技术有限责任公司）

摘　要　SLon 立环脉动高梯度磁选机采用垂直磁场磁系，它们已在工业上广泛用于氧化铁矿、钛铁矿、非金属矿的选矿。为了探索进一步改进设备的可能性，又研制了水平磁场磁系的脉动高梯度磁选机试验设备，并通过理论分析和试验对 2 种磁场磁系的脉动高梯度磁选机进行了对比，结果表明，垂直磁场磁选机具有磁场强度较高、漏磁系数较小、激磁电耗较低、对细粒弱磁性矿物回收率较高、选矿效率较高和设备处理量易于放大的优点。

关键词　脉动高梯度磁选　垂直磁场　水平磁场　弱磁性矿石选矿

Study on Comparison Between Vertical and Horizontal Magnetic Fields in Pulsating High Gradient Magnetic Separation

Xiong Dahe

(*Ganzhou Vertical Ring Magnetic-electric Equipment Hi-tech Co.,Ltd.*)

Abstract　SLon vertical ring pulsating high gradient magnetic separators utilize vertical field magnet system, which have been widely used in the beneficiation of iron oxide ore, ilmenite and nonmetallic minerals. To explore the possibility of further equipment improvement, the author developed a laboratory pulsating high gradient magnetic separator with horizontal field magnet system. Through the theoretical analysis and comparative test on the pulsating high gradient magnetic separators with vertical or horizontal field magnet system, it is demonstrated that the magnetic separator with vertical magnetic field has advantages of higher magnetic intensity, smaller flux leakage factor, lower exciting power consumption, higher recovery separation efficiency and easiness of equipment throughput scaling-up.

Keywords　Pulsating high gradient magnetic separation; Vertical magnetic field; Horizontal magnetic field; Beneficiation of weakly magnetic minerals

　　脉动高梯度磁选是分选细粒弱磁性矿物的有效方法。目前，采用垂直磁场的 SLon 立环脉动高梯度磁选机已广泛应用于氧化铁矿、钛铁矿和非金属矿的选矿工业中。采用垂直磁场的脉动高梯度磁选机是指分选区矿浆流动的方向和磁力线的方向均垂直于地球表面；采用水平磁场的脉动高梯度磁选机是指分选区矿浆流动方向垂直于地球表面，而磁力线方向平行于地球表面。为了比较这两种磁场应用于脉动高梯度磁选的优劣，作者做了一些理

---

❶ 原载于《金属矿山》，2004（10）：24~27。

论性的探讨,并研制了垂直磁场和水平磁场脉动高梯度磁选试验设备,进行设备性能和选别指标的实际验证,获得了一些有参考价值的资料。

1 垂直磁场与水平磁场高梯度磁选原理

1.1 2种磁场对磁性矿物的作用力

高梯度磁选是利用圆柱形磁介质(或丝状磁介质)在磁场中产生磁力线聚焦的现象而形成高梯度磁场。如图1(a)和(b)所示分别为圆柱形磁介质在垂直磁场和水平磁场中产生的磁力线聚焦现象。

由图1可见,圆柱形磁介质在垂直磁场中,其上下表面附近为磁力线密集区,磁场强度和磁场梯度最高,磁性矿物在这2个区域内受到磁场引力;在其左右表面附近为磁力线稀疏区,磁场强度较低且磁场梯度为负值,磁性矿物在这2个区域内受到磁场斥力[1]。在水平磁场中,磁介质表面磁力线的聚焦方向与垂直磁场相差90°。图2(a)和(b)分别为垂直磁场和水平磁场中圆柱形磁介质吸引矿石的示意图。

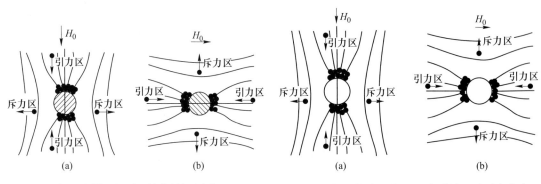

图1 圆柱形磁介质周围的磁力场
(a)垂直磁场;(b)水平磁场
H_0—磁场方向

图2 磁性矿物在圆柱形磁介质周围受磁力状态
(a)垂直磁场;(b)水平磁场
H_0—磁场方向

1.2 流体力对矿物的作用

在湿法高梯度磁选中,一般以水为载体,磨细的矿物与水混合成矿浆。假定磁介质为圆柱形,则水在磁介质表面的流态如图3所示。

无论是垂直磁场还是水平磁场,矿浆的流动一般都要依靠地球引力,因此矿浆流动的方向大体上都是从上至下的。矿浆在磁介质上、下表面附近区域的流速较慢,在磁介质左右表面附近区域的流速较快。

在高梯度磁选中,磁性矿粒被磁介质捕收的必要条件是 $f_{磁力} > f_{竞争力}$,其中 $f_{竞争力}$ 包括

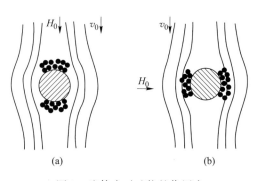

图3 流体力对矿物的作用力
(a)垂直磁场;(b)水平磁场
v_0—矿浆流动方向;H_0—磁场方向

重力、流体力、脉动流体力等。在雷诺系数较小时,矿粒所受流体力的大小与矿浆流速成正比。从图3可见,当$f_{磁力}$一定时,在垂直磁场中被吸附在磁介质上下表面的矿粒受到的流体力较小,有利于提高细粒弱磁性矿物的回收率;在水平磁场中被吸附在磁介质左右表面的矿粒受到的流体力较大,不利于细粒弱磁性矿物的捕收。

1.3 2种磁场对设备处理量的影响

工业生产中的高梯度磁选机是采用成千上万根圆柱形磁介质(或丝状磁介质)按一定的规律排列组合成磁介质矩阵(简称磁介质组),对矿浆中数以亿计的细粒弱磁性矿粒(若将1kg密度为5g/cm³的铁矿石磨至直径为0.05mm的颗粒,其数量大约为3.8亿颗)进行捕收[2]。如图4(a)所示,在垂直磁场中,由于圆柱形磁介质上下表面捕收磁性矿物,磁性矿物的堆积对矿浆流动通道影响较小,因此采用垂直磁场的高梯度磁选机有较大的处理

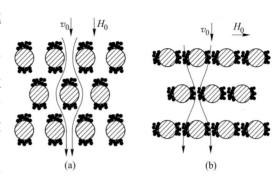

图4 磁场方向对设备处理量的影响
(a) 垂直磁场; (b) 水平磁场
v_0— 矿浆流动方向; H_0— 磁场方向

能力;如图4(b)所示,在水平磁场中,由于圆柱形磁介质左右表面捕收磁性矿物,随着磁性矿物的堆积,矿浆通道越来越小,严重时会导致后续矿浆流不下去,非磁性矿粒无法通过,造成精尾不分,选矿指标恶化。因此,采用水平磁场的高梯度磁选机具有设备处理量较小,选矿指标不稳定的缺点。

2 2种磁场磁系比较

磁系是高梯度磁选机重要的组成部分,可分为电磁磁系和永磁磁系。电磁磁系具有磁场强度高、处理量大、选矿指标好、设备安装和检修方便的优点。目前工业生产中应用的高梯度磁选机绝大多数为电磁磁系。图5(a)和(b)分别为垂直磁场磁系和水平磁场磁系示意图。这2种磁系均由激磁线圈和铁轭组成,当激磁线圈通以直流电时,在气隙空间(即分选区)产生感应磁场。根据安培环路定理,线圈的安匝数之和等于闭合磁力线的磁场强度与路径的积分,即:

$$NI = \oint H \cdot dL = H_0 L_0 + \sum H_i L_i = (1+K)H_0 L_0$$

式中,N为激磁线圈匝数;I为激磁电流;H为磁场强度;H_0为气隙磁场强度;L_0为气隙路径;H_i为铁轭内部各路径上的磁场强度;L_i为铁轭内部各处路径;K为漏磁系数。

上式还可写成:

$$H_0 = \frac{NI}{(1+K) \cdot L_0}$$

上式说明在安匝数NI一定的条件下,漏磁系数越小,在气隙空间(即分选区)的磁场强度H_0越高;或者说在气隙空间磁场强度一定的条件下,K值越小,所需安匝数越少,磁

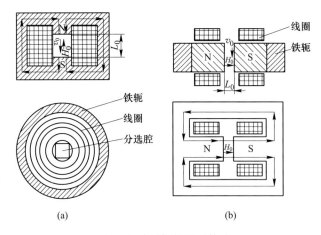

图 5　2 种磁场的磁系构造
(a) 垂直磁场磁系；(b) 水平磁场磁系

选机电耗越低。因此，漏磁系数的大小是衡量磁系设计水平的一个非常重要的参数。

垂直磁场的磁系采用闭合式磁系（或称马斯顿磁路），激磁线圈外围铁轭包裹的比例较大，其气隙（分选区）安排在激磁线圈的核心部位，具有磁路短、漏磁系数小、场强高、电耗低的优点。而水平磁场磁系磁路较长，激磁线圈外围铁轭包裹的比例较小，因此水平磁场磁系具有漏磁系数较大、场强较低、电耗较高的缺点。

垂直磁系的另一个优点是有利于制造大型的电磁选矿设备。如图 5（a）所示，当分选区高度一定时，矿浆的通过量与激磁线圈内径的平方成正比，即激磁线圈的内径放大 1 倍，则矿浆的通过量增加至 4 倍；而由于激磁线圈的周长只增加 1 倍，因此激磁功率也只增加 1 倍。可见，采用垂直磁系的高梯度磁选机，设备规格越大，处理每吨矿石的电耗越低。

与此相反，水平磁系则不利于制造大型的电磁选矿设备。由于水平磁系的矿浆通过量与分选区（气隙）的长度或磁极头的宽度成正比，无论是增加气隙的长度，还是增加磁极头的宽度，其激磁功率的增加均与处理量的增加成正比。此外，水平磁系的气隙（分选区）长度越大，其漏磁系数越大，要达到一定磁场强度的电耗显著增加。因此，采用水平磁系制造的高梯度磁选机，难以实现依靠设备大型化而省电的目的。

3　2 种磁场的脉动高梯度磁选试验设备

为了验证垂直磁场与水平磁场脉动高梯度磁选机的电磁性能和选矿性能，设计并制造了图 6 所示的垂直磁场脉动高梯度磁选机和水平磁场脉动高梯度磁选机，其磁场特性实测数据如图 7 所示。

如图 7 所示，垂直磁场脉动高梯度磁选机具有漏磁系数小、背景磁感强度高的优点，当其背景磁感强度达到 0.7T 时，漏磁系数为 0.18；当其背景磁感强度达到 1.0T 时，漏磁系数为 0.56；该机最高背景磁感强度可达到 1.75T，相应的漏磁系数为 1.17。而水平磁场脉动高梯度磁选机当其背景磁感强度达到 0.7T 时，漏磁系数高达 1.34，这时其铁轭已趋近饱和，随着激磁电流继续增加，激磁功率急剧增加，而背景磁感强度增长缓慢，尽管其磁极头之间气隙长度只有60mm，当背景磁感强度达到0.78T时，漏磁系数已达到2.16，若想继续提高背景磁感强度非常困难。

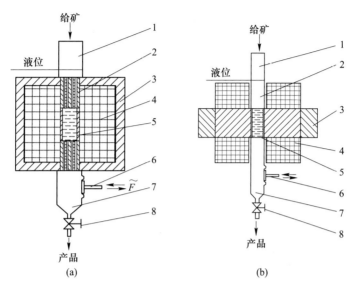

图6 实验室型周期式脉动高梯度磁选机
（a）垂直磁场脉动高梯度磁选机；（b）水平磁场脉动高梯度磁选机
1—给矿盒；2—磁极头；3—磁轭；4—激磁线圈；5—磁介质；6—脉动机构；7—接矿斗；8—阀门

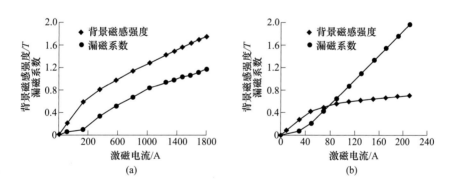

图7 2种磁选机的磁场特性实测数据
（a）垂直磁场脉动高梯度磁选机；（b）水平磁场脉动高梯度磁选机

4 选矿对比试验

为了检验垂直磁场和水平磁场脉动高梯度磁选机在选矿方面的性能，以海南昌江赤铁矿为试样在实验室进行了对比试验。

海南省昌江县是海南钢铁公司所在地，其铁矿资源主要以赤铁矿为主，其中含少量磁铁矿。本次样品从昌江光大矿业有限公司生产流程中采取，采样点为强磁选给矿，经化验，铁品位为46.89%，粒度为小于0.074mm（200目）占90%。

如图6所示的垂直磁场和水平磁场脉动高梯度磁选机进行对比试验。固定脉动冲程冲次，矿浆流速相等，磁介质均采用ϕ2mm导磁不锈钢棒，给矿量按分选腔水平截面积比例配给。试验时,除了磁感应强度为变量之外,其余参数均固定在相同的水平。对比试验数据见表1和图8。

表1 2种磁选机分选昌江赤铁矿对比试验结果

磁 场	磁感强度/T	产品	产率/%	品位/%	回收率/%	选矿效率/%
垂直磁场	0.4	精矿	31.41	64.24	43.03	12.72
		尾矿	68.59	38.94	56.97	
		给矿	100.00	46.89	100.00	
	0.5	精矿	37.06	64.81	51.22	17.45
		尾矿	62.94	36.34	48.78	
		给矿	100.00	46.89	100.00	
	0.6	精矿	39.05	64.59	53.79	18.52
		尾矿	60.95	35.76	46.21	
		给矿	100.00	46.89	100.00	
	0.7	精矿	50.57	64.62	69.69	29.68
		尾矿	49.43	28.75	30.31	
		给矿	100.00	46.89	100.00	
	0.8	精矿	54.35	65.43	75.84	37.77
		尾矿	45.65	24.82	24.16	
		给矿	100.00	46.89	100.00	
	0.9	精矿	56.60	64.25	77.55	36.26
		尾矿	43.40	24.25	22.45	
		给矿	100.00	46.89	100.00	
水平磁场	0.2	精矿	25.77	63.60	34.95	8.94
		尾矿	74.23	41.09	65.05	
		给矿	100.00	46.89	100.00	
	0.3	精矿	33.73	63.93	45.99	13.64
		尾矿	66.27	38.22	54.01	
		给矿	100.00	46.89	100.00	
	0.4	精矿	38.85	65.19	54.01	19.63
		尾矿	61.15	35.26	45.99	
		给矿	100.00	46.89	100.00	
	0.5	精矿	42.54	64.17	58.22	20.40
		尾矿	57.46	37.86	41.78	
		给矿	100.00	46.89	100.00	
	0.6	精矿	45.34	63.58	61.48	21.33
		尾矿	54.66	33.05	38.52	
		给矿	100.00	46.89	100.00	
	0.7	精矿	47.76	63.36	64.53	22.88
		尾矿	52.24	31.83	35.47	
		给矿	100.00	46.89	100.00	

从对比试验数据可见,垂直磁场脉动高梯度磁选所获得的最高精矿铁品位和最高选矿效率在背景磁感强度为0.8T处,分别为65.43%和37.77%。而水平磁场脉动高梯度磁选的最高精矿铁品位为65.19%,此时背景磁感强度为0.4T,相应的选矿效率仅为19.63%;选矿效率最高为22.88%,相应的精矿铁品位为63.36%,背景磁感强度为0.7T(如前所述,水平磁场磁系漏磁系数大,其背景磁感强度达到0.7T以后,很难继续增加,因此本试验中因设备的限制而未做更高的背景磁感强度试验)。垂直磁场的尾矿铁品位最低为

24.25%，最高作业回收率为77.55%；而水平磁场的尾矿铁品位最低为31.83%，最高作业回收率为64.53%，前者明显优于后者。

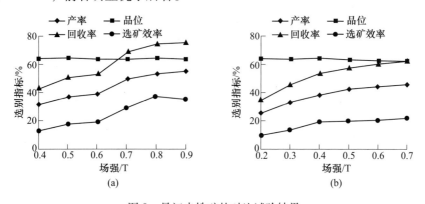

图 8 昌江赤铁矿的对比试验结果
(a) 垂直磁场脉动高梯度磁选机；(b) 水平磁场脉动高梯度磁选机

5 结论

（1）在垂直磁场中，磁介质捕收弱磁性矿物的表面区域是矿浆流体竞争力最小的地方，有利于捕收微细粒的弱磁性矿物；而水平磁场磁介质捕收弱磁性矿物的表面区域是矿浆流体竞争力最大的地方。因此，垂直磁场脉动高梯度磁选机具有对弱磁性矿物回收率较高的优点。

（2）垂直磁场中的磁介质不易产生磁封堵现象，垂直磁场的磁系结构有利于设备处理量的放大，因此，垂直磁场脉动高梯度磁选机比水平磁场脉动高梯度磁选机易于增加设备处理量。

（3）垂直磁场磁系结构具有漏磁系数小、场强高、电耗低的优点，采用垂直磁场的脉动高梯度磁选机在电磁性能方面明显优于水平磁场的磁选机。

（4）选矿试验证明，采用垂直磁场的脉动高梯度磁选机在铁精矿品位略显优势的前提下，其选矿回收率、选矿效率、尾矿品位均明显优于水平磁场脉动高梯度磁选机。

参 考 文 献

[1] 熊大和. 弱磁性矿粒在棒介质高梯度磁场中的动力学分析 [J]. 金属矿山，1998（8）：19~22.
[2] 熊大和. SLon-1750 立环脉动高梯度磁选机的研制与应用 [J]. 金属矿山，1999（10）：23~26.

写作背景 出于学术上的探讨发表了此论文。当时有人认为垂直磁系好，也有人认为水平磁系好。为了比较这两种磁系，我们研制了垂直磁系和水平磁系的小型实验磁选机。通过理论分析和小型实验表明，垂直磁系磁选机具有场强较高、能耗较低和选矿指标较好的优点。

SLon型磁选机在红矿选矿工业中的应用

熊大和[1,2]

（1. 赣州金环磁选设备有限公司；2. 赣州有色冶金研究所）

摘　要　SLon立环脉动高梯度磁选机和反浮选等高效选矿设备和技术促进了我国红矿选矿工业的高速发展。介绍了SLon磁选机在强磁—反浮选和全磁选等红矿选矿流程中的工业应用，也探索了强磁—离心机精选分选细粒铁矿以获得较高铁精矿品位的新途径。

关键词　SLon立环脉动高梯度磁选机　氧化铁矿　强磁选　反浮选　离心选矿机

Application of SLon Magnetic Separator in Oxidized Iron Ore Processing Industry

Xiong Dahe[1,2]

(1. Ganzhou Jinhuan Magnetic Separation Equipment Co., Ltd.
2. SLon Magnetic Separator Ltd.)

Abstract　High-efficiency equipment and technology such as SLon vertical ring pulsating high gradient magnetic separators and reverse flotation has promoted the rapid development of China's red ore (oxidized iron ore) processing industry. The paper describes the industrial application of SLon magnetic separators in the flowsheets of high intensity magnetic separation-reverse flotation and sole magnetic separation for red ore beneficication and investigates the new way to obtain high grade iron concnetrate by treating fine oxidized iron ore with high intensity magnetic separator-centrifugal separator.

Keywords　SLon vertical ring pulsating high gradient magnetic separator; Oxidized iron ore; High intensity magnetic separation; Reverse flotation; Centrifugal separator

　　随着国民经济的发展，我国钢铁工业近年得到飞速的发展，2003年我国钢产量达到2.2亿吨，钢铁工业对铁矿石的巨大需求促使了我国铁矿选矿技术水平的迅速提高。经过几十年的攻关，我国红矿（氧化铁矿）选矿工艺和设备有了很大的发展，其中SLon立环脉动高梯度磁选机和反浮选等高效选矿设备和技术的应用使我国红矿选矿技术达到了国际领先水平。

　　我国红矿资源的特点是储量大、品位低、嵌布粒度细、绝大部分难磨难选，而我国的钢铁冶炼工业又普遍需要高品位的铁精矿，SLon立环脉动高梯度磁选机就是在这种需求旺盛而供给不足的环境下成长起来的。通过20多年的研制、创新与改进，SLon立环脉动高梯度磁选机已形成系列化产品，具有优异的选矿性能和设备性能，并已广泛应用于红矿

---

❶ 原载于《全国选矿新技术及其发展方向学术研讨与技术交流会会议论文集》，2004：166~169。

的选矿工业中,该机已成为我国新一代高效强磁选设备。

1 SLon 磁选机在强磁—反浮选中的应用

强磁—反浮选是当前红矿选矿最有效的技术,可获得相当高的铁精矿品位和较高的铁回收率。例如,鞍钢集团鞍山矿业公司在 2001 年对齐大山选矿厂二选车间进行技术改造,将煤气焙烧—弱磁选分选鞍山式贫赤铁矿的生产工艺改为阶段磨矿—重选—强磁—反浮选流程,其中细粒级部分用滚筒式弱磁选机选出磁铁矿,弱磁选机的尾矿用 6 台 SLon-1750 立环脉动高梯度磁选机进行粗选抛尾,弱磁选精矿和强磁选精矿混合后用反浮选精选。2003 年,该车间又采用 6 台 SLon-1500 立环脉动中磁选机(背景磁感应强度 0.4T)取代原有的 6 台滚筒式永磁中磁机,作为螺旋溜槽中矿的抛尾设备,使该流程选矿指标进一步提高。改进后的数质量流程见图 1。该流程中,6 台 SLon-1750 立环脉动高梯度磁选机将细粒级尾矿品位控制在 8.91% 左右的水平,6 台 SLon-1500 立环脉动中磁机将螺旋溜槽尾矿品位控制在 8.05% 左右的水平,这两种设备抛弃的尾矿量占全流程尾矿量的 77.82%(另有 22.18% 由浮选抛尾),有效地保证了全流程铁回收率达到 78% 以上。

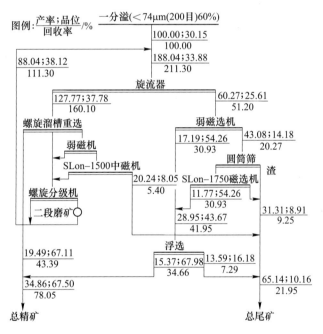

图 1 齐大山选矿厂二选车间数质量原则流程

目前,齐大山选矿厂一选车间也采用 4 台 SLon-1750 立环脉动高梯度磁选机和 4 台 SLon-1500 立环脉动中磁选机组成图 1 所示的选矿流程。该厂一选和二选车间年处理鞍山式贫赤铁矿共 720 万吨,全厂的综合指标为原矿品位 30.15%,铁精矿品位 67.50%,尾矿品位 10.16%,铁回收率 78.05%,其选矿指标再创历史新高。

强磁—反浮选技术在鞍山地区的成功应用,带动了我国红矿选矿技术的高速发展,目前我国采用 SLon 磁选机进行强磁—反浮选的部分厂矿见表 1。

表1 我国部分采用SLon磁选机—反浮选分选红矿的厂矿

厂矿名称	选矿流程	设备型号	分选指标/%	备 注
鞍钢齐大山选矿厂	阶段磨矿、分级重选—强磁—反浮选	SLon-1750强磁机11台 SLon-1500中磁机11台 SLon-1500强磁机1台	$\alpha=30.15$, $\beta=67.50$, $\theta=10.15$, $\varepsilon=78.05$	工业生产
鞍钢东鞍山烧结厂一选车间	阶段磨矿、分级重选—强磁—反浮选	SLon-1750强磁机8台 SLon-2000强磁机2台 SLon-1750中磁机10台	$\alpha=32.50$, $\beta=64.50$, $\theta=14.50$, $\varepsilon=71.45$	工业生产
鞍钢调军台选矿厂	连续磨矿—强磁—反浮选	SLon-2000强磁机6台 Shp-3200磁选机9台	$\alpha=30.50$, $\beta=67.50$, $\theta=11.00$, $\varepsilon=76.38$	工业生产
鞍钢弓长岭选矿厂	阶段磨矿、分级重选—强磁—反浮选	SLon-2000强磁机4台 SLon-2000中磁机4台	$\alpha=27.78$, $\beta=67.19$, $\theta=9.96$, $\varepsilon=76.29$	在建工程设计指标
唐钢司家营铁矿	阶段磨矿、分级重选—强磁	SLon-1750强磁机12台 SLon-1750中磁机8台		在建工程
安阳钢铁公司舞阳矿业公司	阶段磨矿—强磁(一粗一扫)—强磁精矿再磨—反浮选	SLon-2000强磁机12台		在建工程
海南钢铁公司	连续磨矿—强磁一粗一扫—反浮选	SLon-1750强磁机8台		在建工程

注：表中α为入选原矿品位，β为铁精矿品位，θ为尾矿品位，ε为铁回收率。

2 SLon磁选机在全磁选流程中的应用

尽管强磁—反浮选工艺流程可获得较高的铁精矿品位，分选鞍山式铁矿石取得了很大的成功。但是该工艺流程较适应于矿石性质比较简单，铁矿物与脉石解离较好，脉石以石英为主的矿石。此外，由于浮选工艺较复杂，药剂成本和环保投资较高等原因，对于一些中小型矿山或矿石性质比较复杂的矿石，或早期已建设好的矿山要实现强磁—反浮选有一定的难度。因此目前有不少厂矿仍采用全磁选流程。

例如马钢姑山铁矿分选赤铁矿采用6台SLon-1750立环脉动高梯度磁选机和4台SLon-1500立环脉动高梯度磁选机组成的全磁选流程，见图2。该流程中，一段磨矿的分级溢流经圆筒筛隔渣后用3台SLon-1750立环脉动高梯度磁选机粗选，抛弃产率为27%左右的尾矿，其粗精矿进二段磨矿至小于74μm（200目）85%，然后用3台SLon-1750立环脉动高梯度磁选机精选和4台SLon-1500立环脉动高梯度磁选机对精选尾矿进行扫选。全流程的选矿指标大致为给矿品位43.15%，铁精矿品位60.17%，尾矿品位22.41%，铁回收率76.59%。该流程的特点是：（1）阶段磨矿、强磁粗选可提前抛弃产率为27%左右的尾矿，减少二段磨矿量；（2）全部是开路流程，没有中矿循环，矿石处理量大，生产流程稳定易控制；（3）磁选生产成本低。

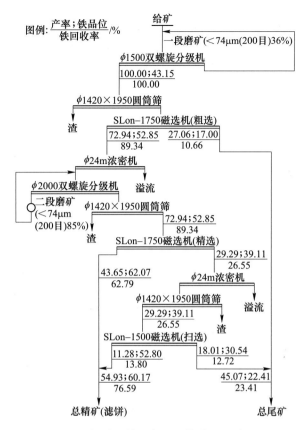

图 2 马钢姑山铁矿全磁选数质量原则流程

目前我国采用 SLon 磁选机用于全磁选流程的部分厂矿见表 2。

表 2 我国采用 SLon 磁选机用于全磁选流程的部分厂矿

厂矿名称	选矿流程	设备型号	分选指标/%	备注
马钢姑山铁矿	一段磨矿、强磁粗选,二段磨矿、强磁(一精一扫)	SLon-1750 强磁机 6 台 SLon-1500 强磁机 4 台	$\alpha=43.15$, $\beta=60.17$, $\theta=22.41$, $\varepsilon=76.59$	工业生产
宝钢梅山铁矿	弱磁—强磁(一粗一扫)降磷流程	SLon-1500 强磁机 18 台	$\alpha=\text{Fe }52.77$, $\alpha=\text{P }0.399$, $\beta=\text{Fe}56.08$, $\beta=\text{P }0.246$, $\theta=\text{Fe}26.13$, $\varepsilon=\text{Fe }94.51$	工业生产
内蒙古苏尼特宏鑫选厂	连续磨矿、弱磁—强磁(一粗一精)	SLon-2000 强磁机 2 台 SLon-1750 强磁机 2 台	$\alpha=38.36$, $\beta=60.19$, $\theta=22.34$, $\varepsilon=66.40$	工业生产, 铁精矿含 锰约 4%
海南钢铁公司选矿厂	连续磨矿—强磁(一粗一精一扫)	SLon-2000 强磁机 4 台	$\alpha=51.09$, $\beta=64.20$, $\theta=24.50$, $\varepsilon=84.16$	工业生产
海南昌江光大公司	连续磨矿—弱磁—强磁(一粗一精)	SLon-1750 强磁机 2 台	$\alpha=49.01$, $\beta=63.50$, $\theta=25.60$, $\varepsilon=80.03$	工业生产
昆钢大红山铁矿	阶段磨矿—弱磁—强磁阶段选别	SLon-1500 强磁机 4 台	$\alpha=37.69$, $\beta=64.24$, $\theta=12.14$, $\varepsilon=83.59$	工业生产
昆钢罗次铁矿	连续磨矿—弱磁—强磁(一粗一精)	SLon-1500 强磁机 3 台	$\alpha=45.86$, $\beta=62.51$, $\theta=24.74$, $\varepsilon=76.22$	工业生产

续表2

厂矿名称	选矿流程	设备型号	分选指标 /%	备注
宁夏中卫铁矿（褐铁矿）	一段磨矿—强磁（一粗一扫）	SLon-1500 强磁机 5 台	$\alpha = 46.36$, $\beta = 53.06$, $\theta = 28.43$, $\varepsilon = 83.32$	工业生产
首钢秘鲁铁矿	阶段磨矿—弱磁—强磁—浮硫	SLon-1750 强磁机 2 台	$\alpha = 55.20$, $\beta = 65.09$, $\theta = 28.10$, $\varepsilon = 86.49$	工业生产

3 SLon 磁选机在强磁—离心机流程中的探索

上述全磁选流程分选红矿具有流程较简单，生产成本较低的优点，但铁精矿品位普遍不如强磁—反浮选。为了探索提高细粒红矿铁精矿品位的新途径，我们开展了强磁粗选—离心机精选的探索试验。

从选矿原理上分析，磁铁矿的磁性率（比磁化系数）大约是赤铁矿的 100 倍，SLon 磁选机对细粒铁矿物的作用力大约是弱磁选机的 100 倍。因此，在强磁选过程中，对于含有少量磁铁矿的贫连生体，强磁选机均捕收至铁精矿中。如果这种铁精矿不经过磨矿再用强磁选机精选，这些贫连生体仍然捕收至铁精矿中。因此，强磁精选对提高铁精矿品位的作用不大。

强磁粗选的细粒铁精矿若用离心机精选，含少量磁铁矿的石英等贫连生体因密度较轻可进入到尾矿之中去，因此离心机精选可获得较高的铁精矿品位。

以海南昌江铁矿为例，当地矿石含磁铁矿产率占 10%～15%，其余为赤铁矿，对这种矿样分别进行强磁粗选—强磁精选和强磁粗选—离心机精选的试验结果如图 3 和表 3 所示。

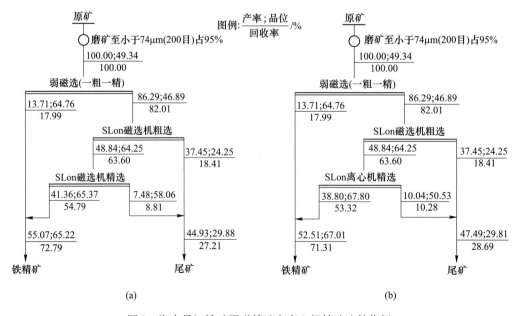

图 3 海南昌江铁矿强磁精选和离心机精选比较指标
(a) 弱磁—强磁粗选—强磁精选；(b) 弱磁—强磁粗选—离心机精选

表3 细粒铁矿强磁精选和离心机精选对比指标　　　　　　　　　　　　（%）

项目名称	作业指标				全流程指标			
	给矿品位	精矿品位	尾矿品位	铁回收率	给矿品位	精矿品位	尾矿品位	铁回收率
强磁机精选	64.25	65.37	58.06	86.15	49.34	65.22	29.88	72.79
离心机精选	64.25	67.80	50.53	83.83	49.34	67.01	29.81	71.31
差　值	0	+2.43	-7.53	-2.32	0	+1.79	-0.07	-1.48

强磁机精选作业仅将铁精矿品位从64.25%提高至65.37%，提高幅度为1.12%；而离心机精选将铁精矿品位从64.25%提高到67.80%，提高幅度为3.55%。强磁机精选的尾矿品位为58.06%，而离心机精选的尾矿品位为50.53%。因此，离心机精选的指标明显优于强磁机精选指标。

尽管离心机的精选指标较强磁机精选指标好，但由于离心机处理量小，设备故障率较高，在生产上较难操作维护。因此，目前生产上分选铁矿用离心机精选的几乎没有。

4 结论

（1）通过多年不懈的研究、创新与发展，SLon立环脉动高梯度磁选机已具有优异的机电性能和选矿性能。该机具有选矿效率高、磁介质不堵塞、设备作业率高、工作性能稳定、易于操作维护的优点，在我国红矿选矿工业中得到广泛的应用。

（2）SLon立环脉动高梯度磁选机在强磁—反浮选工艺流程中的应用，促进了我国红矿选矿技术的飞跃性发展，目前已有一大批红矿选矿厂已建成或正在建设强磁—反浮选流程。强磁—反浮选工艺分选鞍山式铁矿石可获得的选矿生产指标为原矿品位30%左右，铁精矿品位64.5%~67.5%，铁回收率70%~80%的优良指标，使我国的红矿选矿技术达到国际领先水平。

（3）全磁选流程已在我国众多中小型红矿选矿厂大量的应用。全磁选流程具有流程简单、运行稳定、生产成本低、铁回收率较高的优点，其缺点是难以获得很高的铁精矿品位。

（4）强磁—螺旋溜槽重选已在我国很多红矿选矿厂应用于分选较粗粒级的铁精矿，但对细粒级红矿，重选几乎没有工业应用。通过探索试验表明，强磁—离心机重选分选细粒红矿可获得较高的铁精矿品位。由于目前工业离心机存在处理量小、设备故障率较高的缺点，因此这种工艺流程的应用前景将取决于离心机设备的发展。

写作背景　本文总结了SLon立环脉动高梯度磁选机分选氧化铁矿的几种流程，即SLon磁选机-反浮选流程、全磁选流程和SLon磁选机-离心选矿机流程。其中SLon磁选机-反浮选流程在全国已得到广泛的应用，而SLon磁选机-离心选矿机流程则刚处于试验阶段。

SLon Magnetic Separators Applied in the Ilmenite Processing Industry[0]

Xiong Dahe

(Ganzhou Nonferrous Metallurgy Research Institute, 36 Qingnian Road, Ganzhou 341000, Jiangxi Province, P. R. China)

(Received 10 September 2004; Accepted 18 September 2004)

Abstract SLon vertical ring pulsating high gradient magnetic separators possess the advantages of a large beneficial ratio, high processing efficiency, strong adaptability, high resistance of the magnetic matrix to clogging, and high equipment availability. In recent years, these separators have been successfully applied to the beneficiation of ilmenite ores at the Panzhihua Ilmenite concentrator, and several other ilmenite processing plants, promoting the rapid development of the ilmenite industry in China.

Keywords SLon high-gradient magnetic separator; Ilmenite; Matrix

1 Introduction

The ilmenite deposit in Panzhihua region, Sichuan Province of China accounts for 35% of the titanium resource in the world, and for approximately of 92% in China. Efficient utilization of the ilmenite resources is very important for development of the titanium industry.

Because a SLon vertical ring pulsating high gradient magnetic separator (SLon VPHGMS) possesses the advantages of high efficiency, a high beneficiation ratio, a high resistance of the matrix to clogging and high reliability[1~7], they have been widely applied, in recent years, in the ilmenite processing industry in China.

2 Mineralogical features

The Panzhihua deposit consists of vanadium, titanium and iron minerals and other gangue minerals. Magnetite is the major iron mineral and vanadium exists in magnetite. Ilmenite is the major titanium mineral. A typical elemental assay and mineralogical analysis of ilmenite-containing minerals are shown in Tables 1 and 2, respectively.

Table 1 Elemental assay of Panzhihua Ore (%)

TFe	FeO	Fe_2O_3	TiO_2	V_2O_5	SiO_2	Al_2O_3	CaO	MgO	Na_2O
27.10	18.51	18.46	6.51	0.27	26.64	9.95	3.90	9.15	0.51
K_2O	S	P	MnO	Cu	Co	Ni	Cr_2O_3	Roasting lost	
0.28	0.35	0.038	0.32	0.027	0.020	0.024	0.025	4.80	

❶ 原载于《Physical Separation in Science and Engineering》,2004,9。

Table 2 Mineralogical analysis of Panzhihua Ore   (%)

Mineral	Magnetite	Hematite	Ilmenite	Pyrite	Olivine	Titaniferous augite	Plagioclase
Mass	30.6	3.5	6.1	0.7	16.2	13.1	15.8
Mineral	Hornblende	Mica	Chlorite	Cyphoite	Kaolin	Spinel	Others
Mass	2.3	0.2	4.2	3.0	3.6	0.5	0.2

All of the mineral processing plants in the Panzhihua region recover magnetite with low-intensity magnetic separation (LIMS), while the tailings which contain about 10% TiO_2, serve as the feed to various ilmenite processing flowsheet. The typical elemental assay and mineral properties in the tailings of LIMS are shown in Tables 3 and 4.

Table 3 Elemental assay of the tails of LIMS(%)

TFe	FeO	Fe_2O_3	TiO_2	V_2O_5	SiO_2	Al_2O_3	CaO	MgO
13.40	10.56	7.37	9.51	0.072	35.76	11.67	11.22	9.26
Na_2O	K_2O	S	P	Cu	Co	Ni	Cr_2O_3	
0.68	0.85	0.564	0.021	0.015	0.019	0.012	0.008	

Table 4 Typical mineral properties in the tailings of LIMS(%)

Mineral	Molecule	Mass /%	Grade (TiO_2)/%	Specific density /g·cm^{-3}	Magnetic susceptibility /cm^3·g^{-1}	Specific resistance /Ω·cm
Vanadium-titanium magnetite	$(V,Ti,Fe)_3O_4$	2-5	13.4	4.6-4.8	$>1\times10^{-3}$	1.38×10^6
Ilmenite	$FeTiO_3$	15.3-17.5	51.89	4.6-4.7	0.24×10^{-3}	1.75×10^5
Titaniferous Augite	$Ca(Mg,Fe,Ti)[(Si,Al)_2O_6]$	46-46.5	1.85	3.2-3.3	0.1×10^{-3}	3.13×10^{13}
Pyrite	FeS_2	2-2.5		4.4-4.9	4.1×10^{-3}	1.25×10^4
Plagioclase	$(100-n)Na[AlSi_3O_8]\cdot nCa[Al_2Si_2O_8]$	31-34	0.097	2.7	0.014×10^{-3}	$>10^{14}$

3 SLon VPHGMS applied to the recovery of fine ilmenite

The Panzhihua Ilmenite Processing Plant is the biggest producer of ilmenite concentrate in China. The iron processing plant recovers magnetite and vanadium before the ilmenite processing plant. The tailings from the iron processing plant represent the feed into the ilmenite processing plant.

The previous flowsheet of the ilmenite processing plant was based on classification into >0.1mm(coarse), 0.1-0.045mm (middle) and <0.045mm (fine) fractions. The coarse and middle particle fractions were treated with spirals and electrical separators. The fine fraction was discarded as the final tailings for the lack of efficient beneficiation technology. The overall recovery of ilmenite had been only 17%-20% of the feed to the ilmenite processing plant.

In recent years, the mass of the fine fraction has reached about 50% as the iron processing plant grinding became finer and finer for higher iron concentrate grade. The TiO_2 grade of the fine fraction is almost the same as that of the feed to the ilmenite plant. The recovery of ilmenite from the fine particle fraction is getting more and more important in order to increase the overall recovery of the ilmenite plant.

From 1997 to 2004, the Panzhihua Ilmenite Processing Plant built two circuits for the recovery of <0.045mm ilmenite using the SLon VPHGMS-Flotation flowsheet. Twelve SLon-1500 VPHGMS are applied as roughers, followed by flotation for cleaning as shown in Figs. 1 and 2.

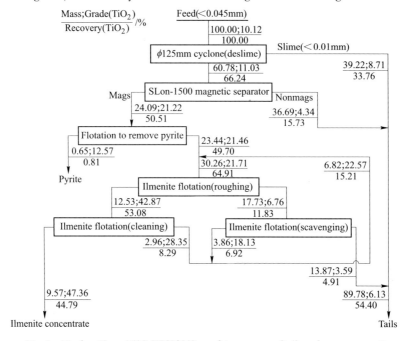

Fig. 1 Twelve SLon-1500 VPHGMS machines are applied to the recovery of <0.045mm ilmenite in the Panzhihua Ilmenite Processing Plant

Fig. 2 A photograph of SLon-1500 VPHGMS in the Panzhihua Ilmenite Processing Plant

The SLon VPHGMS operation is very important. It discards most of the gangue and slimes before flotation. Only a small mass portion containing about 21.22% TiO_2 enters flotation. It creates a very good condition for flotation, which guarantees that the flotation operation can get good results at low cost. The total results of the flowsheet are: feed grade 12.12% TiO_2, ilmenite concentrate grade 47%-48% TiO_2, recovery 44.79% TiO_2.

Now The Panzhihua Ilmenite Processing Plant produces 150000 tonnes of <0.045mm ilmenite concentrate with average 47.5% TiO_2 each year. The TiO_2 overall recovery of the plant was increased from 17%-20% to about 35%.

4 Application of SLon VPHGMS for the recovery of ilmenite from the Panzhihua tailings

The Panzhihua Iron and Steel Company treats 10 million tones of iron ore annually. Approximately 5.7 million tonnes of tailings from the low-intensity magnetic separation circuit, containing about 9.5% TiO_2 serve as the feed to the Panzhihua Ilmenite Processing Plant as mentioned above. After a portion of ilmenite is recovered into the magnetic concentrate, the tailings of the plant still contain about 7% TiO_2, in about 5.4 million tonnes that are discharged and pumped to the tailings dump.

In the past few years, a mineral processing plant has been built at the middle of the pumping pipeline to recover the remaining iron and ilmenite minerals from the discarded tailings. The flowsheet is shown in Fig. 3. Low intensity magnetic separators recover the remaining magnetite. Two SLon-1500 VPHGMS are applied as the roughing stage and one SLon-1500 VPHGMS is used as the cleaning stage to recover ilmenite. As the Panzhihua tailings are enormous, the feed into each SLon-1500 at the roughing stage is about 45t/h (nominally 20 to 30t/h), so that as much ilmenite as possible is recovered. The magnetic fraction of the cleaning SLon-1500 containing 21.55% TiO_2 are condensed and classified into > 0.1mm (coarse), 0.1-0.019mm (fine) and <0.019mm (slime) fractions. The coarse fraction is cleaned further by shaking tables and the fine fraction is cleaned further by flotation. The slime is discharged directly as the final tailings. The advantages of this flowsheet are that three SLon-1500 separators discharge 81.5% of the mass fraction to the final tailings at very low cost. Only 18.50% mass fraction enters the further cleaning

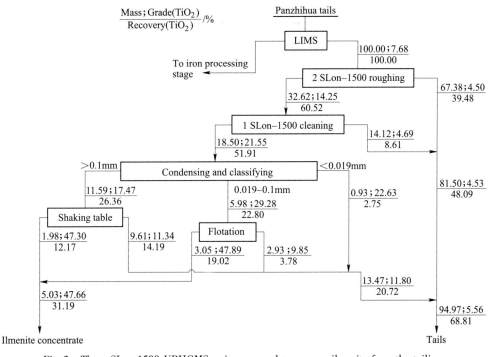

Fig. 3　Three SLon-1500 VPHGMS units are used to recover ilmenite from the tailings

stages. The results of the total flowsheet are: feed grade 7.68% TiO_2, ilmenite concentrate grade 47.66% TiO_2, TiO_2 recovery 31.19%. This flowsheet recovers 30000 tonnes of the ilmenite concentrate annually, at good quality, and about 50000 tonnes of the iron concentrate is also recovered each year.

5 Application of the SLon VPHGMS to the recovery of ilmenite in the Taihe Iron Mine

The Taihe Iron Mine is located in Xichang of the Sichuan Province, close to the Panzhihua Region. Its ore is similar to the Panzhihua ore. A low-intensity magnetic separation (LIMS) circuit was installed to recover magnetite before 1994. In 1994 to 1995, an ilmenite processing circuit was introduced to produce ilmenite concentrate from the tailings from the LIMS circuit. At first spirals and shaking tables were used as roughers and flotation was used as cleaner, however spirals and shaking tables were not able to recover fine ilmenite particles. Their roughing concentrate contained only coarse particles, and the flotation cleaning stage was not able to recover coarse ilmenite particles. Therefore, high losses of TiO_2 from the coarse and fine fractions were incurred. The TiO_2 recovery was, therefore, very low, only about 10% TiO_2 from the LIMS tailings. The ilmenite concentrate production was only 2000 tonnes per annum.

In 2000, three SLon-1500 VPHGMS units were installed at the Taihe Iron Mine to recover ilmenite. Two SLon-1500 separators replaced the previously used spirals and one SLon-1500 separator replaced the previously used shaking tables. These three SLon-1500 separators are incorporated, together with the flotation circuit, into a magnetic-flotation flowsheet as shown in Fig. 4.

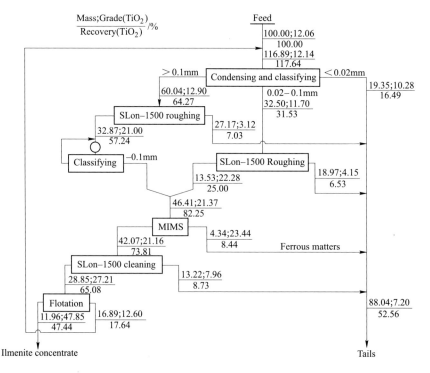

Fig. 4 Three SLon-1500 high-gradient magnetic separators are applied at the Taihe ilmenite processing flowsheet

In the first stage, the feed (the tailings from LIMS) is thickened and classified into >0.1mm(coarse), 0.1-0.02mm (fine) and <0.02mm (slime) fractions. The coarse and the fine fractions are roughed separately by two SLon-1500 separators. The coarse magnetics from the SLon machine are milled to <0.1mm, and mixed with the fine magnetics from the SLon-1500 unit. A medium-intensity magnetic separator (MIMS) of the drum type is used to remove the remaining magnetite and other ferrous materials. The non-magnetics from the MIMS machine are cleaned by a SLon-1500 separator, the magnetics of which are fed into the flotation stage for further cleaning.

The advantages of this flowsheet are as follows:

(1) The coarse and the fine fractions are treated separately by SLon VPHGMS, so that different magnetic parameters can be chosen for optimal results.

(2) The grade of the feed into the flotation circuit is relatively high, about 27.21% TiO_2, which saves flotation reagents and reduces the production costs.

(3) The flotation tailings containing about 12.60% TiO_2 are returned to the feed of the flowsheet which helps to increase TiO_2 recovery.

This flowsheet recovers effectively coarse and fine ilmenite fractions. It guarantees that the grade of the ilmenite concentrate is greater than 47% TiO_2 and the overall recovery of TiO_2 is increased to approximately 47%. Presently the production of the ilmenite concentrate at the Taihe Mine reaches 40000 tonnes per annum, compared to the previous production of 2000 tonnes per annum. In 2003 to 2004, this mine ordered additional three SLon-1750 VPHGMS machines to scale-up the ilmenite processing flowsheet. The work is expected to be completed in 2005.

6 Conclusion

(1) SLon vertical ring pulsating high-gradient magnetic separators possess the advantages of large beneficiation ratio, high mineral processing efficiency, high resistance of the matrix to clogging, high throughput, high equipment reliability and availability. In recent years, these separators have been successfully applied in eight Chinese ilmenite processing plants. Three typical examples of ilmenite processing flowsheet are described in this article.

(2) Twelve SLon-1500 VPHGMS units are applied at Panzhihua Ilmenite Processing Plant to recover <0.045mm ilmenite. These separators are used to beneficiate the feed containing 10.12% TiO_2 to 21.22% TiO_2, at low cost. The mass of SLon magnetic concentrate is only 24.09%, which is very favorable for the following flotation cleaning stage. The overall results of the magnetic-flotation flowsheet are: feed grade 10.12% TiO_2, ilmenite concentrate grade 47.36% TiO_2, recovery 44.79% TiO_2. Each year 150000 tonnes of the ilmenite concentrates of good quality are produced by this flowsheet.

(3) Three SLon-1500 VPHGMS machines are applied to recover ilmenite from the Panzhihua tailings; two of them are used as roughers and the other is used as a cleaner. These machines enrich the feed, with wide size distribution, from 7.68% TiO_2 to 21.55% TiO_2, The SLon magnetics are then classified into >0.1mm (coarse), 0.019-0.1mm (fine), <0.019mm (slime) fractions.

The coarse fraction is further cleaned by shaking tables and the fine fraction is further cleaned by flotation. The overall results of this magnetic—gravity-flotation flowsheet are: feed grade 7.68% TiO_2, ilmenite concentrate grade 47.66% TiO_2, recovery 31.19% TiO_2. Annually, 30000 tonnes of good quality ilmenite concentrate are produced.

(4) Three SLon-1500 VPHGMS units are applied in the Taihe ilmenite processing flowsheet. The feature of this flowsheet is that the feed is first classified into >0.1mm (coarse), 0.02-0.1mm (fine), <0.02mm (slime) fractions. The coarse and the fine fractions are then roughed by the SLon-1500 separators separately for optimized results. The coarse magnetics of a SLon-1500 separator is ground to 0.1mm, mixed with the fine magnetics from the other SLon-1500 separator, and then cleaned by the third SLon-1500 machine, the magnetics of which are further cleaned by flotation. The overall results of this magnetic-magnetic-flotation flowsheet are: feed grade 12.06% TiO_2, ilmenite concentrate grade 47.85% TiO_2, recovery 47.44% TiO_2. Annually 40000 tonnes of the ilmenite concentrates of good quality are produced.

(5) In all of the ilmenite processing flowsheet mentioned in this article, SLon VPHGMS machines are used as roughers or first cleaners. The separators successfully remove most of the gangue at low cost and leave a small mass portion for the final cleaning stage. SLon VPHGMS machines also discard all of the <10μm slimes which otherwise would be harmful to flotation. As the ilmenite flotation reagents are expensive, SLon VPHGMS units greatly reduce the total cost of the ilmenite concentrate production.

(6) The applications of SLon VPHGMS separators successfully solved the difficult problem of beneficiation of the <0.045mm ilmenite fraction, greatly increased the ilmenite recovery and greatly reduced the production cost. The successful applications of SLon VPHGMS separators in the ilmenite processing industry significantly promoted the rapid development of Chinese titanium industry.

References

[1] Xiong Dahe, New development of the SLon vertical ring and pulsation HGMS separator[J]. Magn. Electr., 1994(5):211-222.

[2] Xiong Dahe. Development and commercial test of SLon-2000 vertical ring pulsating high gradient magnetic separator[J]. Magn. Electr.,1997(8):89-100.

[3] Xiong Dahe. Development and applications of SLon vertical ring pulsating high gradient magnetic separators[J]. Proc. XX IMPC, Aachen, Germany, 1997,21-26: 621-630.

[4] Xiong Dahe, Shuyi Liu, Jin Chen. New technology of pulsating high gradient magnetic separation[J]. Int. J. Miner. Process,1998(54):111-127.

[5] Xiong Dahe. Research and commercialisation of treatment of fine ilmenite with SLon magnetic separators[J]. Magn. Electr.,2000(10):121-130.

[6] Xiong Dahe. A large scale application of SLon magnetic separator in Meishan Iron Ore Mine[J]. Magn. Electr.,2002(11):1-8.

[7] Xiong Dahe. SLon magnetic separator applied to upgrading the iron concentrate[J]. Phys. Sep. Sci. Engn., 2003(12):63-69.

写作背景 本文介绍了在攀钢选钛厂应用 12 台 SLon-1500 立环脉动高梯度磁选机分选小于 45μm 粒级钛铁矿，在攀钢尾矿库应用 3 台 SLon-1500 立环脉动高梯度磁选机从尾矿中回收钛铁矿，在西昌太和铁矿应用 3 台 SLon-1500 立环脉动高梯度磁选机分选全粒级钛铁矿的流程。SLon 磁选机的应用大幅度提高了这些厂矿的钛铁矿回收率。

SLon磁选机在红矿强磁反浮选流程中的应用[①]

熊大和[1,2]

(1. 赣州金环磁选设备有限公司; 2. 赣州有色冶金研究所)

摘　要　强磁—反浮选是当今红矿选矿的最有效技术。SLon立环脉动高梯度磁选机具有优异的选矿性能和高度的可靠性,在强磁—反浮选流程中广泛应用于粗选或抛尾作业,可有效地脱除细泥和控制尾矿铁品位,为反浮选作业创造良好的条件和提高全流程铁回收率,促使我国红矿选矿技术高速发展。

关键词　SLon立环脉动高梯度磁选机　氧化铁矿　强磁选　反浮选

SLon Magnetic Separators Applied in Oxidized Iron Ore Processing Industry in High Intensity Magnetic-Reverse Flotation Flowsheets

Xiong Dahe[1,2]

(1. SLon Magnetic Separator Ltd.; 2. Ganzhou Nonferrous Metallurgy Research Institute)

Abstract　High intensity magnetic-reverse flotation is the most efficient technology for oxidized iron ore processing nowadays. SLon vertical ring and pulsating high gradient magnetic separators possess excellent mineral processing ability and high reliability. They are widely applied for roughing or for discharging tails in high intensity magnetic-reverse flotation flowsheets. They can efficiently discharge slims and control the iron grade in the tails, create good conditions for flotation and raise the final iron recovery. They are promoting Chinese oxidized iron ore processing technology developing fast.

Keywords　SLon vertical ring pulsating high gradient magnetic separator; Oxidized iron ore; High intensity magnetic separation; Reverse flotation

我国的红矿(赤铁矿、褐铁矿、镜铁矿、菱铁矿等氧化铁矿)资源丰富,但原矿铁品位低,嵌布粒度细,需要细磨深选才能满足冶炼要求。在探索红矿选矿技术的过程中,广大选矿科技人员通过数十年的努力,强磁—反浮选这种先进的选矿技术在我国得到迅速发展,这种技术可获得很高的铁精矿品位和较高的铁回收率。近几年SLon立环脉动高梯度磁选机在强磁—反浮选流程中得到广泛的应用,将我国红矿选矿工业生产水平推向了世界先进水平的行列。

[①] 原载于《全国选矿高效节能技术及设备学术研讨与成果推广交流会论文集》,2005;79~82。

1 SLon 磁选机简介

SLon 立环脉动高梯度磁选机是一种新型高效的强磁选设备，是国内外第一代成功应用于大规模工业生产的连续式高梯度磁选机。该机成功地解决了平环强磁选机和平环高梯度磁选机磁介质容易堵塞的问题。具有选别效率高、适应性强、设备作业率高、运行费用低、维护工作小的优点。至今有 400 多台 SLon 立环脉动高梯度磁选机在国内外广泛应用于细粒弱磁性金属矿的选矿和非金属矿的提纯。为了满足强磁—反浮选流程的需要，我们研制了多种型号的 SLon 立环脉动高梯度磁选机（见表1）。

表 1 SLon 磁选机主要技术参数

机型	转环外径/mm	给矿粒度/mm	给矿浓度/%	矿浆通过能力/m³·h⁻¹	干矿处理量/t·h⁻¹	额定背景磁场强度/T	额定激磁功率/kW	脉动冲程/mm	脉动冲次/r·min⁻¹	主机质量/t
SLon-500 强磁机	500	-1.0	10~40	0.5~1.0	0.05~0.25	1.0	13.5	0~50	0~400	1.5
SLon-750 强磁机	750	-1.0	10~40	1.0~2.0	0.1~0.5	1.0	22	0~50	0~400	3
SLon-1000 强磁机	1000	-1.3	10~40	12.5~20	4~7	1.0	28.6	0~30	0~300	6
SLon-1250 强磁机	1250	-1.3	10~40	20~50	10~18	1.0	35	0~20	0~300	14
SLon-1500 强磁机	1500	-1.3	10~40	50~100	20~30	1.0	44	0~30	0~300	20
SLon-1750 强磁机	1750	-1.3	10~40	75~150	30~50	1.0	62	0~30	0~300	35
SLon-2000 强磁机	2000	-1.3	10~40	100~200	50~80	1.0	74	0~30	0~300	50
SLon-1500 中磁机	1500	-1.3	10~40	75~150	30~50	0.4	16	0~30	0~300	15
SLon-1750 中磁机	1750	-1.3	10~40	75~150	30~50	0.6	38	0~30	0~300	35
SLon-2000 中磁机	2000	-1.3	10~40	100~200	50~80	0.6	42	0~30	0~300	40

表 1 中的 SLon 强磁选机指额定背景磁感应强度达到 1.0T 的 SLon 立环脉动高梯度磁选机，SLon 中磁机指额定背景磁感应强度为 0.4~0.6T 的 SLon 立环脉动高梯度磁选机。在阶段磨矿、重选—强磁—反浮选流程中，SLon 强磁机用于粗选作业或分选细粒的（小于 74μm（200 目）90% 以上）红矿，SLon 中磁机主要用于粗粒级（如螺旋溜槽尾矿）的抛尾作业。在这些作业中 SLon 磁选机具有处理量大、尾矿品位低、可调性好、工作稳定、生产成本低的优点。

2 SLon 磁选机在鞍山式红矿强磁—反浮选中的应用

强磁—反浮选流程在鞍山地区已得到迅速发展和成功应用，目前齐大山选矿厂、东鞍山烧结厂一选车间、调军台选矿厂、弓长岭选矿厂三选车间及正在建设的胡家庙选矿厂均采用强磁—反浮选流程分选红矿。

2.1 SLon 磁选机在鞍钢齐大山选矿厂的应用

鞍山矿业公司齐大山选矿厂是年处理 720 万吨贫赤铁矿的大型选矿厂，原选矿流程为原矿经粗、中两段破碎后筛分成块矿和粉矿。块矿在二选车间用焙烧磁选工艺处理，粉矿在一选车间用阶段磨矿、重选—强磁—弱酸性正浮选工艺处理。由于选厂的工艺和装备较落后，其铁精矿品位只能达到 63% 左右。

从 2001 年至 2004 年，齐大山选矿厂一选和二选车间全部改为阶段磨矿、重选—强磁—反浮选的选矿流程（见图1），其中采用11台SLon-1750强磁机控制细粒级尾矿品位，另采用11台SLon-1500中磁机控制螺旋溜槽尾矿品位。SLon磁选机在该流程中为降低尾矿品位和提高铁回收率发挥了关键作用。该机脱泥效果好，为反浮选提高铁精矿品位和降低药剂消耗创造了良好的条件。由于SLon磁选机的机电性能良好，磁介质长期不堵塞，设备作业率在99%以上，设备的水电消耗和维护费用低，为选厂降低生产成本、长期稳定地生产起到了重要作用。新流程的铁精矿品位达到67.50%以上，铁回收率达到78%，创我国红矿工业选矿的历史最高水平。

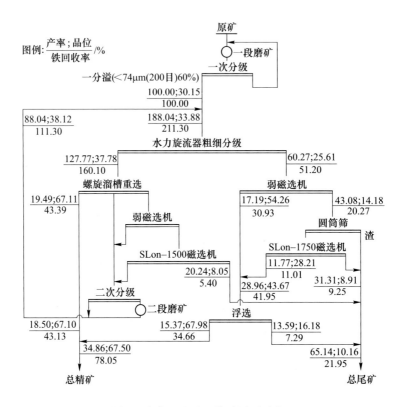

图1 齐大山选矿厂数质量原则流程

2.2 SLon磁选机在东鞍山烧结厂的应用

东鞍山烧结厂一选车间年处理430万吨东鞍山贫赤铁矿，原采用两段连续磨矿、单一正浮选流程。由于该矿石矿物结晶粒度细，氧化程度深，矿物组成极为复杂，所以是鞍山地区最难选的铁矿石。国家始终作为重点项目组织国内选矿界几代人开展攻关，都没能在关键技术上取得突破，精矿品位一直徘徊在60%左右，面对市场的挑战，采选都面临关停的考验。

经历近几年的试验研究，对于鞍山式贫赤铁矿在重大关键技术上取得突破，特别是2001年齐大山选厂技术改造的实践为东鞍山烧结厂一选加速改造提供了技术支撑和成功范例。在此基础上，鞍山矿业公司于2002年10月完成了东鞍山烧结厂一选两段连续磨矿，粗细分选中矿再磨、重选—强磁—阴离子反浮选工艺改造。作为反浮选前的抛尾脱泥设备选用8台SLon-1750立环脉动高梯度磁选机（1.0T）和2台SLon-2000立环脉动高梯度磁选机（1.0T）作为细粒抛尾设备，选用10台SLon-1750立环脉动中磁机（0.6T）作为粗粒抛尾作业。新工艺改造投产后，在保持回收率不降低的前提下，铁精矿品位由60.00%提高到64.50%以上，技术经济指标实现历史性跨越。

SLon立环脉动高梯度磁选机在东鞍山烧结厂的应用，大幅度降低强磁作业尾矿品位为全流程取得较高回收率起到关键作用。该机用于粗、细粒抛尾作业，其尾矿品位均达到12.50%左右，可降低最终尾矿品位1.2%，最终回收率提高3%。所以该项技术和设备在解决东鞍山贫赤铁矿选矿关键技术难题上起到了极其重要的作用（见图2）。

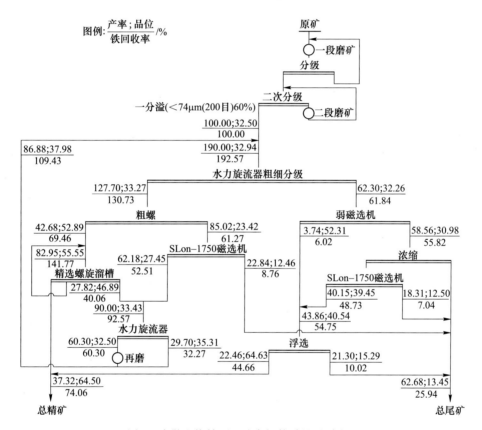

图2 东鞍山烧结厂一选车间数质量原则流程

2.3 SLon磁选机在鞍钢调军台选矿厂的试验与应用

调军台选矿厂是鞍钢于20世纪90年代新建的1座具有现代水平的大型选矿厂，于1998年3月投产。设计规模为年处理鞍山式氧化铁矿900万吨。采用连续磨矿—弱磁—中磁—强磁反浮选的选矿流程（见图3），其中强磁选采用15台Shp-3200平环强磁机。这些平环强磁选机存在着齿板易于堵塞，设备故障率较高、检修和维护较困难的问题，其设备作业率只能达到80%左右。

为了提高调军台选矿厂选矿指标和解决Shp-3200平环强磁选机存在的问题，该厂与我公司合作开展了SLon-2000立环脉动高梯度磁选机在调军台选矿厂的工业试验。

该厂拆去 1 台 Shp-3200 平环强磁选机,将 1 台 SLon-2000 磁选机安装在该 Shp-3200 磁选机的基础上,其给矿系统、精矿和尾矿排放系统均采用原有的设施,因此具有安装费用低、给料性质和给矿量一致、工业试验可比性好的优点。工业试验于 2002 年 11 月 17 日开始至 2002 年 12 月 26 日结束,工业试验期间 SLon-2000 强磁选机与现场现存 14 台 Shp-3200 平环强磁选机工业状况最好的 1 台进行对比试验,其对比指标见表 2。

工业对比试验期间,SLon-2000 磁选机的处理量为 60~80t/h,最大处理量达 100t/h,其处理量与同系统的 Shp-3200 平环强磁选机相同。在给矿条件相同的情况下,SLon-2000 磁选机脉动冲程为 20mm 时,其精矿品位高 1.19%,尾矿品位低 1.56%,精矿产率高 3.07%,精矿回收率高 8.19%,设备作业率高达 99% 以上,各项技术性能指标全面超过 Shp-3200 平环强磁选机。

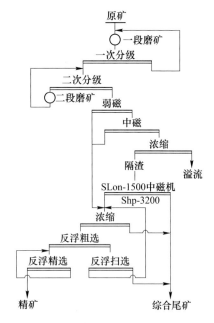

图 3　调军台选矿厂选矿原则流程

表 2　SLon-2000 强磁选机和 Shp-3200 强磁选机对比试验结果　　(%)

设备名称	给矿品位	精矿品位	尾矿品位	精矿产率	精矿回收率
SLon-2000	15.30	29.75	6.42	38.06	74.01
Shp-3200	15.18	28.56	7.98	34.99	65.82
差　值	0.12	1.19	-1.56	3.07	8.19

注:SLon-2000 强磁选机脉动冲程为 20mm。

工业试验完成后,试验的结果得到了齐大山铁矿的一致认可,试验设备直接转换成生产设备。经过 1 年多的连续运转,该机运转平稳,故障率低,选矿指标好,磁介质长期保持清洁无堵塞,设备作业率高达 99% 以上,为提高全流程的回收率和为浮选创造良好的作业条件起到了重要作用,得到现场工程技术人员的普遍好评。

鞍钢集团公司决定用 SLon-2000 立环脉动高梯度磁选机全面替换原来的 Shp-3200 平环强磁选机。第一阶段技术改造已于 2004 年 9 月完成,共采用 6 台 SLon-2000 磁选机替换掉第二、第三系统的 6 台 Shp-3200 磁选机。第二阶段的改造正在准备中,届时将强磁设备全部换成 SLon-2000 强磁选机。

2.4　SLon 磁选机在鞍钢弓长岭选矿厂的应用

弓长岭地区鞍山式红矿储量有数亿吨,弓长岭选矿厂二选车间在 1975 年建成投产,其设计规模为年处理量 300 万吨,设计采用两段连续磨矿、弱磁—重选工艺流程。1985 年至 1988 年将 5~8 系统改为阶段磨矿、强磁—重选工艺流程,生产指标为原矿 29.5%TFe,精矿品位 64%,铁回收率为 68%。由于 20 世纪 90 年代铁精矿价格偏低及当时红矿选矿成本偏高,该车间于 20 世纪 90 年代中期停止分选红矿而改为分选磁铁矿。

随着红矿选矿技术和设备的进步，红矿选矿指标显著提高，生产成本大幅度下降，加上铁精矿价格上涨，鞍钢弓长岭矿业公司决定重新建设弓长岭选矿厂三选车间，设计规模为年处理鞍山式红矿 300 万吨。设计的选矿流程为阶段磨矿，重选—强磁—反浮选流程（见图 4），该流程中采用 4 台 SLon-2000 强磁机作为细粒级的抛尾设备，另采用 4 台 SLon-2000 中磁机作为粗粒级（螺旋溜槽尾矿）抛尾设备，该流程设计的综合选矿指标为原矿品位 28.78%，铁精矿品位 67.19%，铁回收率 76.29%。弓长岭三选车间预计在 2005 年上半年建成投产。

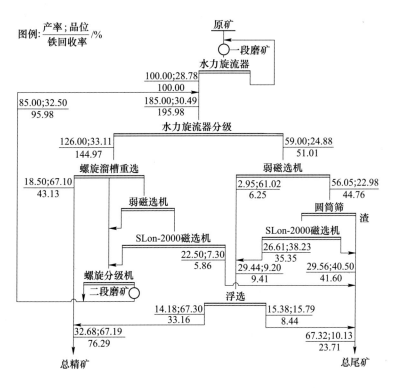

图 4 弓长岭选矿厂三选车间选矿原则流程（设计）

3 SLon 磁选机在我国重点选矿厂强磁—反浮选流程中的应用

由于强磁反浮选可获得很高的铁精矿品位和较高的铁回收率，目前我国已有多家大中型选矿厂采用强磁—反浮选流程。SLon 立环脉动高梯度磁选机在这些流程中的应用统计见表 3。

表 3 SLon 磁选机在红矿选矿厂强磁—反浮选流程中的应用

厂矿名称	设计规模/万吨·a^{-1}	选矿流程	强磁（中磁）机型及台数	铁精矿品位/%	备注
安阳钢铁公司舞阳矿业公司	350	阶段磨矿（二段）、强磁精矿再磨—反浮选	SLon-2000 强磁机 12 台	64.5（设计）	2005 年 4 月进入工业生产调试阶段

续表3

厂矿名称	设计规模/万吨·a^{-1}	选矿流程	强磁（中磁）机型及台数	铁精矿品位/%	备注
海南钢铁公司新建选矿厂	110	连续磨矿（二段）、强磁（一次粗选、一次扫选）—反浮选	SLon-1750 强磁机 8 台	66（设计）	在建工程
唐山钢铁公司司家营选矿厂	600	阶段磨矿（三段）、分级重选—强磁—反浮选	SLon-1750 强磁机 12 台 SLon-1750 中磁机 8 台	66（设计）	在建工程
鞍钢胡家庙选矿厂	900	阶段磨矿（二段）、分级重选—强磁—反浮选	SLon-2000 强磁机 8 台 SLon-2000 中磁机 8 台	67.5（设计）	在建工程
昆明钢铁公司大红山铁矿	400	阶段磨矿（二段）强磁—反浮选	SLon-2000 强磁机 11 台	67.5（设计）	在建工程
鞍钢齐大山选矿厂	720	阶段磨矿（二段）、分级重选—强磁—反浮选	SLon-1750 强磁机 11 台 SLon-1500 中磁机 11 台 SLon-1500 强磁机 1 台	67.5~67.8（生产）	2001 年投产
鞍钢东鞍山烧结厂一选车间	430	阶段磨矿（三段）、分级重选—强磁—反浮选	SLon-1750 强磁机 8 台 SLon-2000 强磁机 2 台 SLon-1750 中磁机 10 台	64.5~64.8（生产）	2002 年投产
鞍钢调军台选矿厂	900	连续磨矿（二段）、弱磁—中磁—反浮选	SLon-2000 强磁机 6 台 SHP-3200 强磁机 9 台	67.5~67.8（生产）	SLon 强磁机 2004 年投产
鞍钢弓长岭选矿厂三选车间	300	阶段磨矿（二段）、分级重选—强磁—反浮选	SLon-2000 强磁机 4 台 SLon-2000 中磁机 4 台	67.19（设计）	2005 年 4 月已进入生产调试阶段

4 结论

（1）通过多年的研究、创新和发展，SLon 立环脉动高梯度磁选机已具有优异的机电性能和选矿性能。该机具有选矿效率高、磁介质不堵塞、设备作业率高、工作性能稳定、易于操作维护的优点，在我国红矿选矿工业中得到了广泛的应用。

（2）在强磁—反浮选流程中，SLon-1750 和 SLon-2000 强磁机主要应用于粗选抛尾和细粒抛尾作业；SLon-1500、SLon-1750 和 SLon-2000 中磁机主要应用于螺旋溜槽尾矿做抛尾作业，它们可有效地控制尾矿品位，提高全流程回收率。

（3）SLon 立环脉动高梯度磁选机用于浮选作业之前具有脱泥效果好、作业铁精矿品位较高的优点，能为浮选作业降低药剂消耗和获得高质量的铁精矿创造良好的条件。

（4）由于强磁—反浮选流程分选红矿可获得很高的铁精矿品位和较高的铁回收率，近年该流程在我国红矿选矿厂得到广泛的应用。SLon 立环脉动高梯度选机的发展和新型浮选药剂的应用使我国红矿选矿水平达到国际先进水平。

写作背景 本文介绍了 SLon 立环脉动高梯度磁选机在鞍钢齐大山选矿厂、鞍钢东鞍山烧结厂、鞍钢调军台选矿厂和鞍钢弓长岭选矿厂分选氧化铁矿的强磁-反浮选流程中的应用。其中弓长岭选矿厂三选车间是新上的氧化铁矿选矿流程。

SLon 立环脉动高梯度磁选机分选红矿的研究与应用[1]

熊大和[1,2]

(1. 赣州金环磁选设备有限公司；2. 赣州有色冶金研究所)

摘 要 经过20多年的持续研究与技术创新，SLon立环脉动高梯度磁选机已发展成为国内外新一代的高效强磁选设备。该设备用于分选红矿具有富集比大、回收率高、磁介质不堵塞、设备作业率高的优点，促进我国红矿选矿工业得到了快速发展。对SLon磁选机在鞍钢、马钢、宝钢、昆钢、海钢及秘鲁铁矿的应用情况及效果进行了全面的介绍。

关键词 SLon立环脉动高梯度磁选机 红矿 强磁选

Research and Application of SLon Vertical Ring Pulsating High Gradient Magnetic Separators in Red Iron Ore Processing

Xiong Dahe[1,2]

(1. SLon Magnetic Separator Ltd.；
2. Ganzhou Nonferrous Metallurgy Research Institute)

Abstract Through more than 20 years' persistent research and innovation, SLon vertial ring pulsating high gradient magnetic separators have been developed into a new generation of high-efficiency high intensity magnetic separators both in and outside China. This equipment, when used in red iron ore processing, has the advantages of high enrichment ratio, great recovery, noclogging of magnetic matrix and high equipment availability, which has greatly accelerated the development of the red iron ore processing industry in China. The paper describes in an all-sided way the application of SLon separators in Anshan Steel, Maanshan Steel, Bao Steel, Kunming Steel, Hainan Steel and Peru Iron Mine.

Keywords SLon vertical ring pulsating high gradient magnetic separator；Red iron ore；High intensity magnetic separation

近几年我国钢铁工业正处于高速发展阶段，钢铁工业对铁矿石的巨大需求和进口铁矿石价格的大幅度上涨促使我国铁矿采选工业快速发展。在磁铁矿资源日益减少的形势下，我国越来越多的厂矿进入或扩大红矿（氧化铁矿）的采矿和选矿。

---

[1] 原载于《金属矿山》2005 (08)：24~29。

我国红矿资源的特点是储量大、品位低、嵌布粒度细、绝大部分难磨难选，而我国的钢铁冶炼工业又普遍需要高品位的铁精矿。经过几十年的技术攻关，我国红矿选矿工业有了很大的发展，其中SLon立环脉动高梯度磁选机和反浮选等高效选矿设备和技术的应用使我国红矿选矿技术达到了国际先进水平。

经过20多年持续的研制、创新与改进，SLon立环脉动高梯度磁选机已形成系列化产品，具有优异的选矿性能和设备性能，已广泛应用于我国红矿选矿工业中，成为我国新一代高效强磁选设备。

1 SLon磁选机设备结构、工作原理和主要技术参数

SLon立环脉动高梯度磁选机结构如图1所示，它主要由脉动机构、激磁线圈、铁轭、转环和各种矿斗、水斗组成，用导磁不锈钢制成的圆棒或钢板网作磁介质。

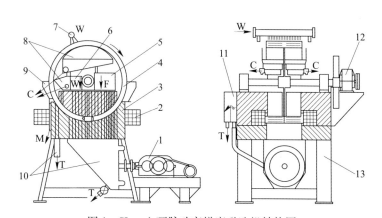

图1 SLon立环脉动高梯度磁选机结构图

1—脉动机构；2—激磁线圈；3—铁轭；4—转环；5—给矿斗；6—漂洗水斗；7—精矿冲洗装置；8—精矿斗；9—中矿斗；10—尾矿斗；11—液位计；12—转环驱动机构；13—机架；
F—给矿；W—清水；C—精矿；M—中矿；T—尾矿

SLon磁选机的工作原理为：激磁线圈通以直流电，在分选区产生感应磁场，位于分选区的磁介质表面产生非均匀磁场即高梯度磁场；转环做顺时针旋转，将磁介质不断送入和运出分选区；矿浆从给矿斗给入，沿上铁轭缝隙流经转环。矿浆中的磁性颗粒吸附在磁介质表面上，被转环带至顶部无磁场区，被冲洗水冲入精矿斗；非磁性颗粒在重力、脉动流体力的作用下穿过磁介质堆，沿下铁轭缝隙流入尾矿斗排走。

该机的转环采用立式旋转方式，对于每一组磁介质而言，冲洗磁性精矿的方向与给矿方向相反，粗颗粒不必穿过磁介质堆便可冲洗出来。该机的脉动机构驱动矿浆产生脉动，可使分选区内矿粒群保持松散状态，使磁性矿粒更容易被磁介质捕获，使非磁性矿粒尽快穿过磁介质堆进入尾矿中去。

显然，反冲精矿和矿浆脉动可防止磁介质堵塞；脉动分选可提高磁性精矿的质量。这些措施保证了该机具有较大的富集比、较高的分选效率和较强的适应能力。

SLon磁选机的主要技术参数见表1。

表1 SLon磁选机主要技术参数

机型	给矿粒度/mm	给矿浓度/%	矿浆通过能力/m³·h⁻¹	干矿处理量/t·h⁻¹	额定背景磁感强度/T	额定激磁功率/kW	脉动冲程/mm	脉动冲次/r·min⁻¹	主机质量/t
SLon-500强磁机	<1.0	10~40	0.5~1.0	0.05~0.25	1.0	13.5	0~50	0~400	1.5
SLon-750强磁机	<1.0	10~40	1.0~2.0	0.1~0.5	1.0	22	0~50	0~400	3
SLon-1000强磁机	<1.3	10~40	12.5~20	4~7	1.0	28.6	0~30	0~300	6
SLon-1250强磁机	<1.3	10~40	20~50	10~18	1.0	35	0~20	0~300	14
SLon-1500强磁机	<1.3	10~40	50~100	20~30	1.0	44	0~30	0~300	20
SLon-1750强磁机	<1.3	10~40	75~150	30~50	1.0	62	0~30	0~300	35
SLon-2000强磁机	<1.3	10~40	100~200	50~80	1.0	74	0~30	0~300	50
SLon-1500中磁机	<1.3	10~40	75~150	30~50	0.4	16	0~30	0~300	15
SLon-1750中磁机	<1.3	10~40	75~150	30~50	0.6	38	0~30	0~300	35
SLon-2000中磁机	<1.3	10~40	100~200	50~80	0.6	42	0~30	0~300	40

在20多年的研制过程中，针对生产中存在的问题，经过无数次的改进，SLon立环脉动高梯度磁选机的选矿和机电性能不断得到提高和发展。迄今为止，SLon磁选机在国内的销售总量已达400多台，并出口南非、秘鲁等国家，广泛应用于红铁矿、钛矿、非金属矿选矿。以下介绍SLon磁选机在红矿选矿工业生产中的应用情况。

2 SLon磁选机分选马钢姑山铁矿细粒赤铁矿的工业应用

姑山铁矿是马钢公司主要原料基地之一，年采选能力约100万吨原矿。铁矿石属于宁芜式赤铁矿，其构造呈块状、网状、浸染状及角砾状，嵌布粒度极不均匀，属难磨难选红铁矿。

磨选车间原流程是将矿石磨至小于74μm（200目）占50%左右，用φ500mm和φ350mm旋流器串联两次分级，沉砂用螺旋溜槽分选，φ350mm旋流器溢流用离心机分选。

1989~2001年，该矿对原流程进行强磁选改造，现行生产工艺流程见图2。该流程的特点为：一段磨矿至小于74μm（200目）占48%左右，采用3台SLon-1750磁选机粗选抛尾，粗精矿进二段磨矿至小于74μm（200目）占85%左右，然后用3台SLon-1750磁选机精选，精选作业的尾矿再用SLon-1500磁选机扫选。该流程与原一段磨矿重选流程选矿指标对比见表2，与重选流程比较，磁选流程精矿品位高4.57%，回收率高14.88%。回收率的提高主要是由于采用SLon磁选机回收了重选方法难以回收的微细粒级弱磁性铁矿物。

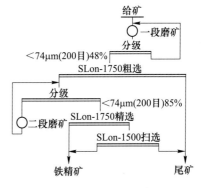

图2 姑山铁矿磨选车间选矿流程

表 2　姑山铁矿重选与磁选流程选别指标对比　　　　　　　　　　　　（%）

流　程	时间	铁品位			铁回收率
		给矿	精矿	尾矿	
重　选	1988	37.16	55.22	24.47	61.32
磁　选	2003	38.02	59.79	17.55	76.20
差　值		0.83	4.57	-6.92	14.88

3　SLon 磁选机在梅山除硫、降磷工程中的应用

梅山铁矿是上海冶金公司的主要钢铁原料基地之一，选厂年处理铁矿石 400 万吨，矿石类型为磁铁矿、赤铁矿、菱铁矿和黄铁矿的混合矿。其铁精矿主要供给上海冶金公司作球团用，但因含磷硫偏高不能单独作球团原料，只能作辅助原料，直接影响该公司经济效益。

为解决上述问题，梅山铁矿联合多家科研院所进行了大量选矿试验，经过分析比较，采用如图 3 所示磁选流程进行降低磷、硫的工业生产。该工艺采用弱磁选机回收磁铁矿，16 台 SLon-1500 强磁选机分别作粗选和扫选，用于回收矿物中的赤铁矿和菱铁矿。通过强磁作业，可有效回收细粒赤铁矿和菱铁矿，降低尾矿品位，铁精矿产品中的磷、硫含量完全符合冶炼要求。全流程选矿指标见表 3。

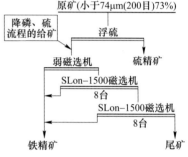

图 3　梅山铁矿降磷硫生产流程

表 3　梅山铁矿降磷硫指标　　　　　　　　　　　　（%）

产品名称	产率	品位			回收率		
		Fe	S	P	Fe	S	P
给　矿	100.00	52.77	0.44	0.399	100.00	100.00	100.00
精　矿	88.90	56.08	0.29	0.246	94.51	59.10	54.76
冶炼要求		>55.0	<0.35	<0.25	>94		
尾　矿	11.10	26.13	1.61	1.626	5.49	40.90	45.24

4　SLon 磁选机在鞍山式红矿选矿中的应用

近年鞍山地区红矿选矿技术得到飞速发展，其中 SLon 立环脉动高梯度磁选机和反浮选技术的普遍应用使该地区的选矿技术水平达到国际领先水平。目前齐大山选矿厂、东鞍山烧结厂一选车间、调军台选矿厂、弓长岭选矿厂三选车间及正在建设的胡家庙选矿厂均采用 SLon 磁选机作为强磁选和中磁选设备。

4.1　SLon 磁选机在齐大山选矿厂的应用

鞍山矿业公司齐大山选矿厂是年处理 720 万吨贫赤铁矿的大型选矿厂。原选矿流程为原矿经粗、中两段破碎后筛分成块矿和粉矿，块矿在二选车间用焙烧磁选工艺处理，粉矿在一选车间用阶段磨矿、重选—强磁—弱酸性正浮选工艺处理。由于选厂的工艺和装备较

落后，其铁精矿品位只能达到 63% 左右。

从 2001 年至 2004 年，齐大山选矿厂一选和二选车间全部改为阶段磨矿、重选—强磁—反浮选的选矿流程（见图 4），采用 11 台 SLon-1750 强磁机控制细粒级尾矿品位，另采用 11 台 SLon-1500 中磁机控制螺旋溜槽尾矿品位。SLon 磁选机在该流程中为降低尾矿品位和提高铁回收率发挥了关键作用，该机脱泥效果好，为反浮选提高铁精矿品位和降低药剂消耗创造了良好的条件。由于 SLon 磁选机的机电性能良好，磁介质长期不堵塞，设备作业率在 99% 以上，设备的水电消耗和维护费用低，对选厂降低生产成本、长期稳定地生产起到了重要作用。新流程的铁精矿品位达到 67.50% 以上，铁回收率达到 78%，创我国红矿工业选矿的历史最高水平。

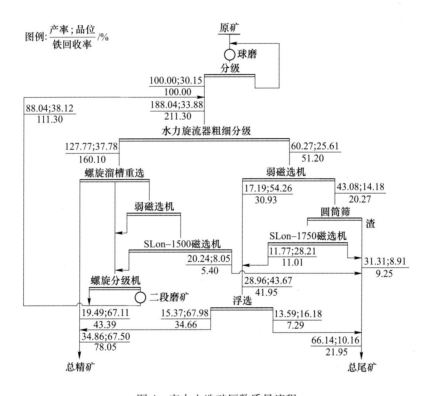

图 4　齐大山选矿厂数质量流程

4.2　SLon 磁选机在东鞍山烧结厂的应用

东鞍山烧结厂一选车间年处理 430 万吨贫赤铁矿，原采用两段连续磨矿单一正浮选流程。由于该矿石矿物结晶粒度细、氧化程度深、矿物组成极为复杂，所以是鞍山地区最难选的铁矿石。国家始终作为重点项目组织国内选矿界几代人开展攻关，都没能在关键技术上取得突破，精矿品位一直徘徊在 60% 左右，面对市场的挑战，采选都面临关停的考验。

经历近几年的试验研究，对鞍山式贫赤铁矿的选矿在重大关键技术上取得了突破，特别是 2001 年齐大山选厂技术改造的实践为东烧厂一选加速改造提供了技术支撑和成功范例。在此基础上，鞍山矿业公司于2002年10月完成了东烧厂一选车间两段连续磨矿、粗细分选、中矿再磨、重选—强磁—阴离子反浮选工艺改造，反浮选前的抛尾脱泥设备选用8台SLon-1750磁选机（1.0T）和2台SLon-2000磁选机（1.0T），另选用10台SLon-1750立环脉动中磁机（0.6T）分选螺旋溜槽尾矿作为粗粒抛尾设备。新工艺改造投产后，在保持回收率不降低的前提下，铁精矿品位由60%提高到64.5%以上，技术经济指标实现历史性跨越。

SLon磁选机大幅度降低强磁作业尾矿品位为全流程取得较高回收率提供了保证，该机用于粗、细粒抛尾作业，其尾矿品位均达到12.50%左右（见图5），为解决东鞍山贫赤铁矿选矿技术难题起到了关键的作用。

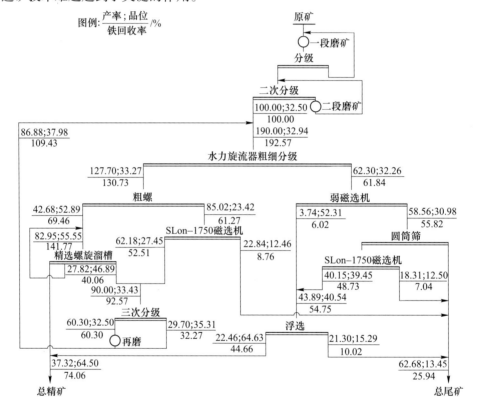

图5　东烧一选车间数质量流程

4.3　SLon磁选机在调军台选矿厂的试验与应用

调军台选矿厂是鞍钢于20世纪90年代新建的一座具有现代化水平的大型选矿厂，于1998年3月投产，设计规模为年处理鞍山式氧化铁矿900万吨。选厂采用连续磨矿、弱磁—中磁—强磁—反浮选流程（见图6），其中强磁选采用15台Shp-3200平环强磁选机。这些平环强磁选机存在着齿板易堵塞、设备故障率较高、检修和维护较困难等问题，其设备作业率只能达到80%左右。

为了提高调军台选矿厂选矿指标和解决Shp-3200平环强磁选机存在的问题，该厂与我公司合作开展了 SLon-2000 立环脉动高梯度磁选机在调军台选矿厂的工业试验。

该厂拆去 1 台 Shp-3200 平环强磁选机，将 1 台 SLon-2000 磁选机安装在该 Shp-3200 磁选机的基础上，其给矿系统、精矿和尾矿排放系统均采用原有的设施，因此具有安装费用低、给料性质和给矿量一致、工业试验可比性好的优点。试验于 2002 年 11 月 17 日开始至 2002 年 12 月 26 日结束，试验期间 SLon-2000 磁选机与现场 14 台 Shp-3200 平环强磁选机中工况最好的一台进行对比试验，其对比指标见表 4。

工业对比试验期间，SLon-2000 磁选机的处理量为 60~80t/h，最大处理量达 100t/h，其处理量与同系统的 Shp-3200 平环强磁选机相同。在给矿条件相同的情况下，SLon-2000 磁选机精矿品位高 1.19%，尾矿品位低 1.56%，精矿产率高 3.07%，精矿回收率高 8.19%，各项技术经济指标全面超过 Shp-3200 平环强磁选机。

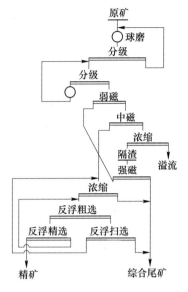

图 6 调军台选矿厂原则流程

表 4　SLon 磁选机与平环强磁机对比试验结果　　　　　　（%）

设备名称	给矿品位	精矿品位	尾矿品位	精矿产率	精矿回收率
SLon-2000	15.30	29.75	6.42	38.06	74.01
Shp-3200	15.18	28.56	7.98	34.99	65.82
差　值	0.12	1.19	-1.56	3.07	8.19

工业试验完成后，试验设备直接转换成生产设备，经过 1 年多的连续运转，该机运转平稳，故障率低，选矿指标好，磁介质长期保持清洁无堵塞，设备作业率高达 99% 以上，为提高全流程的回收率和为浮选创造良好的作业条件起到了重要作用，得到现场工程技术人员的普遍好评。

鞍钢集团公司决定采用 SLon-2000 磁选机全面替换原来的 Shp-3200 平环强磁选机。第 1 阶段技术改造已于 2004 年 9 月完成，共采用 6 台 SLon-2000 磁选机替换掉第 2、第 3 系统的 6 台 Shp-3200 磁选机。

4.4　SLon 磁选机在弓长岭选矿厂的应用

弓长岭地区鞍山式红矿储量有数亿吨。弓长岭选矿厂二选车间在 1975 年建成投产，其设计规模为年处理量 300 万吨，设计采用两段连续磨矿、弱磁—重选工艺流程。1985 年至 1988 年将 5~8 号系统改为阶段磨矿、强磁—重选工艺流程，生产指标为：原矿品位 29.5%、精矿品位 64%、铁回收率 68%。由于 20 世纪 90 年代铁精矿价格偏低及当时红矿选矿成本偏高，该车间于 20 世纪 90 年代中期停止分选红矿而改为分选磁铁矿。

随着红矿选矿技术和设备的进步，红矿选矿指标显著提高，生产成本大幅度下降，加上铁精矿价格上涨，鞍钢弓长岭矿业公司决定新建弓长岭选矿厂三选车间，设计规模为年处理鞍山式红矿 300 万吨。设计的选矿流程为阶段磨矿、重选—强磁—反浮选流程（见图

7)，采用 4 台 SLon-2000 强磁机作为细粒级的抛尾设备，另采用 4 台 SLon-2000 中磁机作为粗粒级（螺旋溜槽尾矿）的抛尾设备。该流程设计的综合选矿指标为：原矿品位 28.78%、铁精矿品位 67.19%、铁回收率 76.29%。弓长岭三选车间预计在 2005 年上半年建成投产。

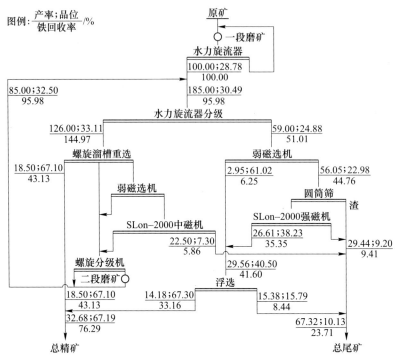

图 7 弓长岭三选车间设计流程

5 SLon 磁选机在海南铁矿的应用

海南铁矿的铁矿石主要为赤铁矿和部分磁铁矿。2003～2005 年，海南钢铁公司采用如图 8 所示选矿流程。该流程将给矿磨至小于 74μm（200 目）占 95% 的粒度，经弱磁选机分选出磁铁矿，再经由 5 台 SLon-2000 磁选机组成的一粗一精一扫工艺选别赤铁矿。SLon 磁选机精选尾矿和扫选精矿浓缩后返回二段球磨，弱磁选机精矿和 SLon-2000 磁选机精选精矿为最终精矿。该流程指标为：给矿品位 51.09%、铁精矿品位 64.57%、铁回收率 88.01%。

目前，海钢又在建设 1 座年处理 110 万吨红矿的选矿厂，采用连续磨矿、强磁—反浮选流程，其中以 8 台 SLon-1750 磁选机作为强磁选设备，预计投产后铁精矿品位可达到 66% 以上。

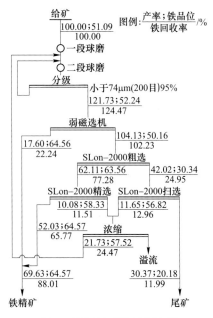

图 8 海钢铁矿数质量流程

6 SLon 磁选机在昆钢大红山铁矿的应用

大红山铁矿是昆钢新建的大型地下开采矿山，其矿石中铁矿物以磁铁矿为主（占60%左右），赤铁矿为辅（占40%左右）；脉石矿物主要为石英和长石。第一期已于2002年建成年处理矿石50万吨的选矿厂，其中采用4台SLon-1500磁选机回收赤铁矿；第二期年处理400万吨矿石的选矿厂正在建设之中，其中采用11台SLon-2000磁选机回收赤铁矿。

大红山铁矿50万吨选厂的数质量流程如图9所示。该流程为阶段磨矿、全磁选流程，原矿破碎后用自磨机磨到小于74μm（200目）55%左右，然后用弱磁选机和SLon-1500磁选机分别选出磁铁矿和赤铁矿，SLon-1500磁选机抛弃产率为26.11%的低品位尾矿，一段弱磁选和强磁选精矿进入二段球磨机磨至小于74μm（200目）85%左右，然后用弱磁选机和SLon-1500磁选机选出最终铁精矿。该流程具有简单易行、生产成本低的优点，其缺点是铁精矿的品位仅达到63%~64%。

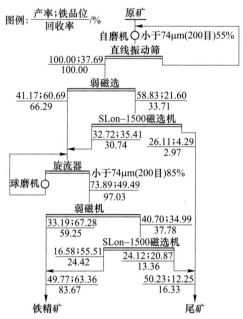

图9 大红山铁矿50万吨选厂数质量流程

大红山铁矿第二期年处理400万吨矿石的选厂在上述流程基础上增加了反浮选作业，其设计的选矿指标为：原矿品位40.03%、铁精矿品位67.00%、尾矿品位16.49%、铁回收率78.00%。该选厂计划2006年投产。

7 SLon 磁选机在秘鲁铁矿的应用

秘鲁铁矿位于秘鲁共和国境内的马尔科纳地区，属古生代寒武纪地层，交代型矿床，岩深达300~500m，蕴含着品位高达50%以上的富铁矿，目前已探明的地质储量为14.3亿吨。矿石的成分以磁铁矿为主，岩体表层多为氧化型赤铁矿。早期秘鲁铁矿由于岩体表层氧化块矿品位较高，无需选矿直接装船外运，而品位较低的氧化矿从表层剥离后直接在采场堆积。近年来随着露天开采规模越来越大，大量堆积的表层氧化铁矿带来的问题日益突出：一方面是造成资源的巨大浪费；另一方面是影响原生矿的进一步开采。

2002 年初，秘鲁铁矿与我公司合作探索用 SLon 磁选机为主体设备分选该矿的表层氧化铁矿，获得了优良的试验指标。2003 年底，经双方研究决定采用如图 10 所示流程处理氧化铁矿（包括过渡矿）。该流程用弱磁选机选出磁铁矿，采用 2 台 SLon-1750 磁选机选出赤铁矿，弱磁精矿和强磁精矿合并后进入浮硫作业。其生产指标为：原矿铁品位 56.43%、精矿铁品位 65.09%、铁回收率 80.32%。

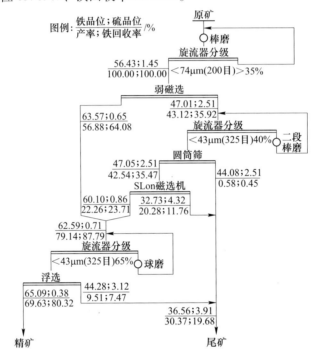

图 10　秘鲁铁矿氧化矿选别数质量流程

该流程分选出的铁精矿用于高炉球团，产品质量满足生产球团的要求。与原生矿精矿相比，氧化矿精矿品位略低，但用于生产高炉球团时可以少加石英，石英添加量由原来的 4.09% 降到 2.50%。该流程的成功应用，不但有效处理了采场大量堆积的氧化铁矿，而且减少了石英的开采、运输、破碎磨矿等费用。流程中 SLon 磁选机优异的设备性能得到现场技术人员的普遍好评。

8　结论

（1）通过 20 多年的持续研究、创新、工业试验、工业应用和改进，发展了系列化新型高效的 SLon 立环脉动高梯度磁选机。该机转环立式旋转、反冲精矿、配有矿浆脉动机构，具有富集比大、对红矿（氧化铁矿）回收率高、磁介质不堵塞、设备作业率高、适应性强、易于维护等优点，现有 300 多台广泛应用在国内外红矿选矿工业生产中。

（2）SLon 磁选机在马钢姑山铁矿、宝钢梅山铁矿、昆钢大红山铁矿 50 万吨选厂和秘鲁铁矿等地的单一磁选流程中，可有效地回收其中的赤铁矿、菱铁矿等细粒弱磁性铁矿石，使全流程获得较高的铁精矿品位和铁回收率，大幅度降低生产成本，获得显著的经济效益。

（3）由于强磁—反浮选流程分选红矿可获得很高的铁精矿品位和较高的铁回收率，近年该流程在我国红矿选矿工业中得到广泛的应用。在强磁—反浮选流程中，SLon 强磁机主要应用于粗选抛尾或细粒抛尾作业，SLon 中磁机主要应用于螺旋溜槽尾矿的抛尾作业，它们可有效地控制尾矿品位，提高全流程回收率。SLon 磁选机用于浮选作业之前具有脱泥效果好、作业铁精矿品位较高的优点，能为浮选作业降低药剂消耗和获得高质量的铁精矿创造良好的条件。

（4）SLon 立环脉动高梯度磁选机已成为国内外新一代高效强磁选设备，在工业上的广泛应用促进了我国红矿选矿工业的高速发展，使我国红矿选矿技术达到国际先进水平。

参 考 文 献

[1] 熊大和. SLon 型立环脉动高梯度磁选机的改进及其在红矿选矿的应用 [J]. 金属矿山，1994（6）：30~34.
[2] 熊大和，等. 提高姑山铁矿生产指标的工业试验研究 [J]. 金属矿山，2000（12）：31~33.
[3] 熊大和. SLon 型磁选机在齐大山选矿厂的应用 [J]. 金属矿山，2002（4）：42~44.
[4] 熊大和. SLon 磁选机分选东鞍山氧化铁矿石的应用 [J]. 金属矿山，2003（6）：21~26.
[5] 赫荣安，等. SLon 强磁选机分选鞍山式贫赤铁矿的试验及应用 [J]. 金属矿山，2003（9）：19~24.
[6] 熊大和，等. SLon-2000 磁选机在调军台选矿厂的工业试验与应用 [J]. 金属矿山，2003（12）：37~39.
[7] 李建设. SLon 立环脉动高梯度磁选机在秘鲁铁矿的应用 [J]. 金属矿山，2004（8）：39~41.

写作背景　本文介绍了 SLon-立环脉动高梯度磁选机在多个选矿厂分选氧化铁矿的应用，其中在海南铁矿的应用、在昆钢大红山铁矿 50 万吨选矿厂的应用和在首钢秘鲁铁矿的应用为新公布的资料。

SLon 磁选机在黑色金属选矿工业应用新进展[1]

熊大和[1,2]

（1. 赣州有色冶金研究所；2. 赣州金环磁选设备有限公司）

摘 要 经过 20 多年的持续研究与技术创新，SLon 立环脉动高梯度磁选机已发展成为国内外新一代的高效强磁选设备，该设备具有优异的选矿性能和高度的可靠性。目前，已有 400 多台 SLon 系列磁选机在鞍钢、马钢、宝钢、昆钢、首钢、海钢、安阳钢铁公司等地广泛应用于氧化铁矿选矿工业，在攀钢选钛厂、重钢太和铁矿、承德黑山铁矿等地广泛应用于钛铁矿选矿工业，近年来在锰矿选矿工业中的试验与应用也显示了良好的应用前景。SLon 磁选机为我国黑色金属选矿工业的发展作出了重要贡献。

关键词 SLon 立环脉动高梯度磁选机　氧化铁矿　钛铁矿　锰矿

1 前言

我国拥有丰富的氧化铁矿、钛铁矿、锰矿等弱磁性金属矿产资源，但它们绝大多数原矿品位低、嵌布粒度细，需要细磨深选才能达到冶炼要求。为了提高我国细粒弱磁性矿物分选效率，于 1981 年开始研究探索振动和脉动高梯度磁选的选矿机理，并于 1986 年开始研制 SLon 立环脉动高梯度磁选机。该机采用转环立式旋转、反冲精矿、配置脉动机构松散矿粒群及磁介质优化组合等措施，显著提高了矿物分选效率，成功地解决了国内外平环强磁选机和平环高梯度磁选机磁介质易堵塞的技术难题。SLon 磁选机分选粒度下限可达 $10\mu m$ 左右，并具有富集比大、分选效率高、磁介质不易堵塞、对给矿粒度、浓度和品位波动适应性强，且工作可靠、操作维护方便等优点。

经 20 多年持续的研究、创新与改进，SLon 立环脉动高梯度磁选机已形成系列化产品（见表 1），该产品具有优异的选矿性能和机电性能，在我国弱磁性矿石选矿工业中得到广泛应用，成为我国新一代强磁选设备。

表 1　SLon 磁选机主要技术参数

机型	给矿粒度 /mm	给矿浓度 /%	矿浆通过能力 /$m^3 \cdot h^{-1}$	干矿处理量 /$t \cdot h^{-1}$	额定背景磁感强度 /T	额定激磁功率 /kW	脉动冲程 /mm	脉动冲次 /$r \cdot min^{-1}$	主机质量 /t
SLon-500 强磁机	<1.0	10~40	0.5~1.0	0.05~0.25	1.0	13.5	0~50	0~400	1.5
SLon-750 强磁机	<1.0	10~40	1.0~2.0	0.1~0.5	1.0	22	0~50	0~400	3
SLon-1000 强磁机	<1.3	10~40	12.5~20	4~7	1.2	28.6	0~30	0~300	6
SLon-1250 强磁机	<1.3	10~40	20~50	10~18	1.0	35	0~20	0~300	14
SLon-1500 强磁机	<1.3	10~40	50~100	20~30	1.0	44	0~30	0~300	20
SLon-1750 强磁机	<1.3	10~40	75~150	30~50	1.0	62	0~30	0~300	35
SLon-2000 强磁机	<1.3	10~40	100~200	50~80	1.0	74	0~30	0~300	50
SLon-1500 中磁机	<1.3	10~40	75~150	30~50	0.4	16	0~30	0~300	15
SLon-1750 中磁机	<1.3	10~40	75~150	30~50	0.6	38	0~30	0~300	35
SLon-2000 中磁机	<1.3	10~40	100~200	50~80	0.6	42	0~30	0~300	40

---

[1] 原载于《采选技术进展报告会论文集》，马鞍山矿业快报，2006：281~286。

2 SLon 磁选机设备结构和工作原理

SLon 立环脉动高梯度磁选机结构如图 1 所示，它主要由脉动机构、激磁线圈、铁轭、转环和各矿斗、水斗组成。采用导磁不锈钢制成的圆棒或板网作磁介质。其工作原理如下。

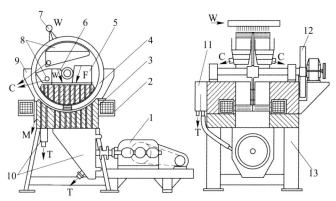

图 1 SLon 立环脉动高梯度磁选机结构图

1—脉动机构；2—激磁线圈；3—铁轭；4—转环；5—给矿斗；6—漂洗水斗；7—精矿冲洗装置；
8—精矿斗；9—中矿斗；10—尾矿斗；11—液位斗；12—转环驱动机构；13—机架；
F—给矿；W—清水；C—精矿；M—中矿；T—尾矿

激磁线圈通以直流电，在分选区产生感应磁场，位于分选区的磁介质表面产生非均匀磁场即高梯度磁场；转环作顺时针旋转，将磁介质不断送入和运出分选区；矿浆从给矿斗给入，沿上铁轭缝隙流经转环。矿浆中的磁性颗粒吸附在磁介质表面上，被转环带至顶部无磁场区，被冲洗水冲入精矿斗，非磁性颗粒在重力、脉动流体力的作用下穿过磁介质堆，沿下铁轭缝隙流入尾矿斗排走。

该机的转环采用立式旋转方式，对于每一组磁介质而言，冲洗磁性精矿的方向与给矿方向相反，粗颗粒不必穿过磁介质堆便可冲洗出来。该机的脉动机构驱动矿浆产生脉动，可使分选区内矿粒群保持松散状态，使磁性矿粒更容易被磁介质捕获，使非磁性矿粒尽快穿过磁介质堆进入尾矿中去。

显然，反冲精矿和矿浆脉动可防止磁介质堵塞；脉动分选可提高磁性精矿的质量。这些措施保证了该机具有较大的富集比、较高的分选效率和较强的适应能力。

在 20 多年的设备研制过程中，针对使用中存在的问题，进行了无数次的改进，SLon 立环脉动高梯度磁选机的选矿和机电性能不断得到提高和发展。迄今为止，SLon 磁选机的销售总量已达 400 多台，并出口南非、秘鲁等国家，广泛应用于铁矿、钛矿、非金属矿的选矿。

3 SLon 磁选机在氧化铁矿选矿中的应用

3.1 SLon 磁选机在鞍钢齐大山选矿厂的应用

鞍山矿业公司齐大山选矿厂年处理 720 万吨贫赤铁矿，由于该选矿厂原选矿流程工艺和装备较落后，其铁精矿品位只能达到 63% 左右。2001~2004 年，齐大山选矿厂一选和二选车间选矿流程全部改为阶段磨矿、重选—强磁—反浮选的选矿流程（见图 2），新流程中 11 台 SLon-1750 强磁机和 11 台 SLon-1500 中磁机有效地控制了细粒级尾矿品位，对提高铁回收率起到了关键作用。该机脱泥效果好，为反浮选作业提高铁精矿品位和降低药剂消耗创造了良好的条件。新流程中 SLon 磁选机机电性能良好、磁介质长期不堵塞、设备作业率达 99% 以上、设备水电消耗和维护费用低，为选厂降低生产成本、长期稳定生产起到了重要作用。新流程的铁精矿品位达到 67.50% 以上，铁回收率达到 78%，创我国红矿选矿工业历史最高水平。

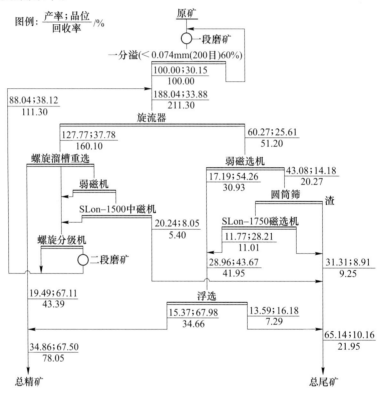

图 2　齐大山选矿厂数质量原则流程图

3.2　SLon 磁选机在鞍钢调军台选矿厂的应用

鞍钢调军台选矿厂设计规模为年处理鞍山式氧化铁矿 900 万吨，为一具有现代化水平的大型选矿厂，采用连续磨矿—弱磁—中磁—强磁—反浮选的选矿流程（见图 3）。原采用 15 台 Shp-3200 平环强磁选机作为强磁选设备。这些平环强磁选机存在齿板介质易堵塞、设备故障率较高、检修和维护较困难等问题，设备作业率只能达到 80% 左右。

为了提高调军台选矿厂选矿指标和解决 Shp-3200 平环强磁选机存在的问题，该选厂与该公司合作开展了 SLon-2000 磁选机在调军台选矿厂的工业试验。工业对比试验期间，在处理量、给矿条件相同的情况下，SLon-2000 磁选机精选精矿品位比 Shp-3200 平环强磁选机高 1.19%，尾矿品位低 1.56%，精矿产率高 3.07%，精矿回收率高 8.19%。SLon 磁

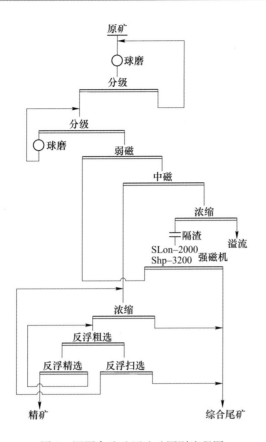

图 3　调军台选矿厂选矿原则流程图

选机机电性能、各项技术经济指标全面超过 Shp-3200 平环强磁选机。

工业试验完成后，鞍钢集团公司决定采用 SLon-2000 磁选机全面替换流程中的 Shp-3200 平环强磁选机，第一阶段技术改造已于 2004 年 9 月完成，共采用 6 台 SLon-2000 磁选机替换掉第二、第三系统的 6 台 Shp-3200 磁选机。新流程中 SLon 磁选机运转平稳、故障率低、选矿指标好、磁介质长期不堵塞、设备作业率高达 99% 以上，为提高全流程的铁回收率和为浮选作业创造良好的条件起到了重要作用。目前，鞍钢集团已批准再用 9 台 SLon-2000 磁选机取代剩余的 9 台 Shp-3200 平环强磁选机，该项目预计将于 2005 年年底完成。

3.3　SLon 磁选机在首钢秘鲁铁矿的应用

秘鲁铁矿位于秘鲁共和国境内的马尔科纳地区，属古生代寒武纪地层，交代型矿床，岩深达 300~500m，蕴含着品位高达 50% 以上的富铁矿，目前已探明地质储量为 14.3 亿吨。矿石成分以磁铁矿为主，岩体表层多为氧化型赤铁矿。早期秘鲁铁矿由于岩体表层氧化块矿品位较高，无需选矿直接装船外运，而品位较低的氧化矿从表层剥离后直接在采场堆积。近年来随着露天开采规模越来越大，大量堆积的表层氧化铁矿带来的问题日益突出：一方面是造成资源的巨大浪费；另一方面是影响原生矿的进一步开采。

2002 年初，秘鲁铁矿与该公司合作探索用 SLon 磁选机为主体设备分选该矿的表层氧化铁矿，获得了优良的试验指标。2003 年底，经双方研究决定采用如图 4 所示流程处理该氧化铁矿（包括过渡矿）。该流程中用弱磁选机选出磁铁矿，采用 2 台 SLon-1750 磁选机选出赤铁矿，弱磁精矿和强磁精矿合并后进入浮硫作业。其生产指标为：给矿品位 56.43%，精矿品位 65.09%，铁回收率 80.32%。

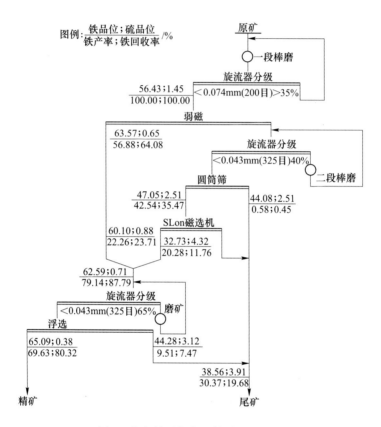

图 4 秘鲁铁矿氧化矿数质量流程图

该流程分选出的铁精矿用于高炉球团，产品质量满足生产球团的要求。与原生矿精矿相比，氧化矿精矿品位略低，但用于生产高炉球团时可以少加石英，石英添加量由原来的 4.09% 降到 2.50%。该流程的成功应用，不但有效处理了采场大量堆积的氧化铁矿，而且减少了石英的开采、运输、破碎磨矿等费用，流程中 SLon 磁选机优异的选矿性能和设备性能得到现场技术人员的普遍好评。2005 年年底秘鲁铁矿向该公司又订购了 3 台 SLon-1750 磁选机用于扩大生产。

3.4 SLon 磁选机国内外大中型氧化铁矿选矿厂应用统计

近年来我国的钢铁工业得到飞速发展，但我国的铁矿资源绝大多数为贫矿，钢铁工业对优质铁精矿的巨大需求促进了以 SLon 磁选机、反浮选等新技术在我国贫、细、杂氧化铁矿选矿工业中的高速发展和应用，使我国氧化铁矿的选矿技术水平达到国际先进水平。表 2 列举了 SLon 磁选机在细粒氧化铁矿选矿工业中的应用情况。

表2 SLon磁选机在大中型氧化铁矿选矿厂的应用统计

厂矿名称	设备型号	使用台数	分选矿种	选矿流程	分选指标/%	备注
鞍钢齐大山选矿厂	SLon-1750强磁机 SLon-1500中磁机	11 11	鞍山式赤铁矿	阶段磨矿—重选—强磁—反浮选	$\alpha=30.15$, $\beta=67.50$, $\theta=10.16$, $\varepsilon=78.05$	全流程选矿指标
鞍钢东烧一选车间	SLon-1750强磁机 SLon-2000强磁机 SLon-1750中磁机	8 2 10	鞍山式赤铁矿	阶段磨矿—重选—强磁—反浮选	$\alpha=32.50$, $\beta=64.50$, $\theta=14.50$, $\varepsilon=71.45$	全流程选矿指标
鞍钢调军台选矿厂	SLon-2000强磁机	6	鞍山式赤铁矿	连续磨矿—强磁—反浮选	$\alpha=29.69$, $\beta=67.50$, $\theta=10.87$, $\varepsilon=75.56$	全流程选矿指标
鞍钢弓长岭三选车间	SLon-2000强磁机 SLon-2000中磁机	5 5	鞍山式赤铁矿	阶段磨矿—重选—强磁—反浮选	$\alpha=28.78$, $\beta=67.19$, $\theta=10.13$, $\varepsilon=76.29$	设计指标,已投产调试
鞍钢胡家庙选矿厂	SLon-2000强磁机 SLon-2000中磁机	8 8	鞍山式赤铁矿	阶段磨矿—重选—强磁—反浮选	$\alpha=28.40$, $\beta=67.50$, $\theta=10.60$, $\varepsilon=74.35$	设计指标,在建工程
安阳钢铁公司舞阳铁矿	SLon-2000强磁机	17	赤铁矿	阶段磨矿—重选—强磁—反浮选	$\alpha=25.95$, $\beta=64.50$, $\theta=10.84$, $\varepsilon=70.00$	设计指标,已投产调试
海南钢铁公司选矿厂	SLon-2000强磁机	5	磁铁矿、赤铁矿混合矿	弱磁—强磁（一粗一精一扫）	$\alpha=51.09$, $\beta=64.20$, $\theta=24.50$, $\varepsilon=84.16$	细粒选矿流程综合指标
	SLon-1750强磁机	8	磁铁矿、赤铁矿混合矿	连续磨矿—强磁—反浮选	$\alpha=47.63$, $\beta=64.50$, $\theta=29.04$, $\varepsilon=71.00$	设计指标在建工程
唐钢司家营铁矿	SLon-1750强磁机 SLon-1750中磁机	8 12	氧化铁矿	阶段磨矿—重选—强磁—反浮选	$\alpha=30.44$, $\beta=66.00$, $\theta=9.65$, $\varepsilon=80.00$	设计指标,在建工程
马钢姑山铁矿	SLon-1750强磁机 SLon-1750强磁机	7 4	宁芜式赤铁矿	阶段磨矿—强磁（一粗一精一扫）	$\alpha=43.15$, $\beta=60.17$, $\theta=22.41$, $\varepsilon=76.59$	磨选主厂房全流程指标
宝钢南京梅山铁矿	SLon-1500强磁机	18	赤铁矿、黄铁矿、菱铁矿	弱磁—强磁（一粗一扫）联合降磷流程	$\alpha=\mathrm{Fe}52.77$, $\alpha=\mathrm{S}0.44$, $\alpha=\mathrm{P}0.399$, $\beta=\mathrm{Fe}56.08$, $\beta=\mathrm{S}0.29$, $\beta=\mathrm{P}0.246$, $\varepsilon=\mathrm{Fe}94.51$	降磷流程综合指标
昆钢大红山铁矿	SLon-1500强磁机	4	磁铁矿和赤铁矿混合矿	阶段磨矿—弱磁—强磁流程	$\alpha=37.69$, $\beta=64.24$, $\theta=12.14$, $\varepsilon=83.59$	全流程选矿指标
	SLon-2000强磁机	11	磁铁矿和赤铁矿混合矿	阶段磨矿—弱磁—强磁—反浮选	$\alpha=40.03$, $\beta=67.00$, $\theta=15.34$, $\varepsilon=80.00$	设计指标,在建工程
首钢秘鲁铁矿	SLon-1750强磁机	2	含硫磁铁矿、赤铁矿	阶段磨矿—弱磁—强磁—浮硫	$\alpha=56.43$, $\beta=65.09$, $\theta=36.56$, $\varepsilon=80.32$	全流程指标

注：分选指标中，α为原矿品位,%；β为精矿品位,%；θ为尾矿品位,%；ε为回收率,%。

4 SLon磁选机在钛铁矿选矿工业中的应用

我国的攀枝花地区和承德地区蕴藏着丰富的钛铁矿资源。过去由于选矿技术和设备落后，我国对钛铁矿回收利用率很低。近年来，随着微细粒级选钛技术的发展和SLon磁选机的广泛应用，使我国选钛水平有了大幅度提高，TiO_2的综合回收率达到35%左右。

4.1 SLon 磁选机在攀枝花选钛厂的应用

攀枝花是我国最大的钛铁矿产地，攀钢（集团）矿业公司选钛厂是一座大型原生钛铁矿选厂，入选原料（原矿）为选铁厂所排出的磁选尾矿。原生产流程采用重选—浮选—电选工艺流程，用于回收粗粒级（>0.1mm）和细粒级（0.1~0.045mm）两级别钛铁矿，而对产率和 TiO_2 金属量均占入选原料45%的小于0.045mm的微细粒级钛铁矿因难于回收而直接丢弃，导致其全厂钛回收率仅有17%~20%，每年生产钛精矿仅5万吨。

为了提高钛回收率，1997~2004 年该厂建成了 3 条磁—浮流程生产线分选小于 0.045mm 粒级的钛铁矿。12 台 SLon-1500 磁选机用于粗选作业，然后用浮选精选。磁—浮流程的综合选矿指标为：给矿品位 10.12% TiO_2，钛精矿品位不小于 47% TiO_2，TiO_2 回收率不小于 44%。该流程 2004 年从细粒尾矿中回收了 14 万吨品位为 47.5% TiO_2 的细粒钛精矿。TiO_2 的综合回收率提高了 10~12 个百分点。第一条工业试验生产线的数质量流程图见图 5，目前攀钢选钛厂每年生产优质钛精矿产量达到 25 万吨。

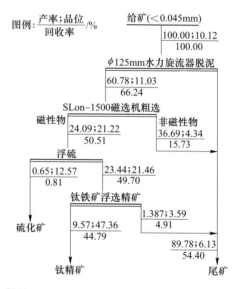

图 5　攀枝花选钛厂小于 0.045mm 钛铁矿选矿数质量流程

4.2 SLon 磁选机在重钢太和铁矿选钛流程中的应用

重庆钢铁公司太和铁矿矿石为磁铁矿与钛铁矿的混合矿。1994 年以前，该矿建立了弱磁选流程分选磁铁矿。1994~1995 年，该矿又建立了以螺旋溜槽、摇床为粗选和浮选为精选的钛铁矿选矿流程从弱磁选尾矿中回收钛铁矿。但是，由于螺旋溜槽和摇床都不能回收细粒钛铁矿，而浮选又难以回收粗粒钛铁矿，TiO_2 从细粒级和粗粒级中大量流失，因此选钛回收率很低，仅占弱磁选尾矿中 TiO_2 的 10% 左右。

2000~2004 年，该矿先后安装了 3 台 SLon-1500 磁选机和 3 台 SLon-1750 磁选机取代螺旋溜槽和摇床，这 6 台磁选机与浮选作业组成磁—浮流程分选钛铁矿。该流程首先采用浓缩分级将给矿大致分为大于 0.1mm（粗粒级）、0.1~0.02mm（细粒级）和小于 0.02mm（溢流）3 个粒级，粗粒级和细粒级分别采用 SLon 磁选机粗选，粗粒级磁选精矿再磨后与细粒级精矿合并，用永磁中磁机除去磁铁矿等强磁性物质后再用 SLon-1500 磁选机精选，其磁选精矿再进入浮选精选。选别流程见图 6，其特点为：

（1）粗、细分级磁选可采用不同的磁选参数以获得最佳的选矿指标。

（2）浮选给矿品位较高，有利于节省浮选药剂降低生产成本。

（3）浮选尾矿返回流程中形成闭路，有利于提高全流程的 TiO_2 回收率。

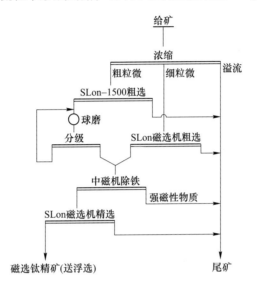

图 6 太和铁矿选钛流程

该流程有效地回收了各粒级的钛铁矿，新流程在保证钛精矿品位不小于 47% 前提下，使 TiO_2 的综合回收率提高至 47% 左右。目前，太和铁矿的钛精矿产量已由改造前的 2000 余吨/a 提高到 7 万吨/a 的能力。

4.3 SLon 磁选机在承德黑山铁矿选钛流程中的应用

承德钢铁公司黑山铁矿矿石类型为磁铁矿和钛铁矿为主的混合矿，其磁铁矿选矿的尾矿中含有 8%~9% TiO_2。过去曾多次尝试回收其中钛铁矿，但均只能生产少量低品位的钛精矿，绝大部分长期作为尾矿排放。2004 年黑山铁矿采用 3 台 SLon-1750 磁选机，成功地解决了其钛铁矿选矿技术难题，至今已实现年产 3 万吨品位不小于 46.5% TiO_2 的设计目标，其钛精矿可作为优质的生产钛白粉原料。黑山铁矿选钛生产数质量原则流程图如图 7 所示。

4.4 SLon 磁选机在钛铁矿选矿工业中的应用统计

随着国民经济的快速发展，我国钛白粉、钛金属的需求与日俱增，钛工业的发展对钛精矿的需求量迅速增加。表 3 列举了 SLon 磁选机在我国选钛工业中的应用状况。

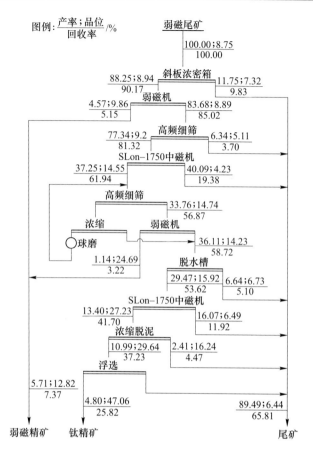

图 7 黑山铁矿选钛生产数质量原则流程图

表 3 SLon 磁选机在我国选钛工业中的应用

厂矿名称	设备型号	使用台数	分选矿种	选钛流程	分选指标（TiO$_2$）/%	钛精矿产量/万吨·a^{-1}
攀钢钛业公司选钛厂	SLon-1500 强磁机	12	<0.045mm 微细粒级钛铁矿	SLon 强磁粗选—浮选精选	$\alpha=10.12$, $\beta=47.36$, $\theta=6.13$, $\varepsilon=44.79$	14（微细粒级）
攀钢兴矿公司综合厂	SLon-1500 强磁机	3	从攀钢总尾矿回收钛资源	SLon 强磁粗选—分级—粗粒重选精选—细粒浮选精选	$\alpha=7.68$, $\beta=47.66$, $\theta=5.56$, $\varepsilon=31.19$	3
重钢太和铁矿	SLon-1500 强磁机 SLon-1750 强磁机	3 3	从钒钛磁铁矿选铁尾矿中回收钛资源	SLon 强磁粗选—粗粒再磨—粗细合并 SLon 再选—浮选精选	$\alpha=12.06$, $\beta=47.85$, $\theta=7.20$, $\varepsilon=47.44$	7
攀枝花永琦商贸公司	SLon-1750 强磁机	1	从钒钛磁铁矿选钛尾矿中回收钛资源	SLon 强磁粗选—摇床精选	$\alpha=14.50$, $\beta=47.60$, $\theta=8.30$, $\varepsilon=51.80$	3.8
攀枝花红发有限公司	SLon-1750 强磁机	1	从钒钛磁铁矿选钛尾矿中回收钛资源	SLon 强磁粗选—摇床精选	$\alpha=14.00$, $\beta=47.50$, $\theta=8.10$, $\varepsilon=50.80$	3.5
四川盐边述伦矿业有限公司	SLon-1750 强磁机	1	从钒钛磁铁矿选钛尾矿中回收钛资源	SLon 强磁粗选—摇床精选	$\alpha=11.00$, $\beta=47.20$, $\theta=7.20$, $\varepsilon=40.76$	3.3
承德黑山铁矿选钛厂	SLon-1750 中磁机 SLon-1750 强磁机	1 2	从磁铁矿选矿尾矿中回收钛资源	弱磁—中磁—强磁粗选—浮选精选	$\alpha=8.75$, $\beta=47.06$, $\theta=6.44$, $\varepsilon=25.82$	3

注：分选指标中，α 为原矿品位 TiO$_2$%；β 为钛精矿品位 TiO$_2$%；θ 为尾矿品位 TiO$_2$%；ε 为 TiO$_2$ 回收率，%。

5 SLon 磁选机在锰矿选矿工业中的应用

我国的锰矿资源绝大多数为贫矿，国产锰精矿远不能满足国内市场需求，每年要从国外进口大量的锰精矿，因此如何开发利用国内低品位的锰矿资源成为当务之急。近几年通过广泛的研究、试验和生产应用，充分证明了 SLon 磁选机能有效地分选低品位锰矿，显示了该设备在锰矿选矿工业中具有良好的应用前景。

5.1 SLon 磁选机在内蒙古分选低品位锰矿的应用

2004~2005 年，内蒙古金水矿业公司先后安装了 SLon-1000、SLon-1250、SLon-1500 立环脉动高梯度磁选机各一台进行分选低品位锰矿的工业生产，其选矿流程见图 8 所示。

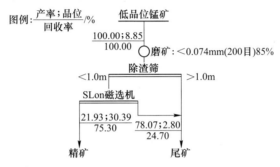

图 8 内蒙古金水矿业锰矿选矿流程图

该矿的低品位锰矿从前作为尾矿排放，经 SLon 磁选机一次粗选即可获得 30%Mn 的锰精矿，实现了变废为宝的目的，获得了良好的经济效益。

5.2 SLon 磁选机分选锰矿的试验与应用统计

近年来对广西、福建、内蒙古、四川、河南等地的低品位锰矿进行了广泛的选矿试验并有一部分已走向工业应用阶段，见表 4。

表 4 SLon 磁选机分选锰矿的试验与应用统计

锰矿产地	设备型号	锰矿类型	选矿流程	分选指标（Mn）/%	备 注
内蒙古金水矿业	SLon-1000 强磁机 1 台 SLon-1250 强磁机 1 台 SLon-1500 强磁机 1 台	菱锰矿	一次粗选	$\alpha=8.85, \beta=30.39, \theta=2.80,$ $\varepsilon=75.30$	工业生产指标
甘肃永登矿业	SLon-1250 强磁机 1 台 SLon-1500 强磁机 1 台	硬锰矿	一次粗选	$\alpha=16.94, \beta=31.88, \theta=9.74,$ $\varepsilon=61.20$	在建项目
内蒙古包头	SLon-100	黑锰矿	一次粗选	$\alpha=12.85, \beta=39.53, \theta=6.69,$ $\varepsilon=57.74$	探索试验指标
辽宁建平	SLon-100	硬锰矿	一次粗选	$\alpha=11.84, \beta=29.27, \theta=5.53,$ $\varepsilon=65.73$	探索试验指标
福建连城锰矿	SLon-100	硬锰矿	一次粗选	$\alpha=28.27, \beta=46.47, \theta=16.32,$ $\varepsilon=65.14$	探索试验指标
广西木圭	SLon-750	松软锰矿	一粗一扫	$\alpha=22.92, \beta=30.38, \theta=13.03,$ $\varepsilon=75.97$	半工业试验指标
四川攀枝花锰矿	SLon-100	菱锰矿	一次粗选	$\alpha=13.36, \beta=38.08, \theta=4.76,$ $\varepsilon=73.54$	探索试验指标

6　结论

（1）SLon 磁选机在国内外大中型氧化铁矿选矿厂的广泛应用，使鞍钢齐大山、调军台等选矿厂的铁精矿品位达到 67.5% 以上，铁回收率达到 78% 左右，选矿指标多次创我国氧化铁矿选矿的最高水平。

（2）SLon 磁选机在我国多个钛铁矿选钛厂的应用，解决了一系列钛铁矿选矿技术难题，显著提高了我国选钛技术水平，使我国作为钛白粉原料的优质钛精矿产量迅速增加。

（3）SLon 磁选机在我国锰矿选矿方面的广泛试验及应用，获得了优良的选矿试验指标和生产指标，显示了该设备在开发低品位锰矿资源方面具有很大的发展潜力。

（4）SLon 磁选机在国内外众多黑色金属选矿工业中的广泛应用，显著地提高了弱磁性矿石的回收率和精矿品位，并使大量的低品位矿石得到回收利用，为提高我国矿产资源的利用率发挥了越来越大的作用。

参 考 文 献

[1] 熊大和. SLon-1000 立环脉动高梯度磁选机的研制 [J]. 金属矿山，1988（10）：39~40.
[2] 熊大和. SLon-2000 立环脉动高梯度磁选机的研制 [J]. 金属矿山，1995（6）：32~34.
[3] 赫荣安，陈平，熊大和. SLon 强磁机选别鞍山式贫赤铁矿的试验及应用 [J]. 金属矿山，2003（9）：5~10.
[4] 李建设. SLon 立环脉动高梯度磁选机在秘鲁铁矿的应用 [J]. 金属矿山，2004（8）：39~41.

写作背景　本文介绍了 SLon 立环脉动高梯度磁选机在多个氧化铁矿和多个钛铁矿选矿流程中的应用，其中在承德钢铁公司黑山铁矿分选钛铁矿的流程为新公布的资料。

SLon 立环脉动高梯度磁选机与《金属矿山》共发展

熊大和[1,2]

(1. 赣州有色冶金研究所;2. 赣州金环磁选设备有限公司)

1 高梯度磁选技术的发展及存在的问题

我国拥有丰富的氧化铁矿、钛铁矿、锰矿、黑钨矿等弱磁性矿石资源,但多数矿床的原矿品位低、嵌布粒度细,而原有的选矿设备和工艺不能满足选矿工业的要求,细粒选矿的回收率低和磁性精矿品位低是一个普遍存在的问题,大量的有用矿物流失到尾矿中,每年对我国矿产资源造成巨大浪费。因此,当前选矿技术研究面临以下课题:研制新一代的强磁选设备和技术以适应日益贫化的入选矿石,提高精矿质量和资源回收率,提高我国弱磁性矿石的选矿水平。

高梯度磁选是 20 世纪 70 年代发展起来的新技术,被公认为具有对细粒弱磁性颗粒回收率较高的优点,美国等国家首先研制了周期式高梯度磁选机用于高岭土提纯。人们很早就想利用这一技术来提高细粒级金属矿物选矿的回收率。20 世纪 70 年代末期和 80 年代初期国内外均研制出平环高梯度磁选机,其中最有影响的是瑞典萨拉磁力公司的 Sala 平环高梯度磁选机,并在瑞典铁选厂进行了分选细粒赤铁矿的工业试验。尔后我国有几个单位也研制了平环高梯度磁选机,分别对我国鞍山式赤铁矿和宁芜式赤铁矿做了工业试验。实践证明,高梯度磁选机分选细粒弱磁性矿物有回收率较高的优点,但是由于平环高梯度磁选机的给矿方向和排磁性精矿的方向一致,其磁介质(网状的导磁材料)极易被矿粒、草渣、木屑、炸药皮等较粗的物质和铁屑、磁铁矿等强磁性物质堵塞。设备结构也很复杂,在选矿区的密封、转环转动、磁介质的清洗等方面都存在很多问题。因此,平环式高梯度磁选机至今没有在工业生产中获得广泛的推广应用。

2 振动、脉动高梯度磁选技术的诞生

针对上述问题,中南工业大学于 1980 年开始在实验室研究振动、脉动高梯度磁选技术,先后对赤铁矿、锰矿、黑钨矿和硫化矿等试样做了大量的探索试验。试验结果表明,振动或脉动高梯度磁选是高梯度磁选技术的进一步发展,能大幅度提高磁性精矿品位,并保持高梯度磁选对细粒磁性矿物回收率高的优点。例如,用振动高梯度磁选机分选赤铁矿,与无振动比较铁精矿品位可提高 8%~12%,铁回收率基本不变。此外,振动或脉动对防止磁介质堵塞也是相当有效的。但是,这项研究当时尚处于小型试验阶段。这种高效的选矿方法能否在工业生产上获得应用,关键在于研制出适用于工业生产的连续振动或脉动高梯度磁选机。

---

❶ 原载于《采选技术进展报告会》,矿业快报,2006:281~286。

3 SLon-1000 型立环脉动高梯度磁选机的发明

SLon-1000 型立环脉动高梯度磁选机是由赣州有色冶金研究所和中南工业大学联合研制的第一台工业型脉动高梯度磁选机。该机转环立式旋转，对于每一组磁介质而言，冲洗精矿的方向与给矿方向相反，粗颗粒不必穿过磁介质堆便可冲洗出来。该机的脉动机构驱动矿浆产生脉动，使位于分选区磁介质堆中的矿粒群保持松散状态，使磁性矿粒易于被磁介质吸附，使非磁性矿粒尽快穿过磁介质堆进入到尾矿中去，减少机械夹杂。反冲精矿和矿浆脉动可防止磁介质堵塞，脉动分选可提高磁性精矿的质量。

这台 SLon-1000 型立环脉动高梯度磁选机于 1987 年 8 月在赣州有色冶金研究所制造出来，然后运至马钢姑山铁矿进行工业考核试验。当时姑山铁矿有 10 台 ϕ1600mm×900mm 离心选矿机分选细粒赤铁矿，由于离心选矿机存在机械故障率很高，选矿效果较差的问题，姑山铁矿决定拆去 1 台离心选矿机装上 SLon-1000 型立环脉动高梯度磁选机。至 1988 年 3 月，该机累计运转了 3000 余小时，其平均生产指标为给矿品位 33.26%，精矿品位 56.78%，尾矿品位 19.31%，作业回收率 63.56%。与采用离心机工艺（1987 年 1~6 月的生产指标）比较，精矿品位提高 4.00% 以上，尾矿品位降低 6.00%，作业回收率提高 20.00% 左右。

1988 年 6 月，SLon-1000 型立环脉动高梯度磁选机在马鞍山市通过了冶金工业部和中国有色金属工业总公司联合主持的技术鉴定，到会的 40 多位专家充分肯定了该机研制成功的重要意义，认为该机具有新颖性和创造性，是国内外一种新型高效的高梯度磁选设备，为提高我国细粒弱磁性矿的选矿水平开创了新的途径，具有国际先进水平。

1988 年 7 月和 10 月《金属矿山》杂志率先刊登了作者撰写的科技论文《脉动高梯度磁选分选细粒氧化铁矿的研究》和《SLon-1000 立环脉动高梯度磁选机的研制》。文章刊登后，该技术及设备引起了国内选矿工程界的关注，为该机的进一步发展奠定了基础。

4 SLon-1500 型立环脉动高梯度磁选机的研制

SLon-1000 型立环脉动高梯度磁选机研制成功，并不意味着就能在生产上大规模应用。为了满足大中型选厂需要，赣州有色冶金研究所与马钢姑山铁矿合作，于 1989 年研制出了 SLon-1500 型立环脉动高梯度磁选机。该机较 SLon-1000 型立环脉动高梯度磁选机处理量增加 3~4 倍，在结构方面作了不少改进，使其稳定性和选矿性能有了进一步提高。首台 SLon-1500 型立环脉动高梯度磁选机于 1989 年 6 月制造出来，运至马钢姑山铁矿做工业试验。该机在马钢姑山铁矿用于从尾矿中回收赤铁矿，当给矿含铁品位为 22.00%~35.00% 时，一次分选获得品位为 55.00% 以上的合格铁精矿，作业回收率平均为 60.67%，每年可从尾矿中多回收 2 万~3 万吨铁精矿。该机在姑山铁矿连续进行了 6 个多月的工业试验，设备运转稳定，单机月作业率达 98% 以上。选矿指标和设备作业率均达到了合同规定的指标。工业试验完成之后，该机在姑山铁矿转化为生产设备，在生产中取代了 5 台 ϕ1600mm×900mm 离心选矿机，这是国内外第一台应用于工业生产的立环脉动高梯度磁选机。随后 SLon-1500 型立环脉动高梯度磁选机又在宝钢集团梅山铁矿、鞍钢弓长岭选厂等地推广应用，有效地推动了这些厂矿的技术进步。

SLon-1500 型立环脉动高梯度磁选机于 1990 年通过了有色金属工业总公司组织的技术鉴定。"立环脉动高梯度磁选机"同年荣获国家发明奖和发明专利。1990 年第 7 期的《金属矿山》刊登了笔者撰写的《SLon-1500立环脉动高梯度磁选机的研制》论文,介绍了该设备的研制及在马钢姑山铁矿的试验情况。

5 SLon-2000型立环脉动高梯度磁选机的研制

鞍钢弓长岭选矿厂在20世纪80年代安装了5台Shp-2000型仿琼斯强磁选机用于一段磨矿分级溢流的粗选作业,因其给矿粒度粗及强磁性矿物较多,该机存在着齿板堵塞频繁、选矿指标不稳定和维修工作量大的缺点。为了解决上述问题,鞍钢矿业公司与赣州有色冶金研究所签订了SLon-2000型立环脉动高梯度磁选机的研制合同,由赣州有色冶金研究所研制1台SLon-2000型立环脉动高梯度磁选机,运至弓长岭选矿厂取代1台Shp-2000型平环强磁选机进行6个月的工业对比试验,若铁精矿品位提高2%,铁回收率提高5%,磁介质不堵塞,工作稳定则由鞍钢矿业公司出资购买该设备。

SLon-2000型立环脉动高梯度磁选机在1992年完成了设计工作,该机的设计吸收了SLon-1000、SLon-1500型立环脉动高梯度磁选机的设计及多次试验、多次改进的经验,采用了一系列的新技术,例如采用ϕ5mm棒介质,采用不锈钢外壳保护激磁线圈,采用滚道式脉动冲程箱等。该机的设计制造取得了成功,1994年7月运至弓长岭选矿厂与Shp-2000型平环强磁选机进行工业对比试验。

为期6个月的工业试验,SLon-2000型磁选机与Shp-2000型平环强磁选机对比,取得铁精矿品位提高7.21%、尾矿品位降低5.41%、铁回收率提高7.36%的优异指标,其处理量平均为60t/h,最大处理量达到104t/h,设备作业率达到98.80%,而Shp-2000型平环强磁选机处理量为40t/h,其设备作业率为78.50%。

SLon-2000型立环脉动高梯度磁选机于1997年通过了中国有色金属工业总公司主持的技术鉴定,到会的专家认为该机研制成功,使贫赤铁矿选矿取得重大突破,是强磁选设备发展的一个质的飞跃,该机的选矿性能和机电性能达到了国际领先水平。

《金属矿山》杂志1995年第6期刊载了《SLon-2000立环脉动高梯度磁选机的研制》以及在鞍钢弓长岭选厂试验和应用的情况,使SLon磁选机的知名度进一步提高。

6 SLon立环脉动高梯度磁选机的系列化

针对马钢姑山铁矿用于赤铁矿粗选作业的SQC-6-2770型湿式强磁选机精矿品位偏低、尾矿品位偏高、齿板较易堵塞的问题,从1998年开始用了3年的时间研制和改进了SLon-1750型立环脉动高梯度磁选机。SLon-1750型立环脉动高梯度磁选机的研制采用了一系列的新技术和新材料,如绕制线圈采用新型绝缘材料,显著提高了激磁线圈的耐腐蚀性和绝缘性能;采用计算机精确定位和优化排列方式研制了ϕ4mm和ϕ2mm棒形磁介质;研制了耐磨矿浆阀、耐磨材料制成冲程箱滚道,使这些零部件的使用寿命大幅度提高;采用了可控硅数字式整流技术提高了整机的控制性能。新技术的应用使SLon-1750型立环脉动高梯度磁选机具有优异的选矿性能和机电性能。6台SLon-1750型立环脉动高梯度磁选机取代了姑山铁矿6台SQC-6-2770型湿式强磁选机,彻底解决了该机存在的问题,使铁精矿品位和回收率有了大幅度的提高,并且增加了设备的处理能力和作业率。在未增加球磨机的情况下,姑山铁矿的精矿产量从16万吨/a增加到38万吨/a。铁精矿品位从57.00%提高到60.00%,铁回收率从70.00%提高到76.00%。SLon-1750型立环脉动高梯度磁选机于2002年通过江西省科技厅主持的技术鉴定,其整体水平达到了国际领先水平。

安徽省来安县境内具有丰富的长石资源，除铁含量较高之外，其他成分均可满足玻璃原料的要求，1997年苏州非金属设计院为来安县设计了一条年产3万吨长石精矿的选矿生产线，其中由赣州有色冶金研究所负责研制1台SLon-1250型立环脉动高梯度磁选机。长石原矿含Fe_2O_3 1.48%左右，要求一次分选达到长石精矿含Fe_2O_3<0.3%。SLon-1250型立环脉动高梯度磁选机采用计算机辅助设计和制图，从设计到制造成功仅用了6个月的时间。首台SLon-1250型立环脉动高梯度磁选机在来安皖东长石厂除铁工艺中起到了关键作用，含Fe_2O_3 1.48%的长石原矿经该机一次选别后使长石精矿含Fe_2O_3 0.26%，成为优质的玻璃原料。

在1986~2005年期间，先后研制了SLon-100型（1.2T和1.75T）周期式脉动高梯度磁选机、SLon-750型立环脉动高梯度磁选机和SLon-500型立环脉动高梯度磁选机等试验型设备，并被国内外众多科研院所、大专院校和厂矿企业的实验室采用，为国内外矿产资源开发利用的前期试验工作打下了良好的基础。

2002~2005年，针对鞍山地区原有的滚筒式中磁机尾矿品位偏高的问题，相继研制出了SLon-1500、SLon-1750、SLon-2000型立环脉动中磁机（额定背景场强0.4~0.6T）用于控制螺旋溜槽尾矿品位，大幅度降低了中矿循环量，显著地优化了流程结构。

经20多年持续的研究、创新与改进，SLon立环脉动高梯度磁选机已形成系列化产品，处理量涵盖0.1~100t/(台·h)。该产品具有优异的选矿性能和机电性能，在我国弱磁性矿石选矿工业中得到广泛应用，成为我国新一代强磁选设备。

7　SLon立环脉动高梯度磁选机的大规模推广应用

迄今我们已将400多台SLon立环脉动高梯度磁选机成功应用在鞍钢、宝钢、首钢、马钢、攀钢、昆钢、包钢、酒钢、海钢等企业中（见图1），并出口秘鲁、印尼、南非等国家，解决了许多弱磁性铁矿、微细粒钛铁矿和多种非金属矿石选矿的技术难题。

近年来我国的钢铁工业得到飞速发展，钢铁工业对优质铁精矿的巨大需求促进了SLon磁选机、反浮选等新技术在我国贫、细、杂氧化铁矿选矿工业中的高速发展和应用，使我国氧化铁矿的选矿技术水平达到国际先进水平。

图1　鞍钢集团采用50多台SLon磁选机应用于齐大山、东鞍山、调军台等选厂技术改造，创我国红矿工业指标最高水平

鞍钢齐大山选矿厂采用23台SLon系列磁选机建成阶段磨矿—重选—强磁—反浮选的新工艺流程，解决了原生产工艺煤气焙烧能耗大、成本高、污染严重的问题，铁精矿品位从63.00%提高到67.50%，铁回收率达到78.00%以上。在齐大山工艺改造获得成功的基础上，鞍钢东鞍山选矿厂、调军台选矿厂、弓长岭选矿厂、胡家庙选矿厂以及安阳钢铁公司舞阳铁矿相继采用了84台SLon磁选机建成阶段磨矿—重选—强磁—反浮选流程处理细

粒氧化铁矿，选矿指标多次创我国氧化铁矿选矿的最高水平。

我国的攀枝花、西昌和承德地区蕴藏着丰富的钛铁矿资源，过去由于选矿技术和设备落后，我国对钛铁矿回收利用率很低。近年来，SLon 磁选机在我国攀钢集团钛业公司选钛厂、重庆钢铁公司太和铁矿、承德钢铁公司黑山铁矿、攀西地区的兴矿、龙蟒、永琦、红发、述伦矿业公司等十多家钛铁矿选钛厂应用，解决了一系列钛铁矿选矿技术难题，显著提高了我国选钛技术水平，使我国作为钛白粉原料的优质钛精矿产量迅速增加（见图2）。

图 2　30 多台 SLon 磁选机应用于攀枝花、西昌、承德等地选别钛铁矿

SLon 立环脉动高梯度磁选机在安徽来安县提纯长石矿、在江西乐平提纯石英砂、在四川南江提纯霞石矿、在安徽淮北提纯煤系高岭土、在四川德昌选别稀土矿，这些应用使很多非金属矿和稀土矿产资源从废石或低级产品提升为优质工业原料，为开发我国矿产资源利用开创了新的途径。

我国的锰矿资源绝大多数为贫矿，国产锰精矿远不能满足国内市场需求，每年要从国外进口大量的锰精矿，因此如何开发利用国内低品位的锰矿资源成为当务之急。目前 SLon 磁选机已在内蒙古金水、甘肃永登等地用于锰矿选别，证明了 SLon 磁选机能有效地分选低品位锰矿。

SLon 磁选机在国内外弱磁性矿物选矿工业中的广泛应用，显著地提高了弱磁性矿石的回收率和精矿品位，并使大量的低品位矿石得到回收利用，为提高我国矿产资源的利用率发挥了越来越大的作用。

8　《金属矿山》的宣传促进了 SLon 磁选机的发展

《金属矿山》作为在矿业工程界的知名媒体见证了 SLon 立环脉动高梯度磁选机的发展历程，《金属矿山》杂志及历年学术会议论文集先后刊登有关 SLon 立环脉动高梯度磁选机研制与应用的论文 50 余篇。由于《金属矿山》多年来的跟踪报道，SLon 立环脉动高梯度磁选机在国内外的影响逐年扩大。SLon 磁选机如今已成为选矿工业知名品牌，市场占有率连续多年居全国第一。

SLon 立环脉动高梯度磁选机的设备研制和应用研究项目多次获得各级部门的表彰和奖励。例如"SLon 立环脉动高梯度磁选机的推广应用"获 1995 年度国家科学技术进步奖三等奖、"SLon-2000 立环脉动高梯度磁选机的研制"获 1998 年度中国有色金属工业总公司科技进步奖二等奖、"攀枝花微细粒级钛铁矿选矿工程技术及选钛装备研究"获 2002 年度四川省科学进步奖一等奖、"SLon-1750 立环脉动高梯度磁选机"获 2003 年度中国有色金属工业科学技术奖一等奖、"弱磁性矿石高效强磁选技术与设备"获 2003 年度江西省科学技术进步奖一等奖、"鞍山贫赤（磁）铁矿选矿新工艺、新药剂与新设备研究及工业应用"获2004年度国家科学技术进步奖二等奖，这些成绩的取得与《金属矿山》的大力宣传也是分不开的。

今年是《金属矿山》杂志创刊40周年的喜庆之年，40年来《金属矿山》为广大科技工作者提供了学习新知识、展示新技术的平台，广告宣传使广大读者了解新材料、新设备，也使企业有了宣传自己的机会。我们相信《金属矿山》在促进行业技术交流，推动科技进步和生产力的发展方面必将作出更大的贡献。

写作背景　本文为答谢《金属矿山》创刊40周而作。SLon-1000、SLon-1500、SLon-2000等立环脉动高梯度磁选机的研制和应用都在《金属矿山》期刊上得到发表，使SLon磁选机在国内外的影响逐年扩大，已成为选矿工业知名品牌，市场占有率连续多年居全国第一。《金属矿山》期刊在推动科技进步和生产力的发展方面作出了重大的贡献。

SLon Magnetic Separator Promoting Chinese Oxidized Iron Ore Processing Industry[1]

Xiong Dahe

(SLon Magnetic Separator Ltd. ,Ganzhou ,341000 ,Jiangxi ,P. R. China)

Abstract In recent years, Chinese iron and steel industry develops rapidly and consumes huge amounts of iron concentrates. There are abundant iron ore deposits in China, but most of them are of low grade. The most important technologies of SLon magnetic separators and reverse flotation have driven Chinese oxidized iron ore processing industry developing fast. Applications of SLon magnetic separators in Chinese oxidized iron ore processing industry are introduced in the paper.

1 Introduction

High gradient magnetic separation (HGMS) has the advantage of high recovery of fine weakly magnetic particles (Oberteuffer et al. ,1974,Svoboda,1981). Cyclic HGMS separators have been successfully applied for kaolin purification since 1970s. The most successfully industrialized HGMS separators are SLon vertical ring pulsating high gradient magnetic separators (SLon VPHGMS or SLon magnetic separator). SLon magnetic separators possess the advantages of high efficiency, high reliability and high availability. Now more than 400 SLon magnetic separators have been widely applied in Chinese oxidized iron ore, ilmenite and non-metallic minerals processing industry.

2 Structure and working princple

SLon magnetic separators mainly consist of the pulsating mechanism, energizing coils, magnetic yoke, separating ring, feeding and product boxes as shown in Fig. 1 (Xiong,1997,1998). Expanded metals or round bars made of magnetic stainless steel are used as matrix. While a direct electric current flows through the energizing coil, a magnetic field is build up in the separating zone. The ring with magnetic matrix rotates around its horizontal axis. When slurry from the feeding boxes enters into matrix located in the separating zone, magnetic particles are attracted from the slurry onto the surface of the matrix, brought to the top of the ring where the magnetic field is negligible, then flushed out into the concentrate box. Non-magnetic particles, however, pass through the matrix and enter into the tailings box under the combined actions of gravity and pulsating hydrodynamic force.

As the ring rotates vertically, the flushing direction of the mags is opposite to that of feeding so that coarse particles can be flushed out without having to pass through the entire depth of the matrix pile. The pulsating mechanism drives the slurry in the separating zone up and down, keeping particles in the matrix pile in a loose state. therefore magnetic particles can be more easily captured

---

[1] 原载于《土耳其第23届国际选矿会议论文集》,2006。

by the matrix and non-magnetic particles can be more easily dragged to the tailings box through the matrix pile. Obviously, opposite flushing and pulsating prevent matrix from being clogged, and most importantly, pulsating helps to purify magnetic product.

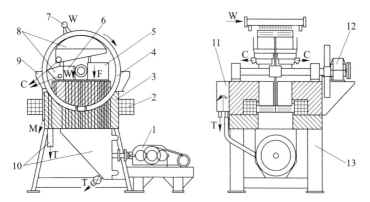

Fig. 1 SLon vertical ring and pulsating high gradient magnetic separator
1—pulsating mechanism; 2—energizing coils; 3—magnetic yoke; 4—working ring; 5—feeding box; 6—wash water box; 7—concentrate flush; 8—concentrate box; 9—middling chute; 10—tailings box; 11—slurry level box; 12—ring driver; 13—support frame; f—feed; w—wate; c—concentrate; M—middlings; T—tails

The main parameters of SLon magnetic separators of different models are listed in Table 1.

Table 1 Main parameters of SLon magnetic separators of different models

Parameters	Ring diameter /mm	Back-ground field /T	Energizing power /kW	Driving power /kW	Pulsating stroke /mm	Pulsating frequency /min^{-1}	Ore-feeding size(mm) <0.074mm (200mesh)/%	Solid density of feeding /%	Slurry through-hput /m$^3 \cdot$ h^{-1}	Ore through-hput /t $\cdot$ h^{-1}	Mass of machine /t
SLon-2500	2500	0-1.0	0-94	11+11	0-30	0-300	<1.3(30-100)	10-40	200-400	80-150	105
SLon-2000	2000	0-1.0	0-74	5.5+7.5	0-30	0-300	<1.3(30-100)	10-40	100-200	50-80	50
SLon-1750	1750	0-1.0	0-62	4+4	0-30	0-300	<1.3(30-100)	10-40	75-150	30-50	35
SLon-1500	1500	0-1.0	0-44	3+4	0-30	0-300	<1.3(30-100)	10-40	50-100	20-30	20
SLon-1250	1250	0-1.0	0-35	1.5+2.2	0-30	0-300	<1.3(30-100)	10-40	20-50	10-18	14
SLon-1000	1000	0-1.2	0-30	1.1+2.2	0-30	0-300	<1.3(30-100)	10-40	10-20	4-7	6
SLon-750	750	0-1.0	0-22	0.55+0.75	0-50	0-400	<1.3(30-100)	10-40	1.0-2.0	0.1-0.5	3
SLon-500	500	0-1.0	0-13.5	0.37+0.37	0-50	0-400	<1.3(30-100)	10-40	0.5-1.0	0.05-0.25	1.8

3 Application in Qidashan iron ore processing plant

Qidashan mineral processing plant is a large oxidized iron ore-processing base of Anshan Iron and Steel Company in Liaoning Province, Northeast China(Xiong, 2003). It processes eight million tons of oxidized iron ore and produces 2.7 million tons of iron concentrate annually.

As a poor iron deposit, Qidashan iron ore mainly consists of hematite, magnetite and quartz. The sizes of iron particles are from 0.005mm to 1.0mm, 0.05mm on average. The sizes of quartz particles are 0.085mm on average. The harmful elements as S and P are very low in the ore, thus obtaining excellent iron concentrate with efficient processing technology being possible.

The previous processing flowsheet of the ore is coal gas roasting-low intensity magnetic separation with which only low grade of 63.22%Fe iron concentrate can be obtained. In recent years, the plant developed a new flowsheet of spiral-SLon magnetic separating-reverse flotation as shown in Fig. 2. In the new flowsheet, 11 SLon-1750 (magnetic field: 1.0T) magnetic separators are applied to control the tailings of the fine fraction (cyclone overflow) and 12 SLon-1500 (magnetic field: 0.4T) magnetic separators are applied to control the tailings of the coarse fraction (spiral tailings). The results of the flowsheet are: feed grade 30.15%Fe, concentrate grade 67.50%Fe, tailings grade 10.16%Fe, iron recovery 78.05%Fe, being a historical high record of the plant.

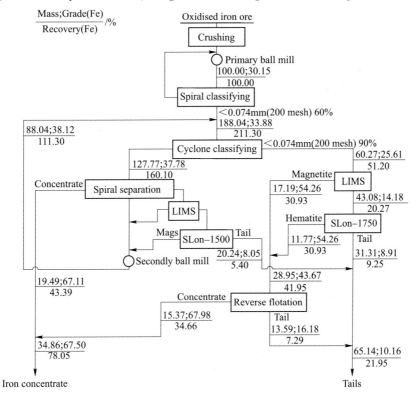

Fig. 2 The technical data of the new processing flowsheet of Qi Dashan

4 Application in Donganshan iron ore processing plant

Donganshan iron ore processing plant of Anshan Iron and Steel Company treats 4.3 million tons of oxidized iron ore of mainly hematite each year. The crystal sizes of the ore are much finer and the iron minerals are much deeply oxidized than that of Qidashan ore. The previous flowsheet was two-stage grinding and flotation, and the iron concentrate grade could only reach about 60%Fe.

In 2002, the plant innovated its flowsheet into a spiral-SLon magnetic separation-reverse flotation. In the flowsheet, 8 SLon-1750 magnetic separators (magnetic field: 1.0T) and 2 SLon-2000 magnetic separators (magnetic field: 1.0T) are applied to control the tailings of the fine fraction (Cyclone overflow), and 10 SLon-1750 magnetic separators (magnetic field: 0.6T) to control the tailings of the coarse fraction (spiral tailings). The average results of the flowsheet are: feed grade 32.50% Fe, concentrate grade 64.50% Fe, tails grade 13.45% Fe, iron recovery 74.06% Fe.

5 Application in Diaojuntai iron ore processing plant

Diaojuntai Iron Ore Processing Plant of Anshan Iron and Steel Company treats 9 million tons oxidized iron ore consisting of magnetite and hematite each year and the iron ore processing flowsheet is two stage grinding-high intensity magnetic separation-reverse flotation. In the past, 15 Shp-3200 horizontal ring wet high intensity separators (Shp-3200 WHIMS) were used as high intensity magnetic separators, but their grooved plates as magnetic matrix were easy to be clogged, and difficult to maintain with low availability of only 80%-90%.

As SLon magnetic separators possess advantages of higher beneficiating ability, matrix not easy to be clogged and availability as high as 98%-99%, by the end of 2005, Diaojuntai Iron Ore Processing Plant has applied 15 SLon-2000 magnetic separators to replace the 15 Shp-3200 WHIMS. The present flowsheet is shown in Fig. 3, which shows that the 15 SLon-2000 magnetic separators

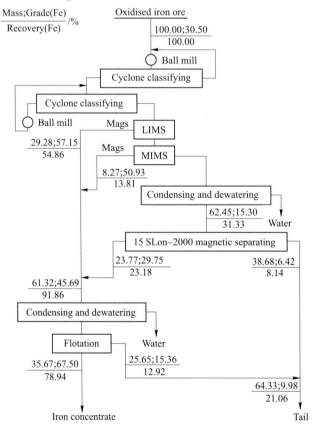

Fig. 3 The technical data of processing flowsheet of Diaojuntai

control the tails grade on a very low level of only 6.42% Fe, throwing most of the slimes into tails. They have created very good conditions for the following flotation stage.

For the excellent performance of the SLon-2000 magnetic separators, the plant gets very good results as: feed grade 30.50% Fe, concentrate grade 67.50% Fe, tails grade 9.98% Fe, iron recovery 78.94% Fe.

6 Application in Peru Iron Mine of Shougang Hierro Peru S. A. A

Peru Iron Mine was bought by Shougang (Beijing) Iron and Steel Company ten years ago, with its deposit of iron ore about 1.43 billion tons in Malkedia region. In the early years, the Mine only mined and processed magnetite ores of high grade, with huge lower grade oxidized iron ores discarded and untreated, occupying large amounts of land.

Since early 2004 to late 2005, 4 SLon-1750 magnetic separators are applied in the mine to beneficiate oxidized iron ores and a magnetic separation-sulfur flotation flowsheet is built as shown in Fig. 4. The flowsheet processes about 1.3 million tons iron ore and produces about 0.9 million tons of iron concentrates (65% Fe) annually.

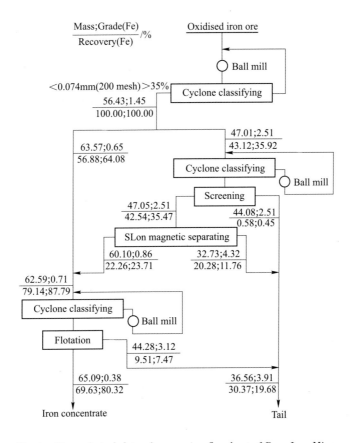

Fig. 4 The technical data of processing flowsheet of Peru Iron Mine

7 Application in Gushan Iron Mine

The deposit of Gushan Iron Mine of Maanshan Iron and Steel Company is mainly refractory hematite ore. In 1980s, the Mine applied a spiral-centriafugal separation processing flowsheet to beneficiate the iron Ore, the concentrate grade could only reach 55% Fe and iron recovery about 61% Fe. Since 1990, the mine gradually changed the flowsheet into SLon magnetic separation. At present, 9 SLon-1750 and 4 SLon-1500 magnetic separators are applied in the flowsheet and the iron concentrate grade reaches 60% Fe and recovery 76% Fe.

8 Application in Hainan Oxidized Iron Ore

Hainan Iron and Steel Company owns a big oxidized iron ore deposit and the ore mainly consists of finely crystallized hematite and about 10% magnetite. From 2003 to 2005, the company built a magnetic separation flowsheet as shown in Fig. 5, in the flowsheet 6 SLon-2000 magnetic separators are applied to recover fine hematite, in which 2 SLon-2000 separators for roughing, 2 SLon-2000 separators for cleaning and 2 SLon-2000 separators for scavenging. The results of the flowsheet are: feed grade 51.09% Fe, concentrate grade 64.57% Fe, iron recovery 88.01% Fe.

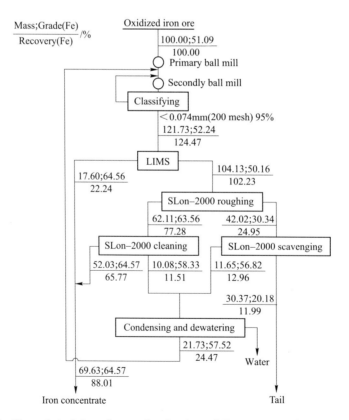

Fig. 5 The technical data of processing flowsheet of Hainan Iron and Steel Company

As SLon magnetic separators successfully applied to beneficiate the Hainan iron ore, now another 25 SLon magnetic separators have been or are to be stalled in the region to treat oxidized iron ore.

9 Conclusions

(1) SLon magnetic separators possess advantages of large beneficial ratio, high processing efficiency, strong adaptability, magnetic matrix not easy to be blocked and high equipment availability. They have become the new generation of high intensity magnetic separator in China.

(2) Most Chinese iron ore are of low grade and finely disseminated. Many Chinese iron mines use SLon separators to achieve iron concentrate of high grade, with low cost.

(3) Nowadays, more than 400 SLon magnetic separators are applied in industry to beneficiate oxidized iron ores and many other iron ore processing plants are being built with SLon magnetic separators in China, with designed iron concentrate grades commonly 66%Fe higher. SLon magnetic separators are creating new records and promoting Chinese oxidized iron ore processing industry developing fast.

References

[1] Oberteuffer J A. Magnetic separation: a review of principals, devices, and applications[J]. IEEE Trans. on Mag, 1974, 2: 223-238.
[2] Svoboda J. High gradient magnetic separation: a research for a matrix material[J]. Int. J. of Miner. Process, 1981, 8: 165-175.
[3] Xiong Dahe. SLon magnetic separator applied to upgrading the iron concentrate[J]. Physical Separation in Science and Engineering, 2003, 2: 63-69.
[4] Xiong Dahe. Development and applications of SLon vertical ring and pulsating high gradient magnetic separators [J]. Int. Proceedings. of XX IMPC, Aachen, 1997: 21-26.
[5] Xiong Dabe, Liu Shuyi, Chen Jin. New technology of pulsating high gradient magnetic separation [J]. Int. J. Miner. Process. , 1998, 54: 111-127.

写作背景　本文向国外的同行介绍了 SLon 立环脉动高梯度磁选机在我国氧化铁矿选矿工业中的应用，列举了该机在鞍钢齐大山选矿厂、鞍钢调军台选矿厂、首钢秘鲁铁矿和海南铁矿的应用。使该机在国际上的知名度有了进一步的提高，为 SLon 磁选机出口起到了一定的宣传作用。

SLon 脉动与振动高梯度磁选机新进展

熊大和[1,2]　刘建平[3]

(1. 赣州有色冶金研究所；2. 赣州金环磁选设备有限公司；
3. 鞍钢集团鞍山矿业公司供销公司)

摘　要　脉动与振动高梯度磁选是细粒弱磁性矿物的高效选矿技术。介绍了高梯度磁选机的有关基本理论及 SLon 周期式和连续式脉动高梯度磁选机、SLon 周期式和连续式振动高梯度磁选机的发展现状。并对 SLon-2500 立环脉动高梯度磁选机和 SLon-1000 立环振动高梯度磁选机新设备作了简介。目前，SLon 型高梯度磁选机已有 500 多台在国内外使用，显著地提高了高梯度磁选技术水平。

关键词　脉动高梯度磁选　振动高梯度磁选　SLon 磁选机　强磁选

New Advance in SLon Pulsating and Vibrating High Gradient Magnetic Separators

Xiong Dahe[1,2]　Liu Jianping[3]

(1. Ganzhou Nonferrous Metallurgy Research Institute; 2. SLon Magnetic Seperator Ltd.;
3. Supply and Marketing Co., Anshan Mining Co. of An Steel (Group))

Abstract　Pulsating and vibrating high gradient magnetic separators are high efficiency separation technology for feebly magnetic minerals. The paper describes the basic theory of high gradient magnetic separators and the state quo of the development of SLon periodic and continuous pulsating high gradient magnetic separators and SLon periodic and continuous vibrating high gradient magnetic separators. Brief introduction is also made of new SLon-2500 vertical ring pulsating high gradient magnetic separators and SLon-1000 vertical ring vibrating high gradient magnetic separators. At present, over 500 SLon high gradient magnetic separators are used both in and outside China, noticeably raising the level of high gradient magnetic separation.

Keywords　Pulsating high gradient magnetic separation; Vibrating high gradient magnetic separation; SLon magnetic separators; High intensity magnetic separators

　　高梯度磁选的基本原理是利用软磁材料制成的丝状介质，在磁场中产生高梯度强磁力，捕收细粒弱磁性矿物，其特点是对微细弱磁性矿物有很强的捕收能力。

　　现有的高梯度磁选机按其工作方式可分为周期式和连续式两类。周期式高梯度磁选机

---

❶ 原载于《金属矿山》2006 (07)：4～7。

可获得较高的磁场强度和较大的分选空间，其缺点是只能间断工作。这种设备在工业生产中主要用于高岭土等非金属矿的提纯和废水处理。连续式高梯度磁选机是针对金属矿选矿而发展起来的，其适应范围较广泛，例如氧化铁矿、黑钨矿、锰矿、钛铁矿、铬铁矿等细粒弱磁性矿物的分选。

为了提高选矿效率，解决磁介质堵塞等技术难题，我们将磁力和脉动流体力或机械振动力有机地结合起来，发展了新一代的高梯度磁选设备——SLon型高梯度磁选机。该机已有500多台在国内外广泛应用，显著地提高了高梯度磁选技术水平。

1 SLon周期式脉动高梯度磁选机简介

SLon-100周期式脉动高梯度磁选机结构示意如图1所示。SLon-100周期式脉动高梯度磁选机与普通周期式高梯度磁选机的不同之处是它的下部装有一脉动机构，由偏心连杆机构产生的交变力 $\tilde{F}$ 推动橡胶鼓膜往复运动，使分选腔内的矿浆产生脉动。调节下部阀门可控制矿浆的流速和液位高度；改变流经激磁线圈的电流值可调节分选区的背景场强。用导磁不锈钢制成的钢板网或圆棒作磁介质。

选矿时，先在分选腔内注满水，使脉动能量传递到分选腔，然后从给矿盒给入矿浆。分选腔内的磁性矿物和非磁性矿物在磁力、脉动流体力、重力的综合力场作用下得到分离，磁性矿粒被吸着在磁介质表面上，非磁性矿粒随矿浆从下部排走。给矿完毕后，放干水分，切断激磁电流，然后用干净水将磁性物冲洗出来。

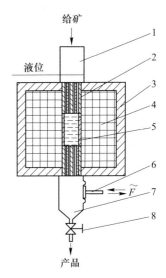

图1 周期式脉动高梯度磁选机结构示意
1—给矿盒；2—磁极头；3—磁轭；4—激磁线圈；
5—磁介质；6—脉动机构；7—接矿斗；8—阀门

2 脉动高梯度磁选原理

在脉动高梯度磁选过程中，当脉动机构工作时，脉动能量从下磁极头的通孔传入到分选区，驱使分选区矿浆产生脉动。矿浆的脉动速度 $\tilde{v}$、最大脉动速度 $\tilde{v}_{max}$ 和平均脉动速度 $\bar{v}$ 分别为：

$$\tilde{v} = \frac{1}{2}S\omega\sin\omega t$$

$$\tilde{v}_{max} = \frac{1}{2}S\omega$$

$$\bar{v} = \frac{1}{2\pi}\int_0^{2\pi} |S\omega\sin\omega t| d(\omega t) = \frac{S\omega}{\pi}$$

式中，S 为选矿区有效脉动冲程；ω 为脉动波的角速度；t 为时间变量。

矿浆的实际流速 v 为给矿速度 v_0 和脉动速度的叠加：

$$v = v_0 + \tilde{v} = v_0 + \frac{1}{2}S\omega\sin\omega t = v_0 + \tilde{v}_{\max}\sin\omega t$$

脉动矿浆的加速度 $\tilde{a}$、最大加速度 $\tilde{a}_{\max}$ 和平均加速度 $\bar{a}$ 分别为：

$$\tilde{a} = \frac{1}{2}S\omega^2\cos\omega t$$

$$\tilde{a}_{\max} = \frac{1}{2}S\omega^2$$

$$\bar{a} = \frac{1}{2\pi}\int_0^{2\pi}\left|\frac{1}{2}S\omega^2\cos\omega t\right|\mathrm{d}(\omega t) = \frac{S\omega^2}{\pi}$$

当 $\tilde{v}_{\max} > v_0$，一个脉动周期内矿浆实际流速如图 2 所示，图中水平轴 ωt 下方的阴影部分表示矿浆的实际流速与给矿方向相反，此时流体对停留在磁介质上方的矿粒产生一个反向推力，使图 3 所示被截住的非磁性矿粒脱离约束状态而进入尾矿。

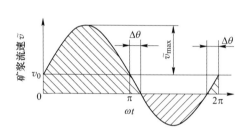
图 2 矿浆在选矿区的流速

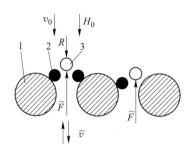
图 3 脉动松散原理
1—磁介质；2—磁性矿粒；3—非磁性矿粒

如矿浆从上至下流动时（图 3），当 $\tilde{v}_{\max} \leqslant v_0$ 或给矿方向不变时，部分脉石会被其他矿粒或介质丝架住，因流体力 R 方向朝下，故这些脉石不能脱离，导致精矿品位下降，严重时还会堵塞磁介质。当 $\tilde{v}_{\max} > v_0$ 时，分选区的矿浆不断变换流速方向，可有效降低介质中非磁性矿粒的机械夹杂，反向脉动力的存在能产生一个松散力，使非磁性矿有更多的机会进入尾矿，从而提高磁性精矿的品位，避免堵塞。无脉动和有脉动时矿粒穿过磁介质的轨迹见图 4。

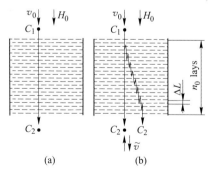
图 4 无脉动和有脉动时矿粒穿过磁介质的轨迹
(a) 无脉动；(b) 有脉动

当有脉动时，只要满足 $\tilde{v}_{max} > v_0$，矿浆便不断地改变流动方向，时而从上至下，时而从下至上，脉动使矿粒可能不止一次地穿过每一层磁介质，即矿粒的穿过次数可能比无脉动时高，有利于提高弱磁性矿石回收率。

3 周期式脉动高梯度磁选机的应用

目前 SLon 周期式脉动高梯度磁选机已在国内外众多科研院所应用于实验室探索试验，例如赣州金环磁选设备有限公司实验室有 2 台 SLon-100 周期式脉动高梯度磁选机，1 台用于弱磁性金属矿的选矿，另 1 台用于非金属矿的除铁提纯，每年约完成 300 个矿样的探索试验。采用 SLon 周期式脉动高梯度磁选机的有武汉工业大学、江西理工大学、西安冶金建筑学院、昆明冶金研究院、梅山铁矿、姑山铁矿、齐大山选矿厂等 20 多家科研院所、厂矿和大学。

4 SLon 立环脉动高梯度磁选机

4.1 整机工作原理

SLon 立环脉动高梯度磁选机结构如图 5 所示，它主要由脉动机构、激磁线圈、铁轭、转环和各种矿斗、水斗组成。用导磁不锈钢制成的圆棒或钢板网作磁介质。其工作原理：激磁线圈通以直流电，在分选区产生感应磁场，位于分选区的磁介质表面产生非均匀磁场即高梯度磁场；转环作顺时针旋转，将磁介质不断送入和运出分选区；矿浆从给矿斗给入，沿上铁轭缝隙流经转环。矿浆中的磁性颗粒吸附在磁介质表面上，被转环带至顶部无磁场区，被冲洗水冲入精矿斗，非磁性颗粒在重力、脉动流体力的作用下穿过磁介质堆，沿下铁轭缝隙流入尾矿斗排走。

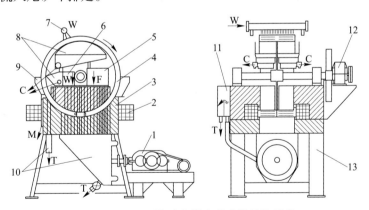

图 5 SLon 立环脉动高梯度磁选机结构示意

1—脉动机构；2—激磁线圈；3—铁轭；4—转环；5—给矿斗；6—漂洗水斗；7—精矿冲洗装置；
8—精矿斗；9—中矿斗；10—尾矿斗；11—液位斗；12—转环驱动机构；13—机架；
F—给矿；W—清水；C—精矿；M—中矿；T—尾矿

该机的转环采用立式旋转方式，对于每一组磁介质而言，冲洗磁性精矿的方向与给矿方向相反，粗颗粒不必穿过磁介质堆便可冲洗出来。该机的脉动机构驱动矿浆产生脉动，可使分选区内矿粒群保持松散状态，使磁性矿粒更容易被磁介质捕获，使非磁性矿粒尽快穿过磁介质堆进入尾矿中去。

显然，反冲精矿和矿浆脉动可防止磁介质堵塞，脉动分选可提高磁性精矿的质量。这些措施保证了该机具有较大的富集比、较高的分选效率和较强的适应能力。

4.2 应用概况

经过20多年的持续研究与技术创新，SLon立环脉动高梯度磁选机已发展成为国内外新一代的高效强磁选设备。该设备用于分选红矿具有富集比大、回收率高、磁介质不堵塞、设备作业率高的优点。目前，已有500多台SLon立环脉动高梯度磁选机在鞍钢、马钢、宝钢、昆钢、首钢、海钢、包钢、安钢等地应用于赤铁矿、镜铁矿、菱铁矿等氧化铁矿的选矿工业，在山西、河南、江西等地应用于褐铁矿的选矿，在攀钢选钛厂、重钢太和铁矿、承德黑山铁矿等地应用于钛铁矿选矿工业，在内蒙古用于铬铁矿、黑钨矿的分选，在内蒙古、南京栖霞山等地用于锰矿的分选，在四川南江、安徽来安、四川乐山、陕西洋县、安徽淮北等地应用于霞石、长石、石英、高岭土等非金属矿的除铁提纯，多次创造了我国弱磁性铁矿、微细粒钛铁矿和多种非金属矿选矿历史最高水平。

4.3 最新研制的SLon-2500立环脉动高梯度磁选机简介

我国矿产资源日益减少，贫矿日益增多。因此，设备大型化、优质高效及运行低成本是今后的主要发展方向。我们从2004年开始研制SLon-2500立环脉动高梯度磁选机，目前已研制出样机，即将开展工业试验。该机主要技术参数见表1。

表1 SLon-2500立环脉动高梯度磁选机主要技术参数

转环直径 /mm	转环转速 /r·min^{-1}	给矿粒度 /mm (<0.074mm (200目)/%)	给矿浓度 /%	矿浆通过能力 /m^3·h^{-1}	干矿处理量 /t·h^{-1}	额定背景场强 /T	额定激磁电流 /A	额定激磁电压 /V	额定激磁功率 /kW
2500	2.5~3	<1.3 (30~100)	10~40	180~450	80~150	1.0	1700	55	94

转环电动机功率 /kW	脉动电动机功率 /kW	脉动冲程 /mm	脉动冲次 /r·min^{-1}	供水压力 /MPa	耗水量 /m^3·h^{-1}	主机质量 /t	最大部件质量 /t	外形尺寸 (长×宽×高) /mm×mm×mm
11	11	0~30	0~300	0.2~0.4	200~400	105	15	5800×5000× 5400

SLon-2500立环脉动高梯度磁选机主要优点如下：

（1）设备处理量大。该机台时处理量可达150t（处理鞍山式贫赤铁矿台时处理量上限可达200t）。该机是迄今我国台时处理量最大的强磁选设备。设备大型化可节约用地，降低操作成本。

（2）高效节能。SLon-2500磁选机处理每吨矿石的电耗为0.63kW·h，比SLon-2000磁选机（1.02kW·h/t）节电38%。

（3）自动化程度较高，有利于选厂自动化控制。

5 周期式振动高梯度磁选机简介

振动高梯度磁选是利用机械力对磁介质产生一个振动力，其优点是既可用于干法选矿，也可用于湿法选矿。

如图6所示为周期式振动高梯度磁选机工作原理。磁介质被固定在一根长杆上，调速电动机驱动凸轮旋转，凸轮及弹簧驱动磁介质上下振动。该机振动频率和振幅可调节。

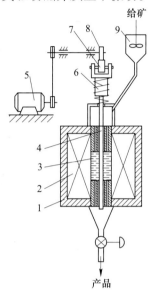

图6　周期式振动高梯度产品磁选机结构示意

1—铁铠；2—激磁线圈；3—磁介质；4—振杆；5—调速电机；6—弹簧；7—凸轮；8—滚动轴承；9—搅拌槽

6　连续式 SLon-1000 振动高梯度磁选机简介

如图7所示为最新研制的SLon-1000振动高梯度磁选机。该机利用振动电机通过转环主轴驱动转环振动，传动机构通过橡胶联轴器驱动转环旋转。该机的场强、振幅、振次、转环转速等多种参数可调，可广泛应用于细粒弱磁性矿的干法分选。该机在非金属矿提纯工业中，与雷蒙磨串联可实现非金属干法除铁的生产过程。目前该机尚处于研制发展阶段。

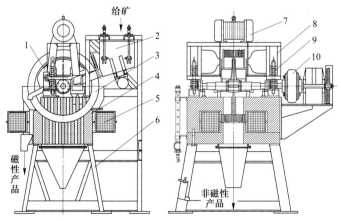

图7　连续式振动高梯度磁选机结构示意

1—磁性矿斗；2—给料系统；3—转环；4—磁轭；5—激磁线圈；6—机架；
7—振动电机；8—振动电机架；9—振动机构；10—橡胶联轴器

7 结论

（1）脉动高梯度磁选与振动高梯度磁选利用流体的振动力或机械振动力对磁选过程中的矿粒群产生松散作用，有效地排除非磁性矿粒的机械夹杂，有利于提高磁性精矿的品位和防止磁介质堵塞。

（2）SLon周期式脉动高梯度磁选机场强高，技术参数易于调节，试验时每周期给料200g左右，试验方便快捷，物料平衡性好，现已在20余家科研院所、厂矿和大学的实验室中应用于弱磁性矿石的探索试验。

（3）SLon立环脉动高梯度磁选机已广泛应用于弱磁性矿石的选矿工业生产中，新研制的SLon-2500立环脉动高梯度磁选机具有处理量大、能耗低、自动化程度高的优点。该机在近年内有望投入工业生产应用。

（4）周期式振动高梯度磁选机可用于干法选矿试验，也可用于湿法选矿试验。该机为弱磁性矿石的选矿试验提供了一种新的试验设备。

（5）SLon-1000振动高梯度磁选机是一种正在发展的设备，该机对非金属矿干法除铁已获得了较好的试验指标。该机可望与雷蒙磨等干法磨矿设备组成磨矿—强磁选除铁的干法生产线。

参 考 文 献

[1] 熊大和. SLon立环脉动高梯度磁选机分离红矿的研究与应用 [J]. 金属矿山, 2005 (8): 24~29.
[2] 熊大和. SLon磁选机在红矿强磁反浮选流程中的应用 [C] //2005年全国选矿高效节能技术及设备学术研讨与成果推广交流会论文集. 金属矿山, 2005: 59~62.
[3] 熊大和. SLon磁选机在淮北煤系高岭土提纯中的研究与应用 [C] //2004年全国矿产资源合理开发、有效利用和生态环境综合整治技术交流会论文集. 金属矿山, 2004: 427~429.
[4] 熊大和. 应用SLon磁选机从尾矿中回收有价矿产品 [C] //2003年全国矿产资源高效开发和固体废物处理处置技术交流会论文集. 金属矿山, 2003: 269~272.
[5] 熊大和. SLon立环脉动高梯度磁选机在提高红矿质量的应用 [C] //2002年全国铁精矿提质降杂学术研讨及技术交流会论文集. 金属矿山, 2002: 238~241.
[6] 熊大和. SLon磁选机在钛铁矿选矿工业中的研究与应用 [C] //2003年全国破碎、磨矿及选别设备学术研讨及技术交流会论文集. 金属矿山, 2003: 248~250.

写作背景 本文介绍了脉动与振动高梯度磁选机原理，首次提出了研制SLon-2500立环脉动高梯度磁选机的技术参数，同时也首次公布了首台SLon-1000连续式振动高梯度磁选机结构。

SLon 磁选机在大型红矿选厂应用新进展

熊大和[1,2]

（1. 赣州有色冶金研究所；2. 赣州金环磁选设备有限公司）

摘 要 经过20多年的持续研究与技术创新，SLon立环脉动高梯度磁选机已发展成为新一代的高效强磁设备。该机具有富集比大、回收率高、磁介质不堵塞、设备作业率高的优点。本文介绍该机在鞍钢胡家庙选厂、调军台选厂、弓长岭选厂、海南钢铁公司选厂、首钢秘鲁选厂、安阳钢铁公司舞阳铁矿等大型红矿选矿厂的应用新进展。

关键词 SLon立环脉动高梯度磁选机 氧化铁矿 磁选

New Progress in Application of SLon Magnetic Separators in Large Oxidized Iron Ore Concentrators

Xiong Dahe[1,2]

(1. Ganzhou Nonferrrous Metallurgy Research Institute;
2. SLon Magnetic Separator Ltd.)

Abstract Through more than 20 years' persistent research and technical innovation, SLon vertical ring high gradient magnetic separators have been developed into a new generation high efficiency high intensity magnetic separators. They are characterized by great enrichment ratio, high recovery, non-blocked magnetic medium and high availability. The paper describes the new progress in their application in large oxidized iron ore concentrators such as An Steel's Hujiamiao Concentrator, Diaojuntai Concentrator and Gongchangling Concentrator, Hainan Steel's concentrator, Capital Steel's Peru Concentrator and Anyang Steel's Wuyang Iron Mine.

Keywords SLon vertical ring high gradient magnetic separators; Oxidized iron ores; Magnetic separation

随着我国钢铁工业的高速发展，我国自产铁矿石远远满足不了自给需求，近几年铁矿石处于较高的价位，拉动了我国低品位氧化铁矿（红矿）开采与选矿工业的快速发展。

SLon立环脉动高梯度磁选机是针对我国量大面广的低品位氧化铁矿、钛铁矿、锰矿等弱磁性矿石选矿而发展起来的。该机利用导磁不锈钢棒在磁场中产生的高梯度强磁力回收细粒弱磁性矿物，其特点是对细粒弱磁性矿物有很强的捕收能力。该机将磁力和脉动流

---

❶ 原载于《全国金属矿节约资源及高效选矿加工利用学术研讨与技术成果交流会论文集》，2006：80~84。

体力有机地结合起来,并采用转环立式旋转、反冲精矿,并配有矿浆脉动机构,显著提高了高梯度磁选的选矿效率,并解决了平环强磁机磁介质易堵塞的技术难题。

经过 20 多年的持续研究与技术创新,SLon 立环脉动高梯度磁选机具有优异的选矿性能和机电性能。目前已有 500 多台广泛应用在我国的氧化铁矿、钛铁矿、锰矿、非金属矿等弱磁性矿的选矿工业中,为提高我国弱磁性矿石的选矿技术水平作出了突出贡献。

1 SLon 磁选机简介

1.1 SLon 磁选机工作原理

SLon 立环脉动高梯度磁选机结构如图 1 所示。它主要由脉动机构、激磁线圈、铁轭、转环和各种矿斗、水斗组成。用导磁不锈钢制成的钢板网或圆棒作磁介质。该机工作原理:激磁线圈通以直流电,在分选区产生感应磁场,位于分选区的磁介质表面产生非均匀磁场即高梯度磁场;转环作顺时针旋转,将磁介质不断送入和运出分选区;矿浆从给矿斗给入,沿上铁轭缝隙流经转环。矿浆中的磁性颗粒吸附在磁介质棒表面上,被转环带至顶部无磁场区,被冲洗水冲入精矿斗,非磁性颗粒在重力、脉动流体力的作用下穿过磁介质堆,与磁性颗粒分离。然后沿下铁轭缝隙流入尾矿斗排走。

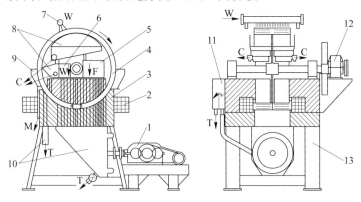

图 1 SLon 立环脉动高梯度磁选机结构示意图

1—脉动机构;2—激磁线圈;3—铁轭;4—转环;5—给矿斗;6—漂洗水斗;7—精矿冲洗装置;8—精矿斗;9—中矿斗;10—尾矿斗;11—液位斗;12—转环驱动机构;13—机架;F—给矿;W—清水;C—精矿;M—中矿;T—尾矿

该机的转环采用立式旋转方式,对于每一组磁介质而言,冲洗磁性精矿的方向与给矿方向相反,粗颗粒不必穿过磁介质堆便可冲洗出来。该机的脉动机构驱动矿浆产生脉动,可使位于分选区磁介质堆中的矿粒群保持松散状态,使磁性矿粒更容易被捕获,使非磁性矿粒尽快穿过磁介质堆进入尾矿中去。

显然,反冲精矿和矿浆脉动可防止磁介质堵塞;脉动分选可提高磁性精矿的质量。这些措施保证了该机具有较大的富集比、较高的分选效率和较强的适应能力。

1.2 SLon 磁选机研制新进展

1.2.1 研制大型的 SLon-2500 立环脉动高梯度磁选机

我国富矿资源日益减少,入选贫矿日益增多。因此,设备大型化、优质高效及运行低成本是今后的主要发展方向。我们从2004年开始研制SLon-2500立环脉动高梯度磁选机，目前已研制出样机，即将开展工业试验，该机主要优点如下：

（1）设备处理量大。该机台时处理量可达150t（处理鞍山式贫赤铁矿处理量上限可达200t/（台·h））。该机是迄今我国台时处理量最大的强磁选设备。设备大型化可节约用地，降低操作成本。

（2）高效节能。SLon-2500磁选机处理每吨矿石的电耗为0.63kW·h，比SLon-2000磁选机（1.02kW·h/t）节电38%。

（3）自动化程度较高，有利于选厂自动化控制。

1.2.2 研制SLon-2000中磁机

目前鞍山式赤铁矿选矿普遍采用阶段磨矿—螺旋溜槽重选—强磁选—反浮选流程，其中重选尾矿要用中磁机扫选。生产实践证明，一般滚筒式永磁中磁机分选螺旋溜槽尾矿得不到合格的最终尾矿。前几年齐大山选厂、东鞍山选厂都采用了SLon-1500、SLon-1750中磁机，获得了良好的选矿指标。但是这两种机型处理量偏小，流程中设备台数较多，占地面积偏大。为了满足新建选厂的需求，近两年我们又新研制了SLon-2000中磁机，目前该机已在鞍钢胡家庙选厂和弓长岭选厂三车间获得成功应用。

1.2.3 研制SLon-500立环脉动高梯度磁选机

为了满足科研单位做选矿试验的需要，近年又研制了SLon-500立环脉动高梯度磁选机。该机处理量为50~250kg/（台·h），其磁场强度、转环转速、脉动冲程、冲次均连续可调，非常适合于科研院所做小型试验和连选试验。目前已在马鞍山矿山研究院、北京矿冶研究院、首钢矿业公司技术中心、武汉理工大学、赣州金环磁选设备有限公司、酒泉钢铁公司研究院，云南锡业股份公司研究所等地应用。

1.3 SLon磁选机主要技术参数

SLon磁选机的工作方式有周期式和连续式两种，现将连续式SLon立环脉动高梯度磁选机和SLon中磁选机的技术参数见表1和表2。

表1 SLon立环脉动高梯度磁选机主要技术参数

机型	给矿粒度/mm	给矿浓度/%	矿浆通过能力/m³·h⁻¹	干矿处理量/t·h⁻¹	额定背景磁感应强度/T	额定激磁功率/kW	脉动冲程/mm	脉动冲次/r·min⁻¹	主机质量/t
SLon-500强磁机	<1.0	10~40	0.5~1.0	0.05~0.25	1.0	13.5	0~50	0~400	1.5
SLon-750强磁机	<1.0	10~40	1.0~2.0	0.1~0.5	1.0	22.0	0~50	0~400	3.0
SLon-1000强磁机	<1.3	10~40	12.5~20	4~7	1.0	28.6	0~30	0~300	6.0
SLon-1250强磁机	<1.3	10~40	20~50	10~18	1.0	35.0	0~20	0~300	14.0
SLon-1500强磁机	<1.3	10~40	50~100	20~30	1.0	44.0	0~30	0~300	20.0
SLon-1750强磁机	<1.3	10~40	75~150	30~50	1.0	62.0	0~30	0~300	35.0
SLon-2000强磁机	<1.3	10~40	100~200	50~80	1.0	74.0	0~30	0~300	50.0
SLon-2500强磁机	<1.3	10~40	180~450	80~150	1.0	97.0	0~30	0~300	105.0

表2 SLon中磁机主要技术参数

机 型	给矿粒度/mm	给矿浓度/%	矿浆通过能力/m³·h⁻¹	干矿处理量/t·h⁻¹	额定背景磁感强度/T	额定激磁功率/kW	脉动冲程/mm	脉动冲次/r·min⁻¹	主机质量/t
SLon-1500 中磁机	<1.3	10~40	75~150	30~50	0.4	16	0~30	0~300	15
SLon-1750 中磁机	<1.3	10~40	75~150	30~50	0.6	38	0~30	0~300	35
SLon-2000 中磁机	<1.3	10~40	100~200	50~80	0.6	42	0~30	0~300	40

2 SLon 磁选机在大型选矿厂应用新进展

2.1 SLon 磁选机在鞍钢调军台选矿厂的应用

鞍钢调军台选矿厂设计规模为年处理鞍山式氧化铁矿900万吨,是一个具有现代化水平的大型选矿厂,采用连续磨矿—弱磁—中磁—强磁—反浮选的选矿流程。为解决原采用的平环强磁选机存在的问题并提高选矿指标,该选厂从2004年至2005年底先后购入15台 SLon-2000 磁选机全面替换15台 Shp-3200 平环强磁选机。在处理量、给矿条件相同的情况下,SLon-2000 磁选机精选精矿品位比 Shp-3200 平环强磁选机高1.19%,尾矿品位低1.56%,精矿产率高3.07%,精矿回收率高8.19%(生产数质量流程如图2所示)。新流程中 SLon 磁选机运转平稳、故障率低、选矿指标好、磁介质长期不堵塞、设备作业率高达99%以上,为提高全流程的铁回收率和为浮选作业创造良好的条件起到了重要作用。

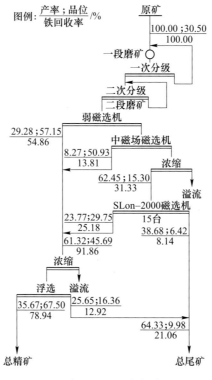

图2 调军台选矿厂生产数质量流程

2.2 SLon 磁选机在鞍钢弓长岭选矿厂的应用

弓长岭选矿厂在 20 世纪 90 年代主要分选磁铁矿。随着红矿选矿技术和设备的进步，鞍钢弓长岭矿业公司决定新建弓长岭选矿厂三选车间分选鞍山式红矿，设计规模为年处理矿石 300 万吨。设计的选矿流程为阶段磨矿、重选—强磁—反浮选流程（见图 3）。该流程中采用 4 台 SLon-2000 强磁机作为细粒级的抛尾设备，另采用 4 台 SLon-2000 中磁机作为粗粒级（螺旋溜槽尾矿）抛尾设备，该流程设计的综合选矿指标为：原矿品位 28.78%，铁精矿品位 67.19%，铁回收率 76.29%。弓长岭三选车间在 2005 年上半年建成投产。

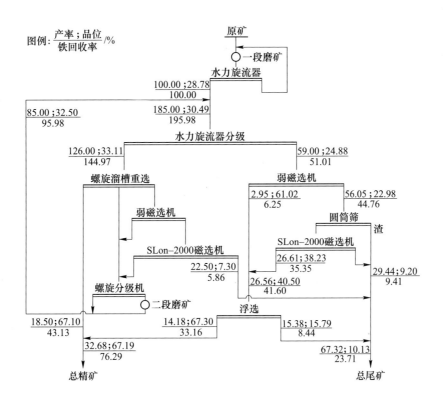

图 3　弓长岭选矿厂三选车间生产数质量流程

2.3 SLon 磁选机在鞍钢胡家庙采选联合企业的应用

鞍钢胡家庙铁矿与齐大山铁矿毗邻，主要以赤铁矿和菱铁矿为主，原矿 TFe 平均含量为 28.40%。2005 年初鞍千矿业有限责任公司投资新建年处理氧化铁矿 900 万吨的胡家庙选矿厂，采用阶段磨矿、重选—强磁—反浮选的选矿流程（见图 4）。其中采用 8 台 SLon-2000 强磁机控制细粒级尾矿品位，另 8 台 SLon-2000 中磁机控制螺旋溜槽尾矿品位。SLon 磁选机为提高全流程的铁回收率和为浮选作业创造良好条件起到了重要作用。该项目第一期工程已于 2006 年初完成，第二期工程计划于 2006 年 6 月完成。

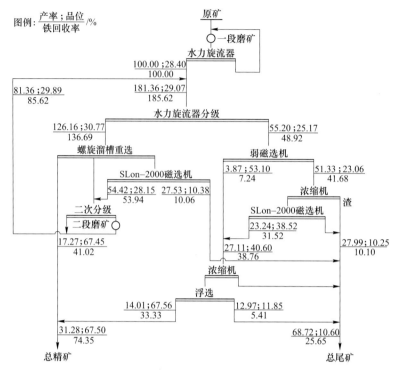

图 4 胡家庙选矿厂数质量流程

2.4 SLon 磁选机在海南铁矿的应用

海南铁矿的铁矿石主要为赤铁矿和部分磁铁矿。2003～2005 年，海南钢铁公司采用 5 台 SLon-2000 立环脉动高梯度磁选机处理铁矿石（见图 5）。该流程中，给矿磨至小于

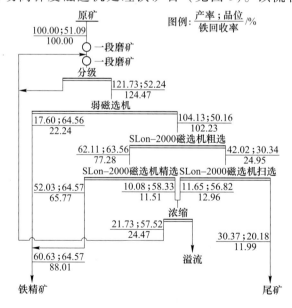

图 5 海钢铁矿选矿厂数质量流程

0.074mm（200目）占95%的粒度，经弱磁选机分选出磁铁矿，再经由 SLon-2000 磁选机组成的一次粗选、一次精选、一次扫选的工艺选别赤铁矿。该流程指标为给矿品位 51.09%，铁精矿品位 64.57%，铁回收率 88.01%。

目前，海钢又在建设 1 座年处理 110 万吨红矿的选矿厂，采用连续磨矿、强磁—反浮选流程，其中采用 8 台 SLon-1750 磁选机作为强磁选设备，预计投产后铁精矿品位可达到 66.00% 以上。

2.5 SLon 磁选机在秘鲁铁矿的应用

首钢秘鲁铁矿的矿石以磁铁矿为主，岩体表层多为氧化型赤铁矿。2003 年底，秘鲁铁矿采用如图 6 所示流程处理该氧化铁矿。流程中用弱磁选机选出磁铁矿，采用 2 台 SLon-1750 磁选机选出赤铁矿，弱磁精矿和强磁精矿合并后进入浮硫作业。其生产指标为给矿品位 56.43%，精矿品位 65.09%，铁回收率 80.32%。该流程分选出的铁精矿用于高炉球团，产品质量满足生产球团的要求。与原生矿精矿相比，石英添加量由原来的 4.09% 降到 2.50%。该流程的成功应用，不但有效处理了采场大量堆积的氧化铁矿，而且减少了石英的开采、运输、破碎磨矿等费用。2005 年底该矿又购入 2 台 SLon-1750 磁选机应用在新建的选矿系统中。

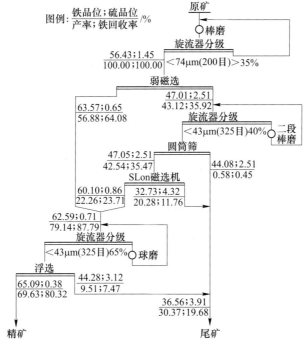

图 6 秘鲁铁矿氧化矿数质量流程

2.6 SLon 磁选机在河南安阳钢铁公司舞阳铁矿的应用

舞阳铁矿是河南安阳钢铁公司的主要原料基地，2004 年以前该矿只有一选车间分选磁铁矿,有大量的贫赤铁矿未得到利用。2004~2005 年,该矿新建了年处理 300 万吨贫赤铁矿的二选车间,其中采用了 23 台 SLon-2000 立环脉动高梯度磁选机,新建生产流程于 2005 年底调试成功,其选矿流程和生产指标如图 7 所示。SLon 磁选机和反浮选在舞阳铁矿的应用,使该矿品位仅 26.00%左右的贫赤铁矿得到利用,工业生产铁精矿品位达 64.00%以上,铁回收率为 50.00%左右。

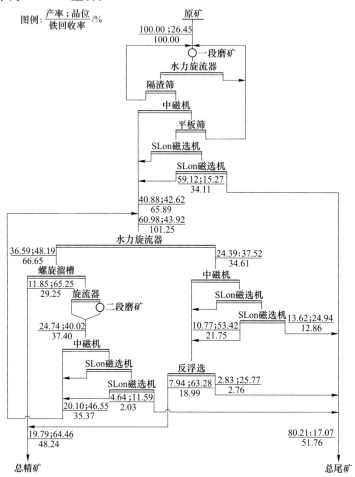

图 7 舞阳铁矿贫赤铁矿选矿数质量流程

舞阳铁矿贫赤铁矿目前开采表土层,入选原矿具有含泥量大、氧化程度深、磁性弱的特点,因此目前选矿回收率偏低。随着采矿向深部延深,矿石氧化程度降低,选铁回收率可望逐步提高。

3 结论

(1) SLon 立环脉动高梯度磁选机在工业上已广泛应用于氧化铁矿(红矿)的选矿,该机具有磁介质不堵塞、选矿效率高、适应性强、设备作业率高、运行成本低的优点。

(2) SLon 磁选机在鞍钢调军台选厂、齐大山选矿厂、东鞍山烧结厂选矿车间、海南钢铁公司选矿厂等地应用,大幅度提高了这些厂矿红矿选矿技术水平,显著地降低了生产成本，选矿指标屡创新高，达到国内外领先水平。

（3）SLon 磁选机在首钢秘鲁铁矿、鞍钢弓长岭三选车间、鞍钢胡家庙选矿厂、安阳钢铁公司舞阳铁矿等地的成功应用，使国内外的大批低品位红矿得到利用，创造了显著的社会经济效益。

（4）目前我国大型红矿选矿厂普遍采用的机型为 SLon-2000、SLon-1750 磁选机。为了进一步提高生产水平，新研制了 SLon-2500 立环脉动高梯度磁选机，该机具有处理量大、高效节能、自动化程度高的优点，可望在近年内在我国大型选矿厂应用。

参 考 文 献

[1] 熊大和，张国庆. SLon-2000 磁选机在调军台选矿厂的工业试验与应用 [J]. 金属矿山，2003（12）：37~39.

[2] 熊大和. SLon 型磁选机在齐大山选矿厂的应用 [J]. 金属矿山，2002（4）：42~44.

[3] 熊大和. SLon 磁选机分选东鞍山红矿的应用 [J]. 金属矿山，2003（8）：21~24.

写作背景 本文介绍了 SLon 立环脉动高梯度磁选机在多家大型氧化铁矿选矿厂应用情况，其中在鞍钢胡家庙选矿厂和安阳钢铁公司舞阳铁矿的应用为新项目。

SLon 磁选机分选锰矿的研究与应用[1]

熊大和[1,2]

（1. 赣州有色冶金研究所；2. 赣州金环磁选设备有限公司）

摘 要 应用 SLon 立环脉动高梯度磁选机对我国多种低品位的锰矿进行了广泛的试验，证明了该机分选锰矿可获得良好的指标。现已在内蒙古、南京等地成功地应用于分选低品位锰矿，显示了该机在锰矿选矿工业中具有良好的应用前景。

关键词 SLon 立环脉动高梯度磁选机 锰矿选矿 磁选

锰是钢铁工业的重要原料，世界上 90% 的锰用于钢铁工业，其余用在轻工、化工、医药等方面。我国的锰矿资源绝大多数为贫矿，国产锰精矿远不能满足国内市场需求，每年要从国外进口大量的锰精矿。因此如何开发国内低品位的锰矿资源成为当务之急。

SLon 立环脉动高梯度磁选机利用磁力、脉动流体力和重力的综合力场选矿，其转环立式旋转、反冲精矿；选矿区下部配有脉动机构，驱动分选区矿浆脉动，使磁介质堆中的矿粒群始终保持松散状态。该机具有适应性强、选矿效率高、工作可靠、设备运转率高的优点。目前已有 500 多台广泛应用于国内外的氧化铁矿、钛铁矿、非金属矿等弱磁性矿石的选矿，已成为国内外新一代高效强磁选设备。近年该机在锰矿选矿工业中也得到较好的应用[1~3]。

1 我国锰矿资源的特点

截至 1995 年末，我国已发现锰矿产 231 处，累计探明锰储量 6.413 亿吨，保有储量 5.8 亿吨，但其中富矿仅占 6.4%。我国锰矿总储量排在南非、乌克兰、加蓬之后，居世界第四位。尽管我国锰矿资源储量较大，但锰矿品位较低，平均品位仅 21%，并且杂质高、加工性能差，可用锰矿资源只占保有资源总量的 40% 左右。折合金属量为 0.48 亿吨。我国锰矿主要分布在中西部地区（约占全国总保有储量的 90%），其中广西和湖南两省的保有储量分别占全国总储量的 38% 和 18%，贵州、云南、四川等西南省份约占全国总储量的 26%[4]。

我国锰矿的基本特点是：

（1）锰矿开采规模小而且矿点分散，地方中、小矿山及民采矿山产量占总产量的 80%。

（2）贫矿多，富矿少。含锰平均品位 21% 的贫矿储量 5.3 亿吨，占保有储量的 93.6%，含锰品位大于 30% 的富矿储量 0.36 亿吨，仅占保有储量的 6.4%。

（3）矿石类型以碳酸锰为主，储量 4.13 亿吨，含锰品位 21.14%，占保有储量的 73%；氧化锰储量 1.19 亿吨，含锰品位 24.56%，占保有储量 21%；铁锰铅锌共生矿储量 0.34 亿吨，含锰品位 14.46%，占保有储量的 6%。

---

[1] 原载于《矿冶工程》，2006（08）：116~118。

2 锰矿的物理化学性质

表1为我国主要锰矿的化学成分。表2为主要锰矿石的比磁化系数。矿物的比磁化系数与其被氧化的程度有关,氧化程度越深,其比磁化系数越小,矿石的磁性越弱,即使是同一种矿物,其比磁化系数也会随氧化程度而改变。由表2可知,大多数锰矿的比磁化系数比赤铁矿小,但与赤铁矿接近,理论上可用强磁选分选。

表1 我国主要锰矿石化学成分

序号	矿山名称	矿石类型	产品名称	化学成分/%							
				Mn	Fe	SiO_2	CaO	Al_2O_3	MgO	P	烧失
1	湖南湘潭锰矿	碳酸矿	原矿	17.0	2.9	26.04	7.32	5.38	2.79	0.124	21.83
2	贵州遵义锰矿	碳酸矿	原矿	18.42	9.00	13.72	5.36	8.86	2.76	0.04	22.45
3	湖南花垣锰矿	碳酸矿	原矿	19.54	2.94	24.86	5.35	3.80	2.90	0.29	27.01
4	广西大新锰矿	氧化矿	I层矿	35.43	8.45	21.06	<0.8	9	0.2	0.15	12.5
5	江西乐华锰矿	混合矿	原矿	24.29	15.90	8.26	10.38	0.876	1.39	0.028	
6	广西平乐锰矿	氧化矿	块矿	27.98	10.42	10.59	0.30	16.19	0.16	0.105	
7	广西木圭锰矿	松软锰	原矿	20.92	9.34	36.73	0.15	3	0.12	0.091	
8	广西天寺锰矿	氧化矿		29.25	7.0	25.83	0.48	7.90	0.34	0.057	
9	福建连城锰矿	氧化矿		32.91	3.7	21.63				0.042	

表2 主要锰矿石的比磁化系数

矿物名称	水锰矿	软锰矿	硬锰矿	褐锰矿	菱锰矿	赤铁矿
比磁化系数/$m^3 \cdot kg^{-3}$	$(0.35\sim1.02)\times10^{-6}$	0.34×10^{-6}	$(0.30\sim0.62)\times10^{-6}$	1.51×10^{-6}	$(1.31\sim1.70)\times10^{-6}$	$(0.50\sim2.16)\times10^{-6}$

3 SLon磁选机分选锰矿的试验

为了充分利用我国低品位锰矿资源,提高锰精矿质量,采用SLon立环脉动高梯度磁选机对各种锰矿进行了广泛试验。

3.1 广西木圭松软锰矿的试验

广西木圭松软锰矿主要矿物成分为偏锰酸矿(又称水锰矿)、软锰矿、黝锰矿及褐铁矿,脉石矿物为黏土和燧石。该锰矿的特点为密度小($1.5g/cm^3$),湿度大(含水约40%),质地松软,一捏即碎。该松软锰矿石的平均品位为Mn 20.90%,Fe 9.43%,P 0.019%,SiO_2 36.73%。

2002年,赣州有色冶金研究所与广西地质测试中心合作,在广西木圭锰矿开展了3个月的半工业试验。

为了避免泥化,该流程首先用振动筛将大于10mm和1.5~10mm粒级筛出来,然后将小于1.5mm细粒级用SLon立环脉动高梯度磁选机进行一次粗选和一次扫选作业。试验流程及指标如图1所示。

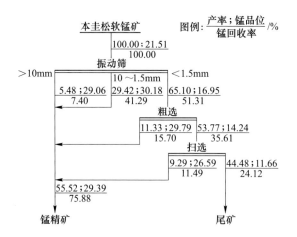

图 1 广西木圭锰矿半工业试验流程

3.2 分选其他锰矿的探索试验

近年用 SLon 磁选机对国内多处低品位锰矿进行了探索试验，其选矿指标见表 3。

表 3 SLon 磁选机分选锰矿的探索试验指标 （%）

锰矿产地	锰矿类型	选矿流程	原矿品位	精矿品位	精矿产率	锰回收率	尾矿品位
内蒙古金水矿业	菱锰矿	一次粗选	8.85	30.39	21.93	75.30	2.80
甘肃永登矿业	菱锰矿	一次粗选	16.94	31.88	32.52	61.20	9.74
甘肃永登矿业	菱锰矿	一次粗选	34.65	43.40	50.23	62.91	25.82
内蒙古包头	黑锰矿	一次粗选	12.85	39.53	18.76	57.70	6.69
辽宁建平	菱锰矿	一次粗选	11.84	29.27	25.58	65.71	5.53
福建连城锰矿	氧化锰矿	一次粗选	28.27	46.47	39.64	65.15	16.32
四川攀枝花锰矿	菱锰矿	一次粗选	13.36	38.08	25.81	73.57	4.76
云南锰矿	氧化锰矿	一次粗选	27.31	39.27	54.32	78.10	13.09
辽宁朝阳锰矿	菱锰矿	一粗一扫	34.42	40.22	70.62	82.51	20.48
安徽锰矿		一次粗选	17.03	27.98	32.62	53.59	11.73

由表 3 可见，内蒙古金水矿业提供的锰矿，原矿 Mn 品位仅为 8.85%，用 SLon 磁选机选一次可达锰精矿 Mn 品位 30.39%，锰回收率 75.30%，尾矿 Mn 品位仅 2.80%。SLon 磁选机分选其他锰矿均可获得良好的选矿指标，显示了该机在锰矿选矿领域中的良好应用前景。

4 SLon 磁选机在锰矿工业中的应用

我国拥有丰富的低品位锰矿资源，此外，我国许多锰矿选矿厂过去用水洗锰矿，只回收粗粒，而含泥量大的细粒锰矿作为尾矿排放或堆存。因此，SLon 磁选机在工业上应用于细粒锰矿的选矿将有利于提高我国锰矿资源的利用率。

4.1 从南京栖霞山铅锌矿尾矿中回收锰矿

南京栖霞山铅锌矿是一家以开采铅锌矿为主的老矿，其铅锌选矿的年尾矿量为 15 万吨，其中含锰 12% 左右，锰金属量约 1.8 万吨。若能将锰精矿品位选至 22% 以上，则可作为制造硫酸锰的原料。2005 年由马鞍山矿山研究院试验、设计，采用 2 台 SLon-1500 立环脉动高梯度磁选机从该尾矿中回收锰精矿。

该锰矿回收流程于 2005 年 11 月顺利投产，生产上达到的选矿指标见图 2 及表 4。

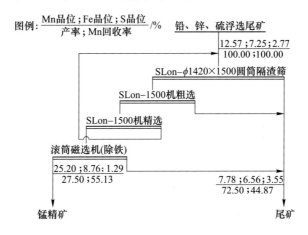

图 2 南京栖霞山锰矿磁选流程

表 4 南京牺霞山铅锌矿锰矿 SLon 磁选机生产指标 （%）

指标	铅锌硫浮选尾矿品位			锰精矿品位			产率	回收率	尾矿品位		
	Mn	Fe	S	Mn	Fe	S			Mn	Fe	S
生产指标	12.57	7.25	2.77	25.20	8.76	1.29	27.50	55.13	7.78	6.56	3.55
设计指标	12.00			24.00	≤6		25	50.00			
差值	+0.57			+1.20	+2.76		+2.50	+5.13			

由表 4 可知，SLon 磁选机选锰主要指标如锰精矿品位、产率和回收率均超过设计指标，但含铁量高于设计要求，说明滚筒磁选机除铁效果不好，建议今后采用除铁效果好的磁选机。

4.2 从内蒙古金水矿业选锰尾矿中回收细粒锰矿

内蒙古金水矿业最初锰矿生产采用水洗方法选出粗粒锰矿出售，洗矿溢流作为尾矿堆放。近几年该公司与我们合作，采用 SLon 立环脉动高梯度磁选机回收尾矿中的细粒锰矿。获得了良好的选矿指标。目前该公司生产上采用 SLon-1000、SLon-1250 和 SLon-1500 磁选机各 1 台回收细粒锰矿，其生产流程和选矿指标如图 3 所示。

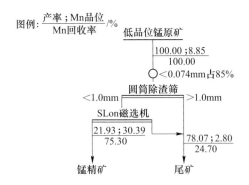

图 3　内蒙古金水矿业锰矿选矿流程

4.3　在甘肃永登矿业应用于分选低品位锰矿

甘肃永登矿业公司于 2005 年采用 1 台 SLon-1250 和 1 台 SLon-1500 磁选机建成一座锰矿选矿厂，用于处理低品位锰矿，获得成功。其选矿流程和指标如图 4 所示。

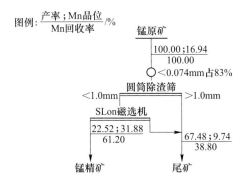

图 4　甘肃永登矿业分选低品位锰矿流程

5　结语

（1）我国拥有丰富的锰矿资源，但绝大多数为贫锰矿，发展新型高效的锰选矿设备和工艺流程，对发展我国的锰业具有重要意义。

（2）SLon 立环脉动高梯度磁选机是新一代高效强磁选设备，具有富集比大，选矿效率高，适应性强，设备运转可靠，可长期稳定工作的优点，该机在国内外的氧化铁矿、钛铁矿、非金属选矿工业中已大量应用，近年在锰矿选矿行业中的应用也得到较快的发展。

（3）用 SLon 立环脉动高梯度磁选机对我国广西、内蒙古、甘肃、辽宁、福建等地的低品位锰矿进行了广泛的选矿探索试验。试验表明，低品位锰矿经该机选别，锰的富集比可达 1.44~3.43，锰回收率可达 53%~83%。

（4）SLon 立环脉动高梯度磁选机已成功在南京栖霞山铅锌矿、内蒙古金水矿业、甘肃永登矿业应用于锰矿选矿工业生产，获得了良好的技术经济指标。随着国民经济的发展和 SLon 磁选机的发展，该机将为提高我国锰矿资源的利用作出更大的贡献。

参 考 文 献

[1] 熊大和. SLon-1000 立环脉动高梯度磁选机的研制 [J]. 金属矿山, 1988 (10): 39~42.
[2] 熊大和. SLon-1500 立环脉动高梯度磁选机的研制 [J]. 金属矿山, 1990 (7): 43~46.
[3] 叶和江. SLon-1250 立环脉动高梯度磁选机研制 [J]. 非金属矿, 2000 (2): 40~43.
[4] 任觉世, 等. 工业矿产资源开发利用手册 [M]. 武汉: 武汉工业大学出版社, 1993.

写作背景 本文介绍了 SLon 立环脉动高梯度磁选机在我国一些锰矿的应用状况, 例如在广西木圭锰矿的半工业试验, 在南京栖霞山铅锌矿从尾矿中回收锰精矿, 在内蒙古金水矿业公司和甘肃永登矿业公司分选低品位锰矿, 使这些锰矿资源的利用率有了显著的提高。

SLon Magnetic Separators Applied in Various Industrial Iron Ore Processing Flow Sheets

Xiong Dahe

(SLon Magnetic Separator Ltd., Shahe Industrial Park,
Ganzhou Jiangxi Province, China. Email: xdh@ slon. com. cn.)

Abstract SLon magnetic separators utilise the combined force fields of magnetism, pulsating fluid and gravity to continuously beneficiate fine weakly magnetic minerals. They possess the advantages of high efficiency, low operative cost and high reliability. Until now there were about 600 SLon magnetic separators widely applied in processing oxidized iron ores, ilmenite, manganese ore and many other minerals. In this paper, several typical oxidised iron processing flow sheets with SLon magnetic separators are introduced.

1 Introduction

Chinese iron and steel industry has been developing very fast during the past 20 years. Most Chinese resources of iron ores are of low grade. They must be beneficiated to high-grade iron concentrate to meet metallurgy requirements. Big market demands and low-grade iron resources have fertilised the development of SLon magnetic separators. They are now widely applied in the oxidised iron ore processing industry, raising Chinese oxidised iron ore processing technology up to a higher level.

2 Structure and working principle

SLon magnetic separators mainly consist of the pulsating mechanism, energizing coils, magnetic yoke, separating ring, feeding and product boxes as shown in Fig. 1. Expanded metals or round bars made of magnetic stainless steel are used as a matrix. While a direct electric current flows through the energizing coils, a magnetic field is built up in the separating zone. The ring with the magnetic matrix rotates around its horizontal axis. When slurry from the feeding boxes enters into matrix located in the separating zone, magnetic particles are attracted from the slurry onto the surface of the matrix, brought to the top of the ring where the magnetic field is negligible, then flushed out into the concentrate box. Non-magnetic particles pass through the matrix and enter into the tailings box under the combined actions of gravity and pulsating hydrodynamic force.

As the ring rotate vertically, the flushing direction of the mags is opposite to that of feeding so that coarse particles can be flushed out without having to pass through the entire depth of the matrix pile. The pulsating mechanism drives the slurry in the separating zone up and down, keeping

---

❶ 原载于 Iron Ore Conference, Australia 20-22 August 2007。

particles in the matrix pile in a loose state. Therefore magnetic particles can be more easily captured by the matrix and non-magnetic particles can be more easily dragged to the tailings box through the matrix pile. Obviously opposite flushing and pulsating prevents the matrix from being clogged. and most importantly, pulsating helps to purify the magnetic product. The major technical data are listed in Table 1.

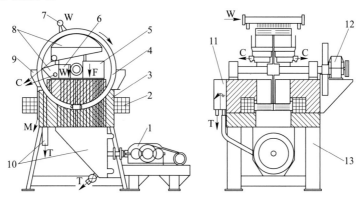

Fig. 1 SLon vertical ring and pulsating high gradient magnetic separator

1—pulsating mechanism; 2—energizing coils; 3—magnetic yoke; 4—working ring; 5—feeding box;
6—wash water box; 7—concentrate flush; 8—concentrate box; 9—middling chute; 10—taillings box;
11—slurry level box; 12—ring driver; 13—support frame; F—feed; W—water; C—concentrate; M—middlings; T—tails

Table 1 Main parameters of SLon magnetic separators of different models

Parameters	Ring diameter /mm	Back ground field /T	Energizing power /kW	Driving power /kW	Pulsating stroke /mm	Pulsating frequency /min^{-1}	Ore-feeding size/mm (<0.074mm (200mesh)/%)	Solid density of feeding /%	Slurry throughput /m$^3 \cdot$h^{-1}	Ore throughput /t$\cdot$h^{-1}	Mass of machine /t
SLon-2500	2500	0-1.0	0-94	11+11	0-30	0-300	<1.3(30-100)	10-40	200-400	80-150	105
SLon-2000	2000	0-1.0	0-74	5.5+7.5	0-30	0-300	<1.3(30-100)	10-40	100-200	50-80	50
SLon-1750	1750	0-1.0	0-62	4+4	0-30	0-300	<1.3(30-100)	10-40	75-150	30-50	35
SLon-1500	1500	0-1.0	0-44	3+4	0-30	0-300	<1.3(30-100)	10-40	50-100	20-30	20
SLon-1250	1250	0-1.0	0-35	1.5+2.2	0-30	0-300	<1.3(30-100)	10-40	20-50	10-18	14
SLon-1000	1000	0-1.2	0-30	1.1+2.2	0-30	0-300	<1.3(30-100)	10-40	10-20	4-7	6
SLon-750	750	0-1.0	0-22	0.55+0.75	0-50	0-400	<1.3(30-100)	10-40	1.0-2.0	0.1-0.5	3
SLon-500	500	0-1.0	0-13.5	0.37+0.37	0-50	0-400	<1.3(30-100)	10-40	0.5-1.0	0.05-0.25	1.8

3 SLon magnetic separators applied in magnetic-reverse flotation flow sheet

SLon magnetic separators beneficiating low-grade iron ore possesses the advantage of low cost and big capacity. They can be applied in the roughing stage to discharge most tails and slimes, upgrading the rough iron concentrate to a higher level, to make good conditions for reverse-flotation.

3.1 SLon in Diaojuntai oxidized iron ore processing plant

Diaojuntai iron ore processing plant of Anshan Iron and Steel Company treats nine million tons of oxidised iron ore. consisting of magnetite and haematite, each year and the iron ore processing flow sheet is two-stage grinding-high intensity magnetic separation-reveres flotation. In the past, 15 Shp-3200 horizontal ring wet high intensity separators (Shp-3200 WHIMS) were used as high intensity magnetic separators. but their grooved plates as magnetic matrix were easy to be clogged, and difficult to maintain with low availability of only 80%-90%.

As SLon magnetic separators possess the advantages of higher beneficiating ability, the matrix not being easily clogged and availability as high as 98%-99%, by the end of 2005, Diaojuntai iron ore processing plant has applied 15 SLon-2000 magnetic separators to replace the 15 Shp-3200 WHIMS. The present flow sheet is shown in Fig. 2, which shows that the 15, SLon-2000 magnetic separators control the tails grade on a very low level of only 6.42%Fe, throwing most of the slimes into tails. They have created very good conditions for the following flotation stage.

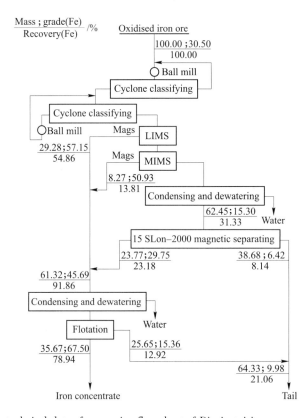

Fig. 2 The technical data of processing flow sheet of Diaojuntai iron ore processing plant

For the excellent performance of the SLon-2000 magnetic separators, the plant gets very good results: feed grade 30.50% Fe, concentrate grade 67.50% Fe, tails grade 9.98% Fe, iron recovery 78.94% Fe.

3.2 SLon in the 1.1 million oxidised iron ore processing line in Hainan

Hainan Iron and Steel Company own a big oxidised Iron ore deposit. The ore mainly consists of finely crystallised haematite and about ten percent magnetite. In 2006, the Company built a yearly 1.1 million iron ore processing line with a magnetic-reverse flotation flow sheet as shown in Fig. 3.

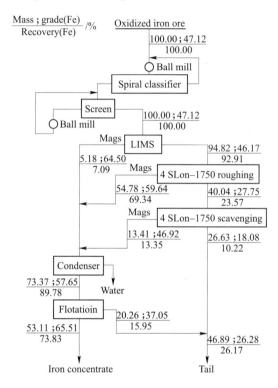

Fig. 3 Hainan 1.1 million oxidised iron ore processing flow sheet

4 SLon magnetic separators applied in magnetic-gravity and reverse flotation flow sheet

Many oxidised iron ores' crystallised particles are different. Coarse iron particles can be recovered by spirals, fine iron particles can be recovered by high-intensity magnetic separators and reverse-flotation. This flow sheet can get high-quality iron concentrate with low cost.

4.1 SLon in Qidashan oxidised iron ore processing plant

Qidashan Mineral Processing Plant is a large oxidised iron ore processing base of Anshan Iron and Steel Company in Liaoning Province, Northeast China. It processes eight million tones of oxidised iron ore and produces 2.7 million tonnes of iron concentrate annually.

As a poor iron deposit, Qidashan iron ore mainly consists of haematite, magnetite and quartz. The sizes of iron particles are from 0.005mm to 1.0mm, 0.005mm on average. The sizes of quartz particles are 0.085mm on average. The harmful elements of S and P are very low in the ore. thus obtaining. excellent iron concentrate with efficient processing technology possible.

The previous processing flow sheet of the ore is coal gas roasting low-intensity magnetic separation with which only a low grade of 63.22% Fe iron concentrate can be obtained. In recent years, the plant developed a new flow sheet of spiral SLon magnetic separating reverse flotation as shown in Fig. 4. In the new flow sheet. 11 SLon-1750(magnetic field: 1.0T) magnetic separators are applied to control the tailings of the fine fraction(cyclone overflow) and 12 SLon-1500(magnetic field: 0.4T) magnetic separators are applied to control the tailings of the coarse fraction(spiral tailings). The results of the flow sheet are: feed grade 30.15% Fe. concentrate grade 67.50% Fe, tailings grade 10.16% Fe. iron recovery 78.05% Fe, being a historical high record for the plant.

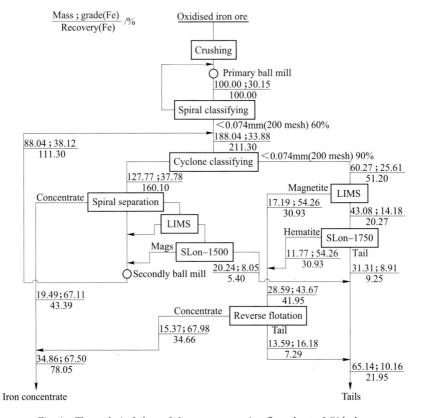

Fig. 4 The technical data of the new processing flow sheet of Qidashan

4.2 SLon in Wuyang oxidised iron ore processing plant

Wuyang Mineral Processing Plant is a large oxidised iron ore processing base of Anyang Iron and Steel Company in Henan Province, Central China. In 2004-2005. the Plant built an oxidised iron ore processing flow sheet as shown in Fig. 5 in which 23 SLon-2000 magnetic separators were applied.

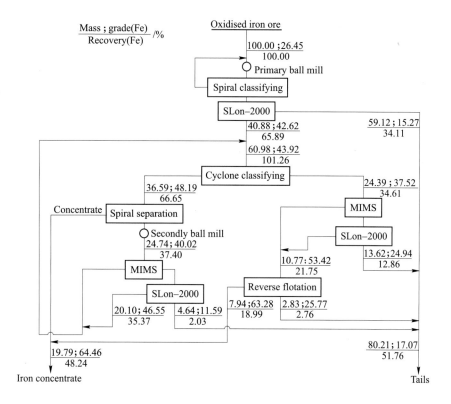

Fig. 5 The technical data of the processing flow sheet of Wuyang

The results of the flow sheet are: feed grade 26.45% Fe, concentrate grade 64.46% Fe, iron recovery about 50%Fe. The recovery is on the low state as the present feed ore from surface soil contains much mud, is deeply oxidised and has weak magnetic susceptibility. which is expected to be improved with deeper mining.

5 SLon magnetic separators applied in pure magnetic separation flow sheet

Many small(<1 million tonnes annually) and medium(one to two million tonnes annually) oxidised iron ore processing plants still use this flow sheet because it is simple and low cost.

5.1 SLon in the first workshop of Hainan iron and steel company

Hainan Iron and Steel Company owns a big oxidised iron ore deposit and the ore mainly consists of finely crystallised haematite and about ten percent magnetite. From 2003 to 2005, the company built a magnetic separation flow sheet as shown in Fig. 6. In the flow sheet six SLon-2000 magnetic separators are applied to recover fine haematite, in which two SLon-2000 separators are used for roughing, two SLon-2000 separators for cleaning and two SLon-2000 separators for scavenging. The results of the flow sheet are: feed grade 51.09% Fe. concentrate grade 64.57% Fe. iron recovery 88.01%Fe.

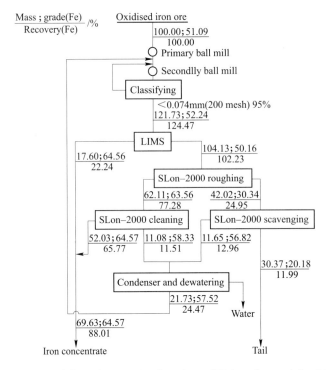

Fig. 6 The technical data of processing flow sheet of Hainan Iron and Steel Company

5.2 SLon in Gushan iron mine

The deposit of Gushan Iron mine of Maanshan Iron and Steel Company is mainly refractory haematilc ore. In the 1980s, the mine applied a spiral-centrifugal separation processing flow sheet to beneficiate the iron ore; the concentrate grade could only reach 55%Fe and iron recovery about 61%Fe. Since 1990, the mine gradually changed the flow sheet into SLon magnetic separation (Fig. 7). At present, nine SLon-1750 and four SLon-1500 magnetic separators are applied in the flow sheet and the iron concentrate grade reaches 60%Fe and recovery 76%Fe.

6 SLon magnetic separators in magnetic-gravity flow sheet

Many small plants use this flow sheet to process oxidised iron ore because it can get good iron concentrate grades with low cost. The disadvantages are that iron recovery is lower and the capacity of the shaking table is small.

6.1 SLon in Dingnan iron ore processing plant

Several small oxidised iron ore processing plants are located in Dingnan, Jiangxi Province, China. Each plant treats about 0.2 million tonnes of oxidised iron ore annually. The ore consists mainly of spiegeleisen, haematite and a small portion of magnetite. The typical flow sheet is shown in Fig. 8.

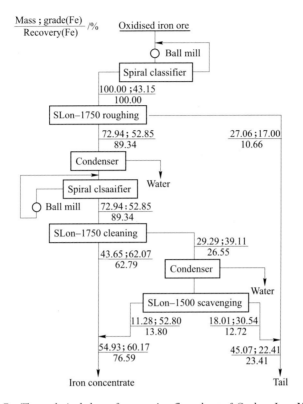

Fig. 7 The technical data of processing flow sheet of Gushan Iron Mine

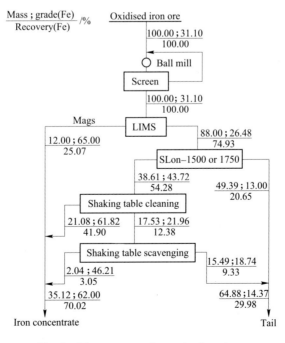

Fig. 8 Dingnan magnetic-gravity flow sheet

7 Conclusions

(1) The SLon magnetic-reverse flotation flow sheet is simple and can get good iron concentrate grade and recovery. But usually the iron grinding size is fine and grinding energy consumption is more than other flow sheets.

(2) SLon magnetic-gravity-reverse flotation flow sheet possesses the advantages that coarse liberated iron ore can be recovered by spirals and part coarse tails can be discharged by SLon magnetic separators (0.2-0.6T) at the primary stage. Only part of fines enters the SLon magnetic separators(1.0T)-reverse flotation stage. This flow sheet can get good iron concentrate grade and recovery with low cost. but the flow sheet is more complex than others.

(3) Pure SLon magnetic separation flow sheet to treat oxidised iron ores possess the advantages of simple, low-cost, good iron recovery. But iron concentrate grade is usually not high enough.

(4) SLon magnetic-gravity flow sheet to treat oxidised iron ores are simple and low cost. But iron recovery is usually not high enough.

(5) Many Chinese iron ore deposits are of low grade. SLon vertical ring pulsating high gradient magnetic separators have become the most advanced high-intensity magnetic separators in China's iron ore processing industry. promoting the develoment of oxidised iron ore processing technology.

写作背景 这是 2007 年在澳大利亚帕斯主办的铁矿学术会上发表的论文。介绍了 SLon 立环脉动高梯度磁选机在几个大中型氧化铁矿选矿厂的应用，其中海南铁矿年处理 100 万吨氧化铁矿、安阳钢铁公司舞阳铁矿年处理 300 万吨氧化铁矿、江西定南氧化铁矿的磁选—摇床选矿流程都是第一次在国外公布。

SLon 磁选机在包钢选厂的试验与应用

熊大和[1,2]

(1. 赣州金环磁选设备有限公司; 2. 赣州有色冶金研究所)

摘 要 包钢选矿厂采用弱磁—强磁—浮选流程回收铁精矿和稀土精矿,其中强磁选设备原为 ϕ3200 平环强磁选机,存在磁介质容易堵塞并引起选矿指标恶化的缺陷。为此,在实验室探索试验的基础上,用 SLon-2000 立环脉动高梯度磁选机与 ϕ3200 平环强磁选机进行了一系列工业对比试验,结果表明,对强磁选作业而言,SLon 磁选机可比平环磁选机提高铁精矿品位 4.05 个百分点、提高铁回收率 16.72 个百分点。根据试验结果,包钢选矿厂用 SLon 磁选机逐步取代平环磁选机,生产实践证实,SLon 磁选机不但可大幅度提高选矿指标,并具有耗水量较小、磁介质不堵塞、工作稳定、维护工作量少、作业率高的优点。

关键词 SLon 立环脉动高梯度磁选机 平环强磁选机 工业试验 应用

Test and Application of SLon Magnetic Separators in Baotou Steels Concentrator

Xiong Dahe[1,2]

(1. SLon Magnetic Separator Ltd.; 2. Ganzhou Nonferrous Metallurgy Research Institute)

Abstract Bao Steel concentrator adopts a flowsheet consisting of low intensity magnetic separation, high intensity magnetic separation and flotation to recover iron concentrate and rare-earth concentrate. The originally used high intensity magnetic separation equipment was ϕ3200 horizontal ring separators, which had the deficiency of easy blockage of magnetic medium that could lead to deterioration of metallurgical performance. Therefore, based on the lab exploratory test, a series of industrial comparative tests were made between SLon-2000 vertical ring pulsating high gradient magnetic separators and ϕ3200 horizontal ring separator. The results indicated that for the high intensity magnetic separation operation, SLon separators, when compared with the horizontal ring separators, can improve the iron concentrate grade by 4.05 percentage points and iron recovery by 16.72 percentage points. According to this, the concentrator has gradually replaced the original separators by SLon magnetic separators. The production practice has proven that SLon magnetic separators can not only greatly improve the metallurgical performance but have the advantages of small water consumption, unblocked medium, stable operation, small maintenance and high availability.

Keywords SLon verical ring pulsating high gradient magnetic separator; Horizontal ring high intensity magnetic separator; Industrial test; Application

---

❶ 原载于《金属矿山》,2007(10):55~58。

1 包钢选矿厂选矿流程简介

包钢选矿厂是我国罕见的大型氧化铁矿选矿厂之一，是全国最大的稀土精矿生产基地，年处理原矿石950万吨，稀土精矿年生产能力可达6.5万吨。我国著名的选矿专家余永富院士为包钢选矿厂的选矿技术发展做出了突出贡献，目前该厂的选矿技术已达到国内领先水平。包钢矿石的矿物组成以磁铁矿、赤铁矿、稀土、铌为主[1]。选厂现共有10个生产系列，采用如图1的弱磁—强磁—浮选流程。该流程用弱磁选机经1次粗选1次精选分选出磁铁矿，用强磁选机1次粗选1次精选分选出赤铁矿，弱磁选精矿和强磁选精矿合并后用反浮选精选出合格铁精矿。强磁精选的尾矿（现场称强磁中矿，下同）作为稀土给矿进入稀土浮选系统。

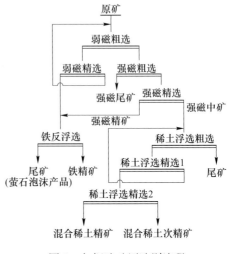

图1 包钢选矿厂选别流程

强磁选作业原采用20台 $\phi3200$ 平环强磁选机进行选别，其中10台粗选，10台精选。这些平环强磁选机在过去的20多年中为包钢选厂作出了重要的贡献。但是平环强磁选机的结构和工作原理存在不合理的地方，例如其给矿方向和冲洗铁精矿的方向都是由上至下，粗粒矿物和树皮、草根、炸药皮等杂物都必须穿过磁介质盒才能被冲洗出来，因此，平环强磁选机存在着磁介质容易堵塞和维护检修工作量大的缺点。此外，平环强磁选机的选矿指标随着齿板磁介质内堵塞物的增加而急剧下降，很不稳定。表1为包钢选厂强磁粗选作业7号 $\phi3200$ 平环强磁选机齿板清洗前后的选矿指标对比情况。

表1 7号 $\phi3200$ 平环强磁选机齿板状况对选矿指标的影响 （%）

齿板状况	取样日期（2005年）	铁品位			精矿产率	铁回收率
		给矿	精矿	尾矿		
清洗前 （齿板内有堵塞物）	9月13日上午	19.30	31.30	14.70	27.71	44.94
	9月13日下午	19.20	29.20	12.30	40.83	62.09
	9月14日下午	18.84	28.80	12.90	37.36	57.11
	9月15日下午	19.60	30.30	12.50	39.89	61.66
	平均	19.24	29.90	13.10	36.52	56.77
清洗后 （齿板干净）	9月16日上午	19.40	32.85	9.45	42.52	72.00
	9月16日下午	19.95	32.85	9.65	44.40	73.10
	9月20日上午	19.75	33.50	10.75	39.56	67.10
	9月20日下午	19.15	33.90	9.80	38.80	68.68
	9月21日上午	19.30	33.30	8.35	43.89	75.72
	平均	19.51	33.28	9.60	41.85	71.39

由表1可知，平环强磁选机齿板清洗前后的铁精矿品位相差3.38个百分点，铁回收率相差14.62个百分点，充分说明了磁介质必须保持干净的重要性。

2 SLon 脉动高梯度磁选机探索试验

SLon 立环脉动高梯度磁选机是20世纪90年代发展起来的新型高效强磁选机，该机采用转环立式旋转，反冲精矿、矿浆脉动机构等多项新技术，具有选矿效率高、磁介质不堵塞、设备作业率高、水电消耗少、作业成本低的优点，目前已在全国绝大多数氧化铁矿选矿厂得到广泛应用[2,3]。

为了解决平环强磁选机齿板容易堵塞、选矿指标不稳定的问题，包钢选矿厂于 2003 年 3 月取样至赣州有色冶金研究所设备研究室（现名为赣州金环磁选设备有限公司），用 SLon-100 周期式脉动高梯度磁选机进行探索试验，并与 ϕ500 平环强磁选机的试验指标（由包钢钢研所提供）进行对比。粗选作业对比结果见表2。

表2 SLon-100 磁选机与 ϕ500 平环强磁选机粗选指标

试验设备	磁感强度/T	铁品位/%			精矿产率/%	铁回收率/%
		给矿	精矿	尾矿		
SLon-100 磁选机	0.9	18.55	36.56	6.04	40.99	80.79
ϕ500 平环磁选机	1.2	18.75	27.90	6.70	56.84	84.58

从表2可看出，SLon 磁选机与 ϕ500 平环强磁选机比较，铁精矿品位高 8.66 个百分点，尾矿品位低 0.66 个百分点，显示了较强的优势。

在粗选试验基础上，又进行了强磁1次粗选1次精选试验指标的对比，结果如图2所示。

图2 两种实验室强磁机1粗1精选别结果
（a）SLon-100 磁选机试验结果（赣州有色冶金研究所）；（b）ϕ500 平环磁选机试验结果（包钢钢研所）

从图2可看出，SLon-100 脉动高梯度磁选机与 ϕ500 平环强磁选机相比，精矿铁品位高 5.11%，铁回收率高 10.08%。以上对比由于试验的时间、地点、部门不同，数据的可靠性不够强，因此有必要开展现场工业对比试验。

3 SLon 磁选机与平环磁选机单机工业对比试验

为了验证采用 SLon 立环脉动高梯度磁选机取代现场 ϕ3200 平环强磁选机的可行性，包钢选矿厂于2005年从赣州金环磁选设备有限公司购入 6 台 SLon-2000 立环脉动高梯度磁选机，用其中3台取代3台强磁粗选作业的φ3200平环强磁选机，另3台取代3台强磁精选作业的φ3200平环强磁选机，进行了单机工业对比试验。

3.1 粗选作业单机对比试验

2005年9月14~19日，采用位于8号位置的SLon-2000立环脉动高梯度磁选机与位于7号位置的φ3200平环强磁选机进行了对比试验，结果见表3。

表3　SLon磁选机与平环强磁选机粗选单机对比试验结果（1）　　　　（%）

日期 （2005年）	设备名称	铁品位			精矿产率	铁回收率
		给矿	精矿	尾矿		
9月14日下午	SLon磁选机 平环磁选机	18.80 18.84	33.80 28.80	9.10 12.90	39.27 32.36	70.60 57.11
9月15日下午	SLon磁选机 平环磁选机	19.80 19.60	35.50 30.30	9.60 12.50	39.38 39.89	70.61 61.66
9月16日上午	SLon磁选机 平环磁选机	19.70 19.40	36.25 32.85	9.75 9.45	37.58 42.52	69.09 72.00
9月16日下午	SLon磁选机 平环磁选机	9.50 19.95	36.00 32.85	10.10 9.65	36.29 44.40	67.00 73.10
9月19日上午	SLon磁选机 平环磁选机	21.10 20.70	37.70 27.40	10.10 12.25	39.86 55.78	71.21 73.83
平均	SLon磁选机 平环磁选机	19.78 19.70	35.85 30.44	9.73 11.35	38.48 43.74	69.74 67.59
差值		+0.08	+5.41	-1.62	-5.26	+2.15

注：SLon磁选机激磁电流1000A，平环磁选机激磁电流160A。

由表3可看出，在给矿相同的条件下，SLon-2000磁选机（1000A）与φ3200平环强磁选机（160A）相比，粗选精矿品位提高5.41个百分点，尾矿品位降低1.62个百分点，铁回收率提高2.15个百分点。

为了探索SLon-2000立环脉动高梯度磁选机进一步提高铁回收率的可能性，又于2005年9月19~21日将该机的激磁电流提高至1200A（相应的背景磁感强度从0.809T提高至0.914T），与φ3200平环强磁选机（该机于2005年9月15日刚清洗完齿板，工作状态正常）进行了对比试验，结果见表4。

表4　SLon磁选机与平环强磁选机粗选单机对比试验结果（2）　　　　（%）

日期 （2005年）	设备名称	铁品位			精矿产率	铁回收率
		给矿	精矿	尾矿		
9月19日下午	SLon磁选机 平环磁选机	19.30 19.70	33.60 30.10	8.80 15.95	42.34 26.50	73.71 40.49
9月20日上午	SLon磁选机 平环磁选机	19.30 19.75	32.50 33.50	8.95 10.75	43.95 39.56	74.01 67.10
9月20日下午	SLon磁选机 平环磁选机	18.80 19.15	32.50 33.90	8.15 9.80	43.74 38.80	75.61 68.68
9月21日上午	SLon磁选机 平环磁选机	19.00 19.30	32.80 33.30	7.90 8.35	44.58 43.89	76.90 75.72

续表4

日期（2005年）	设备名称	铁品位			精矿产率	铁回收率
		给矿	精矿	尾矿		
9月21日下午	SLon磁选机	21.50	35.90	7.95	48.48	80.95
	平环磁选机	21.55	30.10	10.70	55.93	78.12
平均	SLon磁选机	19.58	33.46	8.35	44.72	76.43
	平环磁选机	19.89	32.18	11.11	41.67	67.42
差值		-0.31	+1.28	-2.76	+3.05	+9.01

由表4可见，SLon-2000磁选机激磁电流提高至1200A后，与φ3200平环强磁选机相比，在给矿相同的条件下，粗选铁精矿品位提高1.28个百分点。尾矿品位降低2.76个百分点，铁回收率提高9.01个百分点。

通过强磁粗选作业的对比，充分证明了SLon立环脉动高梯度磁选机可明显提高选矿指标，而且该机工作稳定，磁介质不堵塞，设备作业率高，耗水量大幅度下降，用该机取代φ3200平环强磁选机能取得较大的经济效益。

3.2 精选作业单机对比试验

完成粗选单机对比试验后，包钢选矿厂又在强磁精选作业组织了SLon-2000立环脉动高梯度磁选机与φ3200平环强磁选机的对比试验，试验结果见表5。

表5 SLon磁选机与平环磁选机精选单机对比试验结果

日期（2005年）	设备名称	激磁电流/A	铁品位/%			精矿产率/%	铁回收率/%
			给矿	精矿	尾矿		
11月25日~12月2日	SLon磁选机	250	28.09	39.71	19.50	42.50	60.09
	平环磁选机	20	28.09	36.72	19.61	49.56	64.79
11月21日~11月23日	SLon磁选机	300	26.64	37.89	14.94	50.98	72.51
	平环磁选机	20	26.64	36.36	17.53	48.38	66.03
12月9日~12月22日	SLon磁选机	400	29.07	38.35	19.04	51.94	68.52
	平环磁选机	20	29.07	38.82	20.23	47.55	63.50
平均	SLon磁选机		27.93	38.65	17.83	48.51	67.13
	平环磁选机		27.93	37.30	19.12	48.46	64.72
差值	SLon磁选机		0	+1.35	-1.29	+0.05	+2.41

由表5可见，当SLon-2000磁选机磁电流在250~400A范围变化时，其精选指标与φ3200平环强磁选机相比，铁精矿品位平均提高1.35个百分点，尾矿品位平均降低1.29个百分点，铁回收率平均提高2.41个百分点；当SLon-2000磁选机激磁电流在300~400A范围变化时，其铁精矿品位平均提高0.53个百分点，尾矿品位平均降低1.89个百分点，铁回收率平均提高5.75个百分点。

单机对比试验结果表明，SLon-2000立环脉动高梯度磁选机用于强磁精选作业同样具有明显的优势。

4 强磁作业流程对比试验

在单机试验取得良好结果的基础上，于2005年12月29日至2006年1月6日，用安装在 6 号、7 号、8 号系统原 φ3200 平环强磁选机基础上的 6 台 SLon-2000 磁选机组成 1 次粗选 1 次精选（3 台粗选，3 台精选）的强磁选流程，与 9 号、10 号系统的 4 台 φ3200 平环强磁选机进行了流程对比试验。试验中给矿性质和每个系统的给矿量基本相当，试验结果见图 3。

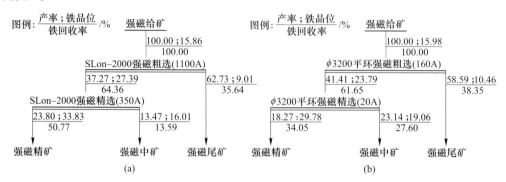

图 3　强磁作业流程对比试验结果
（a）SLon-2000 高梯度磁选机试验结果；（b）φ3200 平环磁选机试验结果

由图 3 可见，SLon-2000 磁选机系统的给矿品位为 15.86%，精矿品位为 33.83%，铁回收率为 50.77%；而 φ3200 平环强磁选机系统的给矿品位为 15.98%，精矿品位为 29.78%，铁回收率为 34.05%。前者比后者精矿品位高 4.05 个百分点，铁回收率高 16.72 个百分点。

与 φ3200 平环强磁机相比，SLon-2000 磁选机在粗选作业中精矿品位高 3.60 个百分点，铁回收率高 2.71 个百分点，为强磁精选作业提供了更优的条件，因此，其系统优势显著。

5　SLon-2000 磁选机在包钢选矿厂的工业应用

大量的试验数据及工业生产上的多方面对比表明，用 SLon-2000 立环脉动高梯度磁选机在包钢选矿厂取代 φ3200 平环强磁选机，可明显提高选矿指标，并具有耗水量较小、磁介质不堵塞、设备工作稳定、维护工作量少、作业率高的优点。2006 年包钢选矿厂又购入 10 台 SLon-2000 磁选机，取代了 10 台 φ3200 平环强磁选机，至今已运转 1 年多，选矿生产指标大幅度提高。2007 年该厂又订购了 4 台 SLon-2000 磁选机，拟取代剩余的 4 台 φ3200 平环强磁选机，预计在 2007 年底完成全部改造。

6　结论

（1）探索试验、现场工业对比试验证明，SLon-2000 立环脉动高梯度磁选机与平环强磁选机相比，可大幅度提高选矿指标，同时，该机磁介质不堵塞、耗水量小、设备作业率高达 99% 以上、检修维护工作量小，在技术指标、机电性能及生产成本等方面都具有明显的优势。

（2）SLon-2000 立环脉动高梯度磁选机用于包钢选矿厂的强磁粗选作业，与 φ3200 平环强磁选机相比，铁精矿品位提高 1.28~5.41 个百分点，尾矿品位降低 1.62~2.76 个百分点，铁回收率提高 2.15~9.01 个百分点。

（3）SLon-2000 立环脉动高梯度磁选机用于包钢选矿厂的强磁精选作业，与 ϕ3200 平环强磁选机相比，铁精矿品位提高 0.53~1.35 个百分点，尾矿品位降低 1.29~1.89 个百分点，铁回收率提高 2.41~5.75 个百分点。

（4）由 SLon-2000 立环脉动高梯度磁选机组成的一粗一精强磁选流程与由 ϕ3200 平环强磁选机组成的一粗一精强磁选流程相比，系统铁精矿品位提高 4.05 个百分点，铁回收率提高 16.72 个百分点。

（5）SLon-2000 立环脉动高梯度磁选机在包钢选矿厂的应用创造了显著的经济效益，为包钢选矿厂氧化铁矿选矿的技术进步作出了新的贡献。

参 考 文 献

[1] 余永富，朱超英. 包头稀土选矿技术进展 [J]. 金属矿山，1999（11）：18~22.
[2] 熊大和. SLon 立环脉动高梯度磁选机分选红矿的研究与应用 [J]. 金属矿山，2005（8）：24~29.
[3] 熊大和. SLon 脉动与振动高梯度磁选机新进展 [J]. 金属矿山，2006（7）：4~7.

写作背景 包钢选矿厂是我国大型氧化铁矿选矿厂之一，原采用了 20 台 ϕ3200 平环强磁选机，后来被 20 台 SLon-2000 立环脉动高梯度磁选机取代。此文介绍了 SLon 磁选机应用于该厂的小型试验和工业试验情况。从对比试验数据可看出，SLon 磁选机矿可显著地提高选矿指标。

SLon 磁选机分选氧化铁矿研究与应用新进展

熊大和

（赣州金环磁选设备有限公司）

摘　要　针对氧化铁矿的特性，研制了多种型号和不同背景磁感应强度的 SLon 立环脉动高梯度磁选机，近年在鞍钢调军台选矿厂、东鞍山烧结厂、胡家庙选矿厂、昆钢大红山选矿厂、唐钢司家营选矿厂等地获得大规模应用，获得了优异的选矿指标和经济效益，显著促进了氧化铁矿选矿技术的发展。

关键词　SLon 立环脉动高梯度磁选机　氧化铁矿　磁选

The New Development of Research and Application of SLon Magnetic Separators in Oxidized Iron Ore Processing

Xiong Dahe

（*SLon Magnetic Separator Ltd.*）

Abstract　According to the features of oxidized iron ores, many models of SLon vertical ring pulsating high gradient magnetic separators with deferent background magnetic induction are developed. In recent years, they have been applied in large scale in the oxidized iron ore processing plants of Diaojuntai, Donganshan, Hujiamiao of Anshan Steel Company, Kunming Steel Mineral Processing Plant and Tang Steel Sijiaying Mineral Processing Plant. Achieve excellent beneficiation-performance and obtain great economic benefits. SLon magnetic separators remarkably promoted the development of oxidized iron ore processing technology.

Keywords　SLon magnetic separator; Oxidized iron ore; Magnetic separation

　　我国氧化铁矿包括赤铁矿、镜铁矿、褐铁矿、菱铁矿等，它们都是弱磁性矿石，其特点是储量大，品位低，嵌布粒度细，绝大部分难磨难选。而我国的钢铁冶炼工业又普遍需要高品位的铁精矿。为了满足氧化铁矿工业的需要，通过持续的创新和研究，近年研制了多种型号规格及不同磁感强度的 SLon 立环脉动高梯度磁选机，在氧化铁矿的选矿工业中得到大规模的应用，取得了最佳的选矿指标和经济效益。

1 SLon 磁选机研制新进展

1.1 SLon 强磁机和中磁机简介

目前我国分选鞍山式氧化铁矿普遍采用的选矿流程为阶段磨矿—分级重选—强磁—反浮选流程，即较粗粒级（旋流器沉砂）采用螺旋溜槽选出部分铁精矿，螺旋溜槽的尾矿采用 SLon 立环脉动中磁机扫选一次，其扫选精矿返回二段磨矿（或三段磨矿）再磨再闭路入选；其尾矿作为最终尾矿抛弃。细粒级（旋流器溢流）采用 SLon 立环脉动高梯度强磁机粗选，其粗选精矿进反浮选精选，其尾矿作为最终尾矿抛弃[1,2]。

上述流程中螺旋溜槽尾矿粒度较粗，利用 SLon 磁选机抛尾实际使用的背景磁感强度为 0.2~0.6T 就够了，因此研制了几种型号的 SLon 立环脉动高梯度中磁机，规格型号见表1，可为用户节约设备投资和减轻设备重量。而旋流器溢流粒度很细，通常在小于 0.074mm 粒级占 90% 左右，利用 SLon 磁选机抛尾需要较高的背景磁感应强度，一般为 0.6~1.0T。用于细粒级抛尾的几种 SLon 立环脉动高梯度磁选机见表2。

表1 SLon 立环脉动高梯度中磁机主要参数

机型项目	转环外径/mm	给矿粒度/mm	<0.074mm 粒级含量/%	给矿浓度/%	矿浆通过能力/m³·h⁻¹	干矿处理量/t·h⁻¹	额定背景场强/T	额定激磁功率/kW	转环电动机功率/kW	脉动电动机功率/kW	脉动冲程/mm	脉动冲次/次·min⁻¹	主机质量/t
SLon-1500 (0.4T)	1500	<1.2	30~100	10~40	75~150	30~50	0.4	16	1.5	4	0~30	0~300	15
SLon-1750 (0.6T)	1750	<1.2	30~100	10~40	75~150	30~50	0.6	38	4	4	0~30	0~300	28
SLon-2000 (0.6T)	2000	<1.2	30~100	10~40	100~200	50~80	0.6	42	5.5	7.5	0~30	0~300	40
SLon-2000 (0.4T)	2000	<1.2	30~100	10~40	100~200	50~80	0.4	32	5.5	7.5	0~30	0~300	38
SLon-2500 (0.6T)	2500	<1.2	30~100	10~40	200~400	80~150	0.6	63	5.5	11	0~30	0~300	60

表2 SLon 立环脉动高梯度强磁机主要参数

机型项目	转环外径/mm	给矿粒度/mm	<0.074mm 粒级含量/%	给矿浓度/%	矿浆通过能力/m³·h⁻¹	干矿处理量/t·h⁻¹	额定背景场强/T	额定激磁功率/kW	转环电动机功率/kW	脉动电动机功率/kW	脉动冲程/mm	脉动冲次/次·min⁻¹	主机质量/t
SLon-750 工业型	750	<1.0		10~40	5~10	2~4	1.0	15	0.75	1.5	0~20	0~300	4
SLon-1000	1000	<1.2	30~100	10~40	12.5~20	4~7	1.0	21	1.1	2.2	0~30	0~300	6
SLon-1250	1250	<1.2	30~100	10~40	20~50	10~18	1.0	35	1.5	2.2	0~20	0~300	14
SLon-1500	1500	<1.2	30~100	10~40	50~100	20~30	1.0	44	3	4	0~30	0~300	20
SLon-1750	1750	<1.2	30~100	10~40	75~150	30~50	1.0	62	4	4	0~30	0~300	35
SLon-2000	2000	<1.2	30~100	10~40	100~200	50~80	1.0	74	5.5	7.5	0~30	0~300	50
SLon-2500	2500	<1.2	30~100	10~40	200~400	100~150	1.0	94	11	11	0~30	0~300	105

1.2 SLon 磁选机的创新与改进

通过多年持续的创新与改进，SLon 磁选机分选氧化铁矿具备如下优点[3]：

（1）设备规格型号多，可根据用户处理量要求和磁感应强度要求配置SLon磁选机，以达到最佳的配置和最低的投资。

（2）该机选矿指标优异，通过大规模工业生产的应用证明，SLon磁选机为用户多次创出历史最高选矿指标。

（3）该机结构持续地得到改进和优化，设备寿命长、作业率高，易于操作维护。

（4）该机采用导磁不锈钢棒作为磁介质，现已研制出$\phi1.0mm$、$\phi1.5mm$、$\phi2mm$、$\phi3mm$和$\phi4mm$等规格的棒介质，它们可采用单一直径的配置，也可采用几种直径的混合配置，采用计算机优化设计、数控机床精确定位实现磁介质最佳有序排列组合，有利于获得优良的选矿指标。

（5）该机转环立式旋转，反冲精矿，并配有矿浆脉动机构，具有选矿效率高，磁介质长期不堵塞的优点。

（6）该机耗水、耗电量小，设备处理量大，生产效率高，作业成本低。

2 SLon磁选机在鞍钢的应用

近几年鞍钢地区氧化铁矿选矿技术发展迅速，各大选厂纷纷采用阶段磨矿—分级重选—强磁—反浮选的流程。SLon磁选机在鞍山地区的应用新进展见表3[4~6]，其中具有代表性的为弓长岭选矿厂三选车间选矿流程，如图1所示。该流程中采用5台SLon-2000强磁机作为细粒级抛尾设备，另采用5台SLon-2000中磁机作为粗粒级（螺旋溜槽尾矿）抛尾设备。该流程设计的综合选矿指标为：原矿品位28.78%，铁精矿品位67.19%，尾矿品位10.13%，铁回收率76.29%。

表3 SLon磁选机在鞍钢各大选厂应用统计

厂矿名称	设备型号	使用台数/台	分选指标/%	选厂原矿处理量/万吨·a^{-1}
调军台选矿厂	SLon-2000强磁机 SLon-2000中磁机	15 15	$\alpha=30.50$, $\beta=67.50$, $\theta=10.50$, $\varepsilon=76.27$	1400
齐大山选矿厂	SLon-1750强磁机 SLon-1500中磁机	11 12	$\alpha=30.50$, $\beta=67.50$, $\theta=10.15$, $\varepsilon=78.53$	800
胡家庙选矿厂	SLon-2000强磁机 SLon-2000中磁机	8 8	$\alpha=28.40$, $\beta=67.50$, $\theta=10.60$, $\varepsilon=74.35$	800
弓长岭三选车间	SLon-2000强磁机 SLon-2000中磁机	5 5	$\alpha=28.78$, $\beta=67.19$, $\theta=10.13$, $\varepsilon=76.29$	300
东鞍山烧结厂	SLon-1750强磁机 SLon-2000强磁机 SLon-1750中磁机 SLon-2000中磁机	8 7 10 6	$\alpha=32.50$, $\beta=64.50$, $\theta=14.50$, $\varepsilon=71.45$	800

注：α为原矿品位；β为精矿品位；θ为尾矿品位；ε为回收率。

调军台选矿厂最初采用了连续磨矿—强磁—反浮选的流程，生产规模为900万吨/a。由于阶段磨矿—分级重选—强磁—反浮选的流程具有磨机台数少、生产成本较低的优点，目前调军台选矿厂也改造成该流程，年生产规模达到1400万吨。

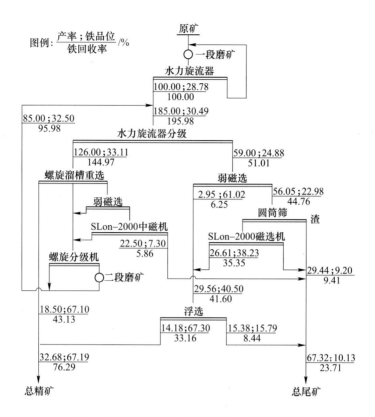

图 1　弓长岭选矿厂三选车间生产数质量流程

3　SLon 磁选机在昆钢大红山选矿厂的应用

昆钢近年在玉溪大红山建成 1 座年处理量为 400 万吨铁矿石的现代化选矿厂。其矿石是磁铁矿和赤铁矿的混合矿。采用阶段磨矿—强磁—反浮选的流程，其设计的原则流程如图 2 所示，设计的全流程选矿指标为：原矿品位 40.03%，精矿品位 67.00%，尾矿品位 16.49%，铁回收率 78%。流程中粗选段采用 8 台 SLon-2000 强磁机，精选段采用 3 台 SLon-2000 强磁机。经近两年的生产运行，强磁选部分运行平稳，而浮选部分存在问题较多，至今浮选作业未开。目前生产上实际达到的铁精矿品位为 64% 左右，铁回收率 85% 左右。

4　SLon 磁选机在唐钢司家营选矿厂的应用

唐山钢铁公司近年在司家营新建一座年处理量为 600 万吨氧化铁矿的现代化选矿厂，其铁矿石具有嵌布粒度细、难磨难选的特点。采用三段磨矿—螺旋溜槽重选—强磁反浮选流程，设计的原则流程如图 3 所示。流程中采用 12 台 SLon-1750 中磁机扫选螺旋溜槽尾矿；采用 8 台 SLon-1750 强磁机对细粒级抛尾，其精矿和弱磁选精矿混合后进入反浮选精选作业。该流程设计的全流程选矿指标为：原矿品位 30.44%，铁精矿品位 66.00%，铁尾矿品位 9.65%，铁回收率 80.00%。该流程已于 2007 年下半年投产[7]。

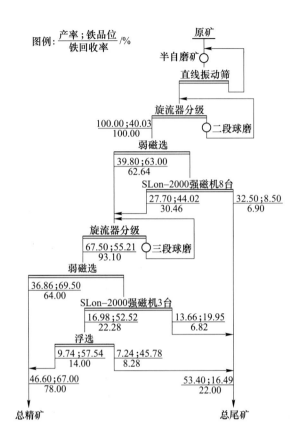

图 2 大红山铁矿 400 万吨选厂原则流程

5 结语

（1）SLon 立环脉动高梯度磁选机的研究近年朝着大型化、多样化发展，以满足多种弱磁性矿石和各种选矿规模的需求。研究人员持续地改进整机和零部件性能，优化选矿指标，提高选矿效率和设备的可靠性，降低选矿生产成本。

（2）我国大多数分选鞍山式氧化铁矿的选矿厂目前采用的选矿流程为阶段磨矿—分级重选—强磁—反浮选流程，该流程具有选矿效率高、铁精矿品位高、生产成本低的优点。其中采用 SLon 立环脉动高梯度中磁机扫选螺旋溜槽尾矿。SLon 立环脉动高梯度强磁机分选细粒级部分。SLon 磁选机脱泥效果好，可有效地控制尾矿品位。为降低全流程的中矿循环量、为浮选作业创造良好的作业条件起到了关键作用。

（3）至今已有 800 多台 SLon 立环脉动高梯度磁选机在工业生产中广泛应用于分选各种类型的氧化铁矿。该机的应用促进了我国氧化铁矿选矿工业的高速发展，使我国氧化铁矿的选矿技术达到国际先进水平。

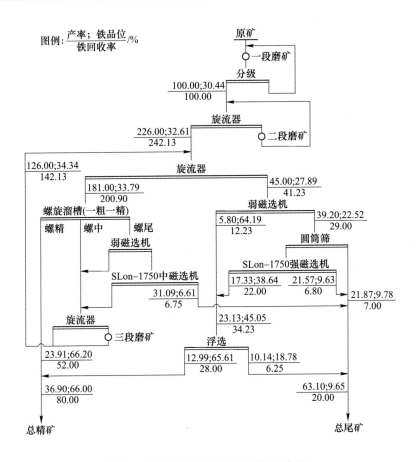

图 3 司家营铁矿 600 万吨现场原则流程

参 考 文 献

[1] 陈占金，李维兵，等. 鞍山地区难选铁矿石选矿技术研究 [J]. 金属矿山，2007 (1)：30~34.
[2] 张丛香，周惠文. 鞍山某铁矿床矿石选矿试验研究 [J]. 矿冶工程，2007，27 (3)：33~36.
[3] 熊大和. SLon-2000 立环脉动高梯度磁选机的研制 [J]. 金属矿山，1995 (6)：32~34.
[4] 熊大和，张国庆. SLon-2000 磁选机在调军台选矿厂的工业试验与应用 [J]. 金属矿山，2003 (12)：1~6.
[5] 熊大和. SLon 型磁选机在齐大山选矿厂的应用 [J]. 金属矿山，2002 (4)：42~44.
[6] 熊大和. SLon 型磁选机分选东鞍山红矿的应用 [J]. 金属矿山，2003 (6)：21~24.
[7] 张久甲，侯吉林. 唐钢司家营氧化铁矿石选矿试验研究 [J]. 金属矿山，2004 (4)：28~40.

写作背景 本文介绍了 SLon 立环脉动高梯度磁选机在多个大型氧化铁矿的应用，其中昆钢大红山铁矿年处理 400 万吨氧化铁矿和唐山钢铁公司司家营铁矿年处理 600 万吨氧化铁矿的选矿流程为新项目。

SLon Magnetic Separators Applied to Beneficiate Low Grade Oxidized Iron Ores[1]

Xiong Dahe

(SLon Magnetic Separator Ltd., Ganzhou, China. Email: xdh@slon.com.cn)

Abstract Low grade oxidized iron ores are usually not economic to utilize. However, the development of SLon magnetic separators and other new technologies make the process cost lower and lower. In this paper, several low-grade oxidized iron ore process flowsheets are introduced. The key procedure is that at the early stage, SLon magnetic separators are applied to discharge as more tails as possible with very low cost, only a small portion of rough concentrate going to the cleaning stage, showing the new technologies are profitable for such iron ores.

Keywords SLon magnetic separator; Oxidized iron ore; Magnetic separation

1 Introduction

Most Chinese iron ores deposits are very poor. The grades of some of them are lower than 30% Fe. In recent years, Chinese iron and steel industry develops rapidly and consumes huge amounts of iron concentrate. The facing problem for iron mining and processing industry is that the production scale is getting bigger and bigger, but the feeding grade is getting lower and lower. To solve this problem, the most important technologies of SLon vertical ring and pulsating high gradient magnetic separators (Xiong, Liu and Chen, 1998) and reverse flotation are widely applied in oxidized iron ores processing industry. Now many low grade oxidized iron ores can be utilized economically (Xiong, 1997, 2003, 2006).

2 SLon applied in Hujiamiao Iron Ore Processing Plant

Hujiamiao Iron Ore Processing Plant was built in 2006. It belongs to Anshan Iron and Steel Company. It treats 8 million low-grade oxidized iron ore annually. 8 SLon-2000 middle intensity magnetic separators (SLon MIMS) and 8 SLon-2000 vertical-ring and pulsating high gradient magnetic separators (SLon VPHGMS) are applied in this plant.

3 The characterization of the ore

The ore consists of magnetite, mag-hematite and hematite. The main gang mineral is quartz. The assay, mineralogical composition, particle size distribution and particle size liberation are shown in

---

[1] 原载于《第24届国际选矿会议论文集》。

Table 1-Table 4, respectively.

Table 1 Assay of Hujiamiao iron ore (%)

Total Fe	FeO	SiO$_2$	CaO	MgO	S	P
23.25	2.51	64.4	0.10	0.56	0.011	0.052

Table 2 Mineralogical compositions of Hujiamiao iron ore (%)

Mineral	Magnetite	Mag-hematite	Hematite	Limonite	Siderite	Silica iron	Total
Grade(Fe)/%	7.95	4.35	9.40	0.45	0.30	0.8	23.25
Ratio/%	34.19	18.71	40.43	1.94	1.29	3.44	100.00

Table 3 Practice crystal size distributions of Hujiamiao iron ore

Mineral	Crystal size/μm						
	2000~589	589~295	295~147	147~74	74~35	35~10	10~0
Iron minerals(mass)/%	2.14	8.26	14.69	22.63	23.27	23.08	5.93
Gang minerals(mass)/%	2.10	13.42	23.22	24.32	18.12	14.63	4.19

Table 4 Praticle size liberation of Hujiamiao iron ore

Particle size(Ground to <74μm)/%	Liberation(mass)/%	
	Iron minerals	Gang minerals
53.59	64.55	57.89
73.32	84.88	68.01
80.37	87.56	74.56
90.54	94.04	85.77

4 The processing flowsheet

The Hujiamiao iron ore processing flowsheet is shown in Fig. 1 and Fig. 2.

Seen from Table 3 and Table 4, the iron crystal size is not even. When ground to <74μm 53.95%, about 64.55% iron minerals and 57.89% gang minerals are liberated. So in the mineral processing flowsheet. The ore is ground to <74μm about 60% with primary ball mill, classified with cyclone to two fractions. The coarse fraction is treated with spirals to take out most of the liberate iron minerals as final concentrate. The spiral tails is scavenged with low intensity magnetic separator(LIMS) and 8 SLon-2000 MIMS(0.4T) which discharge 40.88% mass as final tails containing only 9.57% Fe. The mags of SLon MIMS and the spiral middling are ground further with secondly ball mill. Then returned to the primary cyclone.

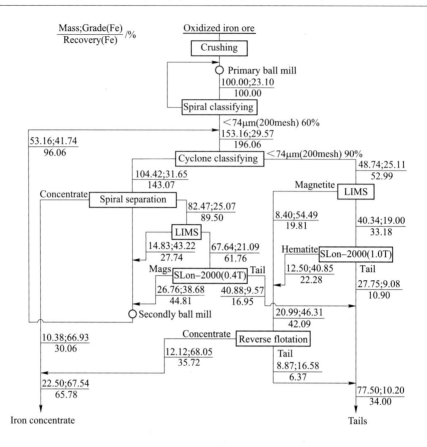

Fig. 1 The technical data of Hujiamiao processing flowsheet

Fig. 2 The photo of SLon-2000 VP HGMS in Hujiamiao processing plant

The fine fraction (<74μm 90%) of the cyclone is treated with LIMS to take out magnetite and with 8 SLon-2000 VPHGMS (1.0T) to take out hematite. The SLon VPHGMS also discharge 27.75% mass as final tails containing only 9.08%Fe.

The mags of the LIMS and SLon VPHGMS are cleaned further by reverse flotation. The collectors of the flotation are very sensitive to quartz and quartz-integrated minerals. Most of the quartz and un-liberated minerals can be removed from the iron concentrate by reverse flotation.

The operational costs of SLon magnetic separators are very low. For example, to treat 1 ton of feed iron ore, the SLon MIMS consumes only about 0.5kW·h electricity and 2m³ recycling water and the SLon VPHGMS consumes only about 1kW·h electricity and 2m³ recycling water. They can discharge about 68.63% mass of low-grade final tails in the early stage. Spiral operational cost is also very low. 10.38% mass of spiral concentrate is also taken out as final iron concentrate.

Flotation operation cost is relatively high. But only about 20.99% mass containing 46.31%Fe goes in to the flotation stage. SLon magnetic separators created very good conditions for flotation. They discharged most of the slimes and gang minerals as final tails, which otherwise would consume a lot of reagents and may damage the flotation process. They help flotation to get good results and greatly reduce reagents consumption. So flotation operation cost is relatively small for the total flowsheet.

Why the LIMS mags and SLon mags of the fine fraction must be cleaned with reverse flotation instead of SLon magnetic separator? The reason is that in the feed of them there are a olt of magnetite and quartz associated particles. For example, if one particle contains 1% or more magnetite and 99% or less quartz, it will be captured by SLon magnetic separator as mags in the roughing stage and re-captured again in the cleaning stage as mags if SLon applied as cleaner. It will enter into the final concentrate and lows the final concentrate grade. But for reverse flotation, the collectors are very sensitive to quartz. For example, if a particle contains 10% quartz and 90% iron or other minerals, the collectors will capture it in to the tails. So reverse flotation can get very pure iron concentrate.

For the excellent performance of SLon magnetic separators and reverse flotation. The plant get very good results as: feed grade 23.10%Fe, concentrate grade 67.54%Fe, tails grade 10.20%Fe, iron recovery 65.78%.

5 SLon applied in Wuyang Iron Ore Processing Plant

Wuyang Iron Ore Processing Plant belongs to Anyang Iron and Steel Company, in Henan Province. A magnetite iron ore-processing workshop has been running for many years to treat magnetite iron ore. But many million tons of low-grade oxidized iron deposit was not utilized. In 2004-2005, a low-grade oxidized iron ore-processing workshop was built to treat 3 million low-grade oxidized iron ore annually. 23 SLon-2000 VPHGMS are applied in this workshop.

6 The characterization of the ore

The ore consists of mainly hematite and a small portion magnetite, mag-hematite, limonite and pyrite. The main gang mineral is quartz, jasper and pyroxene. The iron ore is quite low of grade and deeply oxidized. The mineral crystal sizes are very fine. The ore is relatively difficult to process.

7 The processing flowsheet

The Wuyang iron ore processing flowsheet is shown in Fig. 3 and Fig. 4.

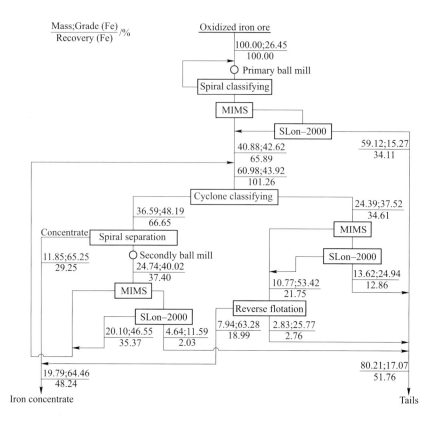

Fig. 3 The technical data of processing flowsheet of Wuyang Iron Mine

Fig. 4 The photo of 23 SLon-2000 VP HGMS in Wuyang Iron Mine

Because the feed iron grade is very low, only 26.45% Fe, it is very important to discharge as more tails as possible with low cost in the primary stage. The ore is ground to <74μm (200mesh) about 50%. Drum type middle intensity magnetic separators (MIMS) are applied to take out magnetite and SLon-2000 VPHGMS are applied for one roughing and one scavenging. 59.12% mass of final tails is rejected immediately with low cost. And the rough iron concentrate reaches 40.88%

Fe, which serves as the feed of the secondly processing stage.

In the second stage, the mixture mags of MIMS and SLon-2000 VPHGMS are classified with cyclone. The coarse fraction is treated with spirals, which take out 11.85% mass of final concentrate containing 65.25% Fe. The spiral tails is classified and the coarse fraction is reground with secondly ball mill. Then MIMS and SLon-2000 VPHGMS are used to discharge tails. Here 4.64% mass of final tails is rejected. And their mags are returned to the second cyclone. The fine fraction of the second cyclone is also treated with MIMS and SLon VPHGMS. Here 13.62% mass of final tails is rejected. Their mags, only 10.77% mass, are cleaned with reverse flotation.

In the flowsheet, three places applied SLon-2000 VPHGMS to discharge tails. Each place SLon-2000 VPHGMS are used for one cleaning and one scavenging due to that the iron ore is deeply oxidized and these arrange can get final tails of lower iron grade.

Because the low cost performance of SLon magnetic separator and reverse flotation, the plant gets profitable results as: feed grade 26.45% Fe, concentrate grade 64.46% Fe, tails grade 17.07% Fe, iron recovery 48.24%.

As the mine is digging the earth suffice ore, the ore is deeply oxidized and contains a lot of slime, the magnetic susceptibilities of the iron minerals are very weak. So the iron recovery in the iron concentrate is not high enough. There is a potential that the iron recovery will be raised further as the mining towards deeper.

8 Tests of abanded iron ore tails with SLon and centrifugal separator

In Jiangxi province, there are several small private iron ore processing plants, each of them treat 0.2 to 0.5 million tons of iron ore annually. Such iron ore consist of mainly magnetite and hematite. In the past years, the plants recovery only magnetite with low intensity magnetic separators (LIMS) and all of the hematite is abanded with tails. As the development of SLon magnetic separator and other new technology and the price of iron concentrate goes up, some people are planning to recover hematite from the abanded tails.

The test flowsheet is shown in Fig. 5.

The abanded tails contain 22.24%. It is roughly grounded to <0.074mm about 60% with ball mill. Then SLon magnetic separator are applied for one roughing and one scavenging. 58.99% mass of final tails is rejected immediately with low cost. Only 41.01% mass of mags containing 41.65% Fe enters the centrifugal separator cleaning stage.

The SLon mags are ground to <0.051mm, and then centrifugal separators are used for one roughing. One leaning and one scavenging. The final concentrate riches: grade 64.26% Fe, iron recovery 38.41%. It is profitable because of there is no mining and crashing costs.

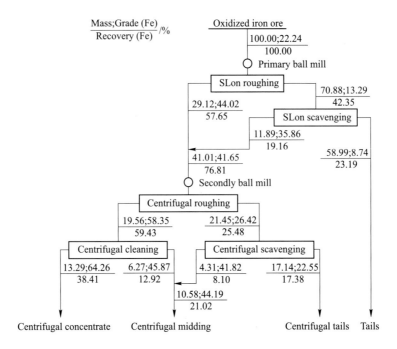

Fig. 5 Test flowsheet with SLon and centrifugal separator

9 Conclusions

(1) SLon magnetic separators are a kind of high efficient high intensity magnetic separator. They can treat weakly magnetic minerals with high capacity, high efficiency and very low cost.

(2) Many low-grade oxidized iron ores now can be profitably treated with the combination flowsheet of SLon magnetic separators, spiral and reverse flotation. The key procedure is that at the early stage, SLon magnetic separators are applied to reject as many as possible final tails with very low cost, and spiral are applied to take out as much as possible final concentrate. Only a small portion pre-richened fine fraction enter into the reverse flotation stage.

(3) The technology applied in Hujiamiao low grade oxidized iron ore processing flowsheet, the plant gets very good results as: feed grade 23.10%Fe, concentrate grade 67.54%Fe, tails grade 10.20%Fe, iron recovery 65.78%.

(4) The technology applied in Wuyang low grade oxidized iron ore processing flowsheet, the plant gets profitable results as: feed grade 26.45%Fe, concentrate grade 64.46%Fe, tails grade 17.07%Fe, iron recovery 48.24%.

(5) The other example is that SLon magnetic separator and centrifugal separators are cooperated to test banded tails containing hematite, the test results are: feed grade 22.24%Fe, concentrate grade 64.26%Fe, tails grade 17.38%Fe, iron recovery 38.41%. This technology may represent a trend for small plants to recover low-grade oxidized iron ores.

References

[1] Xiong D H. SLon magnetic separator applied to upgrading the iron concentrate[J]. Physical Separation in Science and Engineering,2003,2:63-69.
[2] Xiong D H. Development and application of SLon vertical ring and pulsating high gradient magnetic separators [J]. Proceedings of XX IMPC. Anchen,1997:21-26.
[3] Xiong D H, Liu S Y, Chen J. New technology of pulsating high gradient magnetic separation[J]. J. Miner. Process,1998, 54:111-127.
[4] Xiong D H. SLon magnetic separator promoting Chinese oxidized iron ore processing industry[J]. Proceedings of XXIII IMPC,2006(4):276-281.

写作背景 本文向国外同行介绍了 SLon 立环脉动高梯度磁选机在分选低品位氧化铁矿的应用，列举了鞍钢胡家庙铁矿，河南安阳钢铁公司舞阳铁矿的低品位氧化铁矿的矿石性质、SLon 磁选—反浮选流程和选矿指标，江西某地从选磁铁矿的尾矿中用 SLon 磁选—离心机重选回收赤铁矿的流程。

SLon 强磁机在低品位氧化铁矿中的新应用

熊大和[1,2]

（1. 赣州有色冶金研究所；2. 赣州金环磁选设备有限公司）

摘　要　低品位氧化铁矿以往被认为没有利用的价值，SLon 立环脉动高梯度磁选机以及其他新的选矿技术的发展使得这类矿石的选矿成本逐渐降低，从而也能产生良好的经济效益。介绍了几种低品位氧化铁矿的选矿流程，其关键技术是在选矿初期利用 SLon 磁选机低成本多抛尾，仅使一小部分的粗选精矿进入精选作业，发挥了该新技术处理此类铁矿石的突出作用。

关键词　SLon 立环脉动高梯度磁选机　氧化铁矿　磁选

Application of SLon Magnetic Separators to the Beneficiation of Low Grade Oxidized Iron Ores

Xiong Dahe[1,2]

(1. Ganzhou Nonferrous Metallurgy Research Institute;
2. SLon Magnetic Separator Ltd.)

Abstract　It is considered that low grade oxidized iron ores are usually not valuable to utilize. However, the development of SLon magnetic separators and other new technologies make the process costs lower and lower. In this paper, several flow-sheets for low-grade oxidized iron ore process were introduced. The key procedure is that at the early stage, SLon magnetic separators are applied to discharge as much tailings as possible with lower cost. Only a small portion of rough concentrates had gone to the cleaning stage, showing that the new technologies are profitable for such iron ores.

Keywords　SLon vertical-ring pulsating high-gradient magnetic separator; Oxidized iron ore; Magnetic separation

我国大多数铁矿都属于贫矿，某些矿石的铁品位低于 30%。近几年，铁矿山和选矿厂所面临的问题是铁精矿的需求量越来越大，而原矿的品位却越来越低。为了解决这个问题，SLon 立环脉动高梯度强磁选的关键技术[1]和反浮选被广泛的应用于氧化铁矿的选矿工业。现在许多低品位的氧化铁矿也有经济利用的价值[2]。

---

❶ 原载于《2008 年全国金属矿山难选矿及低品位矿选矿新技术学术研讨与技术成果交流暨设备展示会论文集》，2008。

1 SLon 强磁机在胡家庙铁矿选矿厂中的应用

胡家庙选矿厂建于 2006 年,隶属于鞍山钢铁公司。年处理铁矿石 800 万吨。目前,有 8 台 SLon-2000 高梯度中磁机(0.6T)和 8 台 SLon-2000 高梯度强磁机(1.0T)在该选厂应用。

1.1 矿石的主要性质

该矿物主要由磁铁矿、磁赤铁矿、赤铁矿组成。脉石矿物主要为石英。胡家庙铁矿的化学多元素分析、铁物相分析、粒度分布及小于 74μm 粒级不同含量的矿物解离度分别见表 1~表 4。

表 1 胡家庙铁矿的化学多元素分析 (%)

元　素	TFe	FeO	SiO_2	CaO	MgO	S	P
含　量	23.25	2.51	64.4	0.10	0.56	0.011	0.052

表 2 胡家庙铁矿的铁物相分析 (%)

铁物相	磁铁	磁-赤铁	赤铁	褐铁	菱铁	硅酸铁	合计
铁含量	7.95	4.35	9.40	0.45	0.30	0.8	23.25
铁分布率	34.19	18.71	40.43	1.94	1.29	3.44	100.00

表 3 胡家庙铁矿的矿物粒度分布

粒度范围/μm	2000~589	589~295	295~147	147~74	74~35	35~10	10~0
铁矿石含量/%	2.14	8.26	14.69	22.63	23.27	23.08	5.93
脉石矿物含量/%	2.10	13.42	23.22	24.32	18.12	14.63	4.19

表 4 胡家庙铁矿小于 74μm 不同含量下矿物解离情况 (%)

磨矿细度 (<74μm)	解　离　度	
	铁矿石	脉石矿物
53.59	64.55	57.89
73.32	84.88	68.01
80.37	87.56	74.56
90.54	94.04	85.77

1.2 选矿流程

胡家庙铁矿的选矿流程如图 1 所示。从表 3 和表 4 可以看出,该铁矿的粒度分布范围较广,当磨矿至小于 74μm 占 53.95% 时,64.55% 左右的铁矿物和 57.89% 左右的脉石矿物是解离的。因此,在选矿流程中,一段磨矿将矿物磨至小于 74μm 占 60% 左右,采用水力旋流器分级,粗粒部分通过螺旋溜槽将大部分已解离的铁矿物分选出来作为最终的铁精矿。螺旋溜槽的尾矿利用弱磁选机和 8 台 SLon-2000 中磁机(0.6T)抛弃产率为 40.88% 的尾矿,尾矿中铁品位仅为 9.57%。中磁选机的精矿和螺旋溜槽的中矿进入二段磨矿,并返回到一次分级。

细粒级(小于 74μm 占 90%)矿物先经过弱磁选机分选出磁铁矿,然后用 8 台 SLon-2000 高梯度强磁机(1.0T)分选赤铁矿。该机抛弃了产率为 27.75%、铁品位仅为 9.08% 的尾矿。

弱磁选和 SLon 高梯度强磁机的精矿进入反浮选作业,浮选的捕收剂对石英以及石英类矿物非常敏感,大部分的石英以及没有完全解离的矿物能通过反浮选从铁精矿中去除,

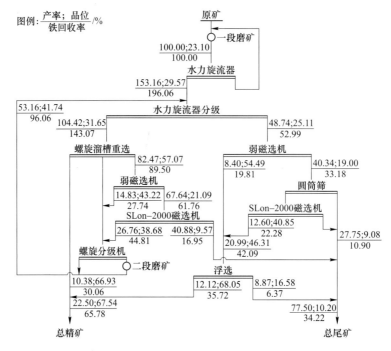

图 1 胡家庙选矿厂生产数质量流程

从而获得高品位的铁精矿。

SLon 高梯度磁选机的作业成本非常低。例如，处理 1t 铁矿石，SLon 高梯度中磁选机仅需消耗 0.5kW·h 的电和 2m^3 可循环利用的水，SLon 高梯度强磁机也仅需消耗 1kW·h 的电和 2m^3 可循环利用的水，而它们却能抛弃大约 68.63% 的低品位的尾矿，螺旋溜槽的使用成本也非常低，它能选出产率为 10.38% 的铁精粉。

浮选的作业成本相对高一些，但进入浮选作业矿物仅占原矿的 20.99%，铁品位为 46.31%，并且先前的 SLon 高梯度强磁机已给浮选作业创造了非常有利的条件，它抛弃了大部分的泥以及脉石矿物。这些泥和脉石矿物如果不在浮选前去除，它们将消耗大量的浮选药剂，并且会影响到整个浮选过程。

SLon 高梯度磁选机能帮助浮选取得好的生产指标并且大幅度的降低了药剂消耗，因此，对整个流程来说，浮选的作业成本相对较小。为什么弱磁选机和 SLon 高梯度强磁机的粗精矿要用反浮选精选，而不再次利用 SLon 高梯度强磁机呢？原因在于原料中有一部分矿物颗粒是以磁性物与石英的连生体形式存在。例如，如果某一个矿粒中含有 1%（甚至还要少）磁铁矿以及 99% 的石英，它能在粗选中作为磁性物被捕集，如果精选也采用 SLon 高梯度强磁机，那么在精选过程中它将再次被捕集上来，进入到精矿中，从而影响了最终精矿的品位。而对于反浮选，捕收剂对石英非常敏感，例如当矿物颗粒中含有 10% 的石英和 90% 的铁或其他矿物，捕收剂就能将它给捕集到尾矿中，所以反浮选能获得高品位的铁精矿。

2 SLon 磁选机在舞阳铁矿选厂中的应用

舞阳铁矿选厂隶属于河南安阳钢铁公司，早期建立的选矿车间都是用来处理磁铁矿，而伴随磁铁矿一起开采出来的数百万吨的低品位氧化铁矿一直得不到有效利用。2004～2005 年，该选厂建立了年处理 300 万吨低品位氧化铁矿的选矿车间。目前，已有 23 台 SLon-2000 高梯度强磁机在该选厂中应用。

2.1 矿石性质

舞阳铁矿的矿石主要成分为赤铁矿、磁铁矿、磁-赤铁矿、褐铁矿和黄铁矿。主要的脉石矿物为石英、碧玉、辉石。该矿石的品位低、氧化程度深、嵌布粒度细，属于难选型铁矿石。

2.2 选矿流程

舞阳铁矿选厂的选矿流程如图 2 所示。

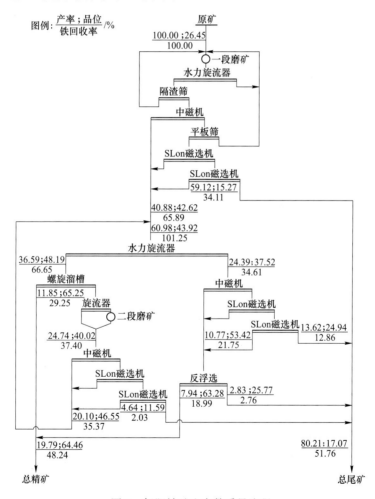

图 2 舞阳铁矿生产数质量流程

由于原矿铁品位仅为 26.45%，因此选别初期尽可能低成本地抛尾就显得十分重要。原矿磨至小于 74μm（200 目）占 50% 左右，采用滚筒中磁机分选出磁铁矿后，再利用 SLon-2000 高梯度强磁选机进行一次粗选一次扫选选别，将产率 59.12% 的尾矿抛弃，得到铁品位为 42.62% 的粗精矿，进入下一个选别作业。

在下一个选别作业中，中磁机与 SLon-2000 高梯度强磁机的混合精矿，通过水力旋流器分级，粗粒级部分进入螺旋溜槽，能得到产率为 11.85% 及铁品位高达 65.25% 的最终铁精矿。螺旋溜槽的尾矿进一步分级，粗粒部分进入二段球磨机再磨。该作业中的中磁机和 SLon-2000 高梯度强磁机都是用来抛尾，在这个过程中，产率为 4.64% 的尾矿被抛弃，它们的磁性产品返回到二次分级，二次分级的细粒部分同样进入中磁机和 SLon-2000 高梯度强磁机，这个过程中又抛弃了产率为 13.62% 的尾矿，最终仅有 10.77% 的磁性产品进入反浮选。整个流程中，有 3 处都是采用 SLon-2000 高梯度强磁机进行抛尾。由于该铁矿的氧化程度较深，使用 SLon-2000 高梯度强磁机时都采用一次精选、一次扫选流程，这样配置能降低最终尾矿的品位。

由于 SLon 强磁机和反浮选技术能降低选矿成本，因此产生了良好的经济效益，其生产指标为原矿品位为 26.45%Fe，精矿品位为 64.46%Fe，尾矿品位为 17.07%Fe，铁回收率 48.24%。

由于矿物的氧化程度较深和含泥量较大，铁矿物的磁性很弱。因此，铁精粉中铁的回收率不够高。随着采矿深度增加，铁矿物的氧化程度以及含泥量都将得到改善，铁的回收率有望进一步提高。

3　SLon 磁选机和离心机处理废弃铁尾矿的试验

江西有很多小型的私营铁矿选厂，年处理铁矿石都在 20 万~50 万吨左右，铁矿的主要成分是磁铁矿和赤铁矿。以往这些选厂都只是利用弱磁选机回收磁铁矿，几乎全部的赤铁矿都是作为尾矿被抛弃。随着 SLon 磁选机及其他新的选矿技术的发展，以及铁精粉价格的不断提高，许多人都想从废弃的尾矿中回收赤铁矿。我们也对此做过些探索性试验，其选矿流程见图 3。

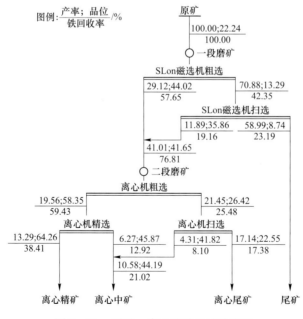

图 3　SLon 磁选—离心选矿机试验流程

废弃的尾矿中，铁的品位为22.24%，将原矿磨至小于0.074μm占60%左右，利用SLon磁选机进行一次粗选、一次扫选两次选别，尽快地以低成本抛弃产率为58.99%的尾矿后，得到的产率为41.01%及铁品位为41.65%的磁选铁精矿，进入离心机的精选作业。

将磁选精矿磨至小于0.051mm后，利用离心选矿机进行一次粗选、一次精选、一次扫选选别，最终可得到铁精矿品位64.26%，铁回收率38.41%，因为无需开采和破碎的成本，也能产生良好的经济效益。

4 结论

（1）SLon高梯度强磁机是一种高效的强磁选矿设备，具有处理能力大、生产效率高、运行成本低的优点，非常适合处理弱磁性矿物。

（2）大多数低品位氧化铁矿，通过SLon强磁机—螺旋溜槽重选—反浮选联合选矿流程，都能够得到很好的利用。该技术的关键是在选别初期利用SLon高梯度磁选机尽早尽可能地多抛尾，然后利用螺旋溜槽尽早尽可能多地选出已解离的较粗铁精矿，仅仅一小部分预富集了的细粒磁选铁精矿进入反浮选作业。

（3）该技术在胡家庙低品位氧化铁矿的选矿工艺应用中，取得了优异的生产指标：原矿铁品位23.10%，精矿铁品位67.54%，尾矿铁品位10.20%，铁回收率65.78%。

（4）该技术在舞阳铁矿选厂低品位氧化铁矿的选矿工艺应用中，为选矿厂创造良好的经济效益，取得了良好的选矿指标：原矿铁品位26.45%，精矿铁品位64.46%，尾矿铁品位17.07%，铁回收率48.42%。

（5）SLon高梯度强磁机—离心选矿机联合工艺在尾矿中回收赤铁矿的试验表明：原矿铁品位22.24%，精矿铁品位64.26%，尾矿铁品位17.38%，铁回收率38.41%。这为小型选厂回收低品位氧化铁矿又增加了一条有效的途径。

参 考 文 献

[1] 熊大和，刘建平. SLon脉动与振动高梯度磁选机新进展[J]. 金属矿山，2006（7）：5~7.
[2] 熊大和. SLon立环脉动高梯度磁选机分选红矿的研究与应用[J]. 金属矿山，2005（8）：24~29.

写作背景 本文向国内同行介绍了SLon立环脉动高梯度磁选机在分选低品位氧化铁矿的应用，列举了鞍钢胡家庙铁矿，河南安阳钢铁公司舞阳铁矿的低品位氧化铁矿的矿石性质、SLon磁选—反浮选流程和选矿指标，江西某地从选磁铁矿的尾矿中用SLon磁选—离心机重选回收赤铁矿的流程。

Application of SLon Magnetic Separators in Modernising the Anshan Oxidised Iron Ore Processing Industry

Xiong Dahe

(SLon Magnetic Separator Ltd. Shahe Industrial Park,
Ganzhou 341000, Jiangxi Province, China. Email: xdh@slon.com.cn)

Abstract The Anshan iron ore deposit is the biggest low-grade oxidised iron deposit in China. Over the past ten years, about 120 SLon vertical ring-pulsating high gradient magnetic separators units have been installed in Anshan to process oxidised iron ore. This paper describes how several plants have upgraded their old technologies to new technologies with SLon magnetic separators as the key equipment for controlling tails grades, including using reverse flotation as the key technology for iron concentrate cleaning. These plants have been modernised with beneficial results at low cost.

1 Introduction

Anshan oxidised iron ore mainly consists of haematite, maghematite and magnetite. The gangue mineral is mainly quartz and the grades of mined ores are only about 26%-34%Fe.

The Anshan Mining Company has five main oxidised iron ore processing plants. Currently their total processing capacity is 41 million tones per annum. Four of them were built up in the early years of operation. In the past, a range of old technologies, equipments and flow sheets were applied to process the oxidised iron ores, but iron concentrate grades and recoveries were not high enough. In order to improve efficiency, many innovations have been implemented in the past 10 years. The most advanced technologies introduced were the SLon vertical ring pulsating high gradient magnetic separators (Xiong, 2007) and reverse flotation. These technologies can produce high-grade iron concentrates up to 67.50%Fe with iron recoveries of up to 78% at low cost. Nowadays, SLon magnetic separators are used in all the oxidised iron ore processing plants in the Anshan area, as well as in many other Chinese oxidised iron ore processing plants.

2 Modernisation of the Qidashan Iron Ore Processing Plant

The Qidashan Iron Ore Processing Plant was built in 1969. Before 2001, the plant used the flow sheet shown in Fig.1. The mined ore was crushed and screened into two size fractions of 0-20mm and 20-75mm. The 0-20mm size fraction was ground and treated using a spiral-wet high intensity magnetic separator-flotation flow sheet. The 20-75mm fraction was roasted to convert haematite into

---

❶ 原载于《Proceedings of Iron Ore 2009》,澳大利亚,2009。

magnetite, then ground and treated with low intensity magnetic separators. The problem with this flow sheet was that the iron concentrate grade reached only 63% and the roasting process was costly, caused heavy pollution and the working conditions were hard.

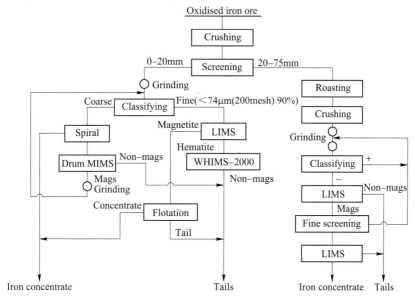

Fig.1 Old flow sheet for Qidashan Iron Ore Processing Plant

Over the period 2001-2005, the plant was modernised by converting it to a staged grinding and spiral-SLon magnetic separator-reverse flotation flow sheet as shown in Fig.2. The major changes were as follows:

(1) The roast-LIMS flow sheet was converted into spiral-SLon magnetic separator-reverse flotation flow sheet;

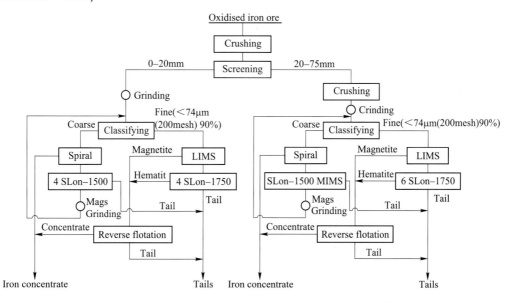

Fig.2 New flow sheet for the Qidashan Iron Ore Processing Plant

(2) The previous WHIMS-2000 horizontal wet high intensity magnetic separators were replaced by SLon-1750 (1.0T) magnetic separators;

(3) The Drum middle intensity magnetic separators (Drum MIMS, 0.4T) were replaced by SLon-1500 vertical-ring pulsating middle intensity magnetic separators (SLon-1500 MIMS, 0.4T).

Using the new flow sheet, the ore treating capacity was raised from 7.5 million tones to 9.6 million tones per annum. In addition, the iron concentrate grade was raised from 63.22%Fe to 67.5%Fe and the iron recovery increased from 74.04% to 77.28%. The detail results are shown in Table 1.

Table 1 Comparison of the penformance of the new Qidashan flow sheet with the previous flow sheet

Items	Feed grade (Fe)/%	Concentrate grade (Fe)/%	Tails grade (Fe)/%	Concentrate mass ratio (Fe)	Iron recovery /%
New flow sheet	29.50	67.50	10.12	33.77	77.28
Previous flow sheet	29.86	63.22	11.92	34.97	74.04
New flow sheet versus previous flow sheet	-0.36	+4.28	-1.8	-1.2	+3.24

3 Modernisation of the Diaojuntai Iron Ore Processing Plant

The Diaojuntai Iron Ore Processing Plant was built in 1998. The flow sheet consisted of two stage continuous grinding and low intensity magnetic separation-mid intensity magnetic separation-high intensity magnetic separation-flotation as shown in Fig. 3. In the flow sheet, 15 Shp-3200 horizontal-ring high intensity magnetic separators were used and their magnetic matrixes were grooved plates made of stainless steel. Because the magnetic matrixes frequently clogged up, the 15 Shp-3200 magnetic separators were difficult to maintain. Hence, in 2003-2005 the plant replaced them with 15 SLon-2000 vertical-ring pulsating high gradient magnetic separators as shown in Fig. 4. Because the SLon magnetic separators have higher efficiency, the final iron concentrate grade was increased from 67.13%Fe to 67.5%Fe and the iron recovery improved from 76.64% to 78.94%.

In order to lower processing costs and increase capacity, in 2005-2007 the flow sheet was further modified to incorporate staged grinding and low intensity magnetic separation-SLon magnetic separation-reverse flotation as shown in Fig. 5. Another 15 SLon-2000 vertical-ring pulsating middle intensity magnetic separators (0.4 T) were added into the flow sheet. Because the SLon magnetic separators discharge part of the coarse tails and the spirals take out part of the coarse final concentrate in the primary grinding stage, the grinding costs are greatly reduced. The ore treating capacity was raised from 9 million tonnes to 14.4 million tonnes per annum using the same ball mill. The production cost of the iron concentrate was lowered about 30 yuan (4.5 US dollar) per tonne. As the mining capacity was subsequently scaled up, the average feed grade was lowered from 30.50% Fe to 29.50% Fe. However, the average concentrate grade remained at 67.5%Fe.

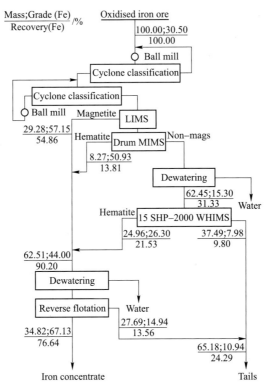

Fig.3 Old flow sheet for the Diaojuntai Iron Ore Processing Plant

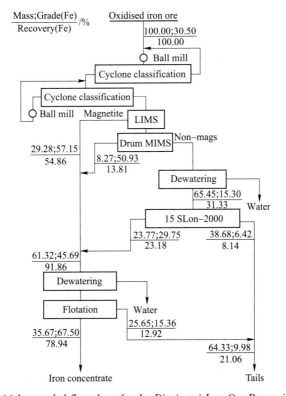

Fig.4 Initial upgraded flow sheet for the Diaojuntai Iron Ore Processing Plant

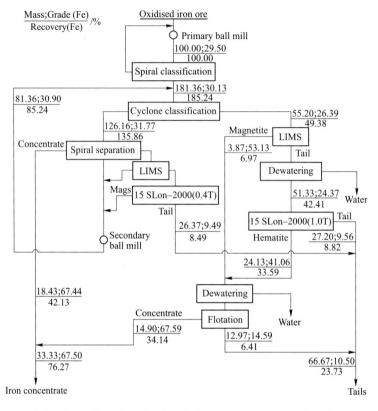

Fig.5 Latest flow sheet for the Diaojuntai Iron Ore Processing Plant

4 Modernisation of the Donganshan Iron Ore Processing Plant

The Donganshan Iron Ore Processing Plant was built in 1958. This iron ore deposit is the most oxidised and the crystal size is the finest in the Anshan region. The old iron ore processing flow sheet consisted of two stages of continuous grinding and flotation as shown in Fig.6. The average final concentrate grade was only 61.3%Fe and the flotation costs were relatively high.

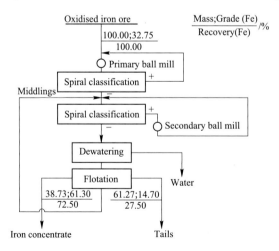

Fig.6 Old flow sheet for the Donganshan Iron Ore Processing Plant

In order to raise the iron concentrate grade and lower the costs, the plant was converted to a staged grinding-spiral-SLon magnetic separator-reverse flotation flow sheet as shown in Fig.7. A total of 31 SLon-2000 and SLon-1750 magnetic separators were installed to reject coarse tails and fine tails. The average iron concentrate grade was raised from 61.3%Fe to 64.50%Fe and the ore treating capacity was raised from four million tones to six million tones per annum. The processing costs were also greatly lowered.

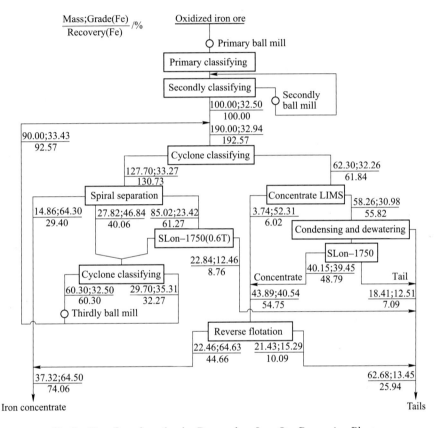

Fig.7 New flow sheet for the Donganshan Iron Ore Processing Plant

5 Modernisation of the Gongchangling Iron Ore Processing Plant

The Gongchangling Iron Ore Processing Plant was built in 1975. It treats three million tonnes of oxidised iron ore annually. The old processing circuit was a two stage grinding-centrifugal separation flow sheet as shown in Fig. 8. The average feed grade was 31.80%Fe, the final concentrate grade 60.50%Fe and the iron recovery 66.35%.

Because high intensity magnetic separation technology was developing quickly in the 1970s to 1980s, the processing circuit was converted into a staged grinding-high intensity magnetic separation-spiral-centrifugal separation flow sheet as shown in Fig.9. In this flow sheet, Shp-2000 wet high intensity magnetic separators (WHIMS) were used to discharge tails and spirals were used to take out part of the final concentrate at the primary grinding stage, so the number of secondary ball

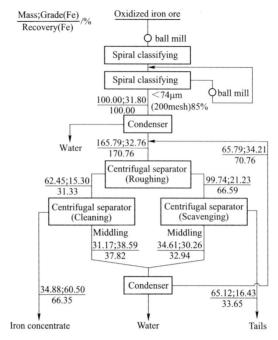

Fig.8 Old flow sheet for the Gongchangling Iron Ore Processing Plant

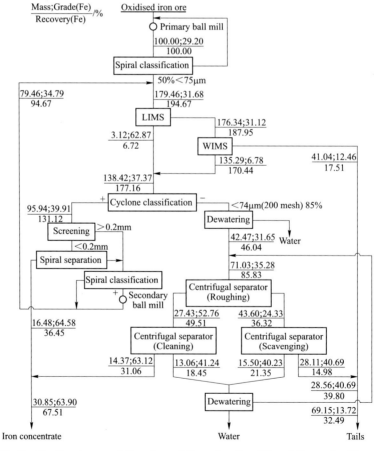

Fig.9 The first innovated flow sheet of Gongchangling Mineral Processing Plant

mills was reduced to half. Even when the feed was lowered from 31.80%Fe to 29.20%Fe, the total concentrate grade was raised from 60.50%Fe to 63.90%Fe and the iron recovery increased from 66.35% to 67.51%.

In 1990-2008, development of SLon magnetic separators and reverse flotation gathered momentum. In order to upgrade plant efficiency, in 2005 the plant flow sheet was converted into a staged grinding-spiral-SLon magnetic separator-reverse flotation circuit as shown in Fig.10. In this flow sheet, 5 SLon-2000 (1.0T) magnetic separators were installed to reject fine particle tails and 5 SLon-2000 (0.6T) magnetic separators were installed to reject coarse particle tails. The final tails grades achieved were much lower than for the previous Shp-2000 WHIMS. This new flow sheet is a major advance over previous folw sheets. In addition, when the plant feed grade dropped to 28.78%Fe, the grade of the iron concentrate increased to 67.19%Fe at an overall iron reeovery of 76.29%.

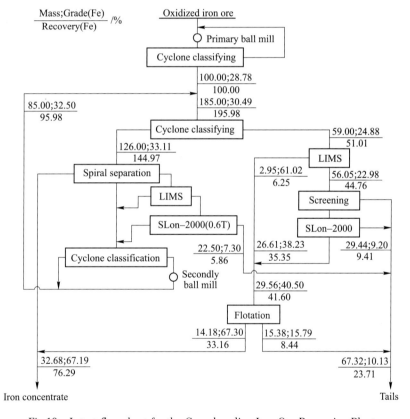

Fig.10 Latest flow sheet for the Gongchangling Iron Ore Processing Plant

6 Conclusions

(1) Anshan has the biggest low-grade oxidised iron ore deposits in China. Over the past 50 years, a huge amount of research work has been conducted to develop processing technologies for oxdidsed iron ores. In this paper, developments at four plants have been discussed as examples of

the application of new technologies. Their initial flow sheets were different from each other, but their latest flow sheets have evolved towards staged grinding-spiral separation-SLon magnetic separation-reverse flotation. The advantages of this flow sheet are that high iron concentrate grades and higher iron recoveries can be obtained at lower cost.

(2) Staged grinding has the advantage that spirals can be used in the early stages of grinding to recover the coarse liberated rion particles and SLon magnetic separators can discharge part of the coarse tails. So the grinding costs are greatly reduced.

(3) SLon magnetic separators are widely applied to the discharge of tails at very low operational cost. They can control tails grades at very low levels, thereby increasing iron recovery. Another important role of SLon magnetic separators is that they remove slimes from the feed to reverse flotation. This is important, because, if the slimes enter the reverse flotation circuit, they impair flotation performance and consume excessive amounts of reagents.

(4) Reverse flotation is widely used to clean up the fine particle fraction of the final product and achieve very high concentrate grades. However, because flotation consumes reagents, its operational costs are much higher than for magnetic and spiral separation. Hence, the mass of the flotation feed should be made as small as possible.

References

[1] Dahe. X, 2007, SLon magnetic separators applied in various industrial iron ore processing flow sheets [J]. Proceedings of Iron Ore, 2007:245-250.

写作背景 这是2009年在澳大利亚帕斯主办的铁矿学术会上发表的论文。此文向国外同行介绍了鞍钢齐大山选矿厂、调军台选矿厂、东鞍山烧结厂和弓长岭选矿厂氧化铁矿选矿流程的演变，SLon立环脉动高梯度磁选机在推动这些选矿流程一步一步走向现代化所起的作用。新技术的应用使这些选矿厂的技术经济指标达到国际领先水平。

SLon 磁选机在鞍山氧化铁矿选矿工业现代化的应用

熊大和

（赣州金环磁选设备有限公司）

摘　要　鞍山拥有我国储量最大的低品位氧化铁矿。在过去的十年中，约有 150 台 SLon 立环脉动高梯度磁选机应用于该地区氧化铁矿的分选。本文介绍了鞍山齐大山、调军台、东鞍山、弓长岭选矿厂利用 SLon 磁选机作为关键设备来控制尾矿品位，以及利用反浮选技术提高精矿质量的工艺流程。目前这些选矿厂已实现了现代化的工业生产，以较低的生产成本获得了优良的选矿指标。

关键词　SLon 立环脉动高梯度磁选机　鞍山　氧化铁矿

1　前言

鞍山氧化铁矿主要成分为赤铁矿、磁赤铁矿和磁铁矿，脉石矿物主要为石英。原矿中的铁品位仅为 26%~34%。鞍山矿业公司下辖五个氧化铁矿选厂，目前综合生产能力为 4100 万吨/a。其中 4 个选厂都是早期建设的，使用的设备、技术包括选矿流程都是早期的技术成果，因此精矿品位和回收率都不是很理想，无法满足现代化钢铁冶炼的需求。为了提高经济效益，在过去的十年里这些选厂进行了多次技术改造，主要采用的选矿技术是 SLon 立环脉动高梯度磁选机—反浮选工艺流程。通过这些技术的运用，最终得到的铁精粉品位高达 67.50%，铁回收率达到了 78% 以上，创造了我国红矿选矿技术的最高水平。如今，SLon 磁选机已广泛应用于鞍山以及全国其他地区的弱磁性氧化铁矿的选厂。

2　齐大山铁矿选厂的工业改造

齐大山选厂建于 1969 年，在 2001 年之前，选厂的选矿流程如图 1 所示。原矿经过破碎和筛分后分为 0~20mm 和 20~75mm 两种粒级的产品。0~20mm 的粒级部分经过磨矿后，通过螺旋溜槽—湿法强磁选—反浮选工艺；20~75mm 的粒级部分经过磁化焙烧，将赤铁矿转化为磁铁矿后，再通过磨矿—弱磁选工艺。这种选别流程，最终的精矿品位只有 63%，而且焙烧工艺投入大，易造成环境污染，生产条件也十分恶劣。

2001~2005 年，选厂为了适应现代生产的需要，将原生产流程替换为阶段磨矿—螺旋溜槽—SLon 磁选机—反浮选流程，见图 2，主要改变如下：

（1）用螺旋溜槽—SLon 磁选机—反浮选流程取代原来的焙烧—弱磁选工艺。

（2）用 SLon-1750 立环脉动高梯度磁选机（背景场强 1.0T）替换原先的 ϕ2000 平环湿式高梯度磁选机。

---

❶ 原载于《现代矿业》，2009（09）：55~58。

（3）用SLon-1500立环脉动高梯度中磁机（0.4T）取代原来的筒式中磁机（0.4T）。

按照新的生产流程，原矿的处理能力从原来的750万吨/a提高到960万吨/a，铁精矿品位从原来的63.22%提高到67.5%，铁回收率也由原来的74.04%提高到了77.28%，新老流程生产指标对比见表1。

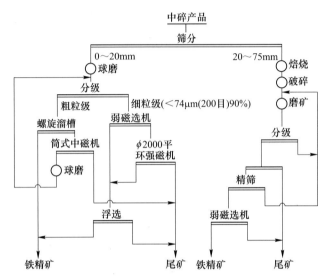

图1 齐大山选厂老流程图

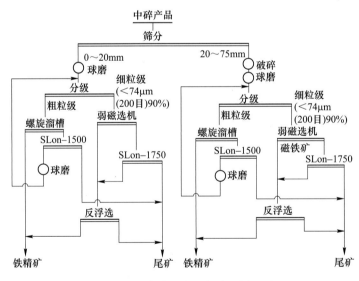

图2 齐大山选厂新流程图

表1 新老流程生产指标对比 （%）

对比项目	给矿铁品位	精矿铁品位	尾矿铁品位	精矿产率	铁回收率
新流程	29.50	67.50	10.12	33.77	77.28
老流程	29.86	63.22	11.92	34.97	74.04
差值	-0.36	+4.28	-1.8	-1.2	+3.24

3 调军台铁矿选厂的工业改造

调军台铁矿选厂建于 1998 年，其原流程为：两段连续磨矿—弱磁选—中磁选—强磁选—浮选，如图 3 所示。新老流程生产指标对比见表 2。

表 2　新老流程生产指标对比　　　　　　　　　　　　　　　　(%)

对比项目	给矿铁品位	精矿铁品位	尾矿铁品位	精矿产率	铁回收率
最新流程	29.50	67.50	10.50	33.33	76.27
第一次改造	30.50	67.50	9.98	35.67	78.94
老流程	30.50	67.13	10.94	34.82	76.64

该流程采用了 15 台 ϕ3200 型平环强磁选机，由于这种磁选机的磁介质是由不锈钢制成的齿板聚磁介质，非常容易堵塞，使得 15 台 ϕ3200 平环强磁选机很难维护。因此，在 2003~2005 年，选厂用 15 台 SLon-2000 立环脉动高梯度磁选机取代了原先的 15 台 ϕ3200 平环强磁选机，如图 4 所示。由于 SLon 型磁选机操作简单，便于维护，作业率高达 99%，并且该机型的综合性能明显优于 ϕ3200 平环强磁选机，使得调军台铁矿选厂最终精矿品位由原来的 67.13% 提高至 67.50%，铁的综合回收率也由原来的 76.64% 提高至 78.94%。

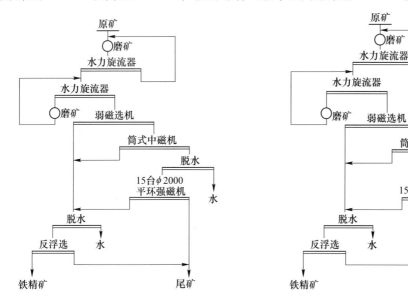

图 3　调军台选厂老流程　　　　图 4　调军台选厂第一次技改流程图

为了进一步扩大生产规模，降低选矿成本，从 2005~2007 年，该选厂又改造成阶段磨矿—弱磁选—SLon 强磁选—反浮选工艺，见图 5，选矿流程得到了进一步优化。新流程增加了 15 台 SLon-2000 立环脉动高梯度中磁机（0.4T）用于粗粒的抛尾作业中：一段磨矿后，螺旋溜槽拿出了部分铁精矿（作为最终精矿），SLon 磁选机又抛弃了部分粗粒尾矿，大大减少了二段球磨的入磨量，磨矿成本也大幅度下降，生产能力由原来的 900 万吨/a 提高至 1440 万吨/a，每吨精矿的选矿成本下降了近 30 元。随着开采能力的加大和开采年数的增加，虽然原矿的平均品位已由原来的 30.50% 下降至 29.50%，但铁精矿铁品位一直保持在 67.50% 左右。

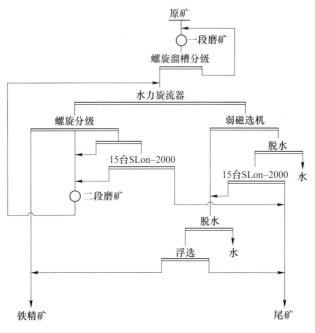

图 5 调军台选厂最近改造的选矿流程图

4 东鞍山铁矿选厂的工业改造

东鞍山铁矿选厂建于 1958 年，该铁矿是鞍山地区氧化程度最深，嵌布粒度最细的铁矿。早期的选矿包括两段连续磨矿和浮选流程，如图 6 所示，最终铁精矿平均品位仅为 61.30%，而且浮选的生产成本较高。

为了提高精矿品位和降低选矿成本，选厂将原流程改为阶段磨矿—螺旋溜槽—SLon 磁选机—反浮选流程，如图 7 所示。选厂共有 31 台 SLon-2000 和 SLon-1750 磁选机用来抛

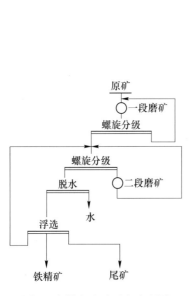

图 6 东鞍山选矿厂老流程图

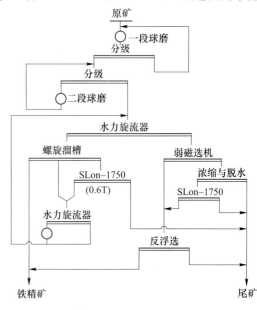

图 7 东鞍山选矿厂新流程图

尾。铁精矿品位由原来的61.30%提高至64.50%,生产能力也由原来的400万吨/a提高至600万吨/a,选矿成本也大幅度下降。新老流程生产指标对比见表3。

表3 新老流程生产指标对比 (%)

对比项目	给矿铁品位	精矿铁品位	尾矿铁品位	精矿产率	铁回收率
新流程	32.50	64.50	13.45	37.32	74.06
老流程	32.75	61.30	14.70	38.73	72.50
差值	-0.25	+3.20	-1.25	-1.41	+1.56

5 弓长岭铁矿选厂的工业改造

弓长岭铁矿选厂建于1975年,年处理氧化铁矿石300万吨,原先采用两段磨矿—离心选矿机选矿流程如图8所示,原矿铁平均品位为31.80%,铁精矿品位为60.50%,铁回收率为66.35%。

由于1970~1985年强磁选技术的快速发展,原选矿工艺逐步转化为阶段磨矿—强磁选—螺旋溜槽—离心选矿机选矿流程,如图9所示,在这个流程中,φ2000平环强磁选机用来抛尾,螺旋溜槽用于一段磨矿后分选出部分产品作为最终的精矿,则进入二段磨矿的矿量就减少了一半。即使当原矿的铁品位由原先的31.80%降至29.20%,铁精矿品位仍能由原来的60.50%提高至63.90%,铁回收率也由原来的66.35%提高至67.51%。

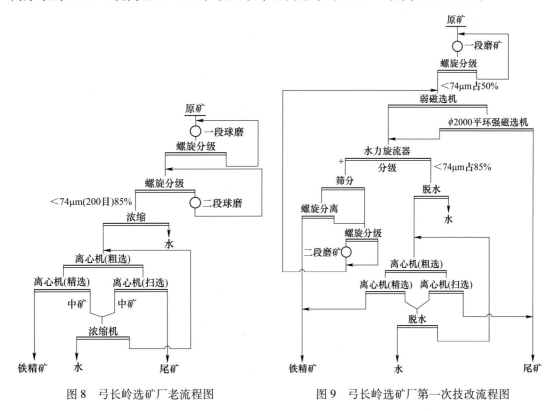

图8 弓长岭选矿厂老流程图　　图9 弓长岭选矿厂第一次技改流程图

1990~2008年,SLon磁选机和反浮选技术发展迅猛,这些技术的优越性逐步得到了业内人士的好评和肯定，具有很好的发展前景。为了进一步提高选厂效益，2005 年选厂第三选矿车间经重建后，又将原先的流程改为阶段磨矿—螺旋溜槽—SLon 磁选机—反浮选流程，如图 10 所示。采用 5 台 SLon-2000 磁选机（1.0T）控制细粒级的尾矿品位，5 台 SLon-2000 中磁机（0.6T）控制粗粒级尾矿品位，最终排出的尾矿品位要比先前用 ϕ2000 平环强磁选机时低得多。新流程比先前的老流程更具优越性，具体体现在：当选厂的入料铁品位降至 28.78% 时，精矿铁品位仍能提高至 67.19%，铁综合回收率达到 76.29%。新老流程生产指标对比见表 4。

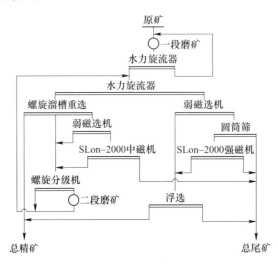

图 10　弓长岭选矿厂三选车间生产数质量流程图

表 4　新老流程生产指标对比　（%）

对比项目	给矿铁品位	精矿铁品位	尾矿铁品位	精矿产率	铁回收率
最新流程	28.78	67.19	10.13	32.68	76.29
第一次改造	29.20	63.90	13.72	30.85	67.51
老流程	31.80	60.50	16.43	34.88	66.35

6　结语

（1）鞍山地区有我国最大的低品位氧化铁矿矿床，在过去的 50 年里，广大科研工作者对该类矿石的选别做过大量的研究工作。本文以该地区的 4 个选厂为例，介绍了一项新技术的应用，尽管各个选厂最初的选矿流程都各不相同，但经过多次技术改造后，最终采取了阶段磨矿—螺旋溜槽—SLon 磁选机—反浮选流程，这种流程的优势在于能以较低的生产成本获得高品位、高回收率的铁精矿。

（2）阶段磨矿的优势在于螺旋溜槽能在二段磨矿前分选出部分粗粒级的铁精矿，SLon 磁选机又能甩弃部分粗粒级的尾矿，这样就能减少二段磨矿的入磨量，大幅度降低了磨矿成本。

（3）SLon 磁选机广泛用于抛尾，它能将尾矿的品位控制在较低的水平，因此能提高产品的回收率。还有一个重要因素就是 SLon 磁选机能脱去很大一部分细泥，为后续的反浮选作业创造更好的作业条件，这一点非常重要，因为细泥一旦进入反浮选作业会干扰浮选作业，额外消耗大量的浮选药剂。

（4）反浮选广泛用于提高细粒级的精矿品位。由于反浮选要消耗浮选药剂，相比于磁选和重选来说，作业成本要高得多，因此，为了尽可能降低生产成本，反浮选的给矿量应尽可能减少。

写作背景 本文向国内同行介绍了鞍钢齐大山选矿厂、调军台选矿厂、东鞍山烧结厂和弓长岭选矿厂氧化铁矿选矿流程的演变，SLon立环脉动高梯度磁选机在推动这些选矿流程一步一步走向现代化所起的作用。新技术的应用使这些选矿厂的技术经济指标达到国际领先水平。

SLon 磁选机与离心机组合技术分选氧化铁矿[❶]

熊大和[1,2]

（1. 赣州有色冶金研究所；2. 赣州金环磁选设备有限公司）

摘 要 SLon 立环脉动高梯度磁选机具有优异的选矿性能，已广泛应用于分选氧化铁矿，SLon 离心选矿机用于强磁精矿的精选作业，可有效地剔除含少量磁铁矿的石英等脉石，这两种设备的优势互补，用它们的组合流程分选某些氧化铁矿可获得良好的选矿指标。

关键词 SLon 立环脉动高梯度磁选机　SLon 离心选矿机　氧化铁矿　选矿

Cooperational Technology of SLon Magnetic Separator and Centrifugal Separator in Processing Oxidized Iron Ores

Xiong Dahe[1,2]

(1. Ganzhou Nonferrous Metallurgy Research Institute;
2. SLon Magnetic Separator Ltd.)

Abstract SLon vertical ring and pulsating high gradient magnetic separators possess excellent mineral processing ability. They have been widely applied to process oxidized iron ores. SLon centrifugal separators are applied to clean the SLon magnetic concentrate. They can remove quartz and other gangue minerals which associated with a small portion of magnetite. Applying the cooperational advantages of the two kinds of equipments to process some oxidized iron ores, good benficial results can be achieved.

Keywords SLon vertical ring and pulsating high gradient magnetic separator; SLon centrifugal separator; Oxidized iron ore; Process

1 问题提出

SLon 立环脉动高梯度磁选机广泛应用于分选弱磁性铁矿。在分选鞍山式氧化铁矿的生产流程中，主要是用于粗选作业，精选作业一般是用螺旋溜槽分选粗粒级和用反浮选分选细粒级。为什么分选鞍山式铁矿的流程中强磁选很少用于精选作业呢？原因是鞍山式氧化铁矿是由磁铁矿、假象赤铁矿、赤铁矿组成，由表 1 可知，磁铁矿的比磁化率是赤铁矿和镜铁矿的 100 倍以上。几种铁矿石的比磁化率见表 1。

---

❶ 原载于《2009 年金属矿产资源高选冶加工利用和节能减排技术及设备学术研讨与技术成果交流设备展示会论文集》。

表1 几种铁矿石的比磁化率 (m³/kg)

矿石名称	磁铁矿	假象赤铁矿	赤铁矿	镜铁矿
比磁化率 λ	$(625\sim1160)\times10^{-6}$	$(6.2\sim13.5)\times10^{-6}$	$(0.6\sim2.16)\times10^{-6}$	3.7×10^{-6}

如图1所示，鞍山式氧化铁矿中含有较多的磁铁矿和石英的连生体，这些连生体在磁场中受到的磁力很大。表2为磁铁矿与石英连生体的视在比磁化率。例如，如果一颗连生体含1%质量的磁铁矿和99%质量的石英，它的视在比磁化率达到了$6.25\times10^{-6}\mathrm{m^3/kg}$，已远远大于赤铁矿单体的比磁化率，它在磁场中所受到的磁力就比相同质量的赤铁矿要大。它很容易被强磁机捕捉到铁精矿中去，从而降低铁精矿品位。

图1 磁铁矿和石英的连生体

表2 磁铁矿和石英连生体的视在比磁化率

连生体中磁铁矿质量/%	连生体中石英的质量/%	连生体的视在比磁化率/m³·kg⁻¹
0	100	0
1	99	6.25×10^{-6}
5	95	31.25×10^{-6}
10	90	62.5×10^{-6}
20	80	125.0×10^{-6}
50	50	312.5×10^{-6}
100	0	625.0×10^{-6}

离心选矿机是一种较好的细粒矿物重选设备，它可提供$20g\sim50g$（g为重力加速度）的离心力，能将细粒矿物按密度分选。石英的密度是$2.6\mathrm{g/cm^3}$，磁铁矿和赤铁矿的密度为$5.0\mathrm{g/cm^3}$，表3为磁铁矿和石英的连生体的视在密度。

表3 磁铁矿和石英连生体的视在密度

连生体中磁铁矿质量/%	连生体中石英的质量/%	连生体的视在密度/g·cm⁻³
0	100	2.60
10	90	2.84
20	80	3.08
30	70	3.32
40	60	3.56
50	50	3.80
60	40	4.04
70	30	4.28
80	20	4.52
90	10	4.76
100	0	5.00

根据生产经验，含30%左右的石英与磁铁矿的连生体（视在密度为$4.28\mathrm{g/cm^3}$）能被离心选矿机排入尾矿中去。因此离心选矿机的精选能力要高于强磁选机的精选能力。SLon立环脉动高梯度磁选机处理量大，用于粗选作业具有富集比大，选矿效率高的优点，而用于精选作业则存在含少量磁铁矿和大部分石英的贫连生体难以剔除的制约因素，而离心选

矿机用于精选作业则可较好地解决这个问题,这两种设备相结合分选某些氧化铁可获得较好的选矿指标。

2 在低品位镜铁矿中的应用

图 2 所示为一座日处理 500t 低品位镜铁矿原矿的选厂生产流程。其原矿品位为 22.63%TFe,铁矿物主要是镜铁矿和少量的磁铁矿,磁铁矿产率占原矿的 1% 左右,因此 SLon 强磁机前面不需用弱磁选机。

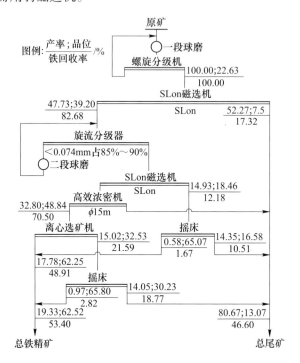

图 2 某低品位镜铁矿选矿流程

采用阶段磨矿—SLon 强磁抛尾—离心机精选—摇床扫选流程。一段磨矿将矿石磨至小于 0.074mm 占 50% 左右,用 SLon 磁选机分选抛去产率为 52.27% 和品位为 7.50%TFe 的低品位尾矿,一次强磁选精矿用水力旋流器分级,水力旋流器沉砂进二段磨矿,二段磨矿的排矿返回到水力旋流器,水力旋流器的溢流粒度为小于 0.074mm 占 85%~95%。水力旋流器溢流进入二次 SLon 高梯度强磁选机分选,二次高梯度强磁精矿浓缩后用离心选矿机精选,二次强磁选尾矿和离心选矿机尾矿用摇床扫选。获得最终综合精矿品位为 62.52%,铁回收率 53.40%。

该流程的特点:

(1) 一次 SLon 高梯度强磁选可抛弃大量的低品位尾矿,较大幅度地节约了二段磨矿的生产成本。

(2) 离心选矿机用于二次强磁选精矿的精选作业,可有效地剔除含磁铁矿的石英连生体,其精矿品位提高幅度较大,从 48.64%TFe 提高到 62.25%TFe。

(3) 采用摇床对二次强磁选尾矿和离心选矿机尾矿扫选,可直接拿出一部分较高品位的铁精矿。

（4）整个流程为开路分选流程，选矿作业不存在循环负荷，生产上很好控制。

上述生产流程具有流程较简单，选矿指标较好，生产成本较低的优点。其缺点是离心选矿机和摇床的台时处理能力较低，目前只适用于中、小型选矿厂，还难以用于大规模的工业生产。

3 海南难选氧化铁矿的试验研究

海南钢铁公司拥有一座储量较大的氧化铁矿，过去长期开采品位50.00%左右的富矿，有一部分铁品位40.00%左右的难选氧化铁暂未得到利用。近年有关单位对这部分氧化铁进行了系统的选矿试验研究。实践证明，若仅用强磁选流程只能得到品位为60.00%左右的铁精矿。而市场上品位为63.00%以上的铁精矿好销且价格较高。因此，选矿试验和生产实践都要求铁精矿品位达到63.00%以上。

经过多次的探索试验和扩大试验，获得图3所示的SLon磁选机与SLon离心选矿机的

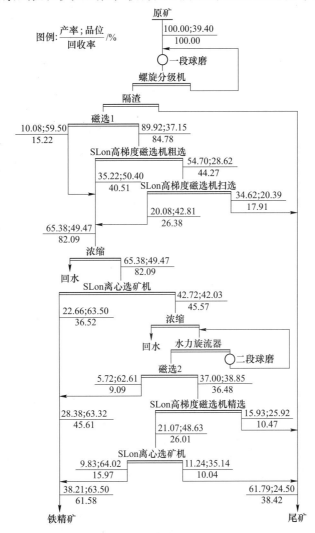

图 3　海南难选氧化铁矿 SLon 磁选机与 SLon 离心选矿机的组合分选流程

组合流程。其入选原矿品位为39.40%，含铁矿物主要是赤铁矿及占原矿产率10%左右的磁铁矿。首先磨矿至小于0.074mm占87%~90%，然后用弱磁选机选出磁铁矿，弱磁选机的尾矿用SLon磁选机一次粗选和一次扫选，磁选精矿合并，浓缩后用离心选矿机精选拿出部分品位为63.00%以上的铁精矿，离心机尾矿经浓缩后用水力旋流器分级，水力旋流器沉砂进入二段球磨机。水力旋流器溢流用弱磁选机选出磁铁矿精矿，弱磁选尾矿用SLon磁选机扫选。SLon磁选机扫选精矿再用离心选矿机精选得出部分品位为64.00%左右的铁精矿。该流程的综合铁精矿品位63.50%，铁回收率61.58%。目前该流程正在向工业生产过渡。

4 鞍山式氧化铁矿的试验研究

目前我国鞍山式氧化铁矿大多数都采用阶段磨矿—分级重选—强磁选—反浮选流程。该流程具有节能、生产成本较低及选矿指标较好的优点。但是，对于一些小型选矿厂来说，反浮选作业存在技术复杂、环保审批时间长等问题，因此，采用离心选矿机代替反浮选作业在一些小型选厂得到应用。

图4所示为鞍山式氧化铁矿的试验流程，该矿石以赤铁矿为主，含有一部分磁铁矿和假象赤铁矿，脉石矿物以石英为主。选矿流程为一段磨矿后用水力旋流器分级，水力旋

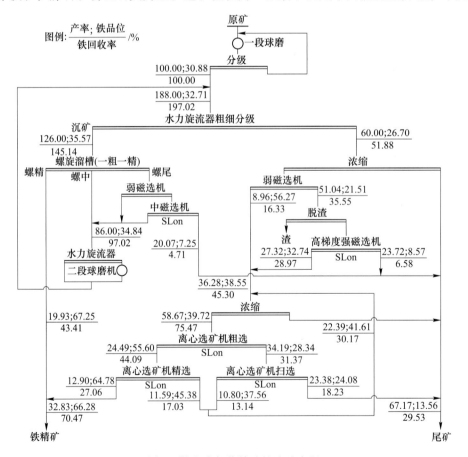

图4 鞍山式氧化铁矿的选矿流程

流器沉砂用螺旋溜槽选出一部分粒度较粗已经单体解离的铁精矿,螺旋溜槽尾矿用 SLon 立环脉动中磁机分选,抛弃产率 20%左右及铁品位 7.25%的粗粒级尾矿。水力旋流器溢流经浓缩后用弱磁选机选出磁铁矿,弱磁选机尾矿用 SLon 立环脉动强磁机分选,该机抛弃产率为 23.72%,品位为 8.57%的细粒尾矿。弱磁选精矿和 SLon 强磁精矿合并,浓缩后用 SLon 离心机进行一次粗选、一次精选、一次扫选,离心选矿机取得大部分的细粒级铁精矿。该流程综合选矿指标为原矿品位 30.88%TFe,综合铁精矿品位为 66.28%,综合铁回收率 70.47%,综合尾矿品位 13.58%。该流程全部采用磁选和重选作业,具有节能、环保、生产成本较低的优点。其缺点是离心机的单机处理能力较小,离心机的作业回收率不如反浮选作业。目前在大规模生产中还难以推广应用。

5 从尾矿中回收铁精矿的应用

我国分选磁铁矿的小厂很多,有的小厂的入选原矿中含有一部分赤铁矿或镜铁矿,这些小选厂用弱磁选机选完磁铁矿后,弱磁选尾矿直接排入尾矿库。采用 SLon 磁选机和 SLon 离心选矿机的组合流程对这种尾矿进行再回收,往往可以获得较好的技术经济指标。图 5 所示为 SLon 磁选机与 SLon 离心机的组合流程分选某尾矿库的堆存尾矿的试验指标。该尾矿中含铁矿物主要是镜铁矿和少量的磁铁矿。该流程的特点:先搅拌,分级磨矿至小于 0.074mm 占 95%,利用弱磁选机分选出少量的磁铁矿,然后利用 SLon 立环脉动高

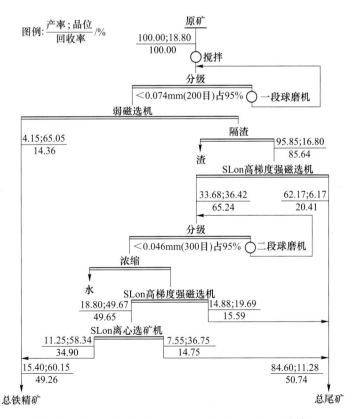

图 5 SLon 磁选机与离心机组合流程从尾矿中回收铁精矿

梯度磁选机处理量大，作业成本低的特点，一次粗选抛弃产率62.17%，品位6.17%TFe的低品位尾矿，SLon磁选机的粗选精矿品位已达到36.42%TFe，而产率只占原矿的33.68%，这部分粗精矿进入二段磨矿分级，二次分级溢流粒度为小于0.052mm占95%，浓缩后再用SLon磁选机精选一次，其精矿用离心选矿机精选。该流程的综合选矿指标：给矿品位18.80%TFe，铁精矿品位为60.15%TFe，铁精矿产率15.40%，铁回收率49.26%，综合尾矿品位为11.28%。该流程目前已在生产中应用，实现了低成本从尾矿中回收铁精矿的目的，使二次资源得到利用。

6 结论

（1）SLon立环脉动高梯度磁选机处理量大，用于低品位氧化铁矿的粗选作业具有富集比大、选矿效率高、生产成本低的优点。

（2）强磁粗选精矿中，若含有磁铁矿和石英的贫连生体，其视在比磁化率远高于赤铁矿的比磁化率，则再利用强磁精选作用不大，例如鞍山式的氧化铁矿是由磁铁矿、假象赤铁矿和赤铁矿组成，这种矿石用强磁粗选后，后续作业一般不再用强磁精选。

（3）SLon离心选矿机是利用矿物密度差异进行分选的，可产生$20g \sim 50g$（g为重力加速度）的离心力，可有效地回收微细粒铁精矿，并可有效地剔除含少量磁铁矿的贫连生体脉石，强磁选精矿用离心机再进行精选往往可获得较高的铁精矿品位。

（4）SLon立环脉动高梯度磁选机和SLon离心选矿机的组合流程具有生产成本低、环境友好、易于操作管理的特点。这种流程已在一部分中小型氧化铁矿选厂得到应用。但是由于离心选矿机的单机处理能力较小，这种组合流程目前还难以在大规模的工业生产中应用。

写作背景 本文首次提出了SLon磁选机与离心机组合技术分选氧化铁矿的原理，介绍了运用该技术分选低品位镜铁矿、海南难选氧化铁矿、鞍山式氧化铁矿、从废弃的老尾矿中回收赤铁矿的试验和生产情况。

SLon 立环脉动高梯度磁选机新技术

熊大和[1,2]

(1. 赣州有色冶金研究所；2. 赣州金环磁选设备有限公司)

摘 要 脉动高梯度磁选具有对细粒弱磁性矿物回收率较高和磁性物精矿品位较高的优点，通过25年持续的创新与发展，研制了SLon立环脉动高梯度磁选机，本文介绍该机的工作原理和最新的工业应用。

关键词 SLon立环脉动高梯度磁选机　铁矿　钛矿　锰矿　长石矿

1 前言

我国拥有丰富的氧化铁矿、钛铁矿、锰矿、黑钨矿等弱磁性矿石资源，但多数矿床的原矿品位低、嵌布粒度细。细粒选矿的回收率较低和磁性精矿品位较低是一个普遍存在的问题，大量的有用矿物流失到尾矿中，每年对我国矿产资源造成巨大浪费。

本研究通过大量试验研究证明：脉动高梯度磁选可显著提高磁性精矿的品位，并保持对细粒磁性矿物回收率高的优点。此外，脉动还具有防止磁介质堵塞的作用。SLon-100周期式脉动高梯度磁选机照片如图1所示。

为了将理论和试验成果转化为生产力，我们用了25年研制成功了同时利用脉动流体力、重力和高梯度磁场力的综合力场选矿的SLon立环脉动高梯度磁选机。其特点是转环立式旋转，反冲精矿并配有矿浆脉动机构，从根本上解决了平环强磁选机磁介质容易堵塞这一世界性技术难题。SLon立环脉动高梯度磁选机照片如图2所示。

 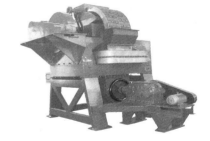

图1　SLon-100周期式脉动高梯度磁选机照片　　图2　SLon立环脉动高梯度磁选机照片

2 SLon磁选机设备结构和工作原理

激磁线圈通以直流电，在分选区产生感应磁场，转环作顺时针旋转，将磁介质不断送入和运出分选区；矿浆从给矿斗给入，沿上铁轭缝隙流经转环。矿浆中的磁性颗粒吸附在磁介质棒表面上，被转环带至顶部无磁场区，被冲洗水冲入精矿斗，非磁性颗粒在重力、

---

❶ 原载于《中国有色金属矿业—2009年昆明学术年会论文集》。

脉动流体力的作用下穿过磁介质堆,与磁性颗粒分离。然后流入尾矿斗排走。SLon 立环脉动高梯度磁选机结构图如图 3 所示。

该机的转环采用立式旋转方式,对于每一组磁介质而言,冲洗磁性精矿的方向与给矿方向相反(图 4),粗颗粒不必穿过磁介质堆便可冲洗出来。该机的脉动机构驱动矿浆产生脉动,可使位于分选区磁介质堆中的矿粒群保持松散状态,使磁性矿粒更容易被捕获,使非磁性矿粒尽快穿过磁介质堆进入到尾矿中去。

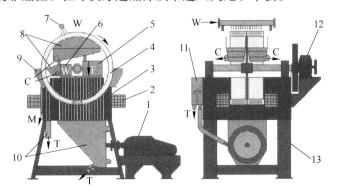
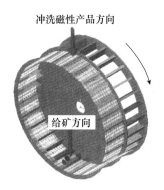

图 3 SLon 立环脉动高梯度磁选机结构图
1—脉动机构;2—激磁线圈;3—铁轭;4—转环;5—给矿斗;
6—漂洗水斗;7—精矿冲洗装置;8—精矿斗;9—中矿斗;
10—尾矿斗;11—液位斗;12—转环驱动机构;13—机架;
F—给矿;W—清水;C—精矿;M—中矿;T—尾矿

图 4 转环冲洗磁性产品方向和给矿方向示意

3 SLon 脉动高梯度磁选机的磁选原理

一个脉动周期内矿浆实际流速如图 5 所示,图中水平轴下方的阴影部分表示矿浆的实际流速与给矿方向相反。

当 $v_{max} > v_0$ 时,分选区的矿浆不断变换流速方向,可有效降低介质中非磁性矿粒的机械夹杂,反向脉动力的存在能产生一个松散力,使非磁性矿有更多的机会进入尾矿,从而提高磁性精矿的品位,避免堵塞。脉动的松散原理如图 6 所示。

图 7(a)和(b)所示分别为无脉动和有脉动时矿粒穿过磁介质堆的运动轨迹。

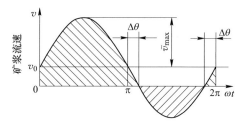

图 5 矿浆在选矿区的流速

SLon 立环脉动高梯度磁选机处理量如下:

型号	处理量/$t \cdot h^{-1}$
SLon-3000	150~250
SLon-2500	80~150
SLon-2000	50~80
SLon-1750	30~50
SLon-1500	20~30
SLon-1250	10~18

SLon-1000   4~7
SLon-750   0.06~0.25
SLon-500   0.03~0.125

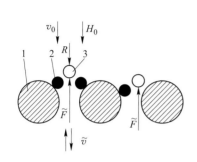

图 6　脉动的松散原理
1—磁介质；2—磁性矿粒；3—非磁性矿粒

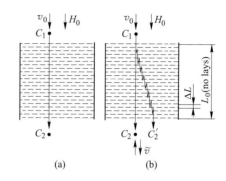

图 7　无脉动和有脉动时矿粒穿过磁介质的轨迹
（a）无脉动；（b）有脉动

4　SLon 磁选机在铁矿的应用

4.1　SLon 磁选机在鞍钢调军台选矿厂的应用

鞍钢调军台选矿厂原年处理鞍山式氧化铁矿 900 万吨，为一具有现代化水平的大型选矿厂，采用连续磨矿—弱磁—中磁—强磁—反浮选的选矿流程。2003~2005 年，选厂用 15 台 SLon-2000 立环脉动高梯度磁选机取代了原先的 15 台 ϕ3200 平环强磁选机。新流程中 SLon 磁选机运转平稳、选矿指标好、作业率高达 99%，使得调军台铁矿选厂最终精矿品位由原来的 67.13% 提高至 67.50%、铁的综合回收率也由原来的 76.64% 提高至 78.94%。

2005~2007 年，该选厂又改造成阶段磨矿—弱磁选—SLon 强磁选—反浮选工艺，选矿流程得到了进一步优化。新增流程增加了 15 台 SLon-2000 立环脉动高梯度中磁机（0.4T）用于粗粒的抛尾作业中。调军台选矿厂生产数质量流程如图 8 所示。

在鞍钢调军台的应用照片如图 9 所示。

生产能力由原来的 900 万吨/a 提高至 1440 万吨/a，每吨精矿的选矿成本下降了近 30 元。随着开采能力的加大和开采年数的增加，虽然原矿的平均品位已由原来的 30.50% 下降至 29.50%，但铁精矿品位一直保持在 67.50% 左右。

4.2　SLon 磁选机在海南铁矿的应用

海南铁矿的铁矿石主要为赤铁矿和部分磁铁矿。2003~2005 年，海南矿业联合有限公司采用 5 台 SLon-2000 立环脉动高梯度磁选机处理铁矿石。

给矿磨至小于 0.074mm（200 目）占 95% 的粒度，经弱磁选机分选出磁铁矿，再经由 SLon-2000 磁选机组成的一粗一精一扫的工艺选别赤铁矿。该流程指标为：给矿品位 51.09%Fe，铁精矿品位 64.57%Fe，铁回收率 88.01%。SLon 立环脉动高梯度磁选机在海南铁矿应用的照片见图 10，海钢铁矿数质量流程如图 11 所示。

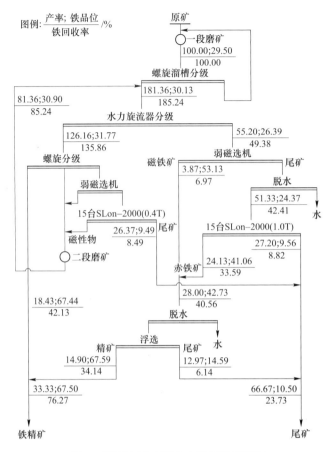

图 8 调军台选矿厂生产数质量流程图

图 9 SLon-2000 立环脉动高梯度磁选机在鞍钢调军台的应用照片

图 10 SLon 立环脉动高梯度磁选机在海南铁矿的应用照片

4.3 SLon 磁选机在秘鲁铁矿的应用

首钢秘鲁铁矿的矿石以磁铁矿为主，岩体表层多为氧化型赤铁矿。2003 年底，秘鲁铁矿采用如图 12 所示流程处理该氧化铁矿。

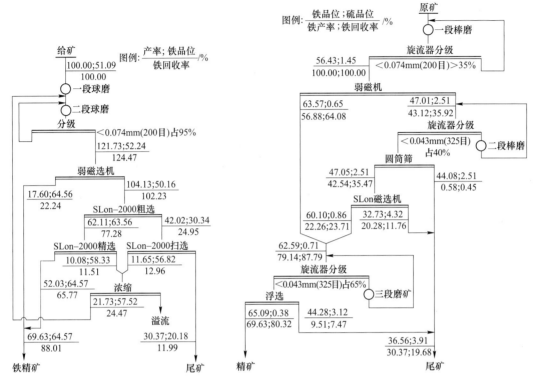

图 11　海钢铁矿数质量流程

图 12　秘鲁铁矿氧化矿数质量流程

在秘鲁铁矿应用的现场照片如图 13 所示。

图 13　SLon 磁选机在秘鲁铁矿现场应用照片

4.4　在昆钢大红山铁矿的应用

昆钢大红山铁矿是磁铁矿和赤铁矿共生的混合矿，目前采用阶段磨矿—弱磁—SLon 强磁选流程（如图 14 所示），获得了给矿品位 40.03%，铁精矿品位 62.97%，铁回收率 87.03% 的良好选矿指标。其中 SLon 磁选机对控制全流程的尾矿品位、对提高铁回收率取得了关键的作用。SLon 磁选机在昆钢大红山的现场应用照片如图 15 所示。

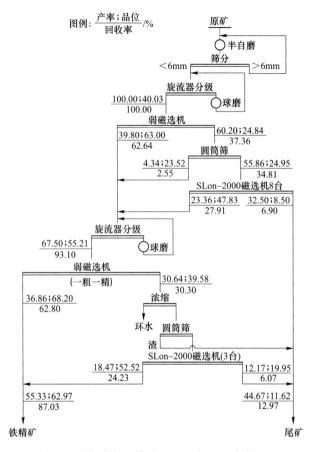

图 14　昆钢大红山铁矿 400 万吨选厂原则流程

图 15　SLon 磁选机在昆钢大红山的现场应用照片

5　SLon 磁选机在钛铁矿的应用

5.1　SLon 磁选机在四川攀钢选钛厂的应用

14 台 SLon-1500 立环脉动高梯度磁选机和 2 台 SLon-1750 中磁机在攀枝花选钛厂用于小于 0.045mm 微细粒级钛铁矿的磁浮选矿流程中（图 16），该机有效地回收了这部分微细粒级钛铁矿，为浮选作业创造了良好的条件。

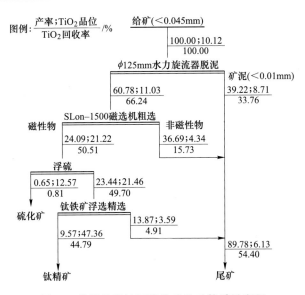

图 16 攀枝花选钛厂钛铁矿选矿数质量流程

全流程在给矿品位 10% 左右的条件下，实现钛精矿品位大于 47%TiO_2，TiO_2 回收率 44% 的良好指标。

5.2 在西昌太和铁矿的应用

SLon-1500、SLon-1750、SLon-2000 立环脉动高梯度磁选机共 11 台在重钢太和铁矿回收钛铁矿，采用分级磁选、粗粒再磨、磁选和浮选精选的流程（见图 17），该流程的选钛指标为：给矿品位 12.06%TiO_2，钛精矿品位 47.85%TiO_2，TiO_2 回收率 47.44%，实现年生产 12 万吨左右优质钛精矿的能力。

6 SLon 磁选机在锰矿的应用

6.1 在南京栖霞山铅锌矿从尾矿中回收锰矿

南京栖霞山铅锌矿是一家以开采铅锌矿为主的老矿，其铅锌选矿的尾矿量为 15 万吨/a，其中含锰 12% 左右，锰金属量 1.8 万吨/a。若能将锰精矿品位选至 22%Mn 以上，则可作为制造硫酸锰的原料。2005 年由马鞍山矿山研究院试验、设计，采用 2 台 SLon-1500 和 1 台 SLon-1750 立环脉动高梯度磁选机从该尾矿中回收锰精矿。南京栖霞山锰矿磁选流程如图 18 所示。

SLon 磁选机给矿（即浮铅、锌的尾矿）12%Mn，采用 SLon 磁选机一粗一精流程，磁选精矿 Mn 品位 24%，锰精矿含铁小于 6%，锰回收率 50%，锰精矿产率 25%，锰精矿产量为 3.75 万吨/a。

6.2 内蒙古金水矿业选锰尾矿中回收细粒锰矿

内蒙古金水矿业最初锰矿生产采用水洗方法选出粗粒锰矿出售，洗矿溢流作为尾矿堆放。近几年该公司与我们合作，采用 SLon 立环脉动高梯度磁选机回收尾矿中的细粒锰矿，

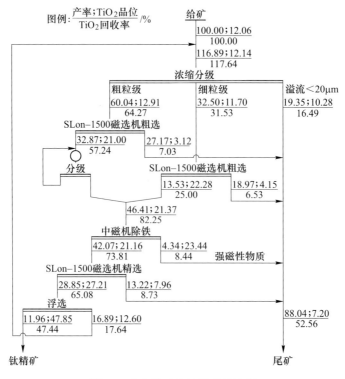

图 17 太和铁矿选钛数质量流程

获得了良好的选矿指标。目前该公司生产上采用 SLon-1000、SLon-1250 和 SLon-1500 磁选机各 1 台回收细粒锰矿,其生产流程和选矿指标如图 19 所示。该流程给矿品位仅 8.85% Mn,经 SLon 磁选机一次选别,锰精矿品位达到 30.39%,锰回收率 75.30%,获得了良好的选矿指标。

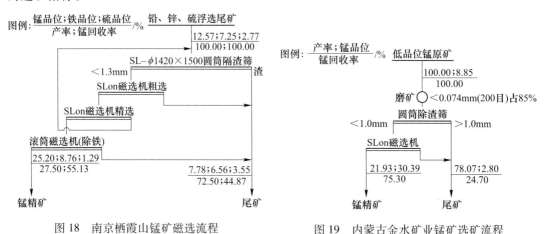

图 18 南京栖霞山锰矿磁选流程   图 19 内蒙古金水矿业锰矿选矿流程

7 SLon 磁选机在开发长石资源中的应用

我国拥有丰富的长石资源,矿石主要由钾长石、斜长石、少量石英、黑云母、角闪石和磁铁矿组成。矿石中 Fe_2O_3 含量普遍偏高。采用 SLon 立环脉动高梯度磁选机除铁工艺，可将矿石中大部分 Fe_2O_3 除去，加工后的长石粉可作为玻璃、陶瓷工业的优质原料。目前已有十几座长石选矿厂采用 SLon 磁选机除铁，其典型的选矿流程如图 20 所示。

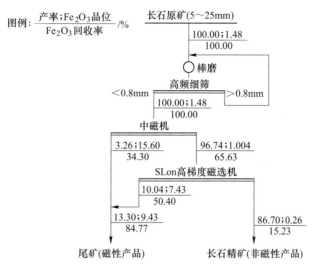

图 20　长石选矿流程

含 Fe_2O_3 1.48% 的原矿经棒磨和高频细筛组成的闭路磨矿筛分至小于 0.8mm，进入滚筒式中磁机除去强磁性铁物质，然后进入 SLon 磁选机除去弱磁性铁矿物。磁选作业的非磁性产品含铁 0.26%，满足玻璃工业和陶瓷工业 Fe_2O_3 含量小于 0.3% 的要求。

8　SLon 磁选机在四川南江霞石矿除铁中的应用

四川南江霞石矿是我国第一个具有较高工业开采价值的霞石矿床。矿石中钠、钾含量较高，具有熔点低和助熔性能强等优点，故在建材、化工等工业中有着广泛的用途。但该霞石矿铁含量高，不能直接开采利用。通过对矿物性质考查及可选性试验研究，确定采用 SLon 立环脉动高梯度磁选机作为除铁主体设备，开展一粗一精全磁流程的提纯工艺，工业试验流程如图 21 所示。

当给矿含 1.69% Fe_2O_3，经一粗一精选别可获得含 0.179% Fe_2O_3 的霞石精矿，回收率 76.36%，除铁率达 91.52%，产品质量达到出口类一级霞石精矿指标。目前该矿采用 SLon 磁选机分选霞石精矿，经济效益可观。

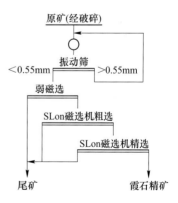

图 21　南江霞石一粗一精流程

9　结论

脉动增加了矿粒与磁介质的碰撞次数，这是对提高磁性矿回收率有益的一面，脉动也增大了作用在磁性矿粒上的竞争力，这是对磁性矿回收率不利的一面，碰撞次数与竞争力的同时增长使磁性精矿的回收率在一定的冲程冲次范围内基本不变或有所增加。

脉动增大了作用在非磁性矿粒上的竞争力，则这有利于消除非磁性矿粒的机械夹杂和因表面力黏附在磁性矿粒上的可能性，因此磁性精矿的品位随冲程冲次的增大而单调上升。

SLon 型立环脉动高梯度磁选机是分选细粒弱磁性矿物的新一代高效强磁选设备，它具有富集比大，磁介质不易堵塞等优点，可广泛应用于赤铁矿、褐铁矿、菱铁矿、锰矿、钛铁矿、黑钨矿等多种弱磁性金属矿和非金属矿如石英、长石、霞石、高岭土等的除铁提纯。

写作背景　本文以较多的图例介绍了 SLon 立环脉动高梯度磁选机的工作原理，在多个厂矿应用于分选赤铁矿、钛铁矿、锰矿、长石和霞石的提纯。

SLon磁选机在云川贵选矿工业中的应用[1]

熊大和[1,2]

(1. 赣州金环磁选设备有限公司; 2. 赣州有色冶金研究所)

摘 要 SLon立环脉动高梯度磁选机是一种新型高效的强磁选设备,在工业上广泛应用于弱磁性矿物,具有富集比大、回收率高、磁介质不堵塞、设备作业率高的优点。本文介绍该机在云南、四川和贵州三省选氧化铁矿、钛铁矿和非金属矿的应用。该机使很多低品位的矿产资源得到利用,生产出大量的优质铁精矿、钛精矿、霞石矿和石英砂,获得了优良的技术经济指标。

关键词 SLon立环脉动高梯度磁选机 氧化铁矿 钛铁矿 霞石 石英

Application of SLon Magnetic Separators in Mineral Processing Industry in Yunnan, Sichuan and Guizhou

Xiong Dahe[1,2]

(1. SLon Magnetic Separator Ltd. ;
2. Ganzhou Nonferrous Metallurgy Research Institute)

Abstract SLon vertical ring pulsating high gradient magnetic separators are a new type of high efficiency high intensity magnetic separator. They are widely applied in industry to beneficiate weakly magnetic minerals. They possess the advantages of large beneficial ratio, high recovery, matrix not easy to be blocked and high availability. This paper introduces their applications in Yunnan Province, Sichuan Province and Guizhou Province for processing oxidized iron ore, ilmenite, nepheline, and quartz. Many low-grade resources are beneficiated and a great quantity of high quality iron concentrate, ilmenite, nepheline, and quartz are produced. Very good technical and economical results are achieved.

Keywords SLon vertical ring pulsating high gradient magnetic separator; Oxidized iron ore; Ilmenite; Nepheline; Quartz

我国云南、四川和贵州拥有丰富的铁矿、钛矿和非金属矿等矿产资源,但是,富矿和易选矿少,贫矿和难选矿多,绝大多数矿产资源都需要经过细磨深选才能得到合格的产品。近年SLon立环脉动高梯度磁选机在云川贵地区的广泛应用,使过去很多不能利用的

---

[1] 原载于《第三届矿业发展高峰论坛会》。

尾矿和低品位的矿石得到回收利用,为这些地区的经济发展作出了一定的贡献。

1 SLon 磁选机简介

SLon 立环脉动高梯度磁选机结构如图 1 所示,它主要由脉动机构、激磁线圈、铁轭、转环和各种矿斗、水斗组成。用导磁不锈钢制成的钢板网或圆棒作磁介质。该机工作原理如下。

激磁线圈通以直流电,在分选区产生感应磁场,位于分选区的磁介质表面产生非均匀磁场即高梯度磁场;转环作顺时针旋转,将磁介质不断送入和运出分选区;矿浆从给矿斗给入,沿上铁轭缝隙流经转环。矿浆中的磁性颗粒吸附在磁介质棒表面上,被转环带至顶部无磁场区,被冲洗水冲入精矿斗,非磁性颗粒在重力、脉动流体力的作用下穿过磁介质堆,沿下铁轭缝隙流入尾矿斗排走,实现与磁性颗粒的分离。

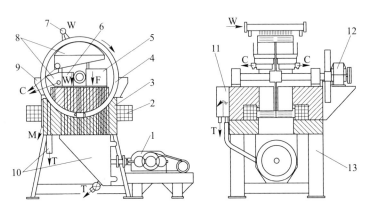

图 1 SLon 立环脉动高梯度磁选机结构

1—脉动机构;2—激磁线圈;3—铁轭;4—转环;5—给矿斗;6—漂洗水斗;
7—精矿冲洗装置;8—精矿斗;9—中矿斗;10—尾矿斗;
11—液位斗;12—转环驱动机构;13—机架;F—给矿;
W—清水;C—精矿;M—中矿;T—尾矿

该机的转环采用立式旋转方式,对于每一组磁介质而言,冲洗磁性精矿的方向与给矿方向相反,粗颗粒不必穿过磁介质堆便可冲洗出来。该机的脉动机构驱动矿浆产生脉动,可使位于分选区磁介质堆中的矿粒群保持松散状态,使磁性矿粒更容易被捕获,使非磁性矿粒尽快穿过磁介质堆进入尾矿中去。

显然,反冲精矿和矿浆脉动可防止磁介质堵塞;脉动分选可提高磁性精矿的质量。这些措施保证了该机具有较大的富集比、较高的分选效率和较强的适应能力。

2 SLon 磁选机在铁矿方面的应用

2.1 在昆钢大红山铁矿的应用

昆钢大红山铁矿是磁铁矿和赤铁矿共生的混合矿,原设计采用阶段磨矿—弱磁—SLon 强磁选—反浮选流程,由于脉石矿物复杂多变,反浮选作业无法稳定而未成功,前几年用摇床取代了反浮选。但是,摇床存在处理量较小,对细粒铁矿回收率较低的缺点。目前大红山铁矿正在进行技术改造，采用赣州金环磁选设备有限公司研制的 SLon-φ2400 离心选矿机取代反浮选作业。工业试验指标及新改造的原则流程如图2所示。该流程预计在2010年投产，预计投产后全流程铁精矿品位可从 62%~63% 提高至 64%~65%。

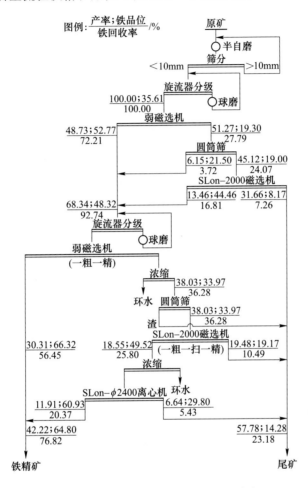

图 2 大红山铁矿 400 万吨选厂工业试验流程

该流程的主要特点为：SLon-2000 磁选机对细粒氧化铁矿回收率较高，可有效地控制强磁选作业的尾矿品位；采用 SLon-φ2400 离心选矿机取代反浮选作业，对强磁选精矿进行精选，2009 年在大红山铁矿的工业试验表明，离心选矿机具有富集比较大，对细粒铁矿回收率较高的优点。这些新技术和新设备的应用，使该流程既可运行顺畅，又能获得良好的选矿指标。预计全流程的选矿指标为原矿品位 35.61%，铁精矿品位 64.80%，铁回收率 76.82%。

2.2 在昆钢罗次铁矿的应用

目前昆钢罗次铁矿共有 11 台 SLon-1500 磁选机组成全高梯度磁选的一粗一精强磁生产流程，如图3所示，采用该机对全流程精矿品位及回收率提高非常显著。由于 SLon-1500 高梯度磁选机具有富集比较大的优点，因此，在保证铁精矿品位大于 59% 的前提下，新流程可降低入选矿的品位，减少富矿的外购量，从而可大幅度降低矿石采购成本。

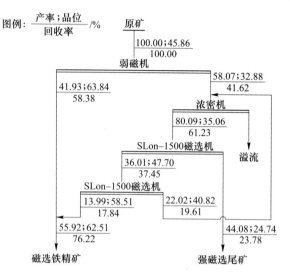

图 3　罗次铁矿全磁选工艺数质量流程

2.3　在昆钢上厂铁矿的应用

在昆钢上厂铁矿尾矿库堆积了含铁 22% 左右的尾矿近千万吨，该矿于 1999 年安装了 3 台 SLon-1500 磁选机回收细粒赤铁矿和褐铁矿（图 4），在给矿品位 22% 的条件下每年回收细粒铁精矿 7 万吨，铁精矿产率 13% 以上，回收率 34% 以上，显著地提高了资源利用率。

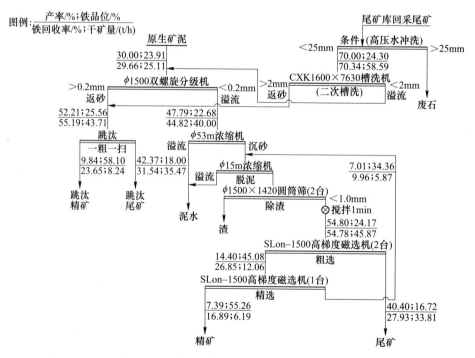

图 4　上厂铁矿细粒尾矿回收生产数质量流程

2.4 在云南北衙黄金选厂的应用

云南北衙黄金选矿厂是一座较大的黄金选厂,其原矿组成主要是金、磁铁矿和褐铁矿,该厂用氧化法回收金后,分别用弱磁选机回收磁铁矿和用 2 台 SLon-2000 磁选机回收褐铁矿(图 5)。目前该矿将回收的磁铁矿和褐铁矿精矿分别销售,用铁矿产生的效益来维持金矿运转,而黄金产品成为纯利润。2009 年该厂又订购了 4 台 SLon-2500 磁选机用于扩大产量,预计 2010 年投产,届时产量可扩大至原有的 4~5 倍。

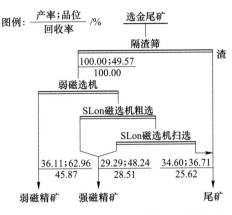

图 5 云南北衙黄金选矿厂褐铁矿回收流程

2.5 在四川满银沟铁矿的应用

四川满银沟铁矿储量有 8000 万吨左右,该矿以赤铁矿和磁铁矿为主,其特点是深度氧化、易泥化、磁性率低,但 P、S 等杂质元素含量极低,经分选后可作为炼铁优质原料。采出的原矿用筛分、手选和干式磁选的方法分选出品位大于 45%Fe 的块矿直接销售,剩余的低品位粉矿以前作为尾矿排弃,而目前进入新建选厂采用 SLon 磁选机回收。其 0~12mm 粉矿选矿流程如图 6 所示。该选矿流程的建成,使过去抛弃的大量低品位铁矿得到了有效的利用。

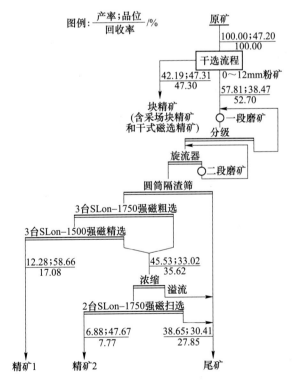

图 6 四川满银沟铁矿细粒部分选矿流程
(精矿 1 单独销售,精矿 2 与块精矿掺和销售)

2.6 分选较粗矿粒的 SLon 磁选机在贵州的应用

SLon 立环脉动高梯度磁选机在细粒级弱磁性矿石的工业生产中得到广泛的应用,其给矿粒度范围为 0~2mm,普遍需要经过磨矿才能入选。然而,现在我国选矿厂入选矿石品位日益下降,如果在入磨前能进行预选抛弃一部分尾矿,则可显著降低后续作业的磨矿成本和选矿成本,通过对 SLon 磁选机的改进,其给矿粒度范围扩展到 0~5mm。

2010 年 3 月,1 台 SLon-1500 立环脉动高梯度磁选机在贵州某选矿厂应用于分选 0~5mm 褐铁矿获得成功。该褐铁矿的地质品位为 35%Fe 左右,在开采过程中混入了较多的围岩,因此入选品位只能达到 26%Fe 左右。经研究表明,该矿石结晶粒度较粗,破碎至 0~5mm 入选即可获得良好的选矿指标。该机的选矿指标见表 1。其入选原矿品位为 26.15%Fe,SLon-1500 磁选机一次分选的精矿品位 53.20%,铁回收率为 64.26%。由于褐铁矿含有结晶水,把结晶水经烧失后相当于品位 63% 的铁精矿。因此该铁精矿可作为商品矿直接销售。

表 1　SLon 磁选机分选 0~5mm 褐铁矿指标　　　　　　　　　　(%)

给矿 TiO₂ 品位	强磁选精矿			强磁选尾矿		
	产率	TiO₂ 品位	回收率	产率	TiO₂ 品位	回收率
26.15	31.59	53.20	64.26	68.41	13.66	35.74

SLon-1500 磁选机在该褐铁矿应用成功,使该选矿厂省去了磨矿作业,大幅度降低了生产成本,获得了显著的经济效益。

3　SLon 磁选机在钛铁矿方面的应用

3.1　在四川攀钢选钛厂的应用

在 2008 年以前,攀钢选钛厂的选钛流程为粗细分选,粗粒级用螺旋流槽重选,重选精矿干燥后用电选机精选;细粒级用图 7 所示的 SLon 磁选机—浮选流程分选。14 台 SLon-

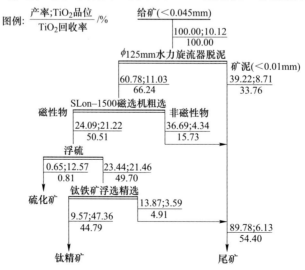

图 7　攀钢选钛厂小于 0.045mm 钛铁矿选矿数质量流程

1500立环脉动高梯度磁选机和2台SLon-1750中磁机在攀枝花选钛厂用于小于0.045mm微细粒级钛铁矿的磁浮选矿流程中，该机有效地回收了这部分微细粒级钛铁矿，为浮选作业创造了良好的条件。全流程在给矿品位10%左右的条件下，实现钛精矿品位大于47% TiO_2，TiO_2回收率44%的良好指标。2006年该厂生产品位为47.5% TiO_2 的微细粒级钛精矿15万吨。

为了进一步提高选钛回收率和降低生产成本，2008~2009年该厂进行了全面的选钛扩能改造，粗粒级和细粒级都分别用强磁—浮选流程。图8所示为采用SLon磁选机的试验流程和指标，该流程预计2010年可以全面投入使用。预计全面投产后该厂钛的回收率可由改造前的40%左右提高到50%左右，钛精矿产量由28万吨增加至60万吨。

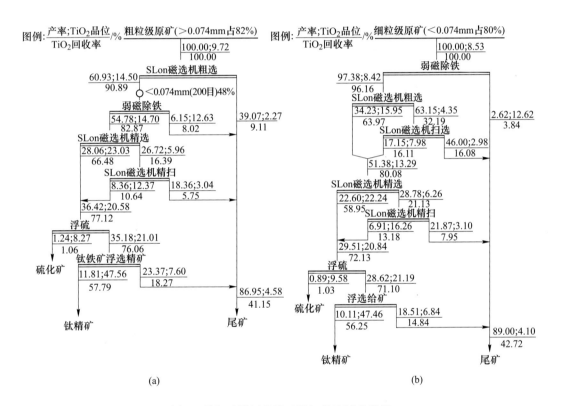

图8 攀钢选钛厂钛铁矿粗细分选试验流程
(a) 粗粒级选钛流程；(b) 细粒级选钛流程

3.2 在西昌太和铁矿的应用

SLon-1500、SLon-1750、SLon-2000立环脉动高梯度磁选机共9台在重钢太和铁矿回收钛铁矿，采用分级磁选、粗粒再磨、磁选和浮选精选的流程（图9），该流程的选钛指标为：给矿 TiO_2 品位12.06%，钛精矿 TiO_2 品位47.85%，TiO_2 回收率47.44%，实现年生产12万吨左右优质钛精矿的能力。

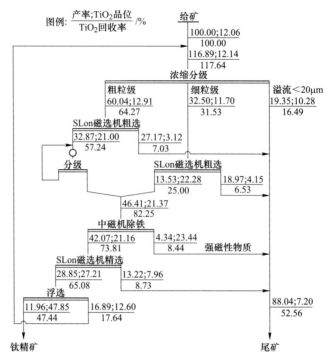

图 9 太和铁矿选钛数质量流程

3.3 在从攀钢总尾矿中回收钛铁矿的应用

3 台 SLon-1500 立环脉动高梯度磁选机在攀枝花综合回收厂用于从攀钢总尾矿中回收钛精矿，选矿流程如图 10 所示，其中两台粗选和一台精选，磁选精矿分级后细粒级用浮

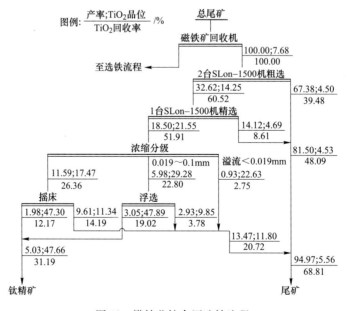

图 10 攀枝花综合厂选钛流程

选精选，粗粒级用摇床精选，可从全粒级中回收钛精矿。该流程的选钛指标为：给矿 TiO_2 品位 7.68%，钛精矿 TiO_2 品位 47.66%，TiO_2 回收率 31.19%，可每年生产 3 万吨左右优质钛精矿。

4 SLon 磁选机在非金属矿提纯的应用

随着我国建筑业飞速发展，对建筑材料的质量和产量迅速提高，近年 SLon 磁选机在石英、长石、霞石、高岭土等非金属矿的提纯方面得到了广泛的应用。

4.1 在四川南江霞石矿除铁的应用

四川南江霞石矿是我国第一个具有较高工业开采价值的霞石矿床。矿石中钠、钾含量较高，具有熔点低和助熔性能强等优点，故在建材、化工等工业中有着广泛的用途。但该霞石矿铁含量高，不能直接利用。通过对矿物性质考查及可选性试验研究，确定采用 SLon 立环脉动高梯度磁选机作为除铁主体设备开展一粗一精全磁的提纯工艺，工业生产流程见图11。

当给矿含 Fe_2O_3 1.69%，经一粗一精选别获得含 Fe_2O_3 0.179% 的霞石精矿，回收率 76.36%，除铁率达 91.91%，产品质量达到出口类一级霞石精矿指标。

4.2 在四川乐山石英砂提纯的应用

石英砂是一种重要的工业原料，广泛应用于玻璃、陶瓷行业，纯度高的石英砂可用于半导体、光纤制造。由于自然界的石英砂含铁量普遍偏高，达不到工业原料的质量要求。近年 SLon 磁选机广泛应用于石英砂的提纯工业。图 12 所示为四川乐山某石英砂厂的除铁流程，石英砂原矿经破碎、筛分、脱泥后，用弱磁选机选出磁铁矿、机械铁等强磁性物质，然后用 SLon 磁选机除去大部分弱磁性矿物质如赤铁矿、褐铁矿等。其非磁性产品为石英砂精矿，整个流程的生产指标为：原矿含量 0.06% Fe_2O_3，非磁性产品精矿产率为 91.20%，品位为 0.03% Fe_2O_3，Fe_2O_3 的除去率为 54.40%。

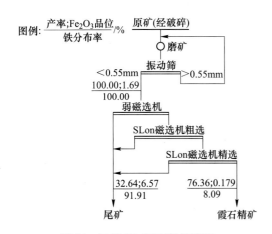

图11 四川南江霞石除铁流程

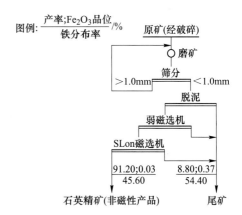

图12 四川乐山石英砂除铁流程

5 结论

(1) SLon 立环脉动高梯度磁选机是新一代高效强磁选设备,具有富集比大、选矿效率高、适应性强、设备运转可靠、可长期稳定工作的优点,该机在国内外的氧化铁矿、钛铁矿、非金属选矿工业中已大量应用。

(2) SLon 磁选机在昆钢大红山铁矿、罗次铁矿、上厂铁矿、四川满银沟铁矿、云南北衙选厂等地应用,大幅度提高了这些厂矿铁矿选矿技术水平,显著地降低了生产成本,选矿指标屡创新高,达到国内外领先水平,创造了显著的社会经济效益。

(3) SLon 磁选机在钛铁矿选矿工业中的成功应用,为解决我国微细粒级钛铁矿选矿技术难题,大幅度提高钛铁矿的回收率和降低选钛生产成本发挥了重要的作用。

(4) SLon 磁选机在霞石、石英等非金属矿提纯的广泛应用,使很多低品质的原矿成为优质的建材原料。

写作背景　本文以较多的图例介绍了 SLon 立环脉动高梯度磁选机在云南、四川、贵州选矿工业中的应用,其中 SLon-ϕ2400 离心机在昆钢大红山铁矿 400 万吨选矿厂的应用,SLon 立环脉动高梯度磁选机在云南北衙黄金选矿厂回收褐铁矿,在四川满银沟分选赤铁矿,首台粗粒级 SLon-1500 立环脉动高梯度磁选机在贵州分选 0~5mm 褐铁矿均为新公布的内容。

SLon-2500立环脉动高梯度磁选机的研制与应用[①]

熊大和[1,2]

(1. 赣州金环磁选设备有限公司;2. 赣州有色冶金研究所)

摘　要　采用最新的先进技术和创新理念,研制了我国目前处理量最大的高梯度磁选机——SLon-2500立环脉动高梯度磁选机。该机具处理量大,选矿效率、设备作业率和自动化程度高,电耗低的优点,用于赤铁矿、钛铁矿的分选及石英砂的提纯,均可获得优良的技术指标。

关键词　SLon-2500立环脉动高梯度磁选机　氧化铁矿分选　钛铁矿分选　石英提纯

Development and Application of SLon-2500 Vertical Ring and Pulsating High Gradient Magnetic Separator

Xiong Dahe[1,2]

(1. Ganzhou SLon Magnetic Separator Ltd.;
2. Ganzhou Nonferrous Metallurgy Research Institute)

Abstract　By adopting the latest advanced technology and creative ideas, the largest capacity high gradient magnetic separator in our China is created, namely SLon-2500 vertical ring and pulsating high gradient magnetic separator. It possesses the advantages of high capacity, high processing efficiency, high equipment operation rate and automation, and low energy consumption, and can be applied into the separation of hematite and ilmenite and the purification of quartz. All can get good technical index.

Keywords　SLon-2500 vertical ring and pulsating high gradient magnetic separators;Separation of oxidized iron ore; Ilmenite separation; Quartz purification

在2004年以前,我国研制的工业型强磁选机台时处理能力不超过80t/h,例如φ3200平环强磁选机和SLon-2000立环脉动高梯度磁选机在工业生产中处理鞍山式赤铁矿,单机给矿量平均为75t/h左右。随着现代选矿工业朝着大规模和高自动化水平的方向不断发展,赣州金环磁选设备有限公司新推出的SLon-2500立环脉动高梯度磁选机台时处理量可达到150t/h,同时该机具有节能、节约占地面积、操作成本低和设备作业率高的优点,可为用户显著降低生产成本,提高经济效益。

[①] 原载于《金属矿山》,2010 (06):133~136。

1 SLon-2500 磁选机的研制

SLon-2500 立环脉动高梯度磁选机是在总结 SLon-1000[1]、SLon-1500[2]、SLon-1750[3]、SLon-2000[4] 立环脉动高梯度磁选机成功经验的基础上研制的,其外形见图 1,主要技术参数见表 1,电磁性能见表 2 和图 2。

图 1 SLon-2500 磁选机外形

表 1 SLon-2500 磁选机主要技术参数

转环外径 /mm	给矿粒度 /mm	给矿浓度 /%	矿浆通过能力 /$m^3 \cdot h^{-1}$	干矿处理量 /$t \cdot h^{-1}$	额定背景磁感应强度/T	额定激磁电流/A	额定激磁电压/V
2500	0~1.2	10~40	200~400	100~150	1.0	1400	45

额定激磁功率/kW	传动功率 /kW	脉动冲程 /mm	脉动冲次 /$r \cdot min^{-1}$	供水压力 /MPa	耗水量 /$m^3 \cdot h^{-1}$	冷却水用量 /$m^3 \cdot h^{-1}$	主机质量 /t
63	11+11	0~30	0~300	0.2~0.4	200~300	6~7	105

表 2 SLon-2500 磁选机电磁性能测定数据

电流/A	电压/V	功率/kW	磁场感应强度/T
200	6.6	1.3	0.168
400	12.7	5.1	0.336
600	19.5	11.7	0.495
800	26.1	20.9	0.654
1000	32.7	32.7	0.793
1200	38.6	46.3	0.911
1400	45.0	63.0	1.001
1600	52.0	83.2	1.074
1700	54.5	92.7	1.107

在研制 SLon-2500 磁选机的过程中,采用了一系列的新技术提升该设备的选矿性能和机电性能。主要措施如下:

(1) 通过优化磁系设计、减少漏磁、降低电阻和提高电效率,使设备具有良好的激磁性能。由表 2 可见,SLon-2500 磁选机的激磁电流为 1400A 时,其背景磁感应强度达到 1.0T,相应的激磁功率仅 63kW。采取低电压大电流激磁,有利于提高激磁线圈的安

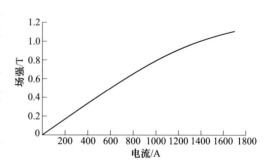

图 2 SLon-2500 磁选机激磁曲线

全可靠性。激磁线圈采用空心铜管绕制，水内冷散热，冷却水直接贴着铜管内壁流动，散热效率高。此外，由于冷却水的流速较高，微细泥沙不易沉淀，可保证激磁线圈长期稳定工作。

（2）采用 ϕ1mm 导磁不锈钢丝作为磁介质，使磁选机对微细粒弱磁性矿物的回收率大幅度提高。表 3 为采用几种不同直径的磁介质对海南钢铁公司排放的总尾矿进行再回收的对比试验结果，可见，ϕ1mm 磁介质与 ϕ2mm 和 ϕ3mm 磁介质相比，在给矿品位相同，精矿品位略高的情况下，铁回收率分别提高 6.41% 和 9.82%。

表 3 SLon-2500 磁选机磁介质对比试验结果

磁介质直径 /mm	给矿铁品位 /%	精矿指标/%			尾矿铁品位 /%
		产率	铁品位	铁回收率	
1	29.45	29.26	49.32	49.01	21.23
2	29.45	25.47	49.26	42.60	22.68
3	29.45	23.96	48.17	39.19	23.55

（3）采用新型数字式脉动冲程箱，具有力量大、平衡性好、可靠性高和能耗低的优点。

（4）采用新型的整流控制柜，激磁电流可根据生产需要调节，并稳定在人工设定点，不会随外电压波动。人机界面友好，自动化程度高，可远程控制。

2 SLon-2500 磁选机工业试验

首台 SLon-2500 立环脉动高梯度磁选机于 2006 年完成制造，其工业试验是在海南钢铁公司（现已更名为海南矿业联合有限公司，下同）红旗尾矿库完成的。

2.1 给矿性质

海南钢铁公司有 3 个选矿车间分选氧化铁矿，选矿流程为一粗一精一扫强磁选或一粗一扫强磁选—反浮选。每年入选原矿约 230 万吨，目前每年大约有 120 万吨尾矿排入红旗尾矿库。

排往红旗尾矿库的总尾矿中铁矿物以赤铁矿为主，有少量的磁铁矿、黄铁矿和褐铁矿；脉石主要是石英。表 4~表 6 分别为该尾矿的化学多元素分析、铁物相分析和粒度分析结果。

表 4 海钢总尾矿化学多元素分析结果　　　　　　　　　　（%）

成分	TFe	FeO	SiO_2	Al_2O_3	CaO	MgO	S	P
含量	29.45	0.93	54.88	6.24	2.90	1.68	1.04	0.043

可见，海钢总尾矿的平均铁品位为 29% 左右，铁主要存在于小于 19μm 粒级中。由于该尾矿氧化程度深，含泥量大，且含有较多的连生体，因此要从中再回收铁精矿技术难度较大。

表5　海钢总尾矿铁物相分析结果　　　　　　　　　　　　　　　　（％）

铁物相	铁含量	铁分布率
赤铁矿	26.93	91.48
磁铁矿	0.54	1.83
黄铁矿	0.99	3.36
褐铁矿	0.98	3.33
合　计	29.44	100.00

表6　海钢总尾矿粒度分析结果

粒级/mm	产率/%	铁品位/%	铁分布率/%
>0.076	6.97	24.00	5.82
0.076~0.037	21.85	29.70	22.56
0.037~0.019	20.41	28.80	20.44
<0.019	50.77	29.00	51.18
合　计	100.00	28.76	100.00

2.2　工业试验流程

首台SLon-2500磁选机安装于红旗尾矿库尾矿输送管的排矿口，从该排矿口截取尾矿入选，处理量大约为72万吨/a。工业试验工艺流程如图3所示。首先用圆筒除渣筛筛除大于0.8mm的粗粒，接着用SLon-2500磁选机粗选，粗选精矿经浓缩后用离心选矿机精选。离心机精矿作为最终精矿，SLon-2500磁选机的尾矿和离心机尾矿作为最终尾矿排入尾矿库。该流程用水取自尾矿库的排水口，流程中浓缩设备的溢流水返回流程再用，因此整个流程具有投资少、生产成本低的优点。

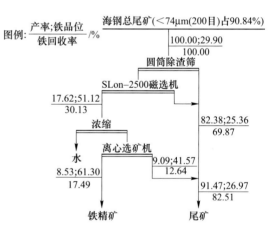

图3　SLon-2500磁选机工业试验流程

2.3　SLon-2500磁选机条件试验

2.3.1　背景磁感应强度试验

SLon-2500磁选机的背景磁感应强度可通过激磁电流调节，表7是以背景磁感应强度为变量的试验结果。可见：激磁电流从600A增大至1300A，相应的背景磁感应强度从0.495T提高至0.960T，精矿品位和尾矿品位都略有下降，精矿回收率增加。当激磁电流达到1000A以上时，激磁功率增加较快，而尾矿品位稳定。兼顾精矿品位和电能消耗，SLon-2500磁选机长期运行时激磁电流以1000A左右为宜，相应的激磁功率为32.7kW左右。

表7 SLon-2500磁选机背景磁感应强度试验结果

激磁电流/A	磁感应强度/T	激磁功率/kW	给矿铁品位/%	精矿指标/%			尾矿铁品位/%
				产率	品位	铁回收率	
600	0.495	11.7	28.6	14.59	52.6	26.83	24.5
700	0.577	15.9	28.5	15.25	52.4	28.04	24.2
800	0.654	20.9	28.8	17.02	52.2	30.85	24.0
900	0.725	26.3	28.4	17.30	52.3	31.86	23.4
1000	0.793	32.7	28.2	17.24	52.2	31.91	23.2
1100	0.853	39.6	28.7	18.90	52.3	34.44	23.2
1200	0.911	46.3	28.8	19.58	51.8	35.22	23.2
1300	0.960	54.0	29.1	20.63	51.8	36.72	23.2

2.3.2 处理量试验

为了核定SLon-2500磁选机对海钢尾矿的处理能力,进行了处理量试验,结果见表8。可见:干矿处理量由97.4t/h增加至148.5t/h,给矿浓度范围为29.76%~36.18%,尾矿品位变化不大;精矿铁品位和回收率略有降低,这与处理量大时给矿品位有所下降有关。试验结果证明了SLon-2500磁选机对给矿浓度和给矿量的变化有很强的适应能力。综合各种因素,判定SLon-2500磁选机对海钢尾矿的处理量以100~125t/h为宜,相应的给矿浓度为30%~35%,给矿体积为260m³/h左右。

表8 SLon-2500磁选机处理量试验结果

给矿浓度/%	给矿体积/m³·h⁻¹	干矿量/t·h⁻¹	给矿铁品位/%	精矿指标/%			尾矿铁品位/%
				产率	铁品位	铁回收率	
29.76	257	97.4	29.0	19.31	54.9	36.56	22.8
31.49	248	100.9	27.6	16.29	53.3	31.45	22.6
35.73	224	107.9	29.6	18.98	53.5	34.31	24.0
32.25	261	109.7	28.9	19.88	54.7	37.62	22.5
36.18	261	127.9	29.8	19.15	52.6	33.80	24.4
35.22	294	138.5	27.2	15.82	52.2	30.37	22.5
34.87	320	148.5	25.6	14.79	52.1	30.10	21.0

2.3.3 给矿粒度对分选指标的影响

本次工业试验SLon-2500磁选机的给矿粒度是不能调节的,通过对121批随机取样化验结果分类统计,给矿粒度对该机分选指标的影响见表9。可见,给矿粒度变细,精矿品位和尾矿品位都有所升高,精矿产率和铁回收率下降。从统计结果看,给矿粒度绝大多数在小于74μm(200目)占90%左右。

2.3.4 给矿品位对分选指标的影响

通过对随机取样化验结果统计,给矿品位对SLon-2500磁选机分选指标的影响列于表10。可见,给矿品位对精矿品位的影响较大,即给矿品位每上升1个百分点,精矿品位大约也上升1个百分点。

表9　给矿粒度对 SLon-2500 磁选机分选指标的影响　　　　　　　　　　（%）

取样批数	给矿指标		精矿指标			尾矿铁品位
	平均小于 74μm（200目）含量	铁品位	产率	铁品位	铁回收率	
3	74.27	28.20	20.75	52.53	38.65	21.83
9	79.00	28.61	20.20	53.10	37.49	22.41
6	84.33	28.18	20.38	52.72	38.12	21.90
55	88.81	27.16	16.32	52.13	31.32	22.29
47	92.65	28.19	16.17	53.74	30.83	23.26
1	96.00	27.50	13.08	55.40	26.36	23.30

表10　给矿品位 SLon-2500 磁选机分选指标的影响　　　　　　　　　　（%）

取样批数	给矿平均铁品位	精矿指标			尾矿铁品位
		产率	铁品位	铁回收率	
30	25.30	15.07	49.99	29.77	20.92
85	28.31	17.27	53.60	32.70	23.03
6	31.80	18.34	56.60	32.64	26.23

2.4　流程试验

完成 SLon-2500 磁选机的条件试验和离心选矿机的调试后，2008年5月13日至26日对全流程进行了连续14天的生产考察，SLon-2500 磁选机单机获得的平均指标为给矿铁品位29.90%，精矿铁品位51.12%，作业铁回收率30.13%，尾矿铁品位25.36%，离心机精选作业平均指标为给矿铁品位51.12%，精矿铁品位61.30%，作业铁回收率58.05%，尾矿铁品位41.57%，综合选别指标为给矿铁品位29.90%，精矿产率8.53%，精矿铁品位61.30%，铁回收率17.49%，尾矿铁品位26.97%（参见图3）。

3　SLon-2500 磁选机的工业应用

3.1　在海钢红旗尾矿库的应用

首台 SLon-2500 磁选机在海钢红旗尾矿库进行工业试验取得成功后，就地转化为生产设备，整个生产线移交给海钢自行管理。至今已运转3年多，SLon-2500 磁选机的单机作业率达到99%以上。

3.2　在赤铁矿选矿工业中的应用

SLon-2500 磁选机研制成功以后，很快在氧化铁矿的选矿工业中得到大规模应用。2008~2009年，印度一公司共采购了10台，构成一次粗选、一次扫选的强磁选流程分选赤铁矿，所获得的选别指标见表11。

3.3　在钛铁矿选矿工业中的应用

2009年，1台 SLon-2500 磁选机在攀枝花地区某选矿厂应用于钛铁矿选矿获得成功。目前该地区的钒钛磁铁矿选矿厂普遍是先用弱磁选机选出磁铁矿精矿，选完磁铁矿的尾矿作为选钛作业的给矿。选钛作业用强磁选机预先抛弃大部分尾矿，然后再通过重选或浮选选出钛精矿。由于选钛流程中入选原料的钛品位低，因此要求强磁选机处理量大，生产成本低，尽可能多抛弃尾矿。SLon-2500 磁选机正好具备这些优势。

表 11 SLon-2500 磁选机分选印度赤铁矿指标　　　　　　　　　　（%）

产品	产率	铁品位	铁回收率
粗选精矿	72.37	66.20	80.15
扫选精矿	13.94	58.77	13.71
综合精矿	86.91	65.00	93.86
综合尾矿	13.69	26.82	6.14
给矿	100.00	59.77	100.00

SLon-2500 磁选机在攀枝花地区某选矿厂用于钛铁矿的强磁粗选作业，其给矿粒度为小于 74μm（200 目）占 41.4%，给矿 TiO_2 品位约为 5%，台时处理量为 100t/h 左右。该机的生产调试指标见表 12。可见，在背景磁感应强度为 0.68T 时，强磁给矿 TiO_2 品位为 4.57%，精矿 TiO_2 品位为 9.86%，TiO_2 回收率为 75.37%，抛弃的尾矿产率达 65.07%，尾矿 TiO_2 品位仅 1.73%。

表 12 SLon-2500 磁选机分选钛铁矿指标

磁感应强度/T	产品	产率/%	TiO_2 品位/%	TiO_2 回收率/%
0.40	精矿	26.82	13.48	70.47
	尾矿	73.18	2.07	29.53
	给矿	100.00	5.13	100.00
0.54	精矿	31.02	11.98	72.43
	尾矿	68.98	2.05	27.57
	给矿	100.00	5.13	100.00
0.68	精矿	34.93	9.86	75.37
	尾矿	65.07	1.73	24.63
	给矿	100.00	4.57	100.00
0.80	精矿	47.64	8.12	84.65
	尾矿	52.36	1.34	15.35
	给矿	100.00	4.57	100.00

SLon-2500 磁选机背景磁感应强度为 0.68T 时，激磁功率仅 22kW，加上 2 台装机容量各为 11kW 的电动机，其总功率为 44kW，即处理每吨给矿的电耗仅 0.44kW·h 左右。

3.4 在石英提纯工业中的应用

2009 年 5 月，1 台 SLon-2500 磁选机在陕西应用于提纯石英砂，获得成功。石英砂给矿的粒度为 0～1mm，Fe_2O_3 含量为 0.15%～0.25%，要求 SLon-2500 磁选机 1 次分选将 Fe_2O_3 含量降到 0.08% 以下。通过生产调试，石英砂精矿 Fe_2O_3 含量降到 0.05%～0.06%，优于设计指标。

4 结论

（1）通过多年的努力和创新，完成了 SLon-2500 立环脉动高梯度磁选机的研制和工业试验，并在工业生产上经过了 3 年多的应用考验。

（2）SLon-2500 磁选机用于赤铁矿和钛铁矿的分选及石英砂的提纯，均可获得良好的技术指标，并具有处理量大，选矿效率、设备作业率和自动化程度高，电耗低的优点。该机的研制成功，为弱磁性矿物的选别提供了一种高效的大型强磁选设备。

参 考 文 献

[1] 熊大和. SLon-1000 立环脉动高梯度磁选机的研制 [J]. 金属矿山，1988（10）：37~40.
[2] 熊大和. SLon-1500 立环脉动高梯度磁选机的研制 [J]. 金属矿山，1990（7）：43~46.
[3] 熊大和，杨庆林，谢金清，等. SLon-1750 立环脉动高梯度磁选机的研制与应用 [J]. 金属矿山，1999（10）：23~26.
[4] 熊大和. SLon-2000 立环脉动高梯度磁选机的研制 [J]. 金属矿山，1995（6）：32~34.

写作背景　SLon-2500 立环脉动高梯度磁选机的研制，从设计、制造、工业试验到成熟的生产应用，大约花了 6 年的时间。此文介绍了该机的研制、工业试验和一些生产应用。该机是当时国内最大的强磁选设备，该机研制成功，是 SLon 磁选机朝着大型化发展的一个重要里程碑。

SLon 磁选机与离心机组合流程分选氧化铁矿新进展[1]

熊大和[1,2]

(1. 赣州金环磁选设备有限公司;2. 赣州有色冶金研究所)

摘 要 SLon 立环脉动高梯度磁选机具有优异的选矿性能,已广泛应用于分选氧化铁矿,SLon 离心选矿机用于强磁精矿的精选作业,可有效地剔除含少量磁铁矿的石英等脉石,这两种设备的优势互补,用它们的组合流程分选某些氧化铁矿可获得良好的选矿指标。本文介绍该技术的最新研究与应用。

关键词 SLon 立环脉动高梯度磁选机 SLon 离心选矿机 氧化铁矿 选矿

New Development of the Cooperational Flowsheet of SLon Magnetic Separator and Centrifugal Separator in Processing Oxidized Iron Ores

Xiong Dahe[1,2]

(1. *SLon Magnetic Separator Ltd.*;
2. *Ganzhou Nonferrous Metallurgy Research Institute*)

Abstract SLon vertical ring and pulsating high gradient magnetic separators possess excellent mineral processing ability. They have been widely applied to process oxidized iron ores. SLon centrifugal separators are applied to clean the SLon magnetic concentrate. They can remove quartz and other gangue minerals which associated with a small portion of magnetite. Applying the cooperation advantages of the two kinds of equipments to process some oxidized iron ores, good beneficial results can be achieved. This paper introduces the latest research and application of this technology.

Keywords SLon magnetic separator; SLon centrifugal separator; Oxidized iron ore; Process

1 SLon 磁选机与离心机组合技术的特点

SLon 立环脉动高梯度磁选机广泛应用于分选弱磁性铁矿。在分选鞍山式氧化铁矿的生产流程中,主要是用于粗选作业,精选作业一般是用螺旋溜槽分选粗粒级和用反浮选分选细粒级。在分选鞍山式铁矿的流程中强磁选很少用于精选作业,原因是鞍山式氧化铁矿

---

[1] 原载于《2009 年金属矿产资源高效选冶加工利用和节能减排技术及设备学术研讨与技术成果推广交流暨设备展示会论文集》。

是由磁铁矿、假象赤铁矿、赤铁矿、镜铁矿组成，由表1可知，磁铁矿的比磁化率是赤铁矿和镜铁矿的100倍以上。

表1　几种铁矿的比磁化率

矿石名称	磁铁矿	假象赤铁矿	赤铁矿	镜铁矿
比磁化率χ /m³·kg⁻¹	$(6.25\sim11.6)\times10^{-4}$	$(6.2\sim13.5)\times10^{-6}$	$(0.6\sim2.16)\times10^{-6}$	3.7×10^{-6}

如图1所示，鞍山式氧化铁矿中含有较多的磁铁矿和石英的连生体，这些连生体在磁场中受到的磁力很大，表2为磁铁矿与石英连生体的视在比磁化率。例如，一颗连生体含1%质量的磁铁矿和99%质量的石英，它的视在比磁化率达到了6.25×10^{-6} m³/kg，已远远大于赤铁矿单体的比磁化率，它在磁

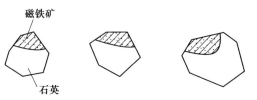

图1　磁铁矿和石英的连生体

场中所受到的磁力就比相同重量的赤铁矿要大。它很容易被强磁机捕捉到铁精矿中去，从而降低铁精矿品位。

表2　磁铁矿和石英连生体的视在比磁化率

连生体中磁铁矿的质量/%	连生体中石英的质量/%	连生体的视在比磁化率 χ/m³·kg⁻¹
0	100	0
1	99	6.25×10^{-6}
5	95	3.125×10^{-5}
10	90	6.25×10^{-5}
20	80	1.25×10^{-4}
50	50	3.125×10^{-4}
100	0	6.25×10^{-4}

离心选矿机是一种较好的细粒矿物重选设备，它可提供$20g\sim50g$（g为重力加速度）的离心力，能将细粒矿物按密度分选。石英的密度是2.6g/cm³，磁铁矿和赤铁矿的密度为5.0g/cm³，表3为磁铁矿和石英的连生体的视在密度。

根据生产经验，含30%的石英与70%的磁铁矿的连生体（视在密度为4.28g/cm³，石英与磁铁矿的体积比为45∶55）能被离心机排入尾矿中去。因此离心机的精选能力要高于强磁选机的精选能力。SLon立环脉动高梯度磁选机处理量大，用于粗选作业具有富集比大、选矿效率高的优点，而用于精选作业则存在含少量磁铁矿和大部分石英的贫连生体难以剔除的制约因素，而离心选矿机用于精选作业则可较好地解决这个问题，这两种设备相结合分选某些氧化铁矿可获得较好的选矿指标。

表3 磁铁矿和石英连生体的视在密度

连生体中磁铁矿质量/%	连生体中石英的质量/%	连生体的视在密度/g·cm⁻³
0	100	2.60
10	90	2.84
20	80	3.08
30	70	3.32
40	60	3.56
50	50	3.80
60	40	4.04
70	30	4.28
80	20	4.52
90	10	4.76
100	0	5.00

2 新设备的研究

2.1 SLon-2500 立环脉动高梯度磁选机的研制

我们在总结以往25年研制SLon立环脉动高梯度磁选机的基础上，研制了SLon-2500立环脉动高梯度磁选机（以下简称SLon-2500磁选机）如图2所示。

该机的主要优点有如下几个方面：

（1）通过优化磁系设计、减少漏磁、降低电阻和提高电效率。使该机有良好的激磁性能。由表4可见，该机的激磁电流为1400A时，其背景场强达到1.0T，相应的激磁功率仅63kW。

图2 SLon-2500立环脉动高梯度磁选机

表4 SLon-2500磁选机电磁性能测定数据

电流/A	电压/V	功率/kW	背景场强/T
200	6.6	1.3	0.168
400	12.7	5.1	0.336
600	19.5	11.7	0.495
800	26.1	20.9	0.654
1000	32.7	32.7	0.793
1200	38.6	46.3	0.911
1400	45.0	63.0	1.001
1600	52.0	83.2	1.074
1700	54.5	92.7	1.107

（2）采用ϕ1mm导磁不锈钢丝作为磁介质，使该机对微细粒弱磁性矿石的回收率大幅度提高。采用ϕ1mm磁介质与ϕ2mm和ϕ3mm磁介质比较，给矿品位相同，精矿品位略高，铁回收率分别提高6.41%和9.82%。

（3）采用新型数字式脉动冲程箱，具有力量大、平衡性好、可靠性高和能耗低的优点。

(4) 采用新型的整流控制柜,激磁电流可根据选矿的需要调节,并稳定在人工设定点,不会随外电压的波动而波动。人机界面友好、自动化高、可远程控制。

2.2 SLon-离心选矿机的研制

针对铁矿选矿要求设备处理量大的特点,我们近年研制出了 SLon-φ1600、SLon-φ2400 离心选矿机(图3)。通过生产应用证明,用它们对强磁粗选的铁精矿进行精选可获得较高的铁精矿品位。表5为 SLon 磁选机和 SLon 离心机应用于分选海钢强磁选精矿的对比指标,离心机的精矿品位比强磁选精矿品位高 2.77%,尾矿品位低 2.51%。因此离心机的精选指标明显优于强磁机的精选指标。

图3 SLon-φ2400 离心选矿机

表5 SLon 磁选机和 SLon 离心机应用于精选作业的对比指标 (%)

名 称	给矿		铁精矿				尾矿铁品位
	<0.074μm(200目)	铁品位	铁品位	产 率	铁回收率		
SLon 磁选机	90.55	55.45	61.75	60.38	67.24		45.85
SLon 离心机	90.55	55.45	64.52	57.18	66.53		43.34

3 工业应用

3.1 在昆钢大红山铁矿的应用

昆钢大红山铁矿是磁铁矿和赤铁矿共生的混合矿,原设计采用阶段磨矿—弱磁—SLon强磁选—反浮选流程,由于脉石矿物复杂多变,反浮选作业无法稳定而未获成功,前几年用摇床取代了反浮选。但是,摇床存在处理量较小,对细粒铁矿回收率较低的缺点。目前大红山铁矿正在进行技术改造,采用 SLon-φ2400 离心选矿机取代反浮选作业。工业试验指标及新改造的原则流程如图4所示。该流程预计在 2010 年投产,预计投产后全流程铁精矿品位可从 62%~63% 提高至 64%~65%。

该流程的主要特点为:SLon-2000 磁选机对细粒氧化铁矿回收率较高,可有效的控制强磁选作业的尾矿品位;采用 SLon-φ2400 离心选矿机取代反浮选作业,对强磁选精矿进行精选,2009 年在大红山铁矿的工业试验表明,离心选矿机具有富集比较大,对细粒铁矿回收率较高的优点。这些新技术和新设备的应用,使该流程既可运行顺畅,又能获得良好的选矿指标。预计全流程的选矿指标为:原矿品位 35.61%,铁精矿品位 64.80%,铁回收率 76.82%。

3.2 在低品位镜铁矿中的应用

如图5所示为一座日处理 500t 低品位镜铁矿的选厂生产流程图。其原矿品位为 22.63%TFe,铁矿物主要是镜铁矿和少量的磁铁矿,磁铁矿产率占原矿的 1% 左右,因此

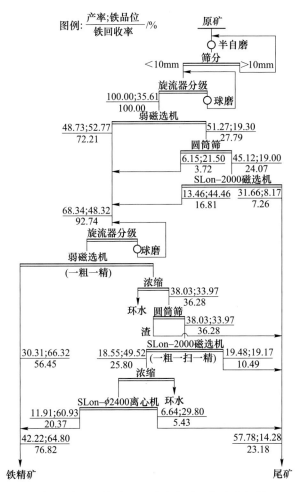

图 4 大红山铁矿 400 万吨选厂工业试验流程

SLon 强磁机前面可不用弱磁选机。

采用阶段磨矿—SLon 强磁抛尾—离心机精选—摇床扫选流程。一段磨矿将矿石磨至小于 0.074mm 粒级占 50%左右，用 SLon 磁选机分选抛去产率为 52.27%和 Fe 品位为 7.50%的低品位尾矿，一段强磁精矿用旋流器分级，旋流器沉砂进二段磨矿，二段磨矿的排矿返回到旋流器，旋流器的溢流粒度为小于 0.074mm 粒级占 85%~95%。旋流器溢流进入二段 SLon 强磁选机分选，二段强磁精矿浓缩后用离心机精选，二段强磁尾矿和离心机尾矿用摇床扫选。获得最终综合精矿品位为 62.52%，铁回收率 53.40%。

该流程的特点是：

(1) 一段 SLon 强磁选可抛弃大量的低品位尾矿，较大幅度地节约了二段磨矿的生产成本。

(2) 离心选矿机用于二段强磁精矿的精选作业，可有效地剔除含磁铁矿的石英连生体，其精矿品位提高幅度较大，TFe 从 48.64%提高到 62.25%。

(3) 采用摇床对二段强磁尾矿和离心机尾矿扫选，可直接拿出一部分较高品位的铁精矿。

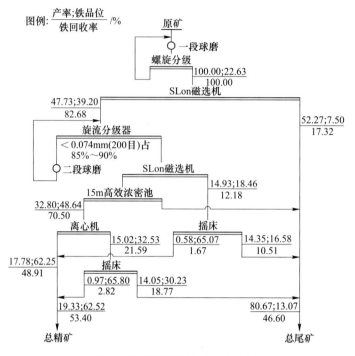

图 5 某低品位镜铁矿选矿流程

（4）整个流程为开路分选流程，选矿作业不存在循环负荷，生产上很好控制。

上述生产流程具有流程较简单，选矿指标较好，生产成本较低的优点。其缺点是离心选矿机和摇床的台时处理能力较低，目前只适用于中、小型选矿厂，还难以用于大规模的工业生产。

3.3 分选海南难选氧化铁矿

海南钢铁公司拥有一座储量较大的氧化铁矿，过去长期开采品位 50% 左右的富矿。现有一部分铁品位 40% 左右的难选氧化铁矿暂未得到利用。近年有关单位对这部分氧化铁矿进行了系统的选矿试验研究，试验证明，若仅用强磁选流程只能得到品位为 60% 左右的铁精矿。而市场上品位为 63% 以上的铁精矿好销且价格较高。因此，选矿试验和生产实践都要求铁精矿品位达到 63% 以上。

经过多次的探索实验和扩大试验，获得图 6 所示的 SLon 磁选机与 SLon 离心选矿机的组合流程。其入选原矿品位为 39.40%，含铁矿物主要是赤铁矿及占原矿产率 10% 左右的磁铁矿。首先磨矿至小于 0.074mm 粒级占 87%~90%，然后用弱磁选机选出磁铁矿，弱磁选机的尾矿用 SLon 磁选机一次粗选和一次扫选，磁选精矿合并，浓缩后用离心选矿机精选拿出部分品位为 63% 以上的铁精矿，离心机尾矿经浓缩后用旋流器分级，旋流器沉砂进入二段球磨。旋流器溢流用弱磁选机选出磁铁矿精矿，弱磁选尾矿用 SLon 磁选机扫选。SLon 磁选机扫选精矿再用离心机精选得出部分品位为 64% 左右的铁精矿。该流程的原矿品位为 39.4%，综合铁精矿品位为 63.50%，铁回收率 61.58%。2009 年底已建设了一条日处理 600t 原矿的生产线。至 2010 年 6 月底，已正常运转 6 个月，生产流程稳定，生产

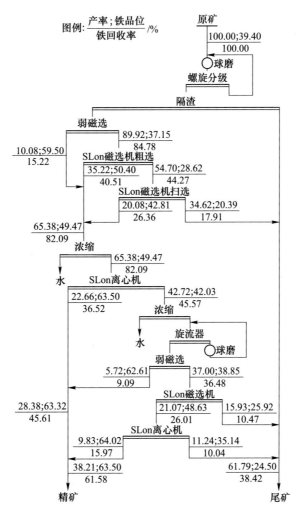

图 6 海南难选氧化铁矿 SLon 磁选机与 SLon 离心选矿机的组合分选流程

指标良好,生产成本较低。目前海南矿业联合有限公司正在按此流程建设一条年处理 200 万吨贫赤铁矿的选矿厂。

3.4 鞍山式氧化铁矿的实验研究

目前我国鞍山式氧化铁矿大多数都采用阶段磨矿—分级重选—强磁—反浮选流程。该流程具有节能、生产成本较低及选矿指标较好的优点。但是,对于一些小型选矿厂来说,反浮选作业存在技术复杂、环保审批时间长等问题,因此,采用离心选矿机代替反浮选作业在一些小型选厂得到应用。图 7 为鞍山式氧化铁矿的试验流程,该矿石以赤铁矿为主,含有一部分磁铁矿和假象赤铁矿,脉石矿物以石英为主。选矿流程为:一段磨矿后用水力旋流器分级,旋流器沉砂用螺旋溜槽选出一部分粒度较粗已经单体解离了的铁精矿,螺旋溜槽尾矿用 SLon 立环脉动中磁机分选,抛弃产率为 20%左右及铁品位为 7.25%的粗粒级尾矿。旋流器溢流经浓缩后用弱磁选机选出磁铁矿,弱磁选机尾矿用 SLon 立环脉动强磁机分选，该机抛弃产率为 23.72%、TFe 品位为 8.57% 的细粒尾矿。弱磁选精矿和 SLon 强磁精矿合并，浓缩后用 SLon 离心机进行一粗一精一扫选，离心机取得大部分的细粒级铁精矿。该流程综合选矿指标为：原矿品位 30.88%TFe，综合铁精矿品位为 66.28%，综合铁回收率 70.47%，综合尾矿品位 13.58%。该流程全部采用磁选和重选作业，具有节能、环保、生产成本较低的优点。

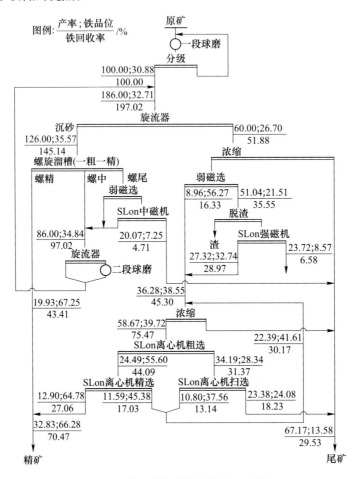

图 7 鞍山式氧化铁矿的选矿流程

3.5 从尾矿中回收铁精矿的应用

我国分选磁铁矿的小厂很多，有的小厂的入选原矿中含有一部分赤铁矿或镜铁矿，这些小选厂用弱磁选机选完磁铁矿后，弱磁选尾矿直接排入尾矿库。采用 SLon 磁选机和 SLon 离心选矿机的组合流程对这种尾矿进行再回收，往往可以获得较好的技术经济指标。图 8 为 SLon 磁选机与 SLon 离心机的组合流程分选某尾矿库的堆存尾矿的试验指标。该尾矿中含铁矿物主要是镜铁矿和少量的磁铁矿。该流程的特点是：先搅拌，分级磨矿至小于 0.074mm 粒级占 95%，利用弱磁选机分选出少量的磁铁矿，然后利用 SLon 立环脉动高梯度磁选机处理量大、作业成本低的特点，一次粗选抛弃产率 62.17%，TFe 品位 6.17% 的低品位尾矿，SLon 磁选机的粗选精矿品位已达到 36.42%TFe，而产率只占原矿的

33.68%，这部分粗精矿进入二段磨矿分级，二段分级溢流粒度为小于 0.054mm 粒级占 95%，浓缩后再用 SLon 磁选机精选一次，其精矿用离心选矿机精选。

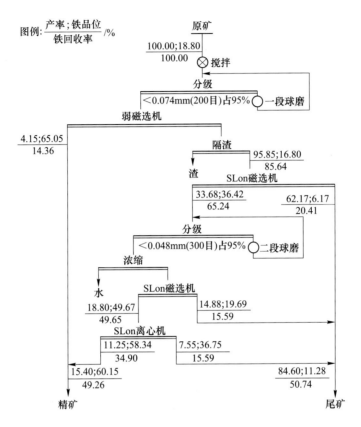

图 8 SLon 磁选机与离心机组合流程从尾矿中回收铁精矿

该流程的综合选矿指标为：给矿 TFe 品位 18.80%，铁精矿 TFe 品位为 60.15%，铁精矿产率 15.40%，铁回收率 49.26%，综合尾矿品位为 11.28%。该流程目前已在生产中应用，实现了低成本从表尾矿中回收铁精矿的目的，使二次资源得到利用。

4 结论

（1）SLon 立环脉动高梯度磁选机具有处理量大、生产成本低，它用于低品位氧化铁矿的粗选作业具有富集比大、选矿效率高的优点。

（2）强磁粗选精矿中，若含有磁铁矿和石英的贫连生体，其视在比磁化率远高于赤铁矿的比磁化率，则再利用强磁精选作用不大，例如鞍山式的氧化铁矿是由磁铁矿、假象赤铁矿和赤铁矿组成，这种矿石用强磁粗选后，后续作业一般不再用强磁精选。

（3）SLon 离心选矿机是利用矿石密度差异进行分选的，可产生 $20g \sim 50g$（g 为重力加速度）的离心力，可有效地回收微细粒铁精矿，并可有效地剔除含少量磁铁矿的贫连生体脉石，强磁选精矿用离心机再进行精选往往可获得较高的铁精矿品位。

（4）SLon 立环脉动高梯度磁选机和 SLon 离心选矿机的组合流程具有生产成本低、环境友好、易于操作管理的特点。这种流程已在一部分中小型氧化铁矿选厂应用成功,并开始在一些较大规模的选矿厂推广应用。

写作背景 本文系统地总结了近几年 SLon 磁选机与离心机组合流程分选氧化铁矿新进展,其中在昆钢大红山铁矿 400 万吨选矿厂已获得工业应用,在海南昌江县已建成年处理 20 万吨难选氧化铁矿的选矿厂(海南昌江光大矿业公司)。这几年是该技术从试验到工业应用快速发展的阶段。

SLon立环脉动高梯度磁选机技术创新与应用

熊大和[1,2]

（1. 赣州金环磁选设备有限公司；2. 赣州有色冶金研究所）

摘　要　近年来对SLon立环脉动高梯度磁选机进行了多方面的技术创新：研制SLon-2500、SLon-3000立环脉动高梯度磁选机，使设备朝着大型化方向发展；通过优化磁系结构，降低激磁电流密度，达到了显著的节能效果；研制了新型高效的磁介质，提高了有用矿物的选矿效率；研制特殊型号的SLon立环脉动高梯度磁选机，可用于0~5mm的较粗矿石选矿作业。通过这些技术创新，进一步提高了SLon立环脉动高梯度磁选机的选矿效率和降低了生产成本，并得到了更加广泛的应用。

关键词　SLon立环脉动高梯度磁选机　大型化　节能　弱磁性矿石选矿

Creative Technologies and Applications of SLon Vertical Ring and Pulsating High Gradient Magnetic Separator

Xiong Dahe[1,2]

(1. *SLon Magnetic Separator Ltd*;
2. *Ganzhou Nonferrous Metallurgy Research Institute*)

Abstract　In recent years many creative technologies are applied to SLon vertical ring and pulsating high gradient magnetic separators. SLon-2500 and SLon-3000 vertical ring and pulsating high gradient magnetic separators are created towards larger scale. Through optimizing the magnetic system and lowering the electric current density, their energy consumption is obviously dropped. New magnetic matrixes are developed to raise the beneficial efficiency of valuable minerals. Special SLon vertical ring and pulsating high gradient magnetic separators are developed which can treat coarser particles of 0-5mm. Through these technology creations, the mineral processing efficiency of SLon magnetic separators is further raised and their processing cost dropped. Therefore, they have been found wider applications.

Keywords　SLon magnetic separator; Scale up; Energy saving; Weak magnetic mineral processing

现代选矿工业的特点是：一方面朝着大规模和自动化水平不断提高的方向发展，另一方面是复杂难选的原矿增多。这就要求选矿设备和选矿工艺向大型化、自动化、高效率和

---

❶ 原载于《中国矿业科技大会论文集》，2010。

低成本方向发展,以满足现代选矿工业的需要。SLon 立环脉动高梯度磁选机在工业上广泛应用于氧化铁矿、钛铁矿、锰矿等弱磁性矿石的选矿及石英、长石、高岭土等非金属矿的提纯。通过多年的持续的技术创新与改进,该机在设备大型化、多样化、自动化、高效、节能、提高可靠性等方面得到了快速的发展,并且得到了更为广泛的应用。

1 设备大型化

设备大型化有利于提高选矿设备的台时处理能力和提高自动化水平,同时大型化设备具有节能、节约占地面积、操作成本低和设备作业率高的优点,可显著为用户降低生产成本和提高用户的经济效益。

1.1 SLon-2500 磁选机

赣州金环磁选设备有限公司在总结以往 25 年研制 SLon 立环脉动高梯度磁选机的基础上,研制了 SLon-2500 立环脉动高梯度磁选机(以下简称 SLon-2500 磁选机),如图 1 所示。

SLon-2500 磁选机的主要优点为:

(1) 能耗低。通过优化磁系设计、减少漏磁、降低电阻和提高电效率。使该机的激磁功率大幅度下降。该机达到额定背景磁感应强度 1.0T 时,相应的激磁功率仅 63kW。

(2) 处理量较大。该机的处理量可达 $100\sim150t/h$,是目前工业生产中实际应用处理量最大的立环脉动高梯度磁选机。

图 1 SLon-2500 磁选机的应用

(3) 采用新型数字式脉动冲程箱,具有力量大、平衡性好、可靠性高和能耗低的优点。采用新型的整流控制柜,激磁电流采用恒流控制,人机界面友好,自动化高,可远程控制。

1.2 SLon-2500 磁选机的工业应用

1.2.1 在海钢红旗尾矿库的应用

首台 SLon-2500 磁选机在海南钢铁公司(现已更名为海南矿业联合有限公司)应用于从尾矿中回收铁精矿。海南钢铁公司有 3 个选矿车间分选氧化铁矿,选矿流程为强磁选一次粗选一次精选一次扫选,或强磁选一次粗选一次扫选—反浮选。每年入选原矿约 230 万吨,目前每年大约有 120 万吨尾矿排入到红旗尾矿库。其总尾矿的平均品位为 28%Fe 左右,其中铁主要是存于 $0\sim19\mu m$ 粒级中。由于该尾矿氧化程度较深,含泥量大,且含有较多的连生体,因此要从该尾矿中再回收铁精矿技术难度较大。

SLon-2500 磁选机安装于该尾矿输送管的排矿口,从该排矿口截取尾矿入选。每年处理大约 72 万吨尾矿。该尾矿再回收系统的选矿工艺流程见图 2。首先用圆筒除渣筛筛出大于 0.8mm 的粗粒,接着用 SLon-2500 磁选机粗选,其粗选精矿经浓缩后用离心选矿机精选,离心机精矿作为最终精矿。

该流程综合选矿指标为给矿铁品位 29.90%，精矿铁品位 61.30%，铁精矿产率 8.53%，铁回收率 17.49%，尾矿铁品位 26.97%。该流程至今已运转 3 年多，该机的单机作业率达到 99% 以上。整个生产流程稳定，指标先进。

1.2.2 在赤铁矿选矿的应用

SLon-2500 磁选机研制成功以后，很快在氧化铁矿的选矿工业中得到大规模应用。2008～2009 年，印度一家大型钢铁公司采购了 10 台 SLon-2500 磁选机分选赤铁矿，采用 SLon-2500 磁选机一次粗选和一次扫选的选矿流程，其入选原矿铁品位为 59.77%，综合铁精矿品位为 65.00%，综合铁回收率为 93.86%，选矿指标优异。

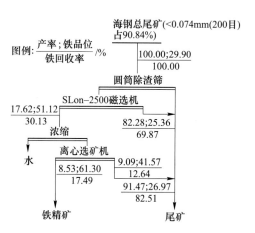

图 2　SLon-2500 磁选机分选海钢尾矿流程

1.2.3 在钛铁矿选矿工业中的应用

2009 年 SLon-2500 磁选机在攀枝花地区应用于钛铁矿选矿获得成功。攀枝花地区的钒钛磁铁矿储量丰富，目前该地区的选矿厂的选矿流程普遍是先用弱磁选机选出磁铁矿精矿，选完磁铁矿的尾矿即作为选钛作业的原矿。选钛作业普遍用强磁选机预先抛弃大部分尾矿，强磁选精矿再用重选或浮选精选。由于选钛流程中入选原料含钛品位低，要求强磁选机处理量大，生产成本低，尽可能多抛弃尾矿。SLon-2500 磁选机正好具备这些优势。

SLon-2500 磁选机在该厂用于钛铁矿的强磁粗选作业，其给矿粒度为小于 0.074mm（200 目）占 41.4%，台时处理量为 100t/h 左右。该机的选矿指标为强磁给矿品位 4.57% TiO_2，强磁精矿品位 9.86% TiO_2，TiO_2 回收率 75.37%。强磁尾矿品位仅 1.73% TiO_2，抛弃的尾矿产率为 65.07%。该机处理每吨给矿的电耗仅 0.42kW 左右。SLon-2500 磁选机在选钛工业的成功应用，显著地降低了选钛生产成本。

1.2.4 在石英提纯工业中的应用

2009 年 7 月 1 台 SLon-2500 磁选机在陕西应用于提纯石英砂，获得成功应用。该机给矿为 0～1mm 粒度的石英砂，给矿中含 0.15%～0.25% Fe_2O_3。要求 SLon-2500 磁选机一次分选的非磁性产品达到 0.08% Fe_2O_3 以下。通过生产调试，该机非磁性产品精矿达到 0.05%～0.06% Fe_2O_3，优于设计指标。

1.3 SLon-3000 磁选机的研制

在完成 SLon-2500 磁选机研制的基础上，近年又开展了 SLon-3000 立环脉动高梯度磁选机的研制。图 3 为该机在组装现场的照片。该机研制成功后，其处理干矿量将达到 150～250t/h。

2 设备节能研究

通过优化磁系结构,减少漏磁,采用优质铜管、增加铜的用量和降低激磁电流密度,使 SLon 立环脉动高梯度磁选机的电耗有了显著的降低。以 SLon-1000 磁选机激磁系统的改进为例,表 1 和图 4 为 SLon-1000 磁选机磁系改进前后的磁场特性。该机在改进前背景磁感应强度达到 1.0T 时,激磁功率为 31kW,改进后背景磁感应强度达到 1.0T 时,激磁功率为 17kW,比改进前节电 45.16%。

图 3 SLon-3000 磁选机

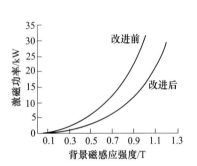

图 4 SLon-1000 磁选机背景磁感应强度与激磁功耗的关系

表 1 SLon-1000 磁选机磁系改进前后的磁场特性

改 进 前				改 进 后			
激磁电流/A	激磁电压/V	激磁功率/kW	磁感应强度/T	激磁电流/A	激磁电压/V	激磁功率/kW	磁感应强度/T
200	4.4	0.9	0.190	200	5.5	1.1	0.289
300	6.6	2.0	0.284	300	8.2	2.5	0.43
400	8.7	3.5	0.378	400	10.6	4.2	0.572
500	10.8	5.4	0.466	500	13.5	6.8	0.707
600	13.1	7.9	0.566	600	16.1	9.7	0.830
700	15.3	10.7	0.658	700	18.7	13.1	0.934
800	17.6	14.1	0.742	800	21.5	17.2	1.026
900	19.9	17.9	0.824	900	24.0	21.6	1.103
1000	22.2	22.2	0.896	1000	26.6	26.6	1.172
1100	24.6	27.1	0.961	1050	28.0	29.4	1.202
1200	26.6	31.9	1.015				

通过技术改进可较大幅度地降低激磁功率,但是制造磁选机的材料消耗也显著增加。以 SLon-1000 磁选机为例,同样达到额定背景磁感应强度 1.0T,额定激磁功率由 31kW 降低至 17kW,激磁线圈铜导线的用量由 640kg 增加至 1130kg,增加幅度为 76.56%,即电耗降低与用铜量大致成反比关系。其他材料如工程纯铁等也要相应增加。

通过技术改进可使该机的激磁电耗较大幅度地降低,但是也不能无限制地降低,否则不但设备耗材量大幅度增加，而且会影响整机的选矿性能。

3 新型磁介质的研究

为了提高 SLon 磁选机的选矿效率，我们针对各种矿石的不同特点，分别研制了 ϕ1mm、ϕ1.5mm、ϕ2mm、ϕ3mm、ϕ4mm、ϕ5mm、ϕ6mm 磁介质组，以及各种磁介质的搭配组合，如 ϕ1mm+ϕ2mm、ϕ2mm+ϕ3mm 等磁介质组合。

磁介质的选择应根据入选矿石的粒度，比磁化系数选取。入选矿石的粒度越细，磁性越弱，磁介质的直径应选得细一些；反之，入选矿石的粒度越粗，磁性越强，磁介质的直径应选得粗一些。

3.1 ϕ1mm 磁介质在海钢尾矿回收中的应用

海南钢铁公司排往红旗尾矿库的尾矿含铁矿物以赤铁矿为主，及少量的磁铁矿、黄铁矿和褐铁矿，脉石主要是石英。表2为该尾矿的粒度分析。

表2 海钢总尾矿粒度分析

粒级/mm	产率/%	铁品位/%	铁分布率/%
>0.076	6.97	24.00	5.82
0.076~0.037	21.85	29.70	22.56
0.037~0.019	20.41	28.80	20.44
<0.019	50.77	29.00	51.18
合　计	100.00	28.76	100.00

该尾矿的平均品位 TFe 为 28%左右，其中的铁主要是存在于 0~19μm 粒级中。表3为用几种不同直径的磁介质对该尾矿进行再回收的对比试验指标。从该表可见，采用 ϕ1mm 磁介质与 ϕ2mm 和 ϕ3mm 磁介质比较，给矿品位相同，精矿品位略高，铁回收率分别提高 6.41%和 9.82%。因此生产上的 SLon-2500 磁选机采用 ϕ1mm 磁介质。

表3 SLon 磁选机磁介质对比试验结果

磁介质直径/mm	给矿品位/%	铁精矿			尾矿		
		品位/%	产率/%	回收率/%	品位/%	产率/%	回收率/%
ϕ1	29.45	49.32	29.26	49.01	21.23	70.74	50.99
ϕ2	29.45	49.26	25.47	42.60	22.68	74.53	57.40
ϕ3	29.45	48.17	23.96	39.19	23.55	76.04	60.81

3.2 ϕ4mm 和 ϕ2mm 磁介质在攀钢选钛的应用

攀钢选钛厂目前是我国最大的钛精矿生产厂家。目前采用粗细分选，强磁—浮选流程分选钛铁矿。为了确定 SLon 磁选机最佳分选磁介质，分别对粗粒级钛铁矿和细粒级钛铁矿做了磁介质的优化试验。表4为该厂粗粒级一段强磁粗选给矿粒度分析，由该表可见，大于 0.074mm 的粒级产率和金属分布率分别占 81.75%和 79.00%。粒度较粗且钛铁矿具有较强的磁性，因此可能需要 ϕ3mm 或 ϕ4mm 磁介质。表5为该物料用 ϕ3mm 和 ϕ4mm 磁介质做对比试验结果。由该表可知，ϕ4mm 磁介质选矿效率高于 ϕ3mm，因此工业生产上的 SLon 磁选机采用 ϕ4mm 磁介质分选该物料。

表 4　攀钢选钛粗粒级一段强磁粗选给矿粒度分析

粒级/mm	产率/%	铁品位/%	钛分布率/%
>0.400	27.50	5.75	15.57
0.400~0.154	30.50	11.00	33.03
0.154~0.074	23.75	13.00	30.40
0.074~0.038	10.50	12.75	13.18
<0.038	7.75	10.25	7.82
合计	100.00	10.16	100.00

表 5　攀钢选钛粗粒一段强磁粗选磁介质对比试验结果

磁介质直径/mm	给矿 TiO_2 品位/%	强磁粗选精矿			尾矿 TiO_2 品位/%	选矿效率/%
		产率/%	TiO_2 品位/%	回收率/%		
$\phi3$	10.05	55.62	16.41	90.81	2.08	11.22
$\phi4$	10.05	54.79	16.75	91.32	1.93	12.04
差值	0	−0.83	+0.34	+0.51	−0.15	+0.82

表 6 为该厂细粒级一段强磁粗选给矿粒度分析，由该表可见，小于 0.074mm 的粒级产率和金属分布率分别占 79.60% 和 86.31%。预计需要 $\phi2mm$ 或 $\phi3mm$ 磁介质。表 7 为该物料用 $\phi2mm$、$\phi3mm$ 磁介质以及 $\phi2mm+\phi3mm$ 混合磁介质做对比试验结果。由该表可知，$\phi2mm$ 磁介质选矿效率最高，选矿指标优于 $\phi3mm$，也优于 $\phi2mm+\phi3mm$ 混合磁介质，因此工业生产上的 SLon 磁选机采用 $\phi2mm$ 磁介质分选该物料。

表 6　攀钢细粒一段强磁粗选给矿粒度分析

粒级/mm	产率/%	铁品位/%	钛分布率/%
>0.400	1.25	3.20	0.43
0.400~0.154	5.25	5.20	2.92
0.154~0.074	13.90	6.95	10.34
0.074~0.038	36.90	11.20	44.25
<0.038	42.70	9.20	42.06
合计	100.00	9.34	100.00

表 7　攀钢选钛细粒一段强磁粗选磁介质对比试验结果

磁介质直径/mm	给矿 TiO_2 品位/%	强磁粗选精矿			尾矿 TiO_2 品位/%	选矿效率/%
		产率/%	TiO_2 品位/%	回收率/%		
$\phi2$	9.47	55.11	15.09	87.82	2.57	8.99
$\phi3$	9.47	64.77	13.14	89.88	2.72	5.75
$\phi2+\phi3$	9.47	68.68	12.69	92.03	2.41	5.27

4　SLon 磁选机与离心选矿机的组合流程研究与应用

目前我国鞍山式氧化铁矿的选矿普遍采用强磁—反浮选流程或强磁选—重选—反浮选流程。鞍山式氧化铁矿矿物组成比较简单，含铁矿物主要是赤铁矿和磁铁矿，脉石矿物主要是石英，因此上述流程可获得较好的选矿指标。对于一些脉石矿物比较复杂的氧化铁矿，反浮选作业可能存在较难实施的问题。此外，对于一些小型选矿厂来说，反浮选作业存在技术复杂、环保审批时间长等问题。近几年，我们致力于 SLon 磁选机—离心选矿机的组合流程选矿的研究与应用，取得了较好的效果。

海南钢铁公司拥有一座储量较大的氧化铁矿，过去长期开采品位 50% 左右的富矿，有一部分铁品位 40% 左右的难选氧化铁矿未得到利用。通过研究和生产表明，若用强磁选流程只能得到品位 60% 左右的铁精矿，而市场上品位为 63% 以上的铁精矿好销且价格较高。因此，选矿生产要尽可能达到铁精矿品位 63% 以上。经过多次的探索试验、扩大试验和生产调试。以 SLon-1750 磁选机和 SLon-2400 离心选矿机组成的选矿流程获得工业应用，见图 5，该流程每小时处理原矿 30t 左右。自 2009 年 8 月投产，至今已运转 8 个月，生产流程稳定，选矿指标良好。

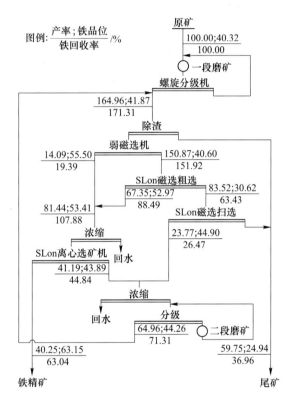

图 5　海南难选氧化铁矿 SLon 磁选机
和 SLon 离心机组合工业流程

5　分选较粗矿粒的 SLon 磁选机的研究与应用

SLon 立环脉动高梯度磁选机在细粒级弱磁性矿石的工业生产中得到广泛的应用，其给矿粒度范围为 0~2mm。普遍需要经过磨矿才能入选。然而，现在我国选矿厂入选矿石品位日益下降，如鞍山式氧化铁矿的入选品位由以前的 30% 左右下降至目前 26% 左右。如果在入磨前能进行预选抛弃一部分尾矿，则可显著降低后续作业的磨矿成本和选矿成本。通过对 SLon 磁选机的改进，其给矿粒度范围扩展到 0~5mm。

2010 年 3 月，1 台新型 SLon-1500 立环脉动高梯度磁选机在工业上应用于分选 0~5mm

褐铁矿获得成功。该褐铁矿的地质品位为 35%Fe 左右，在开采过程中混入了较多的围岩，因此入选 TFe 品位只能达到 26% 左右。经研究表明，该矿石结晶粒度较粗，破碎至 0~5mm 入选即可获得良好的选矿指标。该机的选矿指标见表 8。其入选原矿铁品位为 26.15%，SLon-1500 磁选机一次分选的精矿品位为 53.20%，铁回收率为 64.26%。由于褐铁矿含有结晶水，把结晶水烧失后相当于品位为 63% 的铁精矿。因此该铁精矿可作为商品矿直接销售。

表 8　SLon 磁选机分选 0~5mm 褐铁矿指标　　　　（%）

给矿铁品位	强磁选精矿			强磁选尾矿		
	产率	铁品位	回收率	产率	铁品位	回收率
26.15	31.59	53.20	64.26	68.41	13.66	35.74

SLon-1500 磁选机在该褐铁矿应用成功，使该选厂省去了磨矿作业，大幅度降低了生产成本，获得了显著的经济效益。

6　结论

（1）通过创新和改进显著地提升了 SLon 立环脉动高梯度磁选机的选矿性能和机电性能，拓展了该机的应用范围。

（2）SLon-2500 磁选机的研制与成功应用和 SLon-3000 磁选机的研制代表了该机大型化发展方向。设备的大型化有利于降低生产成本，节约占地面积，提高选矿厂自动化生产水平。

（3）对磁系的优化设计使 SLon 磁选机的激磁功率有了大幅度的下降，以 SLon-1000 磁选机为例，其额定激磁功率由改进前的 31kW 降低至改进后的 17kW，比改进前节电 45.16%。

（4）磁介质的直径和排列组合方式应结合入选矿石的特性，如粒度组成、比磁化系数。通过理论分析和试验研究，确定了 SLon 磁选机分选多种矿石的磁介质优化组合方式。

（5）通过对 SLon 磁选机和离心选矿机组合流程的研究，使这种技术在一些难选氧化铁矿的选矿工业中得到成功应用，为难选氧化铁矿的选矿提供了一条新途径。

（6）SLon 磁选机的给矿粒度范围由原来的 0~2mm 扩展至 0~5mm，可为一些低品位矿石在入磨前增加预选作业或直接选出铁精矿，显著降低磨矿成本。

写作背景　本文介绍了近几年 SLon 立环脉动高梯度磁选机技术创新，SLon-2500 磁选机的研制与应用，首次简要报道了 SLon-3000 磁选机的研制，以及新型磁介质的研究与应用。

A New Technology of Applying SLon-2500 Magnetic Separator to Recover Iron Concentrate from Abandoned Tails[①]

Xiong Dahe

(SLon Magnetic Separator Ltd. Shahe Industrial Park, Ganzhou, 341000, Jiangxi Province, China. Email: xdh@slon.com.cn)

Abstract The first SLon-2500 vertical ring-pulsating high gradient magnetic separator was designed and built up in 2006. It was installed at the tails dam of Hainan Iron Mining Company to recover iron concentrate from the pumping tails. The iron minerals exist mainly in the 0-19 micron fraction and they are deeply oxidised. The average processing results are: the feed of the SLon-2500 magnetic separator is 27.5%Fe, iron concentrate grade 52.40%Fe, iron recovery 35.14%, the final tails grade 21.87%Fe.

In order to further upgrade the iron concentrate, 18 units of SLon-ϕ1600×900 centrifugal separators are applied for the cleaning work. Their average cleaning results are: the feed grade of the centrifugal separator is 52.40%Fe, iron concentrate grade 61.30%Fe, iron recovery 64.19%, tails grade 41.58%Fe.

Up to September 2009, the processing line has been running for 30 months. The average feeding mass of the SLon-2500 is about 100 tons per hour. Each year it treats 720000 tons of abandoned tails, and recovers about 70000 tons of iron concentrate containing 61.30% Fe. This research work made a significant success in recovering valuable iron minerals from abandoned tails.

Keywords SLon magnetic separator; SLon centrifugal separator; Iron ore tails retreat

1 Introduction

Hainan Iron Mining Company locates in the West of Hainan Province, China. It has three workshops to process oxidised iron ore. The iron ore consists of mainly hematite and a small portion of magnetite. Their processing flow sheets are high intensity magnetic separation or high intensity magnetic separation-reverse flotation. Their processing results are: the mined ore grade is 45%Fe, iron concentrate grade 63%Fe; tails grade 29%Fe, iron recovery 66%. Their total processing capacity is 2.3 million tons annually, and about 1.2 million tons of tails containing about 29% Fe are pumped into a tails dam.

The tails grade seems high but difficult to recover. The iron minerals exist mainly in the 0-19 micron fraction and they are deeply oxidised.

In 2004, some one tried to apply ϕ1200 spiral separators at the tails dam to recover iron concentrate from the tails. The average results of five days running were: spiral feed grade 33.75%Fe; spi-

---

① 原载于《第二十五届国际选矿会议》，2010。

ral concentrate grade 33.30%Fe; spiral tails grade 33.82%Fe. The spiral processing results were disappointed and the test failed.

So a new technology need be developed to recover iron concentrate from the abandoned tails. SLon vertical ring-pulsating high gradient magnetic separators are a new generation of high intensity magnetic separator. They have been widely applied in oxidised iron ore processing industry (Xiong,1998,2006,2008). However, if we only use SLon-magnetic separator to treat the tails, we can only get iron concentrate grade about 55%Fe with a one roughing-one cleaning flow sheet. To solve this problem, we developed SLon centrifugal separators, which can reject magnetite associated quartz and other gangue minerals. The final iron concentrate grade can reach 61.30% Fe. The cooperation of the SLon-2500 magnetic separator and SLon centrifugal separators succeeded to recover iron concentrate from the abandoned tails.

2 Laboratory Reseach work

2.1 The characterization of the abandoned tails

The abandoned tails consists of mainly hematite and a small portion magnetite, pyrite and limonite. The gang minerals are mainly quartz. The chemical assay, mineral composition and particle size distribution are shown in Table 1-Table 3 respectively. The data of Table 3 shows that most of the iron exists in the 0-19 micron fraction.

Table 1 Chemical assay of the abandoned tails   (%)

Total Fe	FeO	SiO$_2$	Al$_2$O$_3$	CaO	MgO	S	P
29.45	0.93	54.88	6.24	2.90	1.68	1.04	0.043

Table 2 Mineralogical compositions of the abandoned tails

Mineral	Grade (Fe)/%	Ratio/%
Hematite	26.93	91.50
Magnetite	0.54	1.83
Pyrite	0.99	3.36
Limonite	0.98	3.33
Total	29.44	100.00

Table 3 The particle size analysis of the abandoned tails

Size/mm	Mass ratio/%	Grade(Fe)/%	Fe distribution/%
>0.076	6.97	24.00	5.80
0.076-0.037	21.85	29.70	22.57
0.037-0.019	20.41	28.80	20.45
0.019-0	50.77	29.00	51.18
Total	100.00	28.76	100.00

2.2 The test with SLon pulsating magnetic separator

The laboratory test results with SLon pulsating magnetic separator and centrifugal separator are shown in Table 4 and Table 5 respectively.

Table 4 Results of SLon magnetic separator test for roughing

Items	Mass /%	Size (<0.076mm)/%	Grade (Fe) /%	Iron recovery (Fe) /%
Feed	100.00	95.47	29.45	100.00
Concentrate	29.27		49.32	49.01
Tails	70.23		21.23	50.99

Table 5 Results of centrifugal separator test for cleaning

Items	Mass /%	Size (<0.076mm)/%	Grade(Fe) /%	Iron recovery(Fe) /%
Feed	100.00	91.92	49.32	100.00
Concentrate	46.69		61.49	58.22
Tails	53.31		38.66	41.78

Some one may ask: Can the cleaning work be done by SLon magnetic separator ? To answer this question, a test is done with SLon magnetic separator as shown in Table 6. The data show that the iron concentrate grade of the cleaning SLon magnetic separator is 55.54% Fe, much lower than 61.49% Fe of centrifugal separator. The cleaning tails of SLon magnetic separator is 41.28% Fe against 38.66% Fe of centrifugal separator.

Table 6 Results of SLon magnetic separator test for cleaning

Items	Mass /%	Size (<0.076mm)/%	Grade (Fe)/%	Fe distribution /%
Feed	100.00	91.92	49.32	100.00
Concentrate	56.38		55.54	63.49
Tails	46.69		41.28	36.51

The reason is that there are a lot of magnetite-quartz associated particles in the slurry. The magnetic susceptibilities of several related minerals are shown in Table 7.

Table 7 Magnetic susceptibilities of several related minerals

Mineral	Magnetite	Hematite	Limonite	Quartz
Magnetic susceptibility/$m^3 \cdot kg^{-1}$	$(5.25\text{-}11.6) \times 10^{-4}$	$(0.6\text{-}2.16) \times 10^{-6}$	$(0.31\text{-}0.4) \times 10^{-6}$	0

Fig.1 shows some of the magnetite and quartz associated particles. According to the data of Table 7, their visual magnetic susceptibilities are shown in Table 8.

Fig.1 Magnetite and quartz associated particles

Comparing the data of Table 7 and Table 8, if a particle contains 1% magnetite and 99% quartz by mass, its visual magnetic susceptibility reaches $6.25\times10^{-6}\,\mathrm{m^3/kg}$, much larger than that of a pure hematite particle of $(0.6\text{-}2.16)\times10^{-6}\,\mathrm{m^3/kg}$. In the high gradient magnetic field, the poor magnetite associated particles can be more easily captured by magnetic force than a pure hematite with the same mass. So if a SLon magnetic separator is applied to do the roughing work, a lot of poor magnetite associated quartzes are very easily captured into the magnetic product. If the magnetic product is cleaned by SLon magnetic separator again, they will be captured into the magnetic product again. Iron concentrate grade is difficult to further rise by this method.

Table 8 Visual magnetic susceptibility of magnetite and quartz associated particles

Mass of magnetite/%	Mass of quartz/%	Visual magnetic susceptibility/$\mathrm{m^3 \cdot kg^{-1}}$
0	100	0
1	99	6.25×10^{-6}
5	95	3.125×10^{-5}
20	80	1.25×10^{-4}
100	0	6.25×10^{-4}

Centrifugal separator is a relatively good gravity separator for fine minerals. It can produce a centrifugal force of $20g\text{-}50g$ ($g=9.8\,\mathrm{m/s^2}$). Fine mineral particles can be separated according to their specific gravity. The specific gravity of quartz is $2.6\,\mathrm{g/cm^3}$ and magnetite or hematite is $5.0\,\mathrm{g/cm^3}$. Table 9 shows the visual specific gravity of magnetite and quartz associated particles.

Table 9 Visual specific gravity of magnetite and quartz associated particles

Mass of magnetite/%	Mass of quartz/%	Visual specific gravity/$\mathrm{g\cdot cm^{-3}}$
0	100	2.60
10	90	2.84
30	70	3.32
50	50	3.80
70	30	4.28
100	0	5.00

According to the experience of production, centrifugal separator can reject magnetite and quartz associated particles even containing 70% mass of magnetite and 30% mass of quartz (their volume ratio are 55 : 45). So the cleaning capability of centrifugal separator is better than SLon magnetic separators for such ores.

Combining the test results of SLon magnetic and centrifugal separators as shown in Table 4 and Table 5, the feasible flow sheet to recover iron concentrate from the abandoned tails is shown in Fig. 2.

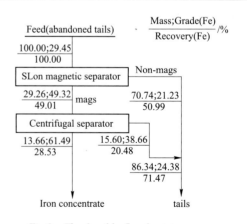

Fig.2 The feasible flowsheet to recover iron concentrate from the abandoned tails

2.3 The test for magnetic matrix optimization

In order to choose the best magnetic matrix for SLon magnetic separator, tests are done with $\phi 1mm, \phi 2mm, \phi 3mm$ wires made of magnetic conductive stainless steel as magnetic matrix. The laboratory test results with a SLon magnetic separator are shown in Table 10. It shows that the $\phi 1mm$ magnetic matrix can get the best beneficial results.

Table 10 SLon magnetic separator test results of matrix optimization

Matrix wire	Feed Grade (Fe)/%	Concentrate			Tails		
		Grade (Fe)/%	Mass /%	Recovery (Fe)/%	Grade (Fe)/%	Mass /%	Recovery (Fe)/%
$\phi 1mm$	29.45	49.32	29.26	49.01	21.23	70.74	50.99
$\phi 2mm$	29.45	49.26	25.47	42.60	22.68	74.53	57.40
$\phi 3mm$	29.45	48.17	23.96	39.19	23.55	76.04	60.81

3 The SLon-2500 magnetic separator

The first SLon-2500 vertical ring-pulsating high gradient magnetic separator is designed and built up in 2006. It utilises the combined force field of magnetism, pulsating fluid and gravity to continuously beneficiate weakly magnetic minerals. It possesses the advantages of high efficiency and energy saving. The major technical data are shown in Table 11.

Table 11 Technical data of SLon-2500 magnetic separator

Ring diameter /mm	Ore feeding size/mm (<200mesh%)	Ore feeding density/%	Slurry put through /m³·h⁻¹	Ore treating capacity /t·h⁻¹	Rated background field(Tesla)	Rated energizing current/A	Rated energizing voltage/V
2500	0-1.2 (30-100)	10-40	200-400	100-150	1.0	1400	45

Rated energizing power/kW	Driving motor power/kW	Pulsating stroke/mm	Pulsating frequency /r·min⁻¹	Water pressure /MPa	Flush water consumption /m³·h⁻¹	Cooling water consumption /m³·h⁻¹	Machine weight/t
63	11+11	0-30	0-300	0.2-0.4	200-300	6-7	105

3.1 Commercial test at the tails dam

In January 2007, The SLon-2500 magnetic separator and 18 units of SLon-$\phi 1600 \times 900$ centrifugal separators were installed at the tails dam of Hainan Iron Mining Company to recover iron concentrate from the pumping tails (Fig. 3).

Most of the iron minerals are exists in the 0-19 micron fraction. $\phi 1mm$ wire matrix boxes are used as the SLon-2500 magnetic matrix. At first, a serious site tests are done to optimize the SLon-2500 operational perimeters.

3.2 Energizing current and magnetic field test

Tests are repeated several times with the SLon-2500 magnetic separator to see the magnetic field effect. The typical processing results are shown in Table 12. When the energizing current goes up, the iron concentrate grade and tail grade go down, but iron recovery goes up. When the energizing current reaches above 1000A, these changes are very slowly. So the best energizing current is about 1000A.

Fig.3　The photo of SLon-2500 magnetic separator in the site

Table 12　The SLon-2500 test results vs. magnetic field

Energizing current /A	Magnetic background induction /T	Feed		Iron concentrate			Tails grade (Fe)/%
		Solid density /%	Grade (Fe)/%	Mass /%	Grade (Fe)/%	Recovery (Fe)/%	
600	0.477	41.31	28.60	14.59	52.60	26.84	24.50
800	0.621	42.29	28.80	16.93	52.35	30.78	24.00
1000	0.755	43.76	28.20	17.24	52.20	31.92	23.20
1200	0.866	41.27	28.80	19.72	51.80	35.47	23.15
1300	0.942	42.73	29.10	21.02	51.64	37.31	23.10

3.3 Feeding capacity test

In order to label the feeding capacity of the SLon-2500 magnetic separator, series tests are done as shown in Table 13. While the feeding ore tonnage changes from 95.01t/h to 147.26t/h, the processing results only slightly change. It seems that the best results appear at the feeding ore tonnage 109.47t/h. The second highest results appear at the feeding ore tonnage 95.01t/h. If we consider that the tests last several ours, the surge of the feeding slurry and the sampling error, we believe that the preferential feeding ore tonnage is around 100t/h.

Table 13　The SLon-2500 test results vs. feeding capacity

Feed				Iron concentrate			Tails grade (Fe)/%
Solid density /%	Slurry volume /$m^3 \cdot h^{-1}$	Ore tonnage /$t \cdot h^{-1}$	Grade (Fe)/%	Mass /%	Grade (Fe)/%	Recovery (Fe)/%	
29.76	257	95.01	28.97	19.31	54.87	36.58	22.77
31.49	248	98.52	27.63	16.30	53.30	31.45	22.63
35.73	224	105.43	29.58	19.00	53.50	34.36	23.97
32.25	261	109.47	28.85	19.75	54.65	37.41	22.50
36.18	261	125.31	29.80	19.15	52.60	33.80	24.40
35.22	294	134.60	27.20	15.72	52.23	30.19	22.53
34.87	320	147.26	25.60	14.79	52.10	30.10	21.00

3.4 The commercial processing flow sheet

The commercial processing flow sheet and the average results of 14 days continuously running are shown in Fig. 4.

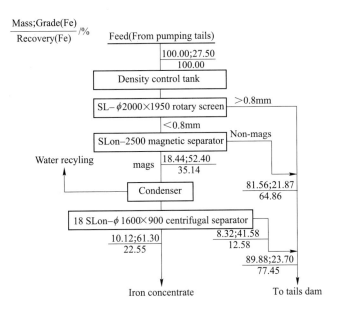

Fig.4 The commercial processing flowsheet

The pumping tails enter a tank and mixed with water to a suitable solid density. A rotary screen is applied to remove >0.8mm matters. The <0.8mm fraction is treated by the SLon-2500 magnetic separator. It gets rough iron concentrate containing 52.4%Fe and discharges 81.56% mass of final tails at very low coast. For example, when it works at the energizing current of 1000A and feeding ore tonnages of 100t/h, its energizing power is only 39kW and its ring and pulsating driving motors are 11+11kW. Each feeding ore tonnage consumes only 0.61kW·h and 2.5t recycling water.

The rough concentrate of SLon-2500 magnetic separator enters a condenser so that the solid density of the centrifugal separator feeding can be controlled. 18 SLon-φ1600×900 centrifugal separators are applied to do the cleaning work. Their average cleaning results are: the feed grade of the centrifugal separator is 52.40%Fe, iron concentrate grade 61.30%Fe, iron recovery 64.19%, tails grade 41.58%Fe.

The final results of the total flow sheet are: feed grade 27.50%Fe, concentrate grade 61.30%Fe, concentrate mass ratio 10.12%, concentrate iron recovery 22.55%, tails grade 23.70%.

Up to September 2009, the processing line has been running for 30 months. The average feeding mass of the SLon-2500 is about 100 tons per hour. Each year it treats 720000 tons of abandoned tails, and recovers about 70000 tons of iron concentrate containing 61.30%Fe.

4 Conclusions

(1) The first SLon-2500 vertical ring-pulsating high gradient magnetic separator is successfully developed and applied to recover iron concentrate from the abandoned pumping tails of Hainan Iron Mining Company. The iron minerals in the pumping tails are deeply oxidised and mainly exist in the 0-19 micron fraction. The SLon-2500 is applied for roughing operation. It can reject most of the final tails with low cost.

(2) Centrifugal separators are applied to do the cleaning work. They can reject the magnetite associated quartz and other gangue minerals. The final iron concentrate grade is more easily raised by centrifugal separator than by high intensity magnetic separation.

(3) The final results of the total flow sheet are: feed grade 27.50% Fe, concentrate grade 61.30% Fe, concentrate mass ratio 10.12%, concentrate iron recovery 22.55%, tails grade 23.70%. Each year the processing line treats 720000 tons of abandoned tails, and recovers about 70000 tons of iron concentrate containing 61.30% Fe. This research work made a significant success in recovering valuable iron minerals from abandoned tails.

References

[1] Xiong D H, Liu S Y, Chen J. New technology of pulsating high gradient magnetic separation [J]. Int. J. Miner. Process, 1998, 54: 111-127.

[2] Xiong D H. SLon magnetic separator promoting Chinese oxidised iron ore processing industry [J]. Int. Proceedings of XXIII IMPC, Istanbul, 2006: 276-281.

[3] Xiong D H. SLon magnetic separators applied to beneficiate low grade oxidised iron ores [J]. Int. Proceedings of XXIV IMPC, 2008: 813-818.

写作背景　本文向国外同行介绍了 SLon-2500 立环脉动高梯度磁选机和离心选矿机在海南铁矿从尾矿中回收赤铁矿的选矿原理，小型试验、工业试验和生产应用。这是赣州金环磁选设备有限公司研制的第一台 SLon-2500 磁选机和第一批 SLon-ϕ1600 离心选矿机成功应用实例。

SLon 磁选机分选氧化铁矿工业应用新进展

熊大和[1,2]

（1. 赣州金环磁选设备有限公司；2. 赣州有色冶金研究所）

摘　要　介绍了新型 SLon 立环脉动高梯度磁选机的研制以及在河北司家营氧化铁矿、安徽李楼铁矿、海南铁矿和云南大红山铁矿的工业应用。司家营氧化铁矿具有结晶粒度细的特点，采用阶段磨矿、螺旋溜槽重选、SLon 高梯度磁选机磁选和反浮选流程，选矿指标为：原矿 Fe 品位 30.44%，铁精矿品位 66%，铁回收率 80%。李楼铁矿是以镜铁矿为主的氧化铁矿，采用 SLon 高梯度磁选和反浮选流程，选矿指标为：原矿 Fe 品位 31.65%，铁精矿品位 65%，铁回收率 80%。海南难选氧化铁矿是以赤铁矿为主，含有少量磁铁矿的混合矿，采用 SLon 高梯度磁选和离心选矿机的组合流程，选矿指标为：原矿 Fe 品位 39.40%，铁精矿品位 63.27%，铁回收率 64.12%。SLon 高梯度磁选和离心选矿机的组合流程还在大红山铁矿应用，解决了该矿强磁选精矿用反浮选精选存在的问题。通过对 SLon 磁选机设备的创新和采用先进的选矿流程，使氧化铁矿的选矿技术得到了新的发展，为用户创造了良好的技术经济指标。

关键词　SLon 立环脉动高梯度磁选机　赤铁矿　镜铁矿　氧化铁矿

经过 26 年持续的技术创新与改进，SLon 立环脉动高梯度磁选机成为新一代的强磁选设备，该机利用磁力、脉动流体力和重力的综合力场分选弱磁性矿石，具有选矿效率高、可靠性高和能耗低的优点，至今已有 2000 余台成功地在工业上应用于赤铁矿、褐铁矿、镜铁矿、钛铁矿、铬铁矿、黑钨矿和锰矿的选矿以及在石英、长石、霞石和高岭土等非金属矿的提纯。近年又研制了 SLon-2500、SLon-3000 立环脉动高梯度磁选机等多种新机型，使 SLon 磁选机得到了更为广泛的应用。

1　新型 SLon 磁选机的研制

SLon 立环脉动高梯度磁选机正朝着大型、高效、节能、可靠性方向发展，近年研发的新型 SLon 磁选机有如下几种。

1.1　SLon-2500 和 SLon-3000 磁选机

近年赣州金环磁选设备有限公司成功研制了 SLon-2500 和 SLon-3000 立环脉动高梯度磁选机，目前 SLon-2500 磁选机已有 80 多台在工业上应用于各种弱磁性矿石的选矿。设备大型化有利于提高选矿设备的台时处理能力和提高自动化水平，并具有节能、节约占地面积、操作成本低和设备作业率高的优点，可显著为用户降低生产成本和提高经济效益。以 SLon-3000 立环脉动高梯度磁选机（图 1）为例，其处理矿石能力达到 150~250t/h，当额定背景磁感应强度为 1.0T 时，相应的激磁功率仅 87kW，整机功率仅为 124kW，处理每吨矿石的电耗仅为 0.3~0.5kW·h。首台 SLon-3000 磁选机于 2010 年 9 月在攀枝花应用于分

---

[1]　原载于《2011 年中国矿业科技大会论文集（金属矿山）》，西安。

选钛铁矿获得成功,至今已运转 9 个月,获得了优良的选矿指标。

1.2 SLon-1500-1.3T 和 SLon-1750-1.3T 磁选机

有些弱磁性矿石磁性很弱,且粒度细,需要较高的磁场力才能捕收上来,为此研制了背景磁感应强度较高的 SLon-1500-1.3T 和 SLon-1750-1.3T 立环脉动高梯度磁选机。它们的额定背景磁感应强度达到 1.3T。该机已在褐铁矿的选矿、氧化铁矿的选矿、尾矿再选和非金属矿的提纯等方面得到应用。

图 1　SLon-3000 立环脉动高梯度磁选机

1.3 分选较粗粒级的 SLon-1500 和 SLon-2000 磁选机研制

SLon 立环脉动高梯度磁选机在细粒级弱磁性矿石的工业生产中得到广泛的应用,其给矿粒度范围为 0~2mm。普遍需要经过磨矿才能入选。然而,现在我国选矿厂入选矿石品位日益下降,如鞍山式氧化铁矿的入选品位由以前的 30% 左右下降至目前 26% 左右。如果在入磨前能进行预选抛弃一部分尾矿,则可显著降低后续作业的磨矿成本和选矿成本。通过对 SLon 磁选机的改进,其给矿粒度范围扩展到 0~5mm。

2010 年新研制的适用于分选较粗粒级的 SLon-1500 和 SLon-2000 立环脉动高梯度磁选机在某褐铁矿选厂应用于分选 0~5mm 褐铁矿获得成功。该褐铁矿的地质品位为 35%Fe 左右,在开采过程中混入了较多的围岩,因此入选 Fe 品位只能达到 26% 左右。经研究表明,该矿石结晶粒度较粗,破碎至 0~5mm 入选即可获得良好的选矿指标。其入选原矿 Fe 品位为 26.15%,SLon 磁选机一次分选的精矿品位为 53.20%,铁回收率为 64.26%。由于褐铁矿含有结晶水,把结晶水烧失后相当于品位为 63% 的铁精矿。因此该铁精矿可作为商品矿直接销售。SLon 在该褐铁矿应用成功,使该选厂省去了磨矿作业,大幅度降低了生产成本,获得了显著的经济效益。

2　SLon 磁选机在选矿工业中的应用

2.1 在司家营氧化铁矿选矿工业中的应用

司家营铁矿是我国的大型铁矿山,浅部以氧化铁矿为主,深部以磁铁矿为主。其氧化铁矿类似于鞍山式氧化铁矿,但是结晶粒度比鞍山式氧化铁矿更细,含铁矿物以赤铁矿为主并含少量的褐铁矿,脉石矿物以石英为主,属低品位难选氧化铁矿。自 1958 年以来,我国多个科研机构对司家营铁矿石进行了大量的研究工作。近年,氧化铁矿选矿技术水平因为选矿新技术、新药剂和新设备的应用得到了飞速的发展,司家营氧化铁矿的选矿技术难题得到解决。2005 年至今该矿分两期建成了年处理 1000 万吨氧化铁矿的选矿厂,其中采用了 39 台 SLon 立环脉动高梯度磁选机,该厂第一期 600 万吨/a 设计的选矿流程如图 2 所示。

该流程中,破碎至 0~12mm 原矿经一段球磨机和螺旋分级机闭路磨矿,螺旋分级机溢流进入二段分级旋流器和二段球磨机闭路磨矿,二段分级旋流器的溢流进入二段粗细分级旋流器,二段粗细分级旋流器的沉砂用螺旋溜槽一次粗选和一次精选,首先得到一部分结晶粒度

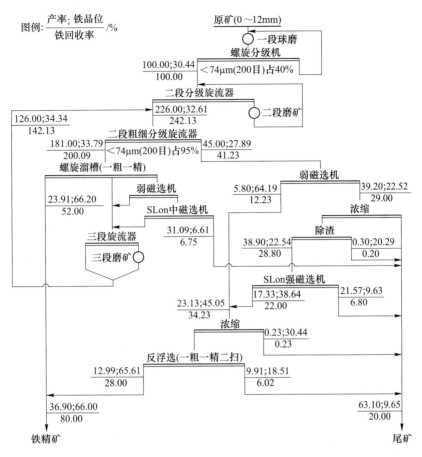

图 2　司家营第一期 600 万吨/a 氧化铁矿设计选矿流程

较粗的铁精矿，螺旋流槽的尾矿用滚筒弱磁选机和 SLon 立环脉动中磁机（背景磁感应强度 0.6T，以下简称 SLon 中磁机）扫选，SLon 中磁机抛弃产率占原矿 31.09%且粒度较粗的低品位尾矿，螺旋溜槽中矿、弱磁选机精矿和 SLon 中磁机的精矿合并后用三段旋流器分级，其沉砂进入三段磨矿，其溢流和三段磨矿的排矿合并后返回与二段分级旋流器的溢流合并进入二段粗细分级旋流器形成闭路；二段粗细分级旋流器的溢流粒度为小于 74μm（200 目）占 95%左右，用弱磁选机选出产率为 5.8%左右的磁铁矿，弱磁选机的尾矿浓缩除渣后用 SLon 立环脉动高梯度磁选机（背景磁感应强度 1.0T，以下简称 SLon 强磁机）选出赤铁矿等弱磁性矿物并抛弃产率占原矿 21.57%粒度较细的尾矿，弱磁选机的精矿和 SLon 强磁机的粗精矿合并，经浓缩后进入反浮选，反浮选精选后得到高品位的细粒铁精矿。

该流程的特点为：采用阶段磨矿，螺旋溜槽拿出了大部分结晶粒度较粗并已单体解离的铁矿物，SLon 中磁机提前抛弃产率占原矿 31.09%、Fe 品位仅 6.61%粒度较粗的尾矿，实现了能收早收，能丢早丢的原则，减少了磨矿量，节约了选矿成本；SLon 强磁机抛弃了产率占原矿 21.57%、品位仅 9.63%的细粒铁尾矿。SLon 中磁机和 SLon 强磁机抛弃的尾矿产率合计占原矿的 52.66%，占尾矿总量的 83.45%，它们有效地控制全流程的尾矿品位，保证了全流程获得较高的选矿回收率。尽管反浮选的成本较高，但该流程反浮选的给矿产率仅占原矿的 23.13%，浮选药剂消耗量较少，因此全流程选矿成本较低。

该流程综合选矿指标为：给矿铁品位 30.44%，精矿铁品位 66.00%，铁精矿产率 36.90%，铁回收率 80.00%，尾矿铁品位 9.65%。如图 3 所示为 SLon 磁选机在司家营铁矿的应用照片。

2.2 在安徽李楼铁矿镜铁矿选矿工业中的应用

安徽省霍邱县铁矿资源丰富，近几年有几个铁矿选矿厂相继建成或正在建设。安徽开发矿业有限公司前几年在当地建成了一座年处理 60 万吨低品位镜铁矿的选矿厂，在该厂建设成功的基础上，目前一座年处理 500 万吨低品位镜铁矿的选矿厂正在建设之中。其 500 万吨/a 设计的选矿流程如图 4 所示。

图 3 SLon 磁选机在司家营铁矿的应用照片

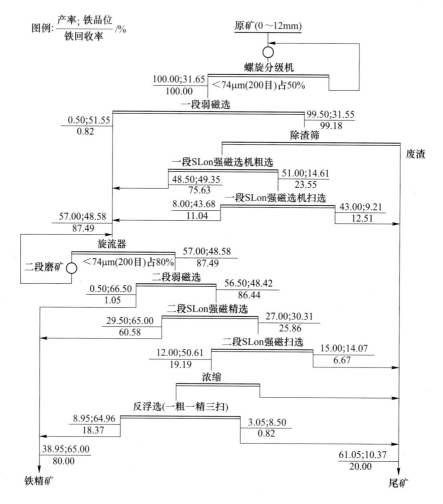

图 4 李楼铁矿 500 万吨/a 镜铁矿设计选矿流程

李楼铁矿镜铁矿的含铁矿物主要是镜铁矿，其他铁矿物含量很少，原矿中磁铁矿产率仅为 0.5% 左右。其选矿流程为：原矿铁品位 31.65%，一段闭路磨矿分级至小于

74μm(200目)占50%左右,用一段弱磁选和一段SLon强磁粗选和一段SLon强磁扫选提前抛弃产率为43.00%,品位为9.21%左右的低品位尾矿。它们的精矿经二段闭路分级磨矿至小于74μm(200目)占80%左右,然后用二段弱磁选和二段SLon强磁精选得到一部分铁品位65%以上的铁精矿成品。二段SLon强磁精选的尾矿再经二段强磁扫选,二段强磁扫选精矿经浓缩后用反浮选精选,反浮选得到另一部分品位为64.96%左右的铁精矿成品。全流程设计的综合选矿指标为:给矿铁品位31.65%,精矿铁品位65.00%,铁精矿产率38.95%,铁回收率80.00%,尾矿铁品位10.37%。

该流程的选矿特点为:一段强磁为一粗一扫,既抛弃了大量的尾矿,又保证了一段强磁尾矿品位处于较低的水平;二段强磁精选直接得到了大部分铁精矿,这部分铁精矿产率占原矿的29.5%,占全流程总精矿的75.74%。由于强磁选作业成本低,强磁选提前抛弃了大部分尾矿和得出了大部分精矿,仅有占原矿12%左右的二段强磁扫选精矿进入反浮选作业,由于反浮选的作业成本较高,进入反浮选的矿量越少,全流程的生产成本越低,因此,SLon磁选机为全流程取得良好的选矿指标和保持较低的生产成本起到了关键的作用。

目前,SLon磁选机还在安徽省霍邱县金日盛矿业公司的年处理原矿450万吨/a选矿厂和安徽省霍邱县环山矿业公司年处理原矿60万吨/a选矿厂应用成功,这两地的铁矿石和选矿流程与李楼铁矿相近。

2.3 SLon磁选机与离心选矿机组合流程分选海南氧化铁矿

海南钢铁公司拥有一座储量较大的氧化铁矿,过去长期开采品位50%左右的富矿。现有一部分铁品位40%左右的难选氧化铁矿暂未得到利用。近年有关单位对这部分氧化铁矿进行了系统的选矿试验研究,试验证明,若仅用强磁选流程只能得到品位为60%左右的铁精矿。而市场上品位为63%以上的铁精矿好销且价格较高。因此,选矿试验和生产实践都要求铁精矿品位达到63%以上。由于该矿石性质复杂多变,经工业试验表明,反浮选用于强磁精矿的精选作业存在选矿指标波动大、流程不稳定的问题。

经过多次的探索实验和扩大试验,获得如图5所示的SLon磁选机与SLon离心选矿机的组合流程。其入选原矿品位为39.40%,含铁矿物主要是赤铁矿及占原矿产率10%左右的磁铁矿。首先磨矿至小于74μm(200目)占87%~90%,然后用弱磁选机选出磁铁矿,弱磁选机的尾矿用SLon磁选机一次粗选和一次扫选,磁选精矿合并,浓缩后用离心选矿机精选拿出部分品位为63%以上的铁精矿,离心机尾矿经浓缩后用旋流器分级,旋流器沉砂进入二段球磨。旋流器溢流用弱磁选机选出磁铁矿精矿,弱磁选尾矿用SLon磁选机扫选。SLon磁选机扫选精矿再用离心机精选得出部分品位为63%以上的铁精矿。该流程的选矿指标为:原矿品位为39.4%,综合铁精矿品位为63.27%,铁回收率64.12%。2009年已建设了一条年处理20万吨原矿的生产线。2010年该生产线已扩大至年处理70万吨原矿,至今生产流程稳定,生产指标良好,生产成本较低。目前海南矿业联合有限公司正在按此流程建设一条年处理200万吨难选氧化铁矿的选矿厂。

2.4 SLon磁选机与离心选矿机组合流程分选昆钢大红山铁矿

昆钢大红山铁矿是磁铁矿和赤铁矿共生的混合矿,原设计采用阶段磨矿—弱磁选—

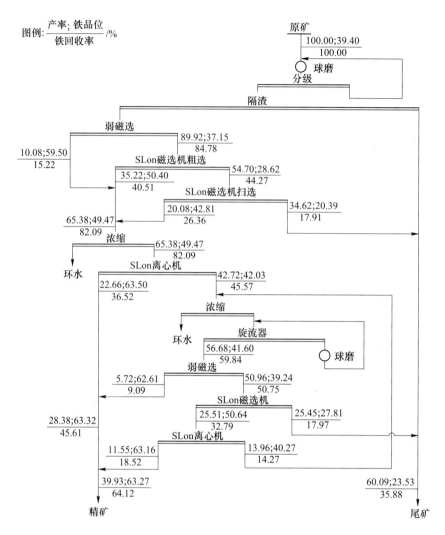

图 5 海南难选氧化铁矿 SLon 磁选机与 SLon 离心选矿机的组合分选流程

SLon 强磁选—反浮选流程，由于脉石矿物复杂多变，反浮选作业无法稳定而未应用成功，前几年用摇床取代了反浮选。但是，摇床存在处理量较小，对细粒铁矿回收率较低的缺点。大红山铁矿采用赣州金环磁选设备有限公司研制的 SLon-2400 离心选矿机取代反浮选作业，该流程于 2010 年顺利投产。新改造的选矿流程见图 6，全流程铁精矿品位从改造前的 62%~63% 提高至目前 64%~65%。

该流程的主要特点为：SLon-2000 磁选机对细粒氧化铁矿回收率较高，可有效地控制强磁选作业的尾矿品位；采用 SLon-2400 离心选矿机取代反浮选作业，对强磁选精矿进行精选，2009 年在大红山铁矿的工业试验结果表明，离心选矿机与螺旋溜槽或摇床比较具有富集比大，对细粒铁矿回收率较高的优点。这些新技术和新设备的应用，使该流程既可运行顺畅，又能获得良好的选矿指标。全流程的选矿指标为：原矿品位 35.61%，铁精矿品位 64.95%，铁回收率 74.06%。

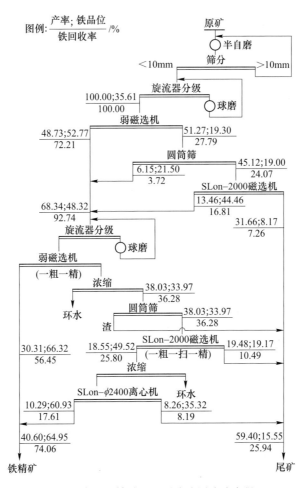

图 6　大红山铁矿 400 万吨选厂选矿流程

3　结论

（1）SLon 磁选机在司家营铁矿年处理 1000 万吨难选氧化铁矿的流程中，主要是用于抛尾作业，可将粗粒级的尾矿品位控制在 6.61% 左右，细粒级尾矿品位控制在 9.63% 左右，为全流程取得较高的铁回收率起到了关键作用。

（2）SLon 磁选机在李楼铁矿分选镜铁矿，由于该镜铁矿含磁铁矿很少，很少有磁铁矿和石英的贫连生体，因此二段强磁精选作业可选出大部分铁精矿产品，进入反浮选作业的矿量仅占原矿产率的 12% 左右，实现了能收早收，能丢早丢的选矿原则。该流程的另一个特点是：由于镜铁矿磁性较弱，一段强磁选采用一粗一扫，二段强磁选采用一精一扫，有效地将强磁选尾矿品位控制在较低水平，保证了全流程获得较高的铁回收率。

（3）强磁粗选精矿中，若含有磁铁矿和石英的贫连生体，其比磁化率远高于赤铁矿的比磁化率，则再利用强磁精选作用不大，离心选矿机可有效地剔除含少量磁铁矿的石英或其他脉石的连生体，强磁选精矿用离心机精选往往可获得较高的铁精矿品位。SLon 立环脉动高梯度磁选机和 SLon 离心选矿机的组合流程分选海南难选氧化铁矿应用成功，获得了良好的选矿指标,并具有生产成本低、环境友好、易于操作管理的特点。

（4）大红山铁矿由于脉石矿物复杂多变,其年处理400万吨氧化铁矿的选厂原设计的反浮选作业暂未应用成功,采用SLon磁选机和SLon离心选矿机的组合流程解决了强磁选粗精矿的精选难题。

（5）通过持续的技术创新,赣州金环磁选设备有限公司研制了多种新型SLon立环脉动高梯度磁选机,它们具有处理量大、选矿效率高、设备可靠性高、能耗低、选矿成本低的优点,它们在低品位弱磁性矿石的选矿工业中得到了更为广泛的应用,为用户创造了良好的技术经济效益。

写作背景 本文介绍了SLon立环脉动高梯度磁选机分选氧化铁矿的新进展,其中SLon-1500-1.3T、SLon-1750-1.3T磁选机,SLon-2000粗颗粒磁选机是新研制的设备,安徽李楼铁矿年处理600万吨镜铁矿的选矿厂,河北唐山钢铁公司司家营铁矿氧化铁矿年处理量从600万吨扩产到1000万吨,海南SLon磁选机与离心机组合流程分选难选氧化铁矿年处理量从20万吨扩产到70万吨均为新项目。

The Integrative Technology of SLon Magnetic Separator and Centrifugal Separator for Processing Oxidised Iron Ores[①]

Xiong Dahe

(*SLon Magnetic Separator Ltd. Shahe Industrial Park Ganzhou* 341000, *Jiangxi Province*, *China. Email*: xdh@slon.com.cn)

Abstract SLon vertical ring and pulsating high gradient magnetic separators possess excellent mineral processing ability and have been widely applied to process oxidised iron ores. SLon centrifugal separators are applied to clean the SLon magnetic concentrate and remove quartz and other gangue minerals associated with a small portion of magnetite. This paper introduces the applications of this technology in Hainan, Dahongshan and Anshan for processing oxidised iron ores.

1 Introduction

SLon vertical ring and pulsating high gradient magnetic separator (SLon VPHGMS or SLon magnetic separator) (Xiong, 2007) is a new generation of continuously working HGMS with high efficiency. It uses the combined force field of magnetism, pulsating slurry and gravity to beneficiate weakly magnetic minerals. It possesses the advantages of large beneficial ratio, high recovery and high reliability. In the past 25 years, We have accelerated the improvement and modernisation of SLon VPHGMS. Currently approximately 15 models and 2000 units of SLon VPHGMS are applied to beneficiate hematite, limonite, specularite, ilmenite, chromite, wolframite and manganese ore to purify nonmetallic minerals such as quartz, feldspar, nepheline and kaolin. Fig. 1 is the latest developed SLon-3000 VPHGMS in operation. Its iron ore treating capacity reaches 150-250t/h.

Fig.1 SLon-3000 SLon VPHGMS

2 Cooperation technology of SLon magnetic separator and centrifugal separator

SLon VPHGMS technology is widely applied to beneficiate weakly magnetic iron ores (Xiong, 2009). They treat oxidised iron ore that contains part of magnetite and is usually applied to the roughing stage, with flotation or other methods applied to the cleaning stage. If SLon technology is applied at the cleaning stage, the effect is ineffective as is cannot remove some of the gang minerals which

---

[①] 原载于《Proceedings of Iron Ore 2011》，Australia 27-29 July 2011。

are associated with magnetite.

Table 1 illustrates the magnetic susceptibility of several iron ores, which shows that the magnetic susceptibility of magnetite is more than 100 times of those of hematite and specularite.

Table 1 Magnetic susceptibility of several iron ores

Iron ore	Magnetite	Mag-hematite	Hematite	Specularite
Magnetic susceptibility/$m^3 \cdot kg^{-1}$	$(6.25\text{-}11.6)\times10^{-4}$	$(6.2\text{-}13.5)\times10^{-6}$	$(0.6\text{-}2.16)\times10^{-6}$	3.7×10^{-6}

Fig. 2 shows some of the quartz associated with a small portion of magnetite that can easily be attracted by SLon magnetic separator while Table 2 shows the visual magnetic susceptibility of them. For example, if a particle contains one per cent mass of magnetite and 99 per cent mass of quartz, its visual magnetic susceptibility reach to $6.25\times10^{-6} m^3/kg$, much larger than that $(0.6\text{-}2.16)\times10^{-6} m^3/kg$ of a pure haematite particle. Therefore, in the high gradient magnetic field, the magnetic attracting force acting on the particle is much larger than that a pure haematite particle. So, if a SLon magnetic separator is applied to do the roughing work, the poor magnetite associated quartzes are very easily captured into the magnetic product. If the magnetic product is cleaned by SLon magnetic separator again, they will be captured into the magnetic product again. Iron concentrate grade is difficult to further rise by SLon magnetic cleaning.

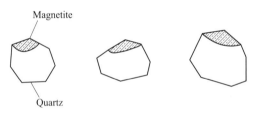

Fig.2 Associated particles of quartz and magnetite

Table 2 Visual magnetic susceptibility of quartz and magnetite associated particles

Mass of magnetite/%	Mass of quartz/%	Visual magnetic susceptibility/$m^3 \cdot kg^{-1}$
0	100	0
1	99	6.25×10^{-6}
2	95	3.125×10^{-5}
10	90	6.25×10^{-5}
20	80	1.25×10^{-4}
50	50	3.125×10^{-4}
100	0	6.25×10^{-4}

In recent years, SLon Magnetic Separator Ltd developed the SLon Centrifugal separator, a gravity separator for fine minerals that can produce a centrifugal force of $20g$-$50g$ (g-Earth gravity acceleration, $g=9.8m/s^2$). Fine mineral particles can be separated according to their specific gravity. The specific gravity of quartz is $2.6g/cm^3$ and magnetite or haematite is $5.0g/cm^3$. Table 3 shows the visual specific gravity of magnetite and quartz associated particles.

Table 3 Visual specific gravity of magnetite and quartz associated particles

Mass of magnetite/%	Mass of quartz/%	Visual specific gravity/g · cm^{-3}
0	100	2.6
10	90	2.84
30	70	3.32
50	50	3.8
70	30	4.28
100	0	5.0

According to production experiment, the centrifugal separator can reject magnetite and quartz associated particles containing up to 70% mass of magnetite and 30% mass of quartz (visual specific gravity is 4.28g/cm^3 while the volume ratio are 55 : 45). So the cleaning capability of centrifugal separator is better than SLon magnetic separators for such ores.

SLon VPHGMS possesses the advantages of large capacity, high benefit efficiency and low operating costs when applied to such iron ore roughing. However, they are limited to the quartz and magnetite associated particles in cleaning. Centrifugal separators can solve this problem for cleaning; by combining the advantages of SLon VPHGMS and centrifugal separator to treat such iron ores, good results can be achieved.

3 Applications in industry

3.1 Application in processing Hainan oxidised iron ore

The Hainan Iron Mine, which contains an oxidised iron ore deposit, is located in the west of the Hainan Province of China. Its iron minerals are primarily haematite with a small portion of magnetite. In the past, the mine mined the rich iron ore upper 50%Fe. A portion of poor iron ore, about 40%Fe, remains unexploited.

In recent years, a lot of mineral processing tests have been done on the poor iron ore. It has been demonstrated that if only SLon VPHGMS are applied for roughing and cleaning, the iron concentrate grade can reach only about 60%Fe. Because the market and price is much better for 63% Fe of iron concentrate, it is required that the iron ore processing test and industrial production should reach 63%Fe of iron concentrate.

Through extensive testing, we developed a flow sheet of SLon VPHGMS and SLon centrifugal separator for processing poor oxidised iron ore, as shown in Fig. 3. The feed iron ore grade is 39.40%. The iron minerals consist of mainly haematite and about 10% mass of magnetite. The process is as follows:

(1) The iron ore is first ground to 87%-90% of <200 mesh, then drum low intensity magnetic separators (LIMS) are used to take out magnetite.

(2) The non-mags of LIMS go to SLon VPHGMS for one roughing and one scavenging to take out haematite.

(3) The mags of the LIMS and SLon VPHGMS are condensed together, and then centrifugal separators are used to take out part of final iron concentrate of 63%Fe.

(4) The tails of SLon centrifugal separator is condensed and classified by hydrocyclone. The underflow of the hydrocyclone goes secondary ball mill.

(5) The overflow of the hydrocyclone is treated by LIMS to take out magnetite.

(6) The non-mags of LIMS are scavenged by SLon VPHGMS.

(7) The mags of SLon VPHGMS is cleaned by centrifugal separator again to get out part final iron concentrate of about 64%Fe.

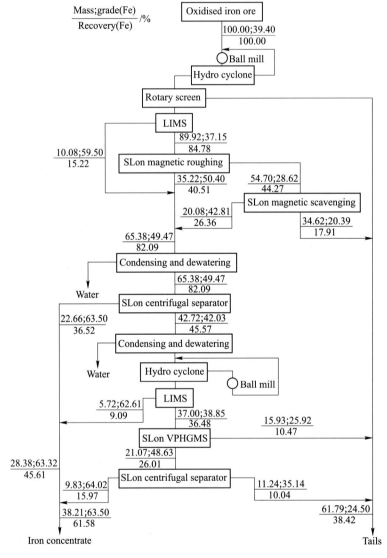

Fig.3 The combined flow sheet of SLon magnetic and centrifugal separators for beneficiating Hainan oxidised iron ore

The total results of the flow sheet are:
(1) Feed grade 39.40%Fe;
(2) Iron concentrate grade 63.5%Fe;

(3) Iron recovery 61.58%.

A yearly production line treating 0.2Mt of such iron ore was built 2009 and was scaled up to yearly 0.7Mt in 2010. Another yearly production line treating 2Mt of such iron ore is currently being built.

3.2 Application in processing Dahongshan oxidised iron ore

Dahongshan Iron Mine belongs Kunming Iron and Steel Company in Yunnan Province of China. The Mine owns a beneficiating plant that annually treats 4.5Mt oxidised iron ore. The mined iron ore grade is about 36%Fe and iron minerals consist of magnetite and haematite.

The flow sheet shown in Fig.3 illustrates stage grinding, LIMS and SLon VPHGMS magnetic separation to take out haematite, and the mags of SLon were cleaned by reverse flotation. However, reverse flotation failed in practice due to the complex of gang minerals. A few years ago, shaking tables were applied to replace the reverse flotation, but the capacity of each shaking table is small and their recovery to fine iron particles is relatively low compared with centrifugal separators. In 2010, the company added 50 units of SLon-2400 centrifugal separators replaced the previous 200 units of shaking table. The new flow sheet and technical date is shown in Fig.4. The final iron concentrate grade is raised from 62%-63%Fe to 64%-65%Fe.

The advantages illustrated in the new flow sheet are:

(1) SLon VPHGMS magnetic separation can effectively recover fine haematite and control the non-mags or final tails grade to a low level.

(2) SLon centrifugal separators replace reverse flotation or shaking table for cleaning the SLon VPHGMS mags. They can achieve higher iron concentrate grade and higher recovery to fine iron particles.

The application of this new technology and equipment improves the operating process, better beneficiating results and lower cost. The total production results of the new flow sheet are:

(1) Feed grade 35.61%Fe;

(2) Iron concentrate grade 64.8% and Fe iron recovery 76.82%.

3.3 The research and test of Anshan type oxidised iron ore

The Anshan, Liaoning Province is the largest oxidised iron deposit location in China. The ore consists of mainly magnetite, mag-haematite, hematite and Quartz. Many similar iron ores located elsewhere are also called Anshan type oxidised iron ore. Many big and middle scale mineral processing plants use the flow sheet of stage grinding, spiral separation, SLon magnetic separation and reverse flotation to beneficiate such ores. However, to some smaller plant, reverse flotation is too complicated and may cause environment problems so they prefer to use centrifugal separators instead of reverse flotation.

Fig.5 is the test flow sheet for beneficiate Anshan type oxidised iron ore. The mined ore grade is only about 30% Fe (most Chinese Anshan type iron ores are around such grades). The ore is ground primarily with ball mill, and then classified by hydrocyclone. The underflow is beneficiated

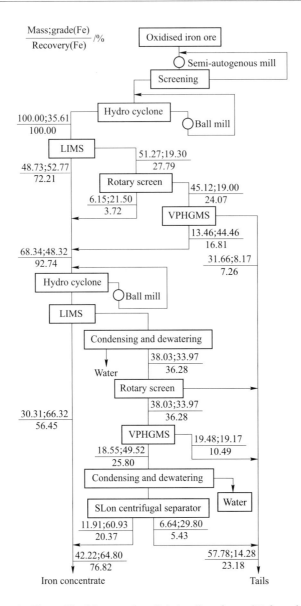

Fig.4 The oxidised iron ore beneficiating flow sheet of Dahongshan

with spiral separators, which can take out the liberated coarser iron concentrate. Spiral tails are scavenged with SLon VPHGMS which can discharge about 20% mass of final tails in the coarser fraction with only 7.25% Fe. The overflow of the hydrocyclone is condensed and LIMS is used to take out magnetite. SLon VPHGMS is used to take out haematite, which discharge 23.72% mass of final tails in the fine fraction with only 8.57% Fe. The mags of LIMS and SLon VPHGMS are mixed and condensed, before entering SLon centrifugal separators for one roughing, one cleaning and one scavenging to achieve a finer iron concentrate.

The total results of this flow sheet are:

(1) Feed grade 30.88% Fe;

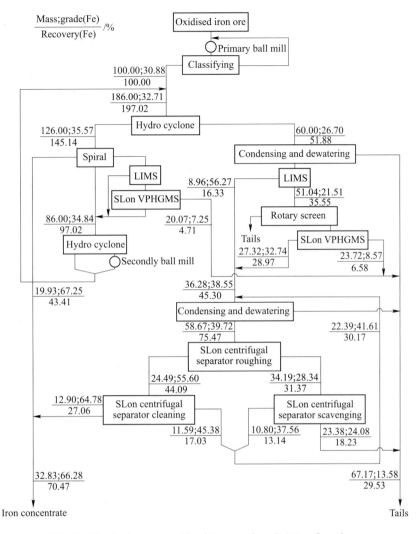

Fig.5 The Anshan type oxidised iron ore beneficiating flow sheet

(2) Iron concentrate grade 66.28%Fe;
(3) Iron recovery 70.47%;
(4) Tails grade 13.58%Fe.

This flow sheet adapts only magnetic and gravity separation. It possesses the advantages of lower energy consumption, which is more environmentally-friendly and has a lower production cost.

4 Conclusions

(1) SLon VPHGMS possesses the advantages of large capacity, low operation cost and high beneficial efficiency for treating low-grade oxidised iron ores.

(2) If the roughing mags of SLon VPHGMS magnetic separation contains quartz or other non-magnetic gang minerals associated with a small portion of magnetite, their visual magnetic susceptibility may be much larger than that of haematite, and therefore it is not effective to use SLon

VPHGMS for cleaning. For example, all of Anshan type oxidised iron ores consist of magnetite, mag-hematite, haematite and quartz. SLon VPHGMS should be used for roughing only, and their mags should be cleaned by reverse flotation or gravity separation.

(3) Centrifugal separator utilises the difference of specific gravity of heavy and light minerals. It produces $20g$-$50g$ (g-Earth gravity acceleration, $g = 9.8m/s^2$) gravity acceleration. It can effectively recover fine iron minerals and reject light gang minerals such as quartz and feldspar. Centrifugal separator can usually achieve a higher iron concentrate if applied to the cleaning the mags of SLon VPHGMS magnetic separation.

(4) The cooperation flow sheet of SLon VPHGMS and SLon centrifugal separators possesses the advantages of lower operational cost, more environmentally-friendly and easier to manage. These new technologies have been successfully applied in several oxidised iron ore beneficiating plants in China.

References

[1] Xiong D H. SLon magnetic separators applied in various industrial iron ore processing flow sheets[C] // Proceedings Iron Ore 2007:245-250 (The Australasian Institute of Mining and Metallurgy:Melbourne).

[2] Xiong D H. Application of SLon magnetic separators in modernising the Anshan oxidised iron ore processing industry[C] // Proceedings Iron Ore, 2009:223-229 (The Australasian Institute of Mining and Metallurgy:Melbourne).

写作背景　这是2011年在澳大利亚帕斯主办的铁矿学术会上发表的论文。向国外同行介绍了SLon磁选机与离心选矿机组合技术分选氧化铁矿的原理，在海南铁矿分选难选氧化铁矿的生产流程，在昆钢大红山铁矿分选氧化铁矿的生产流程，以及分选鞍山式氧化铁矿的试验流程。

SLon 磁选机分选氧化铁矿十年回顾[1]

熊大和[1,2]

(1. 赣州金环磁选设备有限公司；2. 赣州有色冶金研究所)

摘　要　SLon 立环脉动高梯度磁选机利用磁力、脉动流体力和重力的综合力场分选弱磁性矿石，其转环立式旋转、反冲精矿、配有脉动机构松散矿粒群，该机具有富集比大、分选效率高、适应性强、磁介质不易堵塞和运转可靠的优点。2001 年至 2011 年的十年中，该机在鞍钢齐大山选矿厂、东鞍山烧结厂、调军台选矿厂、弓长岭选矿厂、胡家庙选矿厂的氧化铁矿选矿流程中得到应用。在鞍山大规模应用成功后，又在昆钢大红山铁矿、唐钢司家营铁矿等地成功地应用于分选氧化铁矿。该机为我国氧化铁矿提铁降硅的技术进步作出了重要贡献。

关键词　SLon 立环脉动高梯度磁选机　氧化铁矿选矿　提铁降硅

1 SLon 磁选机发展历程

周期式高梯度磁选机在 20 世纪 70 年代已被应用于高岭土提纯和钢铁厂废水处理。高梯度磁选技术对微细粒弱磁性颗粒具有回收率较高的优点，人们很早就想利用这一技术来提高细粒氧化铁矿的回收率。针对氧化铁矿处理量大和磁性物产率较高的特点，20 世纪 70 年代末期和 80 年代初期国内外均研制出了平环高梯度磁选机，其中最有影响的为 Sala HGMS480 平环高梯度磁选机，该机转环外径 7.5m，沿转环一周可配置 4 个磁极头，分选氧化铁矿的处理能力可达 400t/h。该机 1979 年 8 月在瑞典 Strassa 铁矿进行了从尾矿中回收细粒赤铁矿的工业试验，但此后很少见到生产上应用这种设备的报道。

平环式高梯度磁选机的给矿与排矿方向一致，粗粒和木渣草屑都必须穿过磁介质堆才能排走，从而存在磁介质较易堵塞，磁选精矿品位不高，对给矿粒度要求苛刻的缺点。磁介质较易堵塞，维修工作量大是平环高梯度磁选机难以在氧化铁矿选矿工业中推广应用的主要原因。

在实验室周期式高梯度磁选机上做的小型选矿试验研究证明，振动或脉动高梯度磁选可显著提高磁选精矿品位，并保持高梯度磁选对细粒弱磁性矿物回收率较高的优点。例如用振动高梯度磁选分选鞍钢齐大山赤铁矿，与无振动高梯度磁选比较，铁精矿品位可提高 8%～12%，尾矿品位略有降低，铁回收率基本不变。这种高效的选矿方法能否转化为生产力，关键在于研制出适应于工业生产的连续振动或脉动高梯度磁选机。

从 1986 年开始至 2000 年，赣州有色冶金研究所设备研究室（现为赣州金环磁选设备有限公司）先后与中南工业大学、马钢姑山铁矿、赣州有色冶金机械厂、鞍钢矿山公司、弓长岭选矿厂等单位合作研制出了 SLon-1000、SLon-1250、SLon-1500、SLon-1750 和 SLon-2000 立环脉动高梯度磁选机等机型。该机针对铁矿选矿特点，在技术上采取了如下有力的措施。

---

[1] 原载于《2011 年全国选矿学术高层论坛（金属矿山）》。

（1）将设备的可靠性和实用性放在第一位，尽可能做到坚固耐用、便于操作与维护、将理论与实践相结合，不片面地追求理论上预测的最高选矿指标。

（2）将重选理论与磁选理论相结合，设置脉动机构驱动分选区的矿浆脉动，以减少脉石的机械夹杂，提高铁精矿品位。

（3）转环立式旋转、反冲精矿。对于每一组磁介质而言，其给矿方向与排矿方向相反，粗粒和木渣草屑不必穿过磁介质堆便可冲洗出来，可有效地防止磁介质堵塞。

（4）扩大分选粒度范围，简化现场筛分作业。该机的给矿粒度上限可达 2mm，其有效粒度下限为 10μm 左右，比平环高梯度磁选机的适用范围（给矿粒度上限 0.2mm）大得多。

（5）平环高梯度磁选机采用菱形网或编织网作为磁介质，但是网介质容易疲劳折断，维护工作量很大。SLon 磁选机介质采用导磁不锈钢棒作为磁介质，通过多年的生产应用证明，棒介质具有工作稳定、不易堵塞、使用寿命长、维护工作量小和选矿指标稳定等优点。棒介质的应用为该机的大型化和大规模应用打下了良好的基础。

由于 SLon 立环脉动高梯度磁选机具有运转可靠、分选效率高、磁介质不易堵塞、分选粒度范围宽、适应性强的优点，该机首先在马钢姑山铁矿、宝钢梅山铁矿、鞍钢弓长岭选矿厂和昆钢罗次铁矿等地应用于分选赤铁矿和褐铁矿，为各厂矿取得了良好的技术经济指标。

SLon 立环脉动高梯度磁选机大规模推广应用始于 2001 年。当时余永富院士提出铁精矿要提高品位，降低硅含量，以提高冶炼效益。我国的多个铁矿选厂纷纷进行技术改造提高铁精矿质量和提高选矿效率。自 2001~2011 年的 10 年中，SLon 立环脉动高梯度磁选机已有 1500 多台在我国多个氧化铁矿选矿厂应用成功，成为新一代优质高效的强磁选设备（图 1）。

图 1　SLon 立环脉动高梯度磁选机照片

2　SLon 磁选机在齐大山选矿厂的应用

齐大山选矿厂是鞍钢主要铁精矿供应基地，2001 年前后每年处理铁品位 30% 左右的鞍山式氧化铁矿 800 万吨。该厂原矿经粗破和中破后筛分成 0~20mm 和 20~75mm 两个粒级。2001 年以前的选矿工艺为：0~20mm 粒级进一选车间，采用阶段磨矿—重选—强磁—浮选流程分选；20~75mm 粒级进二选车间用竖炉进行还原焙烧成磁铁矿，然后进入磨矿和弱磁选流程。该工艺流程存在着能耗高、污染大等问题。此外该流程的最终铁精矿品位仅 63% 左右，满足不了冶炼的精料方针。

鞍钢矿业公司决定将齐大山二选车间焙烧—弱磁选流程改为阶段磨矿—重选—强磁—反浮选流程。其中强磁选设备的选型是决定该流程改造成功与否的关键之一。

2001 年初，鞍山矿业公司在齐大山选矿厂组织了 SLon 立环脉动高梯度磁选机、永磁立环高梯度磁选机和平环强磁选机的工业对比试验。通过 3 个月的工业对比试验，SLon 立环脉动高梯度磁选机获得了最好的选矿指标，其设备作业率、水耗、电耗及设备价格均具有较大的优势。因此鞍钢矿业公司决定采用 SLon-1750 立环脉动高梯度磁选机作为齐大山选矿厂二选车间技术改造的强磁选设备，尔后又用 SLon-1500 立环脉动中磁机（背景磁感应强度

0.4T）取代了滚筒式永磁机。齐大山选矿厂二选车间改造全面完成后的流程如图2所示。

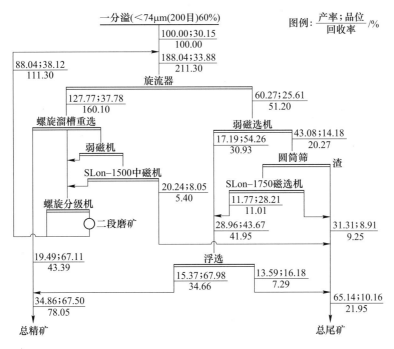

图2 齐大山选矿厂二选车间数质量原则流程

齐大山选矿厂二选车间改造完毕之后，2002年对其一选车间也进行了技术改造，用SLon-1750立环脉动高梯度磁选机取代了原有的永磁立环强磁选机和平环强磁选机。用SLon-1500立环脉动中磁机（背景磁感应强度0.4T）取代了滚筒式永磁机。

齐大山选矿厂的阶段磨矿—重选—强磁—反浮选新流程的改造成功，获得了原矿Fe品位30.15%，铁精矿品位67.50%，铁回收率78.05%的优异指标，将我国氧化铁矿的选矿技术水平推向了一个新的历史高度，使我国的氧化铁矿选矿技术水平从落后于国外的状况转变为达到国际领先的水平。从此该流程成了我国鞍山式氧化铁矿选矿流程的样板。

3 SLon磁选机分选东鞍山氧化铁矿的应用

鞍钢东鞍山矿区的氧化铁矿结晶粒度细、氧化程度深，且含有一定比例的菱铁矿，是鞍山地区最难选的红矿，鞍钢东鞍山烧结厂一选车间处理氧化铁矿470万吨/a。过去长期采用两段连续磨矿，用氧化石蜡皂和塔尔油混合药剂作捕收剂的单一碱性正浮选流程，铁精矿品位一直徘徊在60%Fe左右。由于铁精矿品位较低，随着鞍钢对精矿质量要求的不断提高，东烧车间曾一度面临停产的严峻形势。

为了提高东鞍山氧化铁矿选矿技术指标，鞍钢矿业公司于2002年对东烧一选车间进行了全面的技术改造。新流程（图3）中采用了10台SLon-1750立环脉动高梯度磁选机（额定背景磁感应强度1.0T）作为强磁机控制细粒级尾矿品位；另外采用了10台SLon-1750立环脉动中磁机（额定背景磁感应强度0.6T）扫选螺旋溜槽尾矿。由于SLon-1750中磁机的应用，可及时抛出产率为22.84%，品位为12.46%的粗粒尾矿（注：在此之前用滚筒永磁机做的工业试验表明，因其尾矿品位太高而不能应用，螺旋溜槽尾矿只能作为中矿循环），使中矿的循环量大幅度减少。据测算，粗粒尾矿抛出量与中矿循环减少的量为1∶3的比例关系。SLon-1750中磁机的应用，使中矿循环产率由161.56%（工业试验指标）减少至90.00%（生产运行指标）。

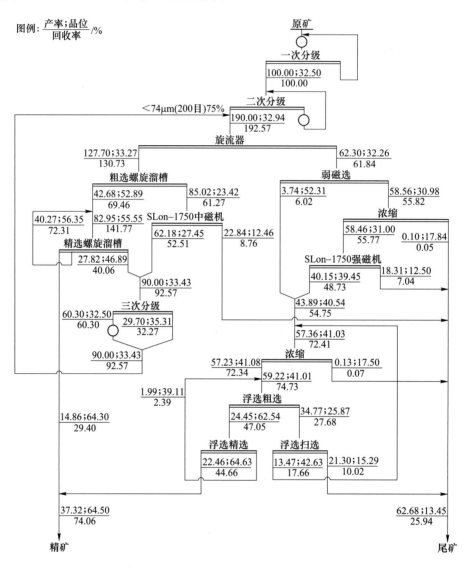

图3 东鞍山阶段磨矿—重选—强磁—反浮选流程

新流程改造成功后，东鞍山烧结厂一选车间铁精矿品位由60%左右提高到64.5%，铁回收率保持在70%左右。使该厂氧化铁矿选矿技术得到了显著的提高。

4 SLon磁选机在调军台选矿厂应用

调军台选矿厂是鞍钢于20世纪90年代新建的一座具有现代化生产水平的大型选矿厂，其设计规模为年处理鞍山式氧化铁矿900万吨。该厂于1998年3月投产，采用连续磨矿—弱磁—中磁—强磁—反浮选流程。

阶段磨矿—重选—强磁—反浮选流程具有生产成本较低的优点，2002~2007 年，调军台选矿厂经过几次技术改造和设备更新，将其选矿流程也改造成为阶段磨矿—重选—强磁—反浮选流程，该厂改造后的选矿流程如图 4 所示。其中采用了 15 台 SLon-2000 立环脉动高梯度磁选机（额定背景磁感应强度 1.0T）作为细粒级抛尾设备，15 台 SLon-2000 立环脉动中磁机（额定背景磁感应强度 0.4T）作为粗粒级抛尾设备。目前该厂生产能力达到年处理鞍山式氧化铁矿 1440 万吨。生产上所获得的选矿指标为：原矿 Fe 品位 29.50%，铁精矿品位 67.50%，尾矿 Fe 品位 10.50%，铁回收率 76.27%。

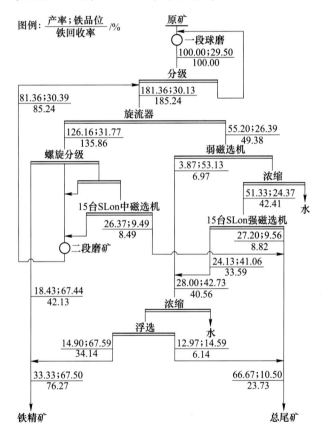

图 4 调军台选矿厂最新的选矿流程

5 SLon 磁选机在胡家庙选矿厂的应用

鞍钢胡家庙选矿厂建于 2006 年，年处理鞍山式氧化铁矿 800 万吨。胡家庙铁矿的特点是原矿品位很低，入选品位只有 24%Fe 左右。采用阶段磨矿—分级重选—强磁—反浮选的选矿流程（图 5）。其中采用了 8 台 SLon-2000 立环脉动高梯度磁选机（额定背景磁感应强度 1.0T）作为细粒级抛尾设备，8 台 SLon-2000 立环脉动中磁机（额定背景磁感应强度 0.4T）作为粗粒级抛尾设备。由于 SLon 磁选机的作业成本非常低，因此用它们尽可能提前抛尾是降低生产成本的关键。它们抛弃的尾矿产率合计占原矿的 68.63%，占总尾量的 88.55%，反浮选的给矿量只占原矿产率 20.99%。全流程的选矿指标为：原矿 Fe 品位 23.10%，精矿品位 67.54%，尾矿 Fe 品位 10.20%，铁回收率 65.78%。

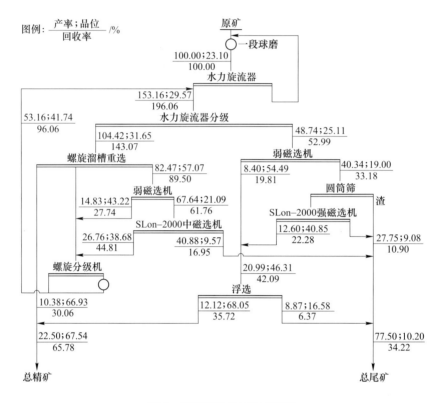

图 5 胡家庙选矿厂选矿流程

6 SLon 磁选机弓长岭选矿厂的应用

鞍钢弓长岭选矿厂建于 1975 年，其二选车间年处理氧化铁矿 300 万吨，最初采用两段磨矿—离心选矿机—粗—精—扫的选矿流程，所获得的选矿指标为：原矿 Fe 品位 31.80%，精矿 Fe 品位 60.50%，铁回收率 66.35%。

由于 1970~1985 年强磁选技术的快速发展，原选矿流程逐步转化为阶段磨矿—强磁选—螺旋溜槽—离心选矿机的选矿流程，所获得的选矿指标为：原矿 Fe 品位 29.20%，铁精矿品位 63.90%，铁回收率 67.51%。1995 年因矿业行情低迷该流程停产。

2005 年弓长岭选矿厂恢复了氧化铁矿的选矿，建设了第三选矿车间，年处理 300 万吨氧化铁矿，采用了阶段磨矿—螺旋溜槽—SLon 磁选机—反浮选流程见图 6。其中采用 5 台 SLon-2000 磁选机（1.0T）控制细粒级尾矿品位，5 台 SLon-2000 中磁机（0.6T）控制粗粒级尾矿品位。全流程的选矿指标为：原矿 Fe 品位 28.78%，铁精矿品位 67.19%，尾矿 Fe 品位 10.13%，铁回收率 76.29%。弓长岭氧化铁矿 3 个阶段不同选矿流程的指标见表 1。

由表 1 可见，弓长岭新流程与以前两种老流程比较，原矿品位下降了，而铁精矿品位和回收率都有大幅度的提高。

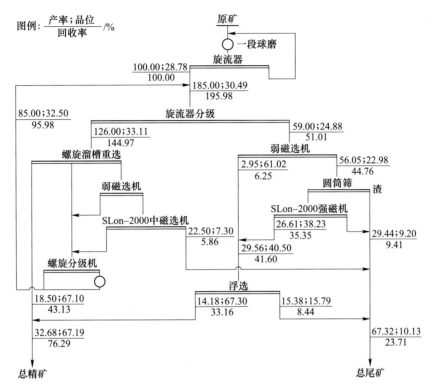

图 6 弓长岭选矿厂三选车间选矿流程

表 1 弓长岭氧化铁矿 3 个阶段不同选矿流程的指标 （%）

选矿流程	原矿铁品位	精矿铁品位	尾矿铁品位	精矿产率	铁回收率
连续磨矿—离心机	31.80	60.50	16.43	34.88	66.35
阶段磨矿—强磁—螺旋溜槽—离心机	29.20	63.90	13.72	30.85	67.51
阶段磨矿—螺旋溜槽—SLon磁选机—反浮选	28.78	67.19	10.13	32.68	76.29

7 SLon 磁选机在司家营氧化铁矿的应用

唐山司家营铁矿是我国大型铁矿山之一，浅部以氧化铁矿为主，深部以磁铁矿为主。其氧化铁矿类似于鞍山式氧化铁矿，但是结晶粒度比鞍山式氧化铁矿更细，含铁矿物以赤铁矿为主并含少量的褐铁矿，脉石矿物以石英为主，属低品位难选氧化铁矿。自 1958 年以来，我国多个科研机构对司家营铁矿石进行了大量的研究工作。近年，氧化铁矿选矿技术水平因为选矿新技术、新药剂和新设备的应用得到了飞速的发展，司家营氧化铁矿的选矿技术难题得到解决。2005 年至今该矿分两期建成了年处理 1000 万吨氧化铁矿的选矿厂，其中采用了 39 台 SLon 立环脉动高梯度磁选机，该厂第一期 600 万吨/a 设计的选矿流程如图 7 所示。

该流程的特点为：采用阶段磨矿，螺旋溜槽拿出了大部分结晶粒度较粗并已单体解离的铁矿物，SLon 中磁机提前抛弃产率占原矿 31.09%、铁品位仅 6.61% 粒度较粗的尾矿，实现了能收早收，能丢早丢的原则，减少了磨矿量，节约了选矿成本；SLon 强磁机抛弃

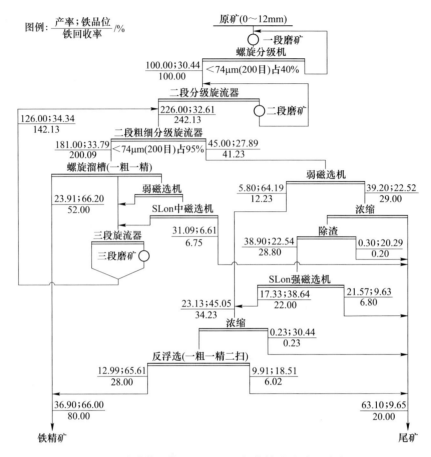

图7 司家营第一期600万吨/a氧化铁矿设计选矿流程

了产率占原矿21.57%、铁品位仅9.63%的细粒尾矿。SLon中磁机和SLon强磁机抛弃的尾矿产率合计占原矿的52.66%，占尾矿总量的83.45%，它们有效地控制全流程的尾矿品位，保证了全流程获得较高的选矿回收率。尽管反浮选的成本较高，但该流程反浮选的给矿产率仅占原矿的23.13%，浮选药剂消耗量较少，因此全流程选矿成本较低。

该流程综合选矿指标为：给矿铁品位30.44%，铁精矿品位66.00%，铁精矿产率36.90%，铁回收率80.00%，尾矿铁品位9.65%。

8 在安徽李楼铁矿镜铁矿选矿工业中的应用

安徽省霍邱县铁矿资源丰富，近年有几个铁矿选矿厂相继建成或正在建设。安徽开发矿业有限公司前几年在当地建成了一座年处理60万吨低品位镜铁矿的选矿厂，在该厂建设成功的基础上，目前一座年处理500万吨低品位镜铁矿的选矿厂正在建设之中。其500万吨/a设计的选矿流程如图8所示。

李楼铁矿镜铁矿的含铁矿物主要是镜铁矿，其他铁矿物含量很少，例如原矿中磁铁矿产率仅为0.5%左右。其选矿流程为：原矿铁品位31.65%，一段闭路磨矿分级至小于74μm（200目）占50%左右，用一段弱磁选和一段SLon强磁粗选和一段SLon强磁扫选提前抛弃产率为43.00%，品位为9.21%左右的低品位尾矿。它们的精矿经二段闭路分级磨

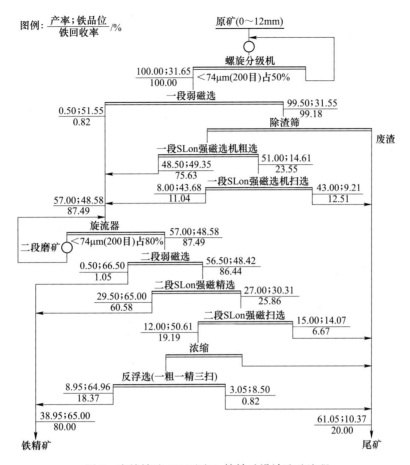

图 8 李楼铁矿 500 万吨/a 镜铁矿设计选矿流程

矿至小于 74μm（200 目）占 80%左右，然后用二段弱磁选和二段 SLon 强磁精选得到一部分品位 65%Fe 以上的铁精矿成品。二段 SLon 强磁精选的尾矿再经二段强磁扫选，二段强磁扫选精矿经浓缩后用反浮选精选，反浮选得到另一部分品位为 64.96%左右的铁精矿成品。全流程设计的综合选矿指标为：给矿品位 31.65%Fe，精矿品位 65.00%Fe，铁精矿产率 38.95%，铁回收率 80.00%，尾矿品位 10.37%Fe。

该流程的选矿特点为：一段强磁为一粗一扫，既抛弃了大量的尾矿，又保证了一段强磁尾矿品位处于较低的水平；二段强磁精选直接得到了大部分铁精矿，这部分铁精矿产率占原矿的 29.5%，占全流程总精矿的 75.74%。由于强磁选作业成本低，强磁选提前抛弃了大部分尾矿和得出了大部分精矿，仅有占原矿 12%左右的二段强磁扫选精矿进入反浮选作业，由于反浮选的作业成本较高，进入反浮选的矿量越少，全流程的生产成本越低，因此，SLon 磁选机为全流程取得良好的选矿指标和保持较低的生产成本起到了关键的作用。

9 在昆钢大红山铁矿的应用

昆钢大红山铁矿是磁铁矿和赤铁矿共生的混合矿，原设计采用阶段磨矿—弱磁—SLon 强磁选—反浮选流程，由于脉石矿物复杂多变，反浮选作业无法稳定而未应用成功，前几年用摇床取代了反浮选。但是,摇床存在处理量较小,对细粒铁矿回收率较低的缺点。大红山铁矿采用赣州金环磁选设备有限公司研制的 SLon-φ2400 离心选矿机取代反浮选作业,该流程于 2010 年顺利投产。新改造的选矿流程如图 9 所示。全流程铁精矿品位从改造前的 62%~63% 提高至目前 64%~65%。

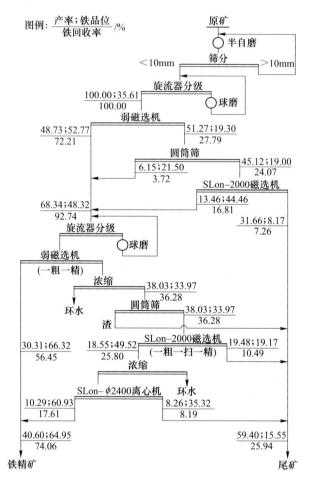

图 9 大红山铁矿 400 万吨选厂选矿流程

该流程的主要特点为:SLon-2000 磁选机对细粒氧化铁矿回收率较高,可有效地控制强磁选作业的尾矿品位;采用 SLon-φ2400 离心选矿机取代反浮选作业,对强磁选精矿进行精选,2009 年在大红山铁矿的工业试验表明,离心选矿机与螺旋溜槽或摇床比较具有富集比较大,对细粒铁矿回收率较高的优点。这些新技术和新设备的应用,使该流程既可运行顺畅,又能获得良好的选矿指标。全流程的选矿指标为:原矿品位 35.61%,铁精矿品位 64.95%,铁回收率 74.06%。

10 结论

(1) 高梯度磁选技术对微细粒弱磁性颗粒具有回收率较高的优点,但是早期针对氧化铁矿选矿研制的平环高梯度磁选机因存在磁介质较易堵塞,磁选精矿品位不高,对给矿粒度要求苛刻的缺点。磁介质较易堵塞，维修工作量大，至今很少在氧化铁矿的选矿工业中推广应用。

（2）SLon立环脉动高梯度磁选机的研制进行了一系列的技术创新，该机利用磁力、脉动流体力和重力的综合力场分选弱磁性矿石，其转环立式旋转、反冲精矿和配有脉动机构松散矿粒群，具有富集比大、分选效率高、适应性强、磁介质不易堵塞和运转可靠的优点。

（3）2001~2011年的10年中，在余永富院士的倡导下，我国许多大型氧化铁矿选矿厂进行了提铁降硅的技术改造或新厂建设。SLon立环脉动高梯度磁选机在鞍钢齐大山选矿厂、东鞍山烧结厂、调军台选矿厂、弓长岭选矿厂、胡家庙选矿厂、昆钢大红山铁矿、司家营铁矿等地成功地应用于分选氧化铁矿，协助这些厂矿获得了优异的技术经济指标，该机为我国氧化铁矿提铁降硅的技术进步作出了重要贡献。

写作背景 本文为纪念余永富院士提出铁矿选矿提铁降硅而作。余永富院士2001年提出铁矿选矿要大力提高铁精矿品位和降低硅的含量。至2011年，在这十年中我国铁矿选矿技术有了突飞猛进的发展，铁精矿品位和回收率达到历史新高，技术经济指标跃居世界领先水平。其中SLon立环脉动高梯度磁选机、反浮选药剂和与它们相结合的选矿流程被公认当时最关键的新技术。文章中回顾了近十年SLon磁选机在我国一些大型氧化铁矿选矿流程中的应用和取得的成果。

SLon 立环脉动高梯度磁选机大型化研究与应用[①]

熊大和[1,2]

（1. 赣州金环磁选设备有限公司；2. 赣州有色冶金研究所）

摘 要 本文介绍 SLon-2500 立环脉动高梯度磁选机和 SLon-3000 立环脉动高梯度磁选机的研制与应用。SLon-3000 磁选机是目前我国自主研发的处理量最大的高梯度磁选机或强磁选机，具有处理量大、选矿效率高、设备可靠性高、电耗低和生产成本低的优点，已成功地在赤铁矿、褐铁矿、钛铁矿选矿和在石英除铁提纯工业中应用。

关键词 SLon 立环脉动高梯度磁选机 大型化 节能 弱磁性矿石选矿

Scaling up and Applications of SLon Vertical Ring and Pulsating High Gradient Magnetic Separators

Xiong Dahe[1,2]

（1. *SLon Magnetic Separator Ltd.*；
2. *Ganzhou Nonferrous Metallurgy Research Institute*）

Abstract The development and applications of SLon-2500 and SLon-3000 vertical ring and pulsating high gradient magnetic separators are introduced. Up to now SLon-3000 magnetic separator is the largest high gradient magnetic separator or high intensity magnetic separator developed in China with the advantages of large capacity, high beneficial efficiency, high reliability, low energy consumption and low operating cost. These separators have been successfully applied in industries to process hematite, limonite and ilmenite and to remove iron impurities for quarz purification.

Keywords SLon magnetic separator; Scale up; Energy saving; Weakly magnetic mineral processing

经济的发展需要更多更好的矿产品作为工业原料，另一方面资源紧缺，复杂难选的原矿增多。这就要求选矿设备和选矿工艺向大型化、自动化、高效率和低成本方向发展，以满足现代选矿工业的需要。SLon 立环脉动高梯度磁选机在工业上广泛应用于氧化铁矿、钛铁矿、锰矿等弱磁性矿石的选矿及石英、长石、高岭土等非金属矿的提纯。通过多年持续的技术创新与改进，该机在设备大型化、多样化、自动化、高效、节能、提高可靠性等

[①] 原载于《第六届全国选矿设备及自动化技术学术会议论文集》。

方面得到了快速的发展,并且得到了更为广泛的应用。

1 设备大型化研制

设备大型化有利于提高选矿设备的台时处理能力和提高自动化水平,同时还具有节能、节约占地面积、操作成本低和设备作业率高等优点,可显著为用户降低生产成本和提高用户的经济效益。

1.1 SLon-2500 磁选机的研制

赣州金环磁选设备有限公司在总结研制 SLon-1000、SLon-1250、SLon-1500、SLon-1750、SLon-2000 立环脉动高梯度磁选机成功的基础上,研制了 SLon-2500 立环脉动高梯度磁选机(以下简称 SLon-2500 磁选机)如图 1 所示。

该机的电磁性能数据分别见图 2 和表 1。

图 1 SLon-2500 磁选机照片

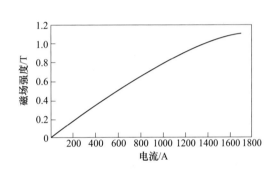

图 2 SLon-2500 磁选机激磁曲线

表 1 SLon-2500 磁选机电磁性能测定数据

电流/A	电压/V	功率/kW	磁感应强度/T
200	6.6	1.3	0.168
400	12.7	5.1	0.336
600	19.5	11.7	0.495
800	26.1	20.9	0.654
1000	32.7	32.7	0.793
1200	38.6	46.3	0.911
1400	45.0	63.0	1.001
1600	52.0	83.2	1.074
1700	54.5	92.7	1.107

1.2 SLon-3000 磁选机的研制

在完成 SLon-2500 磁选机研制的基础上,2010 年又研制出了 SLon-3000 立环脉动高梯度磁选机。如图 3 所示为第一台 SLon-3000 磁选机现场应用照片。该机干矿处理量达到 150~250t/h。

该机的电磁性能数据分别见图 4 和表 2。

图 3　SLon-3000 磁选机应用照片

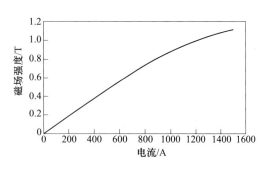

图 4　SLon-3000 磁选机激磁曲线

表 2　SLon-3000 磁选机电磁性能测定数据

电流/A	电压/V	功率/kW	磁感应强度/T
200	8.2	1.6	0.189
400	16.2	6.5	0.375
600	24.6	14.8	0.564
800	33.2	26.6	0.734
1000	42.2	42.2	0.880
1200	51.4	61.7	0.995
1300	55.9	72.7	1.041
1400	60.4	84.6	1.079
1500	64.9	97.4	1.116

在研制 SLon-2500 和 SLon-3000 磁选机的过程中，采用了一系列的新技术提升它们的选矿性能和机电性能，主要措施如下：

（1）通过优化磁系设计、减少漏磁、降低电阻和提高电效率，使其具有优良的激磁性能。采取低电压大电流激磁，更有利于提高激磁线圈的安全可靠性。采用空心铜管绕制，水内冷散热，冷却水直接贴着铜管内壁流动，散热效果更好。此外，由于冷却水的流速较快，微细泥沙不易沉淀，可有效地保证激磁线圈长期稳定工作。

（2）新型磁介质的研究。为了提高 SLon 磁选机的选矿效率，我们针对各种矿石的不同特点，分别研制了 $\phi 1mm$、$\phi 1.5mm$、$\phi 2mm$、$\phi 3mm$、$\phi 4mm$、$\phi 5mm$、$\phi 6mm$ 磁介质组，以及各种磁介质的优化组合，如 $\phi 1mm+\phi 2mm$，$\phi 2mm+\phi 3mm$ 等磁介质组合，可根据入选矿石的粒度，比磁化系数选择，以获得最佳的选矿效果。

（3）采用了具有能耗低、可靠性高、平衡性好和力量大的新型数字式脉动冲程箱。

（4）采用新型的激磁电流可根据选矿需要进行调节的整流控制柜，而可稳定在人工设定点，不会随外电压的波动而波动。可远程控制，自动化程度高，人机界面友好。

SLon-2500 和 SLon-3000 磁选机的技术参数见表 3。

表3　SLon-2500磁选机和SLon-3000磁选机主要技术参数

机型	转环外径/mm	给矿粒度/mm	给矿浓度/%	矿浆通过能力/m³·h⁻¹	干矿处理量/t·h⁻¹	额定背景磁感应强度/T	额定激磁电流/A	额定激磁电压/V
SLon-2500	2500	0~1.2	10~40	200~400	100~150	1.0	1400	45
SLon-3000	3000	0~1.2	10~40	350~650	150~250	1.0	1300	58

机型	额定激磁功率/kW	传动功率/kW	脉动冲程/mm	脉动冲次/r·min⁻¹	供水压力/MPa	耗水量/m³·h⁻¹	冷却水用量/m³·h⁻¹	主机质量/t	最大部件质量/t
SLon-2500	63	11+11	0~30	0~300	0.2~0.4	160~220	5~6	105	15
SLon-3000	75	18.5+18.5	0~30	0~300	0.2~0.4	240~400	7~8	175	25

2　设备节能研究

SLon立环脉动高梯度磁选机的大型化有利于降低处理每吨矿石的能耗单耗，通过优化磁系结构，减少漏磁，采用优质铜管、增加铜的用量和降低激磁电流密度，使电耗有了显著的降低。

图5和表4为几种型号SLon磁选机能耗对比情况。各种机型在背景磁感应强度达到1.0T时，按设备处理量的平均值计算吨矿消耗功率，设备越大，能耗越低，例如SLon-1000磁选机单位能耗为3.69kW·h/t，而SLon-3000磁选机仅为0.56kW·h/t。后者比前者节能85.09%。SLon-3000磁选机与目前工业上大量使用的SLon-2000磁选机比较也节能34.12%。

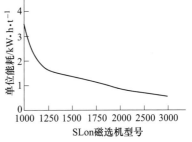

图5　SLon磁选机处理每吨矿石能耗对比

表4　SLon磁选机处理量与能耗对比

机型	额定背景磁感强度/T	额定激磁功率/kW	传动功率/kW	总功率/kW	干矿平均处理量/t·h⁻¹	每吨矿石电耗/kW·h·t⁻¹
SLon-1000	1.0	17	3.3	20.3	5.5	3.69
SLon-1250	1.0	19	3.7	22.7	14.0	1.62
SLon-1500	1.0	27	7.0	34.0	25.0	1.36
SLon-1750	1.0	37	8.0	45.0	40.0	1.13
SLon-2000	1.0	42	13.0	55.0	65.0	0.85
SLon-2500	1.0	63	22.0	85.0	125.0	0.68
SLon-3000	1.0	75	37.0	112.0	200.0	0.56

3　工业应用

3.1　SLon-2500磁选机在海钢红旗尾矿库的应用

首台SLon-2500磁选机在海南钢铁公司（现已更名为海南矿业联合有限公司）应用于从尾矿中回收铁精矿。海南铁矿的铁矿石主要为赤铁矿和部分磁铁矿，建设三个选矿车间分选氧化铁矿，选矿流程为强磁选—粗—扫—反选，或强磁选—粗—精—扫。目前该选矿厂每年入选原矿约 230 万吨，其中有约 120 万吨尾矿排入到红旗尾矿库。其总尾矿的平均铁品位为 28% 左右，其中的铁主要是存在于 0~19μm 粒级。由于该尾矿氧化程度较深，含有较多的连生体，且含泥量大，因此要从该尾矿中再回收铁精矿技术难度较大。

在尾矿输送管的排矿口处安装 SLon-2500 磁选机，尾矿直接从排矿口被截取送入磁选机入选。每年尾矿处理约 72 万吨。该尾矿再回收系统的选矿工艺流程如图 6 所示。首先用圆筒除渣筛筛出大于 0.8mm 的粗粒。接着用 SLon-2500 磁选机粗选，其粗选精矿经浓缩后用离心选矿机精选。离心机精矿作为最终精矿。

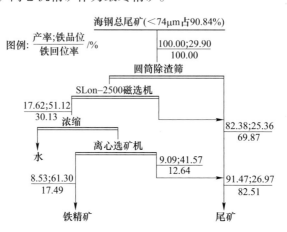

图 6　SLon-2500 磁选机分选海钢尾矿流程

该流程综合选矿指标为：给矿铁品位 29.90%，精矿铁品位 61.30%，铁精矿产率 8.53%，铁回收率 17.49%，尾矿铁品位 26.97%。该流程至今已运转 4 年多，该机的单机作业率达到 99% 以上。整个生产流程稳定，指标先进。

3.2　SLon-2500 磁选机在赤铁矿选矿的应用

SLon-2500 磁选机研制成功以后，很快在铁矿石的选矿工业中得到大规模应用。2008~2010 年，印度一家大型钢铁公司采购了 10 台 SLon-2500 磁选机分选赤铁矿，采用 SLon-2500 磁选机一次粗选和一次扫选的选矿流程，其入选原矿品位为 59.77%Fe，综合铁精矿品位为 65.00%，综合铁回收率为 93.86%，选矿指标优异。

3.3　SLon-2500 磁选机分选褐铁矿的工业应用

云南北衙黄金选矿厂的原矿为金矿与铁矿共生的矿石，其早期的生产仅用浸出法回收金，而选金后的尾矿因粒度很细未得到回收。通过探索试验表明，该选金尾矿中含有磁铁矿和褐铁矿。2006 年该厂建成了一条年处理 50 万吨的选铁生产线，从选金尾矿中回收铁矿物，其中采用 2 台 SLon-2000 磁选机回收褐铁矿。2010 年该厂又建设了一条年处理 150 万吨原矿的选矿生产线，其选金后的尾矿用弱磁选机选出磁铁矿，然后用四台 SLon-2500 磁选机分选褐铁矿（粗选和扫选各 2 台，图 7），该生产线选铁部分的流程如图 8 所示。目前该厂每年可回收 100 多万吨铁精矿，并可根据市场价格安排磁铁矿和褐铁矿分别销售或合并销售。

图 7　SLon-2500 磁选机在北衙黄金选矿厂照片

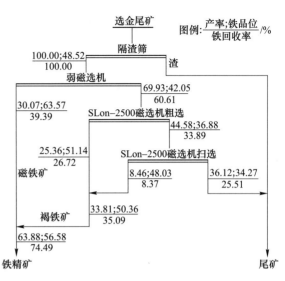

图 8　北衙黄金选矿厂选铁流程

3.4　SLon-2500 磁选机在石英提纯工业中的应用

2009 年 7 月一台 SLon-2500 磁选机在陕西应用于提纯石英砂，获得成功应用。该机给矿为 0~1mm 粒度的石英砂，给矿中含 0.15%~0.25%Fe_2O_3。采用 SLon-2500 磁选机进行除铁。通过生产调试，该机非磁性产品精矿达到 0.05%~0.06%Fe_2O_3，优于设计指标，且石英砂达到了优质标准。

3.5　SLon-3000 磁选机在钛铁矿选矿工业中的应用

2010 年首台 SLon-3000 磁选机成功地应用于攀枝花安宁铁钛有限公司选钛铁矿。攀枝花地区具有丰富的钒钛磁铁矿，目前该地区的选矿厂的选矿流程普遍是先用弱磁选机选出磁铁矿精矿，选完磁铁矿的尾矿即作为选钛作业的原矿。选钛作业普遍用强磁选机预先抛弃大部分尾矿，然后再用重选或浮选精选。由于选钛流程中入选原料含钛品位低，需要强磁选机处理量大，尽可能多抛弃尾矿，同时降低生产成本，SLon-3000 磁选机正好具备这些优势，选矿流程如图 9 所示。

SLon-3000 磁选机在该厂用于钛铁矿的强磁粗选作业，其给矿粒度为小于 74μm 占 40%，台时处理量为 200t/h 左右。该机的选矿指标为：给矿品位 9.52%，TiO_2 精矿品位 17.43% TiO_2，精矿产率 41.01%，TiO_2 回收率 75.09%，

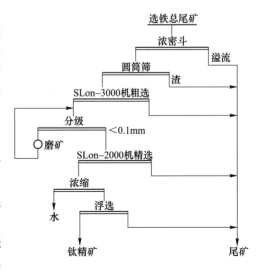

图 9　安宁铁钛有限公司选钛流程

尾矿品位 4.02%TiO$_2$，抛弃的尾矿产率为 58.99%。该机使用的背景磁感强度为 0.7T，处理每吨给矿的电耗仅 0.3kW·h 左右，生产证明该机具有选矿效率高、设备可靠性高和生产成本低的优点。

4 结论

（1）SLon-2500 磁选机和 SLon-3000 磁选机的研制代表了该机型大型化发展方向。设备的大型化有利于降低生产成本，节约占地面积，提高选矿厂自动化生产水平。

（2）对磁系的优化设计使 SLon 磁选机的激磁功率有了大幅度的下降，设备的大型化更有利于降低单位处理量的能耗。以 SLon-3000 磁选机为例，处理每吨矿石的额定电耗为 0.56kW·h（按工作场强 1.0T 计算），处理每吨钛铁矿的实际电耗仅为 0.30kW·h（按工作场强 0.7T 计算）。

（3）SLon-2500 磁选机和 SLon-3000 磁选机具有处理量大、选矿效率高、设备可靠性高、电耗低和生产成本低的优点，已成功地在赤铁矿、褐铁矿、钛铁矿选矿和在石英除铁提纯工业中应用。

写作背景 本文介绍了 SLon-2500、SLon-3000 立环脉动高梯度磁选机的大型化设备研制与应用，其中 SLon-3000 磁选机的一些技术参数及其在安宁铁钛有限公司分选钛铁矿为新公布的内容。文章中表 4 和图 5 列举了 SLon 磁选机大型设备与小型设备耗电量对比数据，表明大型设备比小型设备有显著的节能优势。

应用 SLon 磁选机提高选钛回收率的回顾与展望

熊大和[1,2]

（1. 赣州金环磁选设备有限公司；2. 赣州有色冶金研究所）

摘　要　我国攀枝花和承德等地区拥有丰富的钒钛磁铁矿资源，早期的选钛以重选和电选为主，大量的细粒钛铁矿损失在尾矿中，TiO_2 的选矿回收率只有 10% 左右。自 1994 年以后，SLon 立环脉动高梯度磁选机开始在钛铁矿选矿工业中应用，使细粒级和微细粒级钛铁矿得到了较好的回收。随着 SLon 磁选机的应用和浮选新技术的发展，我国钛铁矿选矿技术水平得到了迅速的提高，目前选钛生产回收率已可达到 40%。然而，我国选钛回收率还有较大的提高潜力，通过优化选矿流程和设备，选钛回收率有可能达到 50%~60%，若能在小于 20μm 钛铁矿选矿技术方面取得突破并从强磁选和浮选尾矿中再选出一部分次钛精矿，则选钛回收率有望达到 70%。

关键词　SLon 立环脉动高梯度磁选机　钛铁矿选矿　TiO_2 回收率

Review and Prospect of SLon Magnetic Separators on Raising Ilmenite Recovery

Xiong Dahe[1,2]

(1. Ganzhou SLon Magnetic Separator Ltd.;
2. Ganzhou Nonferrous Metallurgy Research Institute)

Abstract　There are abundant deposits of magnetite associated with vanadium and titanium in Panzhihua region, Chengde region and other places in China. In the early days, ilmenite is mainly recovered by gravity separation and electric separation. TiO_2 recovery can only reach about 10% and enormous fine ilmenite particles are lost in the tails. From 1994, SLon vertical ring and pulsating high gradient magnetic separators are applied into ilmenite processing industry, which can effectively recover fine ilmenite particles. The development of SLon magnetic separators and flotation technology promoted the rapid development of ilmenite processing technology in China. Nowadays, ilmenite recovery has normally reached 40%. However, there is still great potential to raise ilmenite recovery in China. Through optimizing ilmenite processing flow sheet and equipments, ilmenite recovery still can be raised to 50%-60%. If the breakthrough on beneficiation technology for <20μm ilmenite is made to recover some of titanium ores from the high intensity magnetic separation and tailings flotation, the titanium recovery is expected

---

❶ 原载于《金属矿山》, 2011 (10): 1~8。

to reach 70%.

Keywords　SLon vertical ring and pulsating high-gradient magnetic separator; Beneficiation of ilmenite; Recovery ratio of TiO_2

我国四川省的攀枝花、西昌，河北省的承德及山东、云南等地拥有丰富的钒钛磁铁矿资源。1980 年以前，这些地区的大部分选矿厂将经过破碎、磨矿、弱磁选选出磁铁矿后的磁选尾矿（TiO_2 含量 3%~15%）作为最终尾矿排放，其中的钛铁矿基本上没有回收。1980 年，攀枝花钢铁公司建成了我国第 1 条从钒钛磁铁矿磁选尾矿中回收钛铁矿的生产线[1]，所采用的工艺流程为重选—脱硫浮选—电选，主要回收 0.25~0.074mm 粒级的钛铁矿，每年生产钛精矿 5 万吨，其 TiO_2 品位为 46.2%~46.5%（生产钛白粉的钛精矿要求 TiO_2 品位达到 47%以上），TiO_2 回收率（对磁选尾矿）8.7%左右[2]。从此，我国的钒钛磁铁矿选钛技术和选钛工业得到了较快的发展，而 SLon 型脉动高梯度强磁选机在其中发挥了重要作用。

1　我国早期选钛技术回顾

1980 年至 1995 年期间，我国的选钛技术主要是将螺旋溜槽、摇床等重选设备用于粗选作业回收较粗粒级的钛铁矿[3]，其精矿再用电选或浮选的方法精选，选钛回收率一般占磁选尾矿的 10%左右。由于当时选矿设备和选矿技术的限制，细粒级钛铁矿绝大部分当作尾矿排放。

1.1　攀钢选钛厂早期选钛流程

攀钢选钛厂早期的选钛流程始建于 1980 年，后经过多次技术改造形成了如图 1 所示的流程。该流程将磁选尾矿（即选钛给矿，选钛回收率从这里开始计算，下同）除渣、弱磁选除铁后分成粗粒级（>0.1mm）、细粒级（0.1~0.045mm）和微细粒级（<0.045mm）3 个粒级，其中的粗粒级和细粒级经强磁选、重选、电选后得到 TiO_2 品位为 46.5%左右的钛精矿，而微细粒级因受当时选矿技术和设备的限制作为尾矿排放。

图 1 流程在 1991 年以前每年可生产合格钛精矿 5 万吨，1992 年增加了 1 条生产线以后达到年产钛精矿 10 万吨。该流程的建设成功使我国的钒钛磁铁矿选钛工业实现了从无到有的转变。

图 1 流程存在的主要问题有：

（1）微细粒级未得到回收，而且虽然名义上微细粒级为小于 0.045mm，但由于水力分级设备分级效率不高，该粒级中仍含有较多的大于 0.045mm 钛铁矿。

（2）强磁选机为平环式，磁介质经常发生堵塞，且设备检修困难（强磁选作业后被取消）。

（3）分级、浓缩作业环节多，其溢流中钛铁矿损失率较高。

（4）重选和电选的作业回收率较低。

上述原因导致了图 1 流程选钛回收率低下，按年处理 500 万吨 TiO_2 品位为 10%的磁选尾矿、年产 10 万吨 TiO_2 品位为 47%的钛精矿计算，TiO_2 的回收率仅为 9.4%。

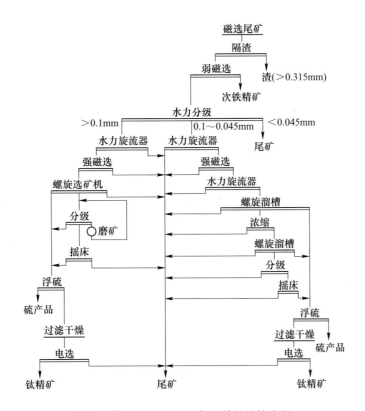

图 1　攀钢选钛厂 1995 年以前的选钛流程

1.2　重钢太和铁矿早期选钛流程

重庆钢铁公司太和铁矿位于四川省西昌市，其矿石为钒钛磁铁矿，性质与攀枝花钒钛磁铁矿类似。该矿为回收磁选尾矿中的钛资源（TiO_2 含量 12% 左右），于 1995 年建成了第 1 条选钛生产线，其工艺流程如图 2 所示。该流程将磁选尾矿分级成粗粒级、细粒级和溢流，粗粒级经螺旋溜槽粗选及磨矿后与细粒级合并，用弱磁选机除去残余的强磁性物质，然后通过摇床重选和浮选得到最终钛精矿。

图 2 流程存在的主要问题有：

（1）由于水力分级设备分级效率不高，有 30% 左右的 TiO_2 损失在分级溢流中。

（2）螺旋溜槽和摇床的分选粒度下限只能达到约 30μm，对细粒级钛铁矿的回收率很低，造成螺旋溜槽的作业回收率仅为 59% 左右，摇床的作业回收率仅为 47% 左右，重选段（包括分级、螺旋溜槽和摇床）的综合回收率仅为 22.42%。

（3）钛铁矿浮选是正浮选，而螺旋溜槽和摇床都不能回收细粒级钛铁矿，使浮选的给矿粒度偏粗，同时钛铁矿密度较大（4.7~4.8 g/cm^3），因此浮选时钛铁矿容易沉槽，造成浮选回收率偏低。

上述原因导致了图 2 中流程的选钛回收率只能达到 10% 左右。1995~2001 年，太和铁矿采用该流程每年只能回收 2000 余吨钛精矿。

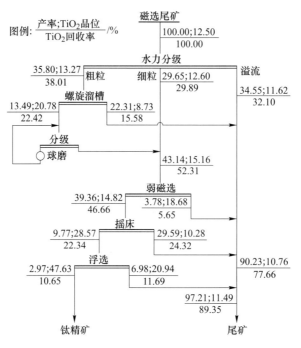

图 2 太和铁矿 2001 年以前的选钛流程

1.3 攀钢综合厂早期选钛流程

攀钢综合厂为攀钢附属集体企业，主要从攀钢排放的总尾矿中分流一部分回收铁精矿和钛精矿，原选钛采用螺旋溜槽粗选、摇床精选的重选流程（图 3），由于螺旋溜槽和摇床对细粒级回收率都很低，因此全流程的选钛回收率仅为 7% 左右。

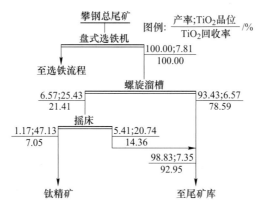

图 3 攀钢综合厂早期选钛流程

2 SLon 磁选机在选钛流程中的应用

SLon 立环脉动高梯度磁选机是我国自行研制的新一代高效强磁选设备[4,5]。该机利用磁力、脉动流体力和重力的综合力场分选弱磁性矿石，转环立式旋转，反冲精矿，配有矿浆脉动机构，具有选矿效率高、磁介质不易堵塞、分选粒度范围较宽、可靠性高和能耗低的优点。该设备首先在我国的氧化铁矿选矿工业中得到广泛应用[6]，从 1994 年开始在国内外的钛铁矿选矿中推广。

2.1 在攀钢选钛厂回收小于 0.045mm 钛铁矿

攀钢选钛厂原先的选钛流程中，小于 0.045mm 粒级作为尾矿排放（图 1），其中 TiO_2 的金属占有率为 40% 左右，因此，攻克这部分物料的选钛难题，对提高全流程的选钛回收率具有重要的意义。1994~1996 年，攀钢选钛厂联合赣州有色冶金研究所等单位开展了从这部分物料中回收钛铁矿的小型试验和工业试验[7]，取得了突破性的研究成果，并于 1996 年在攀钢选钛厂采用 1 台 SLon-1500 立环脉动高梯度磁选机建立了第 1 条微细粒级选钛试验生产线，如图 4 所示。

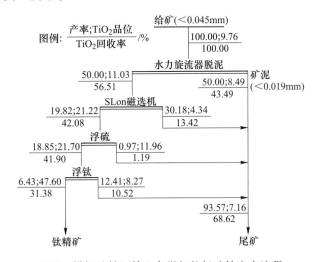

图 4 攀钢选钛厂第 1 条微细粒级选钛生产流程

为避免浮选作业泡沫过多造成跑槽，如图 4 所示流程中首先用旋流器脱除小于 0.019mm 矿泥，然后用 1 台 SLon-1500 磁选机粗选，其粗选精矿再经脱硫浮选和钛浮选，最终得到 TiO_2 品位为 47% 以上的优质钛精矿。

SLon 磁选机在流程中的作业回收率达到 76.24%，浮选的作业回收率也达到了 75% 左右，全流程的选钛回收率为 30% 左右。但是由于小于 0.019mm 粒级未入选，使 TiO_2 金属量损失了 43.49%，严重影响了钛铁矿回收率的进一步提高。

小于 0.045mm 粒级选钛试验生产线每年生产 2 万吨优质钛精矿，随后的几年攀钢选钛厂又分两期建成后八系统和前八系统小于 0.045mm 粒级选钛生产线，至 2004 年，攀钢选钛厂小于 0.045mm 钛精矿的年产量达到 15 万吨左右，加上粗粒级钛精矿，攀钢选钛厂年产钛精矿达到 28 万吨左右。

2.2 在重钢太和铁矿分选钛铁矿

重钢太和铁矿原采用螺旋溜槽和摇床进行钛铁矿的粗选（图 2），由于其作业回收率很低，该矿于 2000 年开始采用 SLon 立环脉动高梯度磁选机进行选钛流程的技术改造，经过几次改造后的选钛流程如图 5 所示。

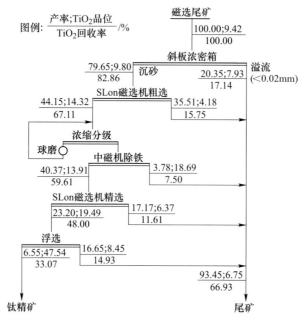

图 5　太和铁矿采用 SLon 磁选机的选钛流程

图 5 流程的主要优点有：

（1）由于 SLon 磁选机的选矿粒度下限可达到 10μm 左右，因此磁选尾矿用水力分级机分级时可减少溢流量，让更多的细粒级钛铁矿进入选别作业。对比图 2 和图 5 可知，改造后溢流部分流失的 TiO_2 金属量由 32% 左右降低至 17% 左右。

（2）由于 SLon 磁选机分选粒度范围较宽，因此入选物料不再分粗粒级和细粒级，从而简化了选钛流程。

（3）SLon 磁选机取代螺旋溜槽和摇床后作业回收率大幅度提高。表 1 为 SLon 磁选机与螺旋溜槽和摇床的选矿指标对比结果。可见，尽管经过几年的开采，选钛给矿的 TiO_2 品位随原矿的 TiO_2 品位大幅度降低，但 SLon 磁选机的作业回收率还是分别比螺旋溜槽和摇床提高了 21.99% 和 32.67%，这使进入浮选作业的 TiO_2 金属量由原来的 22% 左右提高到 48% 左右。

表 1　SLon 磁选机与螺旋溜槽和摇床选别指标对比　　　　　　　　　　（%）

设备	作业	给矿品位	精矿指标			尾矿品位
			产率	品位	作业回收率	
SLon 磁选机	粗选	9.80	55.42	14.32	80.99	4.18
螺旋溜槽	粗粒级粗选	13.27	37.68	20.78	59.00	8.73
SLon 磁选机	二段精选	13.91	57.47	19.49	80.52	6.37
摇床	二段精选	14.82	24.82	28.57	47.85	10.28

（4）改善了浮选作业给矿的粒度组成。由于 SLon 磁选机对 0.15~0.02mm 粒级钛铁矿

的回收率高达 85% 以上，而这一粒级正好是浮选的最佳粒度范围，因此浮选的作业回收率由过去的 48% 左右提高到 69% 左右。

SLon 磁选机在太和铁矿的成功应用大幅度提高了该矿钛铁矿的回收率，尽管选钛系统入选物料的 TiO_2 含量由过去的 12% 左右降低至 9% 左右，TiO_2 的回收率却由过去的 10% 左右提高到 33% 左右，且钛精矿品位保证在 47% 以上。至 2007 年，该矿实现年产优质钛精矿 12 万吨。

2.3 在攀钢综合厂分选钛铁矿

攀钢综合厂从攀钢排放的总尾矿中回收铁精矿和钛精矿，原选钛流程为螺旋溜槽粗选，摇床精选的单一重选流程（图 3），由于螺旋溜槽和摇床对细粒级回收率都很低，全流程的回收率仅为 7% 左右。2002 年该厂采用 SLon 磁选机建立了新的选钛流程（图 6）。该流程的特点是：经盘式选铁机选出残余磁铁矿等强磁性物质的选钛给矿全粒级进入 SLon 磁选机进行 1 次粗选 1 次精选，避免了水力分级机溢流中 TiO_2 的损失（由于水力分级机的分级效率不高，溢流中往往有 15% 左右大于 20μm 的钛铁矿适合于浮选回收）；SLon 磁选机本身具有脱泥的作用，可将绝大部分 10μm 以下的矿泥排入尾矿中；在 SLon 磁选机的精选作业中采用较低的背景磁感应强度，还可将大部分小于 20μm 的矿泥排入尾矿中（小于 20μm 的矿泥在进入浮选之前要尽可能脱除，否则会对浮选造成不利影响）；SLon 磁选机精选作业的精矿产率只有 16% 左右，再用小型的高效浓缩机脱除小于 20μm 的残余矿泥，然后用浮选得到合格钛精矿。该流程使选钛回收率由原来采用重选工艺时的 7% 提高到 27% 左右。

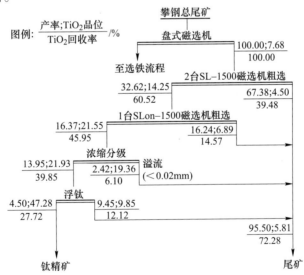

图 6 攀钢综合厂采用 SLon 磁选机的选钛流程

2.4 在承钢黑山铁矿分选钛铁矿

河北省承德地区也有丰富的钒钛磁铁矿资源，其矿石性质类似于攀枝花地区的钒钛磁铁矿，不同的是攀枝花地区的磁选尾矿中 TiO_2 含量一般可达到 10% 左右，而承德地区只能达到 3%~9%，即承德地区的钛铁矿入选品位普遍要低于攀枝花地区。在 2001 年以前，承德地区的选钛工业非常落后,只有很少的企业用螺旋溜槽或摇床回收少量的品位为30%左右的钛铁矿,其用途主要是作为高炉的辅料。

2002~2004年,承德钢铁公司黑山铁矿参考攀枝花地区的选钛流程建设了1条选钛生产线(图7)。该生产线在浮选前采用2台SLon-1750磁选机进行粗选、1台SLon-1750磁选机进行精选,全流程的选钛生产指标为磁选尾矿品位8%左右,钛精矿品位46.5%以上(承德地区生产钛白粉的钛精矿TiO_2品位要求不小于46.5%),TiO_2回收率约25%,每年生产优质钛精矿2万吨左右。

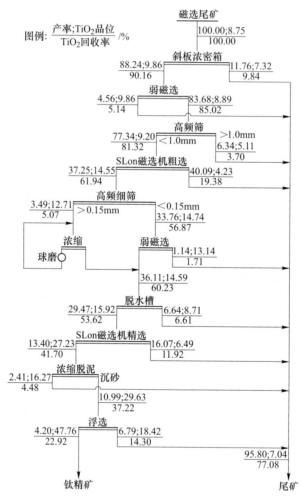

图7 承钢黑山铁矿选钛流程

SLon磁选机在承德地区的成功应用,开创了承德地区生产优质钛精矿的先例。

3 通过优化流程进一步提高选钛回收率

从上述实例中可看出,采用SLon立环脉动高梯度磁选机从钒钛磁铁矿的磁选尾矿中回收钛铁矿,在磁选尾矿TiO_2品位为7%~10%的情况下,生产上全流程的TiO_2回收率在20%~35%。影响钛铁矿回收率的因素主要有如下几方面:

（1）SLon 磁选机（包括其他类似的强磁选机）在生产上的 TiO_2 作业回收率一般为 80% 左右，尽管实验室试验指标可获得更高的回收率，但生产上要求设备处理量大，矿浆流速快，故很难达到实验室的试验指标。现在的选钛流程要求尽可能提高浮选的给矿品位以降低浮选药剂消耗和浮选成本，因此一般采用 1 粗 1 精强磁选。假设强磁粗选和精选的作业回收率均为 80%，则强磁作业的综合回收率为 64%，即强磁选要损失该作业 36% 的 TiO_2 金属量。

（2）隔渣、分级、浓缩环节多，TiO_2 损失大。由于浮选作业要求脱除小于 $20\mu m$ 矿泥，这样在分级浓缩环节中要损失占入选给矿 20%～40% 的 TiO_2，同时由于分级效率不高，又造成了部分大于 $20\mu m$ 钛铁矿的损失。此外，在强磁选前的隔渣过程中，如果隔渣筛效率不高的话，也会从筛上物中损失一部分 TiO_2。

（3）浮选作业的 TiO_2 回收率在生产上一般只能达到 60%～75%[8~10]。尽管实验室试验的浮选作业回收率可达到 85% 以上，但生产上由于浮选设备负荷重、生产环节控制不精准等原因，选钛回收率与实验室试验指标还有较大的差距。

因此，要进一步提高选钛回收率，应从以上几个环节入手。

3.1 攀钢选钛厂扩能改造工程

多年的生产实践证明，强磁选—浮选选钛流程与重选—电选流程相比，具有回收率较高、生产成本较低、生产环节较易控制等优点。2008 年，攀钢选钛厂开始着手进行扩能改造，具体实施计划为：

（1）将粗粒级选钛流程由原来的重选—电选改为强磁选—浮选。

（2）优化细粒级选钛流程。

（3）钛精矿产量由改造前的 28 万吨/a 提高到 40 万吨/a。

赣州金环磁选设备有限公司受攀钢选钛厂委托，进行了选钛扩能改造新工艺试验，结果如图 8 所示。

图 8 流程的主要特点为：

（1）充分发挥 SLon 磁选机本身的脱泥功能，磁选尾矿只分成粗粒级和细粒级入选，避免了分级溢流造成的直接损失。

（2）细粒级强磁选作业由过去的 1 次选别改为 4 次选别，粗粒级强磁选作业采用 SLon 磁选机 1 次粗选，其粗精矿分级再磨并经弱磁选机除铁后，用 SLon 磁选机进行 1 次精选和 1 次扫选，从而保证了强磁选作业有较高的作业回收率和较高的精矿品位（参见表 2）。比较图 4 和图 8 可见，细粒级进入浮选作业的 TiO_2 金属量由 43% 左右提高到 72% 左右（37.65%/52.55% = 72.10%），粗粒级进入浮选作业的 TiO_2 金属量高达该粒级的 77% 左右（36.83%/47.78% = 77.08%）。

（3）高效率的强磁选为浮选创造了较好的条件，在保证钛精矿品位达到 47% 以上的前提下，浮选作业回收率可保证在 70% 左右，全流程回收率达到 52.17%。

2009 年，攀钢选钛厂在图 8 流程基础上，结合现场情况完成了扩能改造工程。2010 年该厂实现了年产优质钛精矿 48 万吨，预计 2011 年可实现年产优质钛精矿 50 万吨以上。目前生产上全流程实际达到的 TiO_2 回收率为 40% 左右，虽然比以前提高了很多，但由于生产条件的限制，许多作业环节还没有达到最佳状况，选钛回收率还有很大的提高潜力。

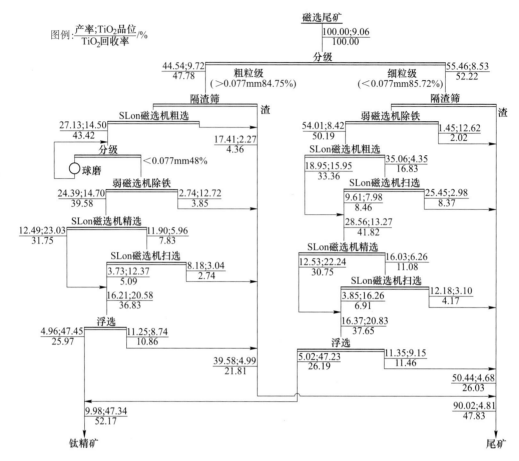

图 8 攀钢选钛厂扩能改造试验流程

表2 图8流程强磁选作业指标 （%）

作业名称	品位			精矿作业回收率
	给矿	精矿	尾矿	
细粒级一段强磁选	8.42	13.27	2.98	83.32
细粒级二段强磁选	13.27	20.83	3.10	90.03
细粒级强磁综合	8.42	20.83	3.02	75.01
粗粒级一段强磁选	9.72	14.50	2.27	90.87
粗粒级二段强磁选	14.70	20.58	3.04	93.07
粗粒级强磁综合	9.72	20.58	2.52	84.57

3.2 重钢太和铁矿新建钛铁矿选矿项目

重庆钢铁公司太和铁矿2007年开始计划将原矿处理量由220万吨/a扩建至630万吨/a，其选钛流程重新建设。该矿于2008年与赣州金环磁选设备有限公司共同进行了选钛流程优化试验，试验结果如图9所示（图中磁选部分由赣州金环公司完成，浮选部分由太和铁矿完成）。

如图9所示流程的特点为：

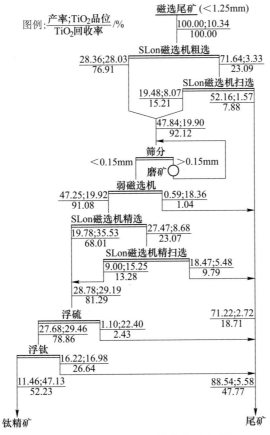

图 9 太和铁矿选钛优化试验流程

(1) 磁选尾矿全粒级进入强磁选，可以避免分级溢流造成的直接损失，另一个好处是矿浆的流动性更好，可减缓矿浆管道和各种矿斗的磨损（攀钢选钛厂粗、细粒级分别入选，粗粒级部分存在流动性较差，矿斗磨损较快的问题）。

(2) 第 1 段强磁选采用 SLon 磁选机 1 次粗选、1 次扫选，TiO_2 作业回收率达到 92.12%，第 2 段强磁选采用 SLon 磁选机 1 次精选、1 次扫选，TiO_2 作业回收率达到 89.25%（81.29%/91.08%），两段强磁选的 TiO_2 综合作业回收率达到 82.22%（92.12%×89.25%），使浮选给矿的 TiO_2 金属量占到了选钛给矿的 81.29%。

(3) SLon 磁选机有良好的脱泥效果。两段强磁选所得强磁选精矿（浮选给矿）的粒度筛析结果见表 3，可见，强磁选精矿含泥量很少，其粒度组成适合浮选，可以直接作为浮选给料而不需再脱泥。

表 3 图 9 流程强磁综合精矿（浮选给矿）粒度筛析结果

粒级/mm	产率/%	TiO_2 品位/%	TiO_2 分布率/%
>0.125	1.0	18.07	0.61
0.125~0.105	3.8	28.44	3.65
0.105~0.074	31.9	28.90	31.17
0.074~0.038	34.9	28.98	34.20
<0.038	28.4	31.62	30.36
合 计	100.0	29.57	100.00

（4）强磁选精矿直接进入浮硫和浮钛作业，浮钛作业采用1次粗选和3次精选，可获得钛精矿 TiO_2 品位为 47.13% 和作业回收率为 69.55% 的指标。

太和铁矿已参考图9流程开始了新建选钛项目的建设，目前尚在进行中。

4 展望

目前，提高我国钛铁矿选矿回收率仍有很大的潜力，有很多工作可做，具体有如下几方面：

（1）浮选作业回收率有待提高。目前我国生产上钛铁矿浮选作业回收率一般在 60%～75%，而实验室试验的钛铁矿浮选作业回收率可达到 85% 以上[10,11]（表4）。如果将来生产上浮选作业回收率能达到 80% 左右，则在选钛给矿品位为 10% 左右、精矿品位达到 47% 以上的前提下，我国钒钛磁铁矿选钛生产的回收率可望达到 60% 左右。

表4 选钛试验浮选作业指标 （%）

试验单位	试验时间	试料	品位			作业回收率
			浮给	浮精	浮尾	
赣州有色冶金研究所	1995年	攀钢 0.045～0.01mm 粒级 SLon-1000 磁选机精矿	23.37	47.60	5.83	85.52
地矿部矿产综合利用研究所[10]	2000年	攀钢小于 0.074mm 粒级强磁选精矿	22.67	48.47	5.08	86.68
长沙矿冶研究院[11]	2002年	攀钢小于 0.045mm 粒级选钛生产线 SLon-1500 磁选机精矿	17.58	47.75	3.68	85.66

注：浮选作业含浮硫和浮钛，本文将硫产品计入尾矿，实际生产中硫产品可能单独回收。

（2）研究小于 $20\mu m$ 粒级钛铁矿的浮选药剂和浮选设备。随着钢铁工业对铁精矿品位要求的提高及入选原矿的日益贫化，钒钛磁铁矿选铁部分的磨矿粒度变细，选钛给矿中小于 $20\mu m$ 粒级含量增多。用 SLon 磁选机分选攀钢选钛厂小于 $20\mu m$ 粒级的试验结果表明，在给矿品位为 7.75% 的情况下，一次选别可获得 TiO_2 品位为 18.73%、TiO_2 回收率为 30.27% 的强磁选精矿。由于 SLon 磁选机有良好的脱泥效果，该强磁选精矿中大部分小于 $10\mu m$ 的矿泥已被脱除，若对其单独进行浮选，并研究出相关的浮选药剂和浮选设备，则钛铁矿的选矿回收率可进一步提高。

（3）生产上二段强磁选尾矿和浮选尾矿的 TiO_2 含量都在 3% 以上，这些物料中连生体较多，返回原流程再选可能不合适，但是可用它们生产一部分品位较低的次钛精矿。赣州金环磁选设备有限公司以太和铁矿选钛流程的二段强磁选尾矿和浮选尾矿为试料进行试验，获得了表5所示的试验结果。

表5 太和铁矿选钛流程二段强磁尾和浮尾再选试验结果 （%）

试料	再选流程	品位			作业回收率
		给矿	次钛精矿	尾矿	
二段强磁尾	SLon-500 磁选机一次分选	6.58	22.79	3.32	57.99
二段强磁尾	SLon-500 磁选机一次分选—浮选	22.85	36.52	9.84	77.94
浮尾	磨至小于 $77\mu m$ 92%，SLon-500 磁选机一粗一精	6.90	19.00	4.76	41.26

综合以上内容，本文作者提出了如图10所示的今后钒钛磁铁矿选钛的设想流程。该流程预计在选钛给矿 TiO₂ 品位为10%左右的情况下，产出 TiO₂ 品位为47%以上、TiO₂ 回收率为60%左右的钛精矿和 TiO₂ 品位为30%左右、TiO₂ 回收率为10%左右的次钛精矿两种钛产品，全流程的 TiO₂ 综合回收率可达到70%左右。

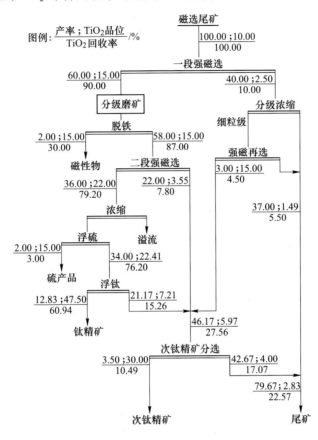

图10 钒钛磁铁矿选钛设想流程

5 结论

（1）我国钒钛磁铁矿的选钛技术在1980年以前几乎是一片空白。1980年后发展了以重选和电选为主的选钛流程，实现了我国钒钛磁铁矿选钛技术从无到有的转变，但是，由于当时细粒级钛铁矿不能回收等原因，选钛流程的回收率只能达到10%左右。

（2）从1994年以后，SLon立环脉动高梯度磁选机在钒钛磁铁矿的选钛流程中得到应用，使 45~20μm 的钛铁矿得到了较好的回收，选钛流程的回收率普遍达到20%~35%，钛精矿品位达到47%以上。

（3）充分利用 SLon 磁选机处理量大和作业成本低的优点，在选钛流程的一段强磁选和二段强磁选作业中增加扫选作业，可较大幅度地提高强磁选的作业回收率。经过优化的选钛流程，对于品位为10%左右的给矿，在保证精矿品位达到47%以上的前提下，目前实验室试验可达到52%左右的回收率，生产上可达到40%左右的回收率。

(4) 随着对选钛流程和选钛设备的进一步优化以及浮选技术水平的提高，我国钒钛磁铁矿选钛生产的回收率有可能达到50%~60%。

(5) 如果在小于20μm钛铁矿选矿技术方面取得突破，并从二段强磁选尾矿和浮选尾矿中再选出一部分品位为30%左右的次钛精矿，我国钒钛磁铁矿选钛的回收率可望在上述基础上进一步提高。

参 考 文 献

[1] 孟长春，刘胜华. 攀钢选钛厂细粒钛铁矿浮选工艺技术研究与发展探讨 [J]. 矿冶工程，2010，30（8）：11~15.
[2] 邹建新，杨成，彭富昌，等. 攀西地区钒钛磁铁矿提钛工艺与技术进展 [J]. 金属矿山，2007（7）：7~9.
[3] 袁国红. 重钢太和铁矿选钛流程技术改造 [J]. 金属矿山，2001（6）：39~40.
[4] 熊大和. SLon-2000立环脉动高梯度磁选机的研制 [J]. 金属矿山，1995（6）：32~34.
[5] 熊大和. SLon-2500立环脉动高梯度磁选机的研制与应用 [J]. 金属矿山，2010（6）：133~136.
[6] 熊大和. SLon立环脉动高梯度磁选机分选红矿的研究与应用 [J]. 金属矿山，2005（8）：24~29.
[7] 袁诗芬，苏树红. SLon高梯度磁选—浮选联合流程选别微细粒级钛铁矿试验研究 [J]. 江西有色金属，1997，11（3）：40~43.
[8] 周友斌. 攀枝花细粒钛铁矿浮选工艺在生产中应用探讨 [J]. 金属矿山，2000（1）：32~36.
[9] 袁国红，余德文. R-2捕收剂选别攀枝花微细粒级钛铁矿试验研究 [J]. 金属矿山，2001（9）：37~39.
[10] 傅文章，张渊，洪秉信，等. 攀枝花微细粒级钛铁矿选矿试验研究 [J]. 金属矿山，2000（2）：37~40.
[11] 谢建国，张泾生，陈让怀，等. 新型捕收剂ROB浮选细粒级钛铁矿的试验研究 [J]. 矿冶工程，2002，22（6）：47~50.

写作背景　本文较系统地总结了过去20年我国钛铁矿选矿的技术进步，选钛流程的进化，SLon立环脉动高梯度磁选机在提高钛铁矿选矿回收率和降低生产成本方面取到的重要作用。展望了采用新设备、新药剂和优化选矿流程进一步提高选钛回收率的前景。

Recent Development and Applications of SLon VPHGMS Magnetic Separator

Xiong Dahe

(*SLon Magnetic Separator Ltd.*, *Ganzhou*, 341000, *Jiangxi*, *P. R. China*)

1 Introduction

The traditional horizontal type Wet High Intensity Magnetic Separators (WHIMS) with grooved plate matrix have been in service for recovering iron ore and other minerals since 1960s. However, it has multiple drawbacks such as matrix blocking. SLon vertical ring and pulsating high gradient magnetic separator (SLon VPHGMS or SLon HGMS) was developed to overcome the multiple drawbacks associated with traditional WHIMS.

SLon HGMS is a new generation of continuously working HGMS with high efficiency. It utilizes the combined force field of magnetism, pulsating slurry, and gravity to beneficiate weakly magnetic minerals. It possesses the advantages of large beneficial ratio, high recovery and high reliability. Till now, over 2000 units of SLon HGMS are applied to beneficiate hematite, limonite, specularite, ilmenite, manganese ore, chromite, wolframite and etc. SLon HGMS are also applied to purify nonmetallic minerals such as quartz, feldspar, nepheline, kaolin and etc.

SLon HGMS is suitable for feed particle size <1.2mm (down to ~10μm), as well as up to 5mm by special configuration.

Fig. 1 is the latest developed SLon-3000 HGMS, it was designed and built up in 2010. Its iron ore treating capacity reaches 150-250t/h.

Fig. 1 SLon-3000 HGMS

2 Structure and working principles

An electromagnetic field is generated within the separating zone. A carousel, rotating on its horizontal axis, houses matrix of paralleled rods. From the feed box, slurry is introduced into the matrix as it passes through the separating zone. The magnetic particles in the slurry are attracted to the surface of the rods in the matrix, carried out of the separation zone to the top of the carousel, where the magnetic field is negligible, and then they are flushed into the concentrate box. Conversely, while the matrix is still in the separation zone, gravity and the force of hydrodynamic pulsing of the slurry drag the non-magnetic particles through the matrix pile into the tailings box.

A schematic of the SLon is shown in Fig. 2.

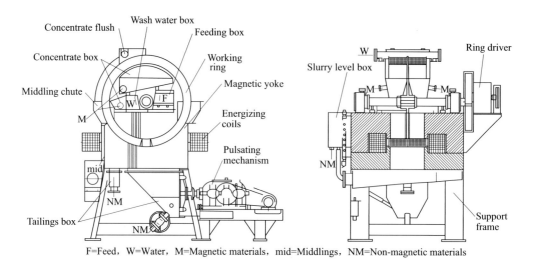

F=Feed, W=Water, M=Magnetic materials, mid=Middlings, NM=Non-magnetic materials

Fig. 2 The Structure of SLon HGMS

3 New development of equipment

3.1 Key technology and practice

A diaphragm actuated by a crankshaft provides the pulsating mechanism. The pulsation drives the slurry in the separating zone up and down keeping particles in a constantly loose status; so fewer non-magnetic particles become trapped. Since the pulsing causes the slurry direction to reverse, particle accumulation occurs on both sides of the matrix. This creates a greater surface area of collection points thus the matrix can process more slurry before becoming saturated, and so capture more magnetic particles in less processing time.

The working ring of the SLon rotates vertically, rather than horizontally like traditional WHIMS. This configuration allows for flushing in the opposite direction of the feed so that strongly magnetic particles and coarse particles can be washed away without having to pass through the entire depth of the matrix. Additionally, the flush zone is designed to have less stray magnetic field and so reduce any residual grip on the magnetic particles.

The SLon matrix is constructed of rods magnetic stainless steel, providing high magnetization. The rods matrix can range in diameter from 1-5mm to accommodate various sizes range of feeds and thus various mineral applications. The rod matrixes have the advantages of stability, reliability, not easy to be blocked and long lifetime. The unique matrix maximizes the physical combination of high background magnetic field and the generated high magnetic flux gradient.

SLon HGMS has a low energizing voltage design. For example, SLon-2500 has a rating DC energizing voltage of 45V, SLon-3000 has a rating DC energizing voltage of 62V. Compared with other HGMS with high DC energizing voltage of 510V, SLon HGMS is much safer in this regards.

3.2 Development history

From 1986 to 2000, SLon Magnetic Separator Ltd. (the former Equipment Division of Ganzhou Nonferrous Metallurgy Research Institute) cooperated with Central South University of Technology, Gushan Iron Mine of Maanshan Iron and Steel Company, and other partners to develop a series of SLon HGMS including model SLon-1000, SLon-1250, SLon-1500, SLon-1750 and SLon-2000.

As SLon HGMS has a very good performance in processing iron ore fines, it has been widely applied in the main oxidized iron mines throughout China. For example, Gushan Iron Mine of Maanshan Iron and Steel Company, Meishan Iron Mine of Bao Steel Group, 5 Processing Plants of Anshan Mining Company, Panzhihua Ilmenite Processing Plant and etc. Till now, over 2000 units of SLon HGMS are applied to beneficiate hematite, limonite, specularite, ilmenite, manganese ore, chromite, wolframite and etc. SLon HGMS are also applied to purify nonmetallic minerals such as quartz, feldspar, nephenline, kaolin and etc.

In recent years a continuous effort was made to scale up the SLon HGMS.

The first SLon-2500 HGMS was designed and built up in 2006. It was installed at the tails dam of Hainan Iron Mining Company to recover iron concentrate from the pumping tails. Till now, over 100 units of SLon-2500 HGMS are in industrial operation.

The first SLon-3000 HGMS was designed and built up in 2010. It has a feed capacity of 150~250 tons per hour. When its rating background magnetic field is at 1.0 Tesla, its energizing power is only 87kW, and total installed power including 2 motors is only 124kW, so it consumes low energy at around 0.5kW per ton of feed capacity. It possesses very good energy saving characteristics. Its specifications are showed in Table 1.

3.3 New improvement/optimization

3.3.1 SLon HGMS 0.6T and 1.3T model

SLon HGMS is usually developed at rating background magnetic field 1.0 Tesla for each size at the beginning. To meet the various requirements of the various minerals, and different particle sizes, we have developed the whole sizes of 0.6T model. In the past 2 years, we also developed the 1.3T model for SLon-1500-1.3T and SLon-1750-1.3T. All of them have found wider and wider use in the industrial application.

Table 1 SLon-3000 specifications

Parameters	SLon-3000
Ring diameter/mm	3000
Ring speed/r · min^{-1}	2~4
Feed size(<200mesh %)/mm	<1.2 (30~100)
Feed solid density/%	10~40
Slurry throughput/m^3 · h^{-1}	350~650
Ore throughput/t · h^{-1}	150~250
Rated Background induction/T	1.0
Energizing current/A	1400
Energizing voltage/V	62
Energizing power/kW	87
Ring motor power/kW	18.5
Pulsating power/kW	18.5
Pulsating stroke/mm	0~26
Pulsating frequency/cycle · min^{-1}	0~300
Water pressure/MPa	0.2~0.4
Washing water consumption/m^3 · h^{-1}	350~530
Cooling water consumption/m^3 · h^{-1}	8~10
Machine weight/t	175
Maximum part weight/t	25
Outside dimension/mm×mm×mm	6600×5300×6400

3.3.2 CE marking certification

To place SLon HGMS into the market in the European Economic Area as well as the specific customers who concern CE marking protection standard, we started to apply the CE marking certification in early 2010. With reference to the model SLon-1000 (Fig. 3), we upgraded all the technical documents, the heath and safety requirements as well as all relevant "essential requirements" of the applicable EC directives. A range of SLon HGMS from SLon-100 to SLon-3000 was certified to CE marking by the end of 2010. the CE marking Certificate is showed in Fig. 4.

Fig. 3 SLon-1000 CE marking sample

Fig. 4 CE marking certificate

Now, the CE marking model of SLon HGMS is available on request.

4 Applications to benificiate oxidized iron ores

4.1 The industrial application in Diaojuntai Iron Ore Processing Plant

The Diaojuntai Iron Ore Processing Plant was built in 1998. It belongs to Anshan Iron and Steel Company. Its ore consists of mainly hematite, magnetite and quartz. Its old flow sheet was consisted of two stage continuous grinding and low intensity magnetic separation→mid intensity magnetic separation→high intensity magnetic separation→floatation. In the flow sheet, 15 SHP-3200 horizontal ring wet high intensity magnetic separators were applied. Their magnetic matrixes were grooved plates made of stainless steel. Because the magnetic matrixes frequently were clogged, the 15 Shp-3200 were difficult to maintain, during 2003-2005, the plant applied 15 units of SLon-2000 vertical ring-pulsating high gradient magnetic separators to replace them as shown in Fig. 5. The final iron concentrate grade was upgraded from 67.13% to 67.50% and iron recovery from 76.64% to 78.94%.

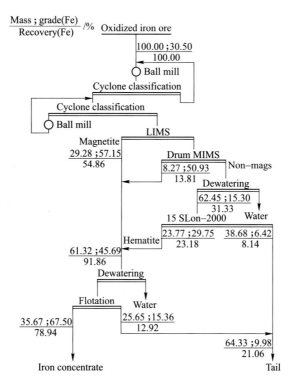

Fig. 5 The Initial upgraded flowsheet of Diaojuntai Mineral Processing Plant

In order to lower processing cost and increase capacity, in 2005-2007, the flow sheet was further modified to incorporate staged grinding with low intensity magnetic separation→SLon magnetic separation→reverse floatation as show in Fig. 6. Another 15 SLon-2000 vertical ring-pulsating middle intensity magnetic separators (0.4T) were added into the flow sheet. Because SLon magnetic separators discharge part of coarse tails and spirals take out part of coarse final concentrate in the primary grinding stage, the grinding costs are greatly reduced. The ore treating capacity is raised from 9 million tonnes to 14.4 million tonnes per annum using the same ball mills. As the mining capacity was subsequently scaled up, the average feed grade was lowered from 30.50%Fe to 29.50%Fe. However, the average concentrate grade remained at 67.50%Fe.

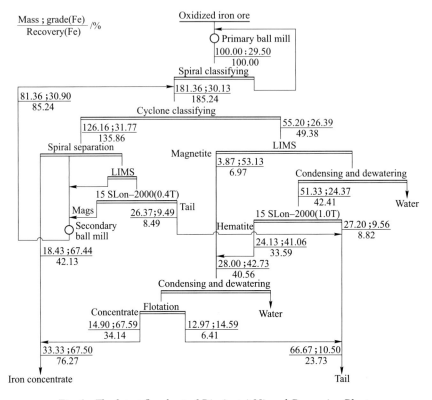

Fig. 6 The latest flowsheet of Diaojuntai Mineral Processing Plant

4.2 The industrial application in Lilou Beneficiation Plant

4.2.1 Flowsheet and technical data

Lilou Beneficiation Plant of Anhui Development Mining Ltd is located in Huoqiu county, Anhui Province. There are abundant iron ore resources there. Lilou Beneficiation Plant built a DEMO Beneficiation Plant with 4 units of SLon-2000 HGMS a few years ago, which capacity was 0.6 million tons per year. With the successful operation of the DEMO plant, Anhui Development Mining Limited has built a new specularite processing plant there in 2010, which capacity is up to 5 million tons per year. The flowsheet is shown in Fig. 7.

 The main iron mineral is specularite, the other iron minerals are very little, such as the mass for magnetite is only around 0.5%. Its flowsheet is: grade for feeding iron ore is 31.65%, after the first stage close circuit milling, the percentage of <200 mesh particles is up to nearly 50%. By the first stage low intensity separator, the first stage SLon HGMS rougher and the first stage SLon

HGMS scavenger, 43% mass of the tailings which grade is 9.21% is discharged. The concentrate goes into the second stage close circuit milling, <200 mesh iron ore is up to properly 80%, then by the second stage low intensity separators and the second stage SLon HGMS rougher, parts of final concentrate which grade is 65% is produced. The tailings from the second stage SLon HGMS rougher goes into the second stage SLon HGMS scavenger, the mags of which is sent to reverse flotation. The flotation concentrate grade is 64.96%. The overall flow sheet data are: grade of feeding: 31.65%Fe, grade of concentrate: 65%Fe, mass of Fe concentrate: 38.95%, Fe recovery: 80%, grade of tailings: 10.37%Fe.

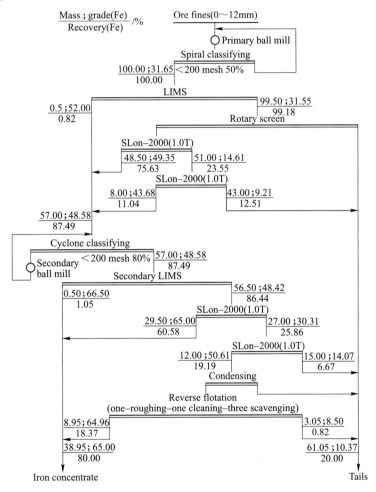

Fig. 7 The technical data of flowsheet of Lilou Beneficiation Plant

SLon HGMS discharged most of the tailings and get most portion of concentrate, only the second stage SLon HGMS scavenger concentrate (around 12% of feeding ore) goes into the reverse flotation stage. Because of the low cost and high efficiency of SLon HGMS, it greatly reduces the cost and load of reverse flotation, the cost of reverse flotation is relatively high, so the less concentrate goes into reverse flotation, the lower cost of whole flowsheet. Therefore, SLon HGMS plays a key role in gaining a very good beneficiation result.

4.2.2 Customer in preference of SLon HGMS

Lilou Beneficiation Plant purchased 4 units of SLon-2000 for DEMO plant in 2006. As several magnetic separation suppliers followed SLon's steps to supply similar HGMS (HGMS COPY) to SLon HGMS in recent years, some end users would like to compare their performance and reliability in industrial test and product. Lilou Beneficiation Plant introduced 1 unit of HGMS COPY for short term industrial comparison in early 2009. Then they purchased 14 units of SLon-2000 and 14 units of HGMS COPY -2000 for their new 5-million-ton plant in late 2009.

After long term production run, Lilou Beneficiation Plant purchased additional 7 units of SLon-2000 for scale up capacity in 2011, and no any HGMS COPY was purchased by Lilou Beneficiation Plant this year. This client finally prefers SLon HGMS.

4.3 SLon HGMS applications in India

India has large reserves of high grade hematite iron ores. However, steady depletion of these iron ores is now a concern forcing to develop beneficiation strategies to utilize low grade iron ores. In India, the iron ore slimes are being rejected in the tailing ponds. These slimes in most cases contain substantial iron values in the range of 50%-58%Fe. Therefore, it is imperative to recover iron values from these slimes because of high demand on the good grade iron ores day-by-day.

The mineralogy of the low grade iron ore sample though indicated that hematite is the major iron bearing phase, goethite also occurs in substantial quantity. As Indian customers' request, we have been engaging in developing strategies for beneficiating low grade iron ores and iron ore slimes since 2004. We have tested hundreds of iron ores samples from India with SLon HGMS. The typical test results of one stage of SLon HGMS are: feed grade from 50%Fe to 58%Fe, magnetic concentrate grade from 60%Fe to 65%Fe, concentrate recovery from 50%Fe to 85%Fe.

Up to now, over 70 SLon HGMS of different sizes have been successfully applied in India. SLon HGMS has become a very good solution for processing Indian low grade iron ore fines/slimes. It makes the previous waste materials to become a valuable resource.

5 Applications to beneficiate manganese ore

Most of Chinese manganese ores are of low grade. In recent years, SLon magnetic separators find applications in manganese ores processing industry, which shows great potential in the field.

5.1 Application in Inner Mongolia

Jinshui Mineral Company, located in Inner Mongolia, installed a SLon-1000, a SLon-1250, and a SLon-1500 magnetic separator to beneficiate low-grade manganese ore and the flowsheet is shown in Fig. 8.

The feed, which had been abandoned as tails in the previous years, only grade 8.85% Mn contained. After one stage grinding and SLon magnetic separating, the concentrate reaches 30.39%Mn and the recovery 75.30%Mn.

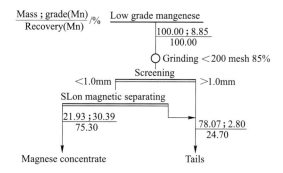

Fig. 8 SLon magnetic separators applied in Inner Mongolia for recovering manganese concentrate

5.2 Application in Nanjing Qixiashan Mine

Qixiashan Mine locates at the city side of Nanjing. The ore consists of Zn, Pb and Mn minerals. In the past years, Zn and Pb are the target minerals and Mn minerals were abandoned as tails. In 2005, the mine added two SLon-1500 magnetic separators into the flowsheet to recover manganese mineral as shown in Fig. 9. The total results are: feed grade 12.57% Mn, concentrate grade 25.20% Mn, recovery 55.13% Mn. Each year, 40,000 tons of manganese concentrate can be recovered.

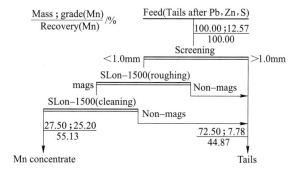

Fig. 9 Two SLon-1500 separators applied in Nanjing Qixiashan Mine

6 Conclusions

New types of SLon-2000, SLon-2500, SLon-3000 have been developed with higher efficiency, larger capacity and low energy consumption and lower operating cost. They have been found more and more applications in weakly magnetic mineral processing industry.

SLon HGMS extends its background magnetic field from 0.6T, 1.0T to 1.3T series; they have found wider and wider use in the industrial applications.

SLon HGMS has become a very good solution for processing Indian low grade iron ore fines and iron ore slimes. It makes the previous waste materials to become a valuable resource.

References

[1] XIONG D. SLon magnetic separator promoting Chinese weakly magnetic ores processing industry[J]. Magnetic and Electrical Separation, 2006, 06, Perth, Australia.

[2] Xiong D. SLon magnetic separators applied to beneficiate low grade oxidised iron ores[C]//Int. Proceedings of XXIV IMPC, 2008:813-818.

[3] Stephen B Hearn, Misty N. Dobbins. SLon magnetic separator: A new approach for recovering and concentrating iron ore fines[J]. Montreal Energy & Mines, Montreal.

写作背景 本文向南非的同行介绍了 SLon 立环脉动高梯度磁选机的优点，最近研制的 SLon-3000、SLon-1500-1.3T、SLon-1750-1.3T 等型号磁选机，SLon 磁选机通过欧盟的 CE 认证，以及 SLon 磁选机在一些氧化铁矿、锰矿和非金属矿中的应用。

SLon 立环脉动高梯度磁选机大型化研究与应用新进展

熊大和[1,2]

(1. 赣州金环磁选设备有限公司；2. 赣州有色冶金研究所)

摘 要 介绍了 SLon-2500、SLon-3000 和 SLon-4000 立环脉动高梯度磁选机的研制与应用。其中 SLon-4000 磁选机是目前我国自主研发的处理量最大的高梯度磁选机或强磁选机。它们具有处理量大、选矿效率高、设备可靠性高、电耗低和生产成本低的优点，它们已成功地在赤铁矿、褐铁矿、钛铁矿选矿和在石英除铁提纯工业中应用。

关键词 SLon 立环脉动高梯度磁选机 大型化 节能 弱磁性矿石选矿

现代选矿工业的特点是：一方面经济的发展需要更多更好的矿产品作为工业原料，另一方面是复杂难选的原矿增多。这就要求选矿设备和选矿工艺向大型化、自动化、高效率和低成本方向发展，以满足现代选矿工业的需要。SLon 立环脉动高梯度磁选机在工业上广泛应用于氧化铁矿、钛铁矿、锰矿等弱磁性矿石的选矿及石英、长石、高岭土等非金属矿的提纯。通过多年持续的技术创新与改进，该机在设备大型化、多样化、自动化、高效、节能、提高可靠性等方面得到了快速的发展，并且得到了更为广泛的应用。

1 设备大型化研制

设备大型化有利于提高选矿设备的台时处理能力和提高自动化水平，同时大型化设备具有节能、节约占地面积、操作成本低和设备作业率高等优点，可显著为用户降低生产成本和提高用户的经济效益。

1.1 SLon-2500 磁选机的研制

赣州金环磁选设备有限公司在总结研制 SLon-1000、SLon-1250、SLon-1500、SLon-1750、SLon-2000 立环脉动高梯度磁选机成功经验的基础上，研制了 SLon-2500 立环脉动高梯度磁选机，如图 1 所示，其电磁性能数据见表 1，该机干矿处理量达到 100~150t/h。

1.2 SLon-3000 磁选机的研制

2010 年研制出了 SLon-3000 立环脉动高梯度

图 1 SLon-2500 磁选机照片

---

[1] 原载于《中国采选技术十年回顾与展望会议论文集》，2012：692~696。

磁选机。如图 2 所示为第一台 SLon-3000 磁选机现场应用照片，其电磁性能数据见表 2，该机干矿处理量达到 150~250t/h。

表 1　SLon-2500 磁选机电磁性能测定数据

电流/A	电压/V	功率/kW	磁感应强度/T
200	6.6	1.3	0.168
400	12.7	5.1	0.336
600	19.5	11.7	0.495
800	26.1	20.9	0.654
1000	32.7	32.7	0.793
1200	38.6	46.3	0.911
1400	45.0	63.0	1.001
1600	52.0	83.2	1.074
1700	54.5	92.7	1.107

表 2　SLon-3000 磁选机电磁性能测定数据

电流/A	电压/V	功率/kW	磁感应强度/T
200	8.2	1.6	0.189
400	16.2	6.5	0.375
600	24.6	14.8	0.564
800	33.2	26.6	0.734
1000	42.2	42.2	0.880
1200	51.4	61.7	0.995
1300	55.9	72.7	1.041
1400	60.4	84.6	1.079
1500	64.9	97.4	1.116

1.3　SLon-4000 磁选机的研制

在完成 SLon-2500 磁选机和 SLon-3000 磁选机研制的基础上，2011 年又研制出了 SLon-4000 立环脉动高梯度磁选机。如图 3 所示为第一台 SLon-4000 磁选机照片，其电磁性能数据见表 3。该机干矿处理量达到 350~550t/h。

图 2　SLon-3000 磁选机应用照片

图 3　SLon-4000 磁选机照片

在研制 SLon-2500、SLon-3000 和 SLon-4000 磁选机的过程中，采用了一系列的新技术提升它们的选矿性能和机电性能，主要措施如下：

（1）通过优化磁系设计、减少漏磁、降低电阻和提高电效率，使它们有良好的激磁性能。采取低电压大电流激磁，有利于提高激磁线圈的安全可靠性。激磁线圈采用空心铜管绕制，水内冷散热，冷却水直接贴着铜管内壁流动，散热效率高。此外，由于冷却水的流速较高，微细泥沙不易沉淀，可保证激磁线圈长期稳定工作。

（2）新型磁介质的研究。为了提高 SLon 磁选机的选矿效率，我们针对各种矿石的不同特点，分别研制了 $\phi1mm$、$\phi1.5mm$、$\phi2mm$、$\phi3mm$、$\phi4mm$、$\phi5mm$、$\phi6mm$ 磁介质组，及各种磁介质的优化组合，如 $\phi1mm+\phi2mm$，$\phi2mm+\phi3mm$ 等磁介质组合，可根据入选矿石的粒度，比磁化系数选择，以获得最佳的选矿效果。

（3）采用新型数字式脉动冲程箱，具有力量大、平衡性好、可靠性高和能耗低的优点。

（4）采用新型的整流控制柜，激磁电流可根据选矿的需要调节，并稳定在人工设定点，不会随外电压的波动而波动。人机界面友好，自动化高，可远程控制。

表3 SLon-4000 磁选机电磁性能测定数据

电流/A	电压/V	功率/kW	磁感应强度/T
200	6.7	1.3	0.15
400	13.6	5.4	0.295
600	20.4	12.2	0.442
800	27.1	21.6	0.579
1000	33.5	33.5	0.709
1200	40.8	48.9	0.825
1400	47.5	66.5	0.921
1600	53.9	86.2	0.997
1800	60.6	109.1	1.056

SLon-2500、SLon-3000 和 SLon-4000 磁选机的主要技术参数见表4。

表4 大型 SLon 磁选机主要技术参数

机型	转环外径/mm	给矿粒度/mm	给矿浓度/%	矿浆通过能力/m³·h⁻¹	干矿处理量/t·h⁻¹	额定背景磁感应强度/T	额定激磁电流/A	额定激磁电压/V
SLon-2500	2500	0~1.2	10~40	200~400	100~150	1.0	1400	45
SLon-3000	3000	0~1.2	10~40	350~650	150~250	1.0	1300	58
SLon-4000	4000	0~1.2	10~40	750~1400	350~550	1.0	1700	60

机型	额定激磁功率/kW	传动功率/kW	脉动冲程/mm	脉动冲次/r·min⁻¹	供水压力/MPa	冲洗水用量/m³·h⁻¹	冷却水用量/m³·h⁻¹	主机重量/t	最大部件重量/t
SLon-2500	63	11+11	0~30	0~300	0.2~0.4	160~220	5~6	105	15
SLon-3000	75	18.5+18.5	0~30	0~300	0.2~0.4	240~400	7~8	175	25
SLon-4000	102	37+37	0~30	0~300	0.2~0.4	560~800	8~9	380	35

2 设备节能研究

SLon 立环脉动高梯度磁选机的大型化有利于降低处理每吨矿石的能耗，通过优化磁系结构，减少漏磁，采用优质铜管、增加铜的用量和降低激磁电流密度，使电耗有了显著的降低。

图 4 和表 5 为几种型号 SLon 磁选机能耗对比情况。各种机型在背景磁感应强度达到 1.0 T 时，按设备处理量的平均值计算吨矿消耗功率，设备越大，能耗越低，例如 SLon-1000 磁选机单位能耗为 3.69 kW·h/t，而 SLon-4000 磁选机仅为 0.41 kW·h/t。后者比前者节能 88.89%。SLon-4000 磁选机与目前工业上大量使用的 SLon-2000 磁选机比较也节能 51.76%。

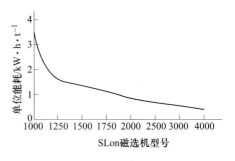

图 4 SLon 磁选机处理每吨矿石能耗对比

表 5 SLon 磁选机处理量与能耗对比

机型	额定背景磁感强度/T	额定激磁功率/kW	传动功率/kW	总功率/kW	干矿平均处理量/t·h^{-1}	每吨矿石电耗/kWh·t^{-1}
SLon-1000	1.0	17	3.3	20.3	5.5	3.69
SLon-1250	1.0	19	3.7	22.7	14	1.62
SLon-1500	1.0	27	7	34	25	1.36
SLon-1750	1.0	37	8	45	40	1.13
SLon-2000	1.0	42	13	55	65	0.85
SLon-2500	1.0	63	22	85	125	0.68
SLon-3000	1.0	75	37	112	200	0.56
SLon-4000	1.0	102	74	176	430	0.41

3 工业应用

3.1 SLon-2500 磁选机在海钢红旗尾矿库的应用

首台 SLon-2500 磁选机在海南钢铁公司（现已更名为海南矿业联合有限公司，下同）应用于从尾矿中回收铁精矿。海南钢铁公司有三个选矿车间分选氧化铁矿，选矿流程为强磁选—粗—精—扫，或强磁选—粗—扫—反浮选。每年入选原矿约 230 万吨，目前每年大约有 120 万吨尾矿排入到红旗尾矿库。其总尾矿的平均铁品位为 28% 左右，其中的铁主要是存在于 0~19μm 粒级中。由于该尾矿氧化程度较深、含泥量大，且含有较多的连生体，因此要从该尾矿中再回收铁精矿技术难度较大。

SLon-2500 磁选机安装于该尾矿输送管的排矿口，从该排矿口截取尾矿入选。每年处理大约 72 万吨尾矿。该尾矿再回收系统的选矿工艺流程见图 5。首先用圆筒除渣筛筛出大于 0.8mm 的粗粒。接着用 SLon-2500 磁选机粗选，其粗选精矿经浓缩后用离心选矿机精选。离心机精矿作为最终精矿。该流程综合选矿指标为：给矿铁品位为 29.90%，精矿铁品位为 61.30%，铁精矿产率 8.53%，铁回收率 17.49%，尾矿铁品位为 26.97%。

该流程至今已运转 5 年，该机的单机作业率达到 99% 以上。整个生产流程稳定，指标先进。

3.2 SLon-2500 磁选机在氧化铁矿选矿的应用

SLon-2500 磁选机研制成功以后，很快在氧化铁矿的选矿工业中得到大规模应用。

2008~2010 年，印度一家大型钢铁公司采购了 10 台 SLon-2500 磁选机分选赤铁矿，采用 SLon-2500 磁选机一次粗选和一次扫选的选矿流程，其入选原矿铁品位为 59.77%，综合铁精矿铁品位为 65.00%，综合铁回收率为 93.86%，选矿指标优异。

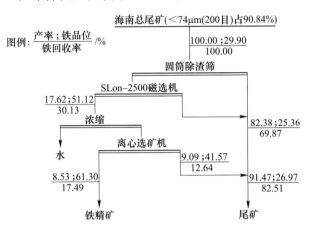

图 5　SLon-2500 磁选机分选海钢尾矿流程

2011 年，11 台 SLon-2500 磁选机出口到非洲塞拉利昂，分选镜铁矿获得成功应用（图 6，图 7），采用 SLon-2500 磁选机一次粗选和一次精选的选矿流程，其入选原矿铁品位为 34.94%，综合铁精矿铁品位为 67.10%，综合铁回收率为 74.74%，获得了优异的选矿指标。

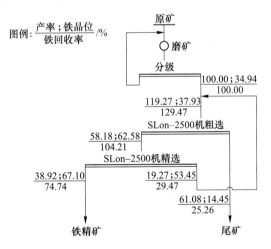

图 6　塞拉利昂镜铁矿选矿原则流程

图 7　SLon-2500 磁选机在塞拉利昂应用照片

3.3　SLon-2500 磁选机分选褐铁矿的工业应用

云南北衙黄金选矿厂的原矿为金矿与铁矿共生的矿石，其早期的生产仅用浸出法回收金，而选金后的尾矿因粒度很细未得到回收。通过探索试验表明，该选金尾矿中含有磁铁矿和褐铁矿。2006 年该厂建成了一条年处理 50 万吨的选铁生产线，从选金尾矿中回收铁矿物，其中采用 2 台 SLon-2000 磁选机回收褐铁矿。2010 年该厂又建设了一条年处理 150

万吨原矿的选矿生产线,其选金后的尾矿用弱磁选机选出磁铁矿,然后用 4 台 SLon-2500 磁选机分选褐铁矿(粗选和扫选各 2 台,图 8),该生产线选铁部分的流程如图 9 所示。目前该厂每年可回收 100 多万吨铁精矿,并可根据市场价格安排磁铁矿和褐铁矿分别销售或合并销售。

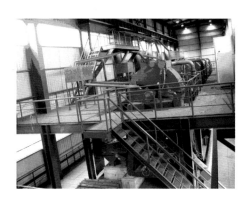

图 8　SLon-2500 磁选机在北衙黄金选矿厂应用照片

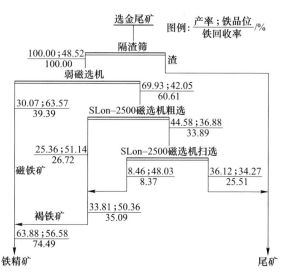

图 9　北衙黄金选矿厂选铁流程

3.4　SLon-2500 磁选机在石英提纯工业中的应用

SLon-2500 磁选机于 2009 年 7 月首先在陕西汉中石英砂提纯工业中应用成功。尔后 SLon-2500 磁选机在凤阳高纯石英砂提纯方面又获得成功应用。凤阳县位于安徽省东北部,处于淮河中游南岸,因石英砂资源丰富,因而享有"中国石英砂之乡"的美誉。SLon-2500 磁选机于 2011 年 7 月在凤阳县远东石英砂有限公司安装完毕,10 月份投入生产(见图 10)。原石英砂矿中 Fe_2O_3 的含量在 0.01%~0.02%,经过 SLon-2500 磁选机磁选后,精矿产率达到了 95%以上,精矿中 $Fe_2O_3 \leq 0.007\%$、$SiO_2 > 99.8\%$,可达到太阳能、光纤通信等高尖端原料的要求。

图 10　SLon-2500 磁选机在安徽凤阳提纯石英砂

3.5　首台 SLon-3000 磁选机在攀枝花的应用

2010 年首台 SLon-3000 磁选机在攀枝花安宁铁钛有限公司应用于钛铁矿选矿获得成功。攀枝花地区的钒钛磁铁矿储量丰富,目前该地区的选矿厂的选矿流程普遍是先用弱磁选机选出磁铁矿精矿,选完磁铁矿的尾矿即作为选钛作业的原矿。选钛作业普遍用强磁选机预先抛弃大部分尾矿，强磁选精矿再用浮选或重选精选。由于选钛流程中入选原料含钛品位低，要求强磁选机处理量大，生产成本低，尽可能多抛弃尾矿，SLon-3000 磁选机正好具备这些优势，该厂选矿流程如图 11 所示。

SLon-3000 磁选机在该厂用于钛铁矿的强磁粗选作业，其给矿粒度为小于 74μm（200 目）占 40%，台时处理量为 200t/h 左右。该机的选矿指标为：给矿品位 9.52%TiO_2，精矿品位 17.43%TiO_2，精矿产率 41.01%，TiO_2 回收率 75.09%，尾矿品位 4.02%TiO_2，抛弃的尾矿产率为 58.99%。该机使用的背景磁感强度为 0.7T，处理每吨给矿的电耗仅 0.3kW·h，生产证明该机具有选矿效率高、设备可靠性高和生产成本低的优点。

3.6 SLon-3000 磁选机在太和铁矿选钛工业中的应用

重庆钢铁公司太和铁矿近年将原矿处理量由 220 万吨/a 扩建至 630 万吨/a，扩建过程中共采用了 3 台 SLon-3000 磁选机用于钛铁矿分选（图 12）。目前该生产流程正处于设备安装及调试阶段。

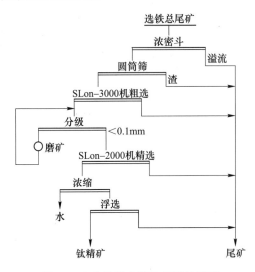

图 11　安宁铁钛有限公司选钛流程　　　　图 12　SLon-3000 磁选机在太和铁矿分选钛铁矿

3.7　首台 SLon-4000 磁选机在攀钢选钛工业试验

攀钢密地选矿厂从钒钛磁铁矿中回收铁精矿和钛精矿，其综合尾矿每年约 700 万吨（每小时约 885t），全部排往尾矿库。但是其综合尾矿中还含有少量的可回收铁矿和钛铁矿。攀钢选矿综合厂主要从攀钢排放的总尾矿中回收铁精矿和钛精矿，最早的选钛流程采用螺旋溜槽粗选，摇床精选的单一重选流程。由于螺旋溜槽和摇床对细粒级回收率都很低，全流程的选钛回收率仅为 7%TiO_2 左右。2002 年该厂采用了 3 台 SLon-1500 磁选机建立新的选钛流程（图 13）。该流程的选钛回收率由原重选作业的 7%TiO_2 提高到 27%TiO_2 左右。但是由于用于粗选作业的 2 台 SLon-1500 磁选机合计只能处理 60t/h 给矿，该厂钛精矿产量只能达到每年 2 万吨左右。随着攀钢密地选钛厂选钛技术水平的提高，目前排往尾矿库的尾矿 TiO_2 品位由原来的 7.68% 左右下降至 6.5% 左右，因此，如果不改造，该厂的钛精矿产量有进一步下降的趋势。2011年年底，首台SLon-4000立环脉动高梯度磁选机在该厂应用于工业试验（图14），取代原有粗选作业的SLon-1500磁选机，该机每小时给矿量为443t（总尾矿的50%）左右，至今已运转4个月，已取得的阶段性工业试验指标为（单机作业指标）：给矿品位6.89% TiO_2，精矿品位13.27% TiO_2，尾矿品位4.02% TiO_2，TiO_2作业回收率59.88%的良好指标。

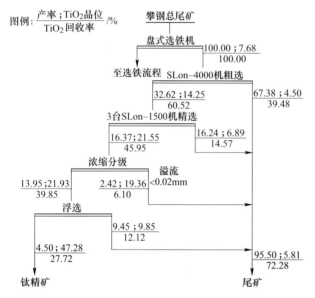

图13 攀钢综合厂采用SLon磁选机的选钛流程

图14 SLon-4000磁选机在攀钢从总尾矿中回收钛铁矿

4 结论

（1）SLon-2500、SLon-3000和SLon-4000磁选机的研制代表了该机型大型化发展方向。设备的大型化有利于降低生产成本，节约占地面积，提高选矿厂自动化生产水平。

（2）对磁系的优化设计使SLon磁选机的激磁功率有了大幅度的下降，设备的大型化更有利于降低单位处理量的能耗。

（3）SLon-2500、SLon-3000和SLon-4000磁选机具有处理量大、选矿效率高、设备可靠性高、电耗低和生产成本低的优点。它们已成功地在赤铁矿、褐铁矿、钛铁矿选矿和在石英除铁提纯工业中应用。

写作背景 本文介绍了SLon-2500、SLon-3000、SLon-4000立环脉动高梯度磁选机的研制、它们的技术性能以及在多个厂矿的应用，其中SLon-4000磁选机的研制、SLon-2500磁选机在云南北衙黄金选矿厂分选褐铁矿、SLon-2500磁选机出口到印度分选赤铁矿、SLon-2500磁选机出口到非洲塞拉利昂分选镜铁矿、SLon-4000磁选机分选钛铁矿均为首次报道。

应用 SLon 磁选机提高弱磁性金属矿回收率

熊大和[1,2]

(1. 赣州金环磁选设备有限公司;2. 赣州有色冶金研究所)

摘 要 SLon-2000、SLon-2500、SLon-3000 和 SLon-4000 立环脉动高梯度磁选等大型设备具有处理量大、选矿效率高、设备可靠性高、电耗低、投资成本低和生产成本低的优点,在氧化铁矿和钛铁矿等弱磁性金属矿的选矿流程中适当增加强磁选作业、优化选矿流程,可显著地提高弱磁性矿物的选矿回收率,使矿产资源的利用率进一步提高。

关键词 SLon 立环脉动高梯度磁选机 氧化铁矿 钛铁矿 选矿回收率

Applications of SLon Magnetic Separators in Raising the Recoveries of Feebly Magnetic Minerals

Xiong Dahe[1,2]

(1. *SLon Magnetic Separator Ltd.*;
2. *Ganzhou Nonferrous Metallurgy Research Institute*)

Abstract Fully utilizing the advantages of large capacity, high beneficiation efficiency, high reliability, low energy consumption, low invest cost and low operating cost of SLon-2000, SLon-2500, SLon-3000 and SLon-4000 magnetic separators, properly adding high intensity magnetic operation stages and optimizing processing flow sheet, The recoveries of feebly magnetic minerals can be greatly raised. Thus the utilization ratio of mineral resources can also be raised further.

Keywords SLon magnetic separator; Oxidized iron ore; Ilmenite; Recovery

在过去的 20 多年中,我国强磁选机技术得到了快速发展,SLon 立环脉动高梯度磁选机得到了广泛应用,由于持续的技术创新和设备的大型化,以台时处理每吨矿石计算的强磁选机的销售价格大约为 2.5 万~3 万元,这个价格维持了 20 年左右并有所下降。而同期的住房价格上涨了十倍左右,铁矿石的价格上涨了 3 倍左右,同期 SLon 磁选机的作业成本降低 50% 左右。如图 1 所示为过去 20 年这几种因素的变化曲线。

由于强磁选机的相对价格和选矿成本的大幅度降低,使选矿厂可采用较多的强磁选机优化选矿流程,提高选矿回收率和选矿效率。

---

[1] 原载于《矿冶工程》,2012(08):75~83。

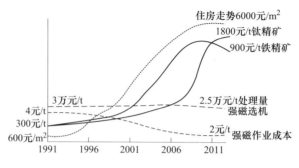

图 1　过去 20 年几种因素的变化曲线

1　设备大型化研制

设备大型化有利于提高选矿设备的台时处理能力和提高自动化水平，同时大型化设备具有节能、节约占地面积、操作成本低和设备作业率高等优点，可显著为用户降低生产成本和提高用户的经济效益。

1.1　SLon-2500 磁选机的研制

赣州金环磁选设备有限公司在总结研制 SLon-1000、SLon-1250、SLon-1500、SLon-1750、SLon-2000 立环脉动高梯度磁选机成功经验的基础上，研制了 SLon-2500 立环脉动高梯度磁选机，如图 2 所示，其电磁性能数据见表 1，该机干矿处理量达到 100~150t/h。

图 2　SLon-2500 磁选机照片

表 1　SLon-2500 磁选机电磁性能测定数据

电流/A	电压/V	功率/kW	磁场强度/T
200	6.6	1.3	0.168
400	12.7	5.1	0.336
600	19.5	11.7	0.495
800	26.1	20.9	0.654
1000	32.7	32.7	0.793
1200	38.6	46.3	0.911
1400	45.0	63.0	1.001
1600	52.0	83.2	1.074
1700	54.5	92.7	1.107

1.2　SLon-3000 磁选机的研制

2010 年我公司研制出了 SLon-3000 立环脉动高梯度磁选机，图 3 为第一台 SLon-3000 磁选机现场应用照片，其电磁性能数据见表 2，该机干矿处理量达到 150~250t/h。

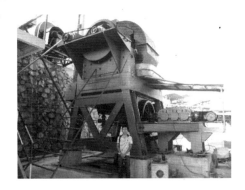

图 3 SLon-3000 磁选机照片

表 2 SLon-3000 磁选机电磁性能测定数据

电流/A	电压/V	功率/kW	磁场强度/T
200	8.2	1.6	0.189
400	16.2	6.5	0.375
600	24.6	14.8	0.564
800	33.2	26.6	0.734
1000	42.2	42.2	0.880
1200	51.4	61.7	0.995
1300	55.9	72.7	1.041
1400	60.4	84.6	1.079
1500	64.9	97.4	1.116

1.3 SLon-4000 磁选机的研制

在完成 SLon-2500 磁选机和 SLon-3000 磁选机研制的基础上，2011 年又研制出了 SLon-4000 立环脉动高梯度磁选机。图 4 为第一台 SLon-4000 磁选机照片，其电磁性能数据见表 3，该机干矿处理量达到 350~550t/h。

在弱磁性金属矿选矿工业中常用的 SLon-2000、SLon-2500、SLon-3000 和 SLon-4000 磁选机的主要技术参数见表 4。

图 4 SLon-4000 磁选机照片

表 3 SLon-4000 磁选机电磁性能测定数据

电流/A	电压/V	功率/kW	磁场强度/T
200	6.7	1.3	0.15
400	13.6	5.4	0.295
600	20.4	12.2	0.442
800	27.1	21.6	0.579
1000	33.5	33.5	0.709
1200	40.8	48.9	0.825
1400	47.5	66.5	0.921
1600	53.9	86.2	0.997
1800	60.6	109.1	1.056

表4 大型SLon磁选机主要技术参数

机 型	转环外径 /mm	给矿粒度 /mm	给矿浓度 /%	矿浆通过能力 /m³·h⁻¹	干矿处理量 /t·h⁻¹	额定背景磁感应强度 /T	额定激磁电流/A	额定激磁电压/V
SLon-2000	2000	0~1.2	10~40	100~200	50~80	1.0	1200	35
SLon-2500	2500	0~1.2	10~40	200~400	100~150	1.0	1400	45
SLon-3000	3000	0~1.2	10~40	350~650	150~250	1.0	1300	58
SLon-4000	4000	0~1.2	10~40	750~1400	350~550	1.0	1700	60

机 型	额定激磁功率/kW	传动功率 /kW	脉动冲程 /mm	脉动冲次 /r·min⁻¹	供水压力 /MPa	冲洗水用量 /m³·h⁻¹	冷却水用量 /m³·h⁻¹	主机质量 /t	最大部件质量/t
SLon-2000	42	5.5+7.5	0~30	0~300	0.2~0.4	80~120	4~5	50	14
SLon-2500	63	11+11	0~30	0~300	0.2~0.4	160~220	5~6	105	15
SLon-3000	75	18.5+18.5	0~30	0~300	0.2~0.4	240~400	7~8	175	25
SLon-4000	102	37+37	0~30	0~300	0.2~0.4	560~800	8~9	380	35

2 设备节能研究

SLon立环脉动高梯度磁选机的大型化有利于降低处理每吨矿石的能耗，通过优化磁系结构，减少漏磁，采用优质铜管、增加铜的用量和降低激磁电流密度等措施，使电耗有了显著的降低。

图5和表5为几种型号SLon磁选机能耗对比情况。各种机型在背景磁感应强度达到1.0T时，按设备处理量的平均值计算吨矿消耗功率，设备越大，能耗越低，例如SLon-1000磁选机单位能耗为3.69kW·h/t，而SLon-4000磁选机仅为0.41kW·h/t。后者比前者节能88.89%。SLon-4000磁选机与目前工业上大量使用的SLon-2000磁选机比较也节能51.76%。

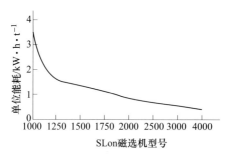

图5 SLon磁选机处理每吨矿石能耗对比

表5 SLon磁选机处理量与能耗对比

机 型	额定背景磁感强度/T	额定激磁功率/kW	传动功率 /kW	总功率 /kW	干矿平均处理量 /t·h⁻¹	每吨矿石电耗 /kW·h·t⁻¹
SLon-1000	1.0	17	3.3	20.3	5.5	3.69
SLon-1250	1.0	19	3.7	22.7	14	1.62
SLon-1500	1.0	27	7	34	25	1.36
SLon-1750	1.0	37	8	45	40	1.13
SLon-2000	1.0	42	13	55	65	0.85
SLon-2500	1.0	63	22	85	125	0.68
SLon-3000	1.0	75	37	112	200	0.56
SLon-4000	1.0	102	74	176	430	0.41

3 工业应用

由于强磁选机价格下降和作业成本的降低，SLon 立环脉动高梯度磁选机得到了更为广泛的应用，许多氧化铁矿和钛铁矿的选矿流程中增加了强磁选作业。

3.1 SLon 磁选机在安徽金日盛铁矿的应用

安徽省霍邱县铁矿资源丰富，近年有几个铁矿选矿厂相继建成。安徽金日盛铁矿的矿石由镜铁矿、赤铁矿、磁铁矿和脉石组成。近几年建成了一座年处理 450 万吨原矿的现代化选矿厂，其选矿流程见图 6。该流程的特点为：

（1）原矿经一段磨矿分级至小于 0.074mm 粒级占 55%，用弱磁选机选出部分品位达到 66%Fe 的磁铁矿作为最终铁精矿，弱磁选尾矿再用螺旋溜槽选出部分品位达到 65%Fe 的氧化铁矿（单体解离度较好的镜铁矿和赤铁矿）。

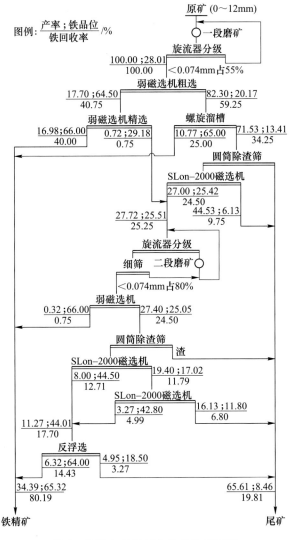

图 6　安徽金日盛氧化铁矿选矿流程

（2）用 SLon-2000 立环脉动高梯度磁选机作为一段强磁选抛尾设备，该机一次选别抛弃了占原矿产率 44.53%，Fe 品位为 6.13% 的尾矿，有效地控制了一段尾矿品位。

（3）一段强磁选精矿和一段弱磁精选的尾矿合并后进入二段分级和磨矿。由于一段成功地分选出了大部分合格铁精矿和抛弃了大部分低品位尾矿，进入二段分级磨矿的矿量只占原矿产率的 27.72%。

（4）二段分级磨矿至小于 0.074mm 占 80%，用弱磁选机分选出剩余的磁铁矿作为最终精矿，弱磁选尾矿用 SLon 磁选机一次粗选和一次扫选分选出镜铁矿和赤铁矿。二段强磁选精矿合并后用反浮选精选，得到 Fe 品位为 64% 的反浮选铁精矿。

该流程综合选矿指标为：原矿品位 28.01%Fe，铁精矿品位 65.32%Fe，铁精矿产率 34.39%，铁回收率 80.19%，尾矿品位 8.46%Fe。整个生产流程稳定，指标优异。

3.2 SLon 磁选机在李楼铁矿的应用

安徽开发矿业有限公司李楼铁矿近几年建成了一座年处理 500 万吨镜铁矿的选矿厂，其设计的选矿流程如图 7 所示。

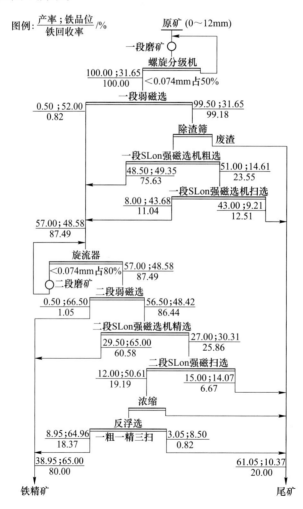

图 7　李楼铁矿 500 万吨/a 镜铁矿设计选矿流程

李楼铁矿镜铁矿的含铁矿物主要是镜铁矿,其他铁矿物含量很少,例如原矿中磁铁矿产率仅为 0.5%左右。其选矿流程的特点为:

(1) 原矿铁品位 31.65%Fe,经一段闭路磨矿分级至小于 0.074mm 占 50%左右,用弱磁选机选出少量的赤铁矿,用强磁选机进行一次粗选和一次扫选提前抛弃产率为 43.00%,品位为 9.21%左右的尾矿,保证了一段强磁尾矿品位处于较低的水平。

(2) 弱磁选精矿和强磁选精矿经二段闭路分级磨矿至小于 0.074mm 占 80%左右,然后用二段弱磁选和二段强磁精选得到一部分 Fe 品位 65%以上的铁精矿成品,其中二段强磁精选的铁精矿产率占原矿的 29.5%,占全流程总精矿的 75.74%。

(3) 二段强磁精选的尾矿再经二段强磁扫选,二段强磁扫选精矿经浓缩后用反浮选精选,反浮选得到另一部分品位为 64.96%左右的铁精矿成品。

(4) 由于强磁选作业成本低,强磁选提前抛弃了大部分尾矿和得出了大部分精矿,仅有占原矿 12%左右的二段强磁扫选精矿进入反浮选作业,由于反浮选的作业成本较高,进入反浮选的矿量越少,全流程的生产成本越低。

(5) 该流程中一段强磁选采用一粗一扫,二段强磁选采用一精一扫,保证了全流程获得较高的选矿回收率。

全流程设计的综合选矿指标为:给矿品位 31.65%Fe,精矿品位 65.00%Fe,铁精矿产率 38.95%,铁回收率 80.00%,尾矿品位 10.37%Fe。

3.3 SLon 磁选机在攀钢选钛厂的应用

3.3.1 SLon 磁选机在攀钢选钛厂回收小于 0.045mm 粒级钛铁矿

攀钢选钛厂原先的选钛流程中,小于 0.045mm 粒级作为尾矿排放,其中 TiO_2 的金属占有率为 40%左右,因此,攻克这部分物料的选钛难题,对提高全流程的选钛回收率具有重要的意义。1994~1996 年,攀钢选钛厂采用一台 SLon-1500 立环脉动高梯度磁选机建立了第一条微细粒级选钛试验生产线(图 8)。

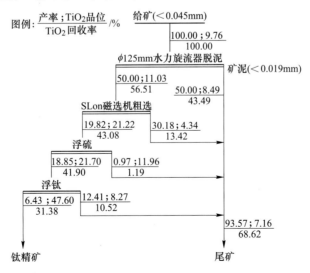

图 8 攀钢选钛厂第一条小于 0.045mm 选钛生产流程

该流程中首先用旋流器脱出小于 0.019mm 矿泥，然后用一台 SLon-1500 磁选机粗选，其粗选精矿再用浮选脱硫和选钛铁矿，最终得到品位为 47%TiO_2 以上的优质钛精矿。

该流程中，SLon 磁选机的作业回收率达到 76.24%TiO_2，浮选的作业回收率也达到了 75%TiO_2 左右，该流程的选钛回收率达到了 30%TiO_2 左右。由于小于 0.019mm 粒级未入选，这部分 TiO_2 金属量损失了 43.49%，影响了钛铁矿回收率的进一步提高。

上述选钛生产线每年生产 2 万吨优质钛精矿。随后的几年攀钢选钛厂又分两期建成后八系统和前八系统小于 0.045mm 粒级选钛生产线，至 2004 年，攀钢选钛厂小于 0.045mm 钛精矿的年产量达到 15 万吨左右，加上粗粒级钛精矿，攀钢选钛厂年产钛精矿达到 28 万吨左右。

3.3.2 攀钢选钛厂钛铁矿选矿的扩能改造工程

2008 年，攀钢选钛厂进行选钛扩能改造工程，具体实施计划有：
（1）将粗粒级选钛流程由原来的重选—电选流程改为强磁选—浮选流程。
（2）优化细粒级选钛流程。
（3）钛精矿产量由改造前的 28 万吨/a 提高到 40 万吨/a。

攀钢选钛厂委托赣州金环磁选设备有限公司所做的选钛试验流程如图 9 所示。

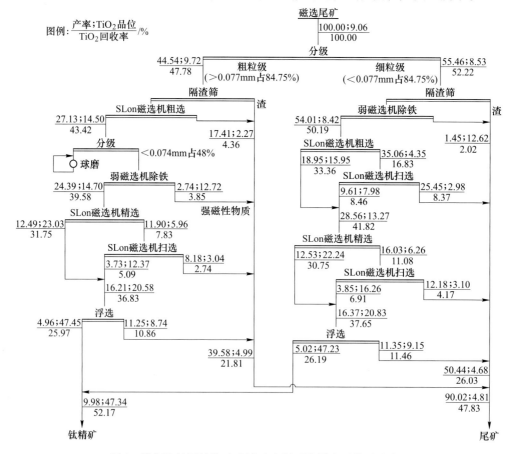

图 9　攀钢选钛厂扩能改造试验流程（赣州金环公司试验）

该流程的主要特点为：

（1）磁选尾矿只分成粗粒级和细粒级入选，充分发挥 SLon 磁选机本身的脱泥功能，尽可能减少分级溢流的损失。

（2）细粒级强磁选作业由过去的 SLon 磁选机一次选别改为四次选别，即 SLon 磁选机一粗一扫和一精一扫；粗粒级强磁选作业采用 SLon 磁选机一次粗选，其粗精矿分级再磨后用弱磁选机除铁，其尾矿用 SLon 磁选机一精一扫选。这些措施保证了强磁选作业对钛铁矿有较高的回收率和较高的精矿品位。比较图 8 和图 9 可见，细粒级进入浮选作业的 TiO_2 金属量由 43% 左右提高到（37.65%/52.55% = 72.10%）72% 左右。粗粒级进入浮选作业的 TiO_2 金属量高达该粒级的（36.83%/47.78% = 77.08%）77% 左右。SLon 磁选机的作业指标见表 6。

表 6　SLon 磁选机分选攀钢钛铁矿作业指标　　　　　　　　　　　　　　　　（%）

作业名称	选别次数	给矿品位	精矿品位	精矿回收率	尾矿品位
细粒级一段强磁选	SLon 一粗一扫	8.42	13.27	83.32	2.98
细粒级二段强磁选	SLon 一精一扫	13.27	20.83	90.03	3.10
细粒强磁综合		8.42	20.83	75.01	3.02
粗粒级一段强磁选	SLon 一粗	9.72	14.50	90.87	2.27
粗粒级二段强磁选	SLon 一精一扫	14.70	20.58	93.07	3.04
粗粒强磁综合		9.72	20.58	84.57	2.52

（3）该选钛流程充分利用了 SLon 磁选机处理量大、生产成本低、泥效果较好的优点，大幅度提高强磁选作业的选钛回收率，为浮选创造了较好的条件。在保证钛精矿品位达到 47% TiO_2 以上的前提下，浮选作业 TiO_2 回收率可保证在 70% 左右，全流程的 TiO_2 回收率达到 52.17%。

2008~2009 年，攀钢选钛厂参考各家的选钛试验流程并结合现场的条件进行了选钛扩能改造，2010 年和 2011 年该厂分别实现了年产优质钛精矿 48 万吨和 50 万吨，比改造前每年提高 20 余万吨优质钛精矿。目前生产上全流程实际达到的 TiO_2 回收率为 40% 左右，虽然选钛回收率比以前提高了很多，由于生产条件的限制，许多作业环节还没有达到最佳状况。选钛回收率的提高还有很大的潜力。

3.4　SLon 磁选机在重钢太和铁矿分选钛铁矿

3.4.1　重钢太和铁矿早期的重选—浮选选钛流程

重庆钢铁公司太和铁矿位于四川省西昌市，其矿石为钒钛磁铁矿，矿石性质与攀枝花钒钛磁铁矿类似。其选矿流程是原矿经破碎、磨矿、永磁磁选机回收磁铁矿。其选铁尾矿中含 TiO_2 12% 左右。为回收该尾矿中的钛资源，太和铁矿于 1995 年建成了第一条选钛生产线（图 10）。该流程首先将磁选尾矿分级成粗粒级、细粒级及溢流。粗粒级用螺旋溜槽粗选，其粗精矿进行分级磨矿后与细粒级合并，用滚筒永磁机选出残余的强磁性物质，然后用摇床分选，摇床精矿用浮选精选，浮选精矿作为最终钛精矿。

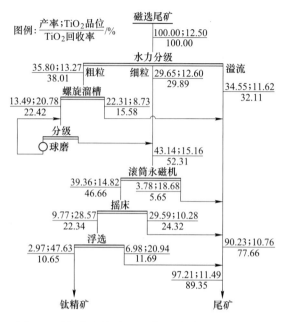

图 10 重钢太和铁矿 2001 年以前采用的选钛流程

该流程存在的主要问题有：

（1）由于当时选用的水力分级设备分级效率不高，有 32% 左右的 TiO_2 损失在溢流中。

（2）螺旋溜槽和摇床的分选粒度下限只能达到 30μm 左右，对细粒级钛铁矿的回收率很低。螺旋溜槽的作业回收率仅为 59% 左右，摇床的作业回收率仅为 47% 左右。

（3）该流程的重选段（包括分级、螺旋溜槽和摇床）的 TiO_2 综合回收率仅为 22.42%。

（4）钛铁矿浮选是正浮选，即泡沫产品为钛铁矿精矿，螺旋溜槽和摇床都不能回收细粒级钛铁矿，造成浮选的给矿粒度偏粗，由于钛铁矿密度（4.7~4.8g/cm³）较大，较粗粒级的钛铁矿容易沉槽，因此造成浮选回收率偏低。

上述原因导致了该流程选钛回收率只能达到 10% TiO_2 左右。1995~2000 年，太和铁矿采用该流程每年只能回收 2000~5000t 钛精矿。

3.4.2 重钢太和铁矿早期的强磁—浮选选钛流程

重钢太和铁矿于 2000 年开始采用 SLon 立环脉动高梯度磁选机进行选钛流程的技术改造，经过几次改造后的选钛流程如图 11 所示。

该流程采用 SLon 磁选机取代了螺旋溜槽和摇床。磁选尾矿首先用水力分级机分成大于 0.02mm（沉砂）和小于 0.02mm（溢流），沉砂用 SLon 磁选机粗选，其精矿分级磨矿，用滚筒永磁中磁机除去磁铁矿等强磁性物质后再用 SLon 磁选机精选，其精矿进入浮选精选。

上述流程的主要优点有：

（1）由于 SLon 磁选机的选矿粒度下限可达到 10μm 左右，磁选尾矿用水力分级机分级时可控制溢流部分少跑一些。让更多的细粒级钛铁矿进入 SLon 磁选机的选别作业。对

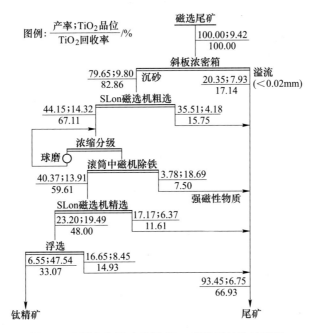

图 11 重钢太和铁矿采用 SLon 磁选机的选钛流程

比图 10 和图 11 可知,溢流部分流失的 TiO_2 金属量由 32% 左右降低至 17% 左右。

(2) 由于 SLon 磁选机分选粒度范围较宽,其入选原矿不再分粗粒级和细粒级,简化了选矿流程。

(3) SLon 磁选机取代螺旋溜槽和摇床后作业回收率大幅度提高。表 7 为它们选矿指标的对比结果,尽管原矿品位大幅度降低了(经过几年开采后,太和铁矿的原矿 TiO_2 品位大幅度降低了),但 SLon 磁选机的作业回收率分别比螺旋溜槽和摇床高 22% 左右和 32% 左右。进入浮选作业的 TiO_2 金属量由原重选作业的 22% 左右提高到 48% 左右。

表 7　SLon 磁选机选与螺旋溜槽和摇床的选矿指标对比　　　　　　（%）

设备	作业名称	给矿 TiO_2 品位	精矿 TiO_2 品位	精矿产率	精矿作业回收率	尾矿 TiO_2 品位
SLon 磁选机	粗选作业	9.80	14.32	55.42	80.99	4.18
螺旋溜槽	粗粒级粗选	13.27	20.78	37.68	59.00	8.73
SLon 与螺旋溜槽差值	粗选作业	-3.47	-6.46	17.74	21.99	-4.55
SLon 磁选机	二段精选	13.91	19.49	57.47	80.52	6.37
摇床	二段精选	14.82	28.57	24.82	47.85	10.28
SLon 磁选机与摇床差值	二段精选	-0.91	-9.08	32.65	32.67	-3.91

(4) 浮选作业的给矿粒度组成更好了。由于 SLon 磁选机对 0.15~0.02mm 钛铁矿粒级回收率高达 85% 以上,这一粒度范围正好是浮选的最佳粒度。因此浮选的作业回收率由过去的 48% TiO_2 左右提高到目前的 69% TiO_2 左右。

SLon 磁选机在太和铁矿的成功应用大幅度提高了该矿钛铁矿的回收率,尽管选钛系

统入选原矿的 TiO_2 由过去的 12% 左右降低至目前的 9% 左右，TiO_2 的回收率由过去的 10% 左右提高到目前的 33% 左右。钛精矿品位保证在 47% TiO_2 以上。至 2007 年，该矿实现年产优质钛精矿 12 万吨左右。

3.4.3 重钢太和铁矿新建钛铁矿选矿项目

重庆钢铁公司太和铁矿前几年计划将原矿处理量由 220 万吨/a 扩建至 630 万吨/a，其选钛流程重新建设。该矿于 2008 年委托赣州金环磁选设备有限公司进行选钛流程优化试验，试验结果如图 12 所示。该选钛流程的特点为：

（1）磁选尾矿（选钛原矿）全粒级进入强磁选，第一段强磁选采用 SLon 磁选机一粗一扫选，实现了第一段强磁选 TiO_2 综合回收率达到 92.12%。磁选尾矿全粒级入选可以避免分级溢流中的 TiO_2 损失，另一个好处是全粒级入选矿浆的流动性更好，细粒级矿物可作为粗粒级矿物的载体，减缓矿浆管道和各种矿斗的磨损（注：攀钢选钛厂粗、细粒级分别入选，粗粒级部分存在流动性较差，矿斗磨损较快的问题）。

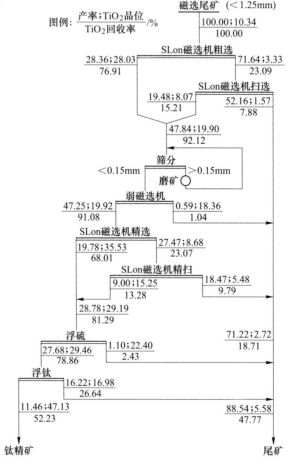

图 12　太和铁矿优化试验流程
（磁选部分由赣州金环公司试验，浮选部分由太和铁矿试验）

(2) 第二段强磁选采用 SLon 磁选机一精一扫选，第二段强磁选 TiO_2 作业回收率达到 (81.29%/91.08% = 89.25%) 89.25%。

(3) 两段强磁选 TiO_2 综合回收率达到 (92.12%×89.25% = 82.22%) 82.22%。实现了浮选的给矿品位 29.19%TiO_2 和 TiO_2 金属量占选钛原矿的 81.29%。

(4) SLon 磁选机有良好的脱泥效果，两段强磁选精矿粒度组成适合浮选，可以直接作为浮选给料。

(5) 强磁选综合精矿直接进入浮硫和浮钛作业，浮选获得钛精矿品位 47.13%TiO_2 和浮选作业回收率 69.55% 的指标。

通过选钛流程的优化，太和铁矿选钛流程试验综合指标为：磁选原矿品位 10.34%TiO_2，钛精矿品位 47.13%TiO_2，TiO_2 回收率 52.23%。

3.4.4 SLon-3000 磁选机在太和铁矿选钛工业中的应用

重庆钢铁公司太和铁矿近年将原矿处理量由 220 万吨/a 扩建至 630 万吨/a，扩建过程中增添了 4 台大型 SLon 磁选机作为一段分选钛铁矿的强磁选设备，其中 3 台 SLon-3000 和 1 台 SLon-2500 磁选机（图 13）。二段沿用现有 SLon-2000 和 SLon-1750 磁选机。目前该生产流程正处于设备安装及调试阶段。

图 13　SLon-3000 和 SLon-2500 磁选机在太和铁矿分选钛铁矿

3.5　首台 SLon-4000 磁选机在攀钢的选钛工业试验

攀钢密地选矿厂从钒钛磁铁矿中回收铁精矿和钛精矿，其综合尾矿每年约 700 万吨（每小时约 885t），全部排往尾矿库。但是其综合尾矿中还含有少量的可回收铁矿和钛铁矿。攀钢选矿综合厂主要从攀钢排放的总尾矿中回收铁精矿和钛精矿，最早的选钛流程采用螺旋溜槽粗选，摇床精选的单一重选流程。由于螺旋溜槽和摇床对细粒级回收率都很低，全流程的选钛回收率仅为 7%TiO_2 左右。2002 年该厂采用了 3 台 SLon-1500 磁选机建立新的选钛流程。该流程的选钛回收率由原重选作业的 7%TiO_2 提高到 27%TiO_2 左右。但是由于用于粗选作业的 2 台 SLon-1500 磁选机合计只能处理 60t/h 给矿，该厂钛精矿产量只能达到每年 2 万吨左右。随着攀钢密地选钛厂选钛技术水平的提高，目前排往尾矿库的尾矿 TiO_2 品位由原来的 7.68% 左右下降至 6.5% 左右。因此，如果不改造，该厂的钛精矿产量有进一步下降的趋势。2011 年年底，首台 SLon-4000 立环脉动高梯度磁选机在该厂应用于工业试验（图 14），取代原有粗选作业的 SLon-1500 磁选机，试验流程如图 15 所示。该机每小时给矿量为 443t（总尾矿的 50%）左右，至今已运转 6 个月，已取得的阶段性工业试验指标为（单机作业指标）：给矿品位 6.89%TiO_2，精矿品位 13.27%TiO_2，尾矿品位 4.02%TiO_2，TiO_2 作业回收率 59.88% 的良好指标。

图 14 SLon-4000 磁选机在攀钢从总尾矿中回收钛铁矿

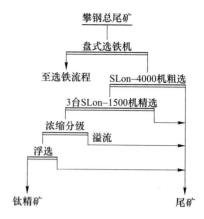

图 15 SLon-4000 磁选机选钛的工业试验流程

4 结论

（1）SLon 磁选机的技术进步及大型化显著地提高了强磁选设备的选矿效率，降低了设备采购成本和弱磁性矿石的选矿作业成本，使弱磁性矿石的选矿流程中可以较多的采用强磁选作业和强磁选设备，优化选矿流程，获得更高的选矿回收率。

（2）弱磁性铁矿选矿流程中，为了有效地降低尾矿品位，一段强磁选可采用强磁一粗一扫作业（如李楼氧化铁矿选矿流程），二段强磁选作业可采用一精一扫作业（如金日盛铁矿和李楼铁矿选矿流程），保证全流程获得较高的选矿回收率。

（3）在钛铁矿选矿流程中，过去一段强磁选和二段强磁选分别只采用一次强磁选作业，现在一段强磁选可以采用强磁一粗一扫（如攀钢选钛厂细粒级部分）；二段强磁选可采用强磁一精一扫作业（如攀钢选钛厂选钛流程，太和铁矿选钛流程）。攀钢选钛厂经扩能改造后，钛精矿产量由 28 万吨/a 增加至 50 万吨/a。

（4）SLon-2500、SLon-3000 和 SLon-4000 等大型强磁选机的研制，有利于降低选矿生产成本，降低电耗，节约占地面积，为优化弱磁性矿石选矿流程提供了条件。

写作背景 本文介绍了 SLon 立环脉动高梯度磁选机的技术进步对降低选矿成本起到的重要作用，SLon 磁选机在大型化方面取得的进步和节能效果，近几年 SLon 磁选机在分选氧化铁矿和钛铁矿的应用新进展，其中 SLon 磁选机在安徽金日盛氧化铁矿选矿流程中的应用、SLon-3000 磁选机在西昌太和铁矿选钛流程中的应用为新项目。

The Creative Technologies of SLon Magnetic Separators in Beneficiating Weakly Magnetic Minerals[1]

Xiong Dahe

(SLon Magnetic Separator Ltd. Shahe Industrial Park, Ganzhou, 341000, Jiangxi Province, China. Email:xdh@slon.com.cn)

Abstract In recent years many creative technologies are applied to SLon vertical ring and pulsating high gradient magnetic separators. SLon-2500, SLon-3000 and SLon-4000 vertical ring and pulsating high gradient magnetic separators are created towards larger scale. Through optimizing the magnetic system and lowering the electric current density, their energy consumption is obviously dropped. New magnetic matrixes are developed to raise the beneficial efficiency of valuable minerals. Special SLon vertical ring and pulsating high gradient magnetic separators are developed which can treat coarser particles of 0~5mm. Through these technology creations, the mineral processing efficiency of SLon magnetic separators is further raised and their processing cost dropped. Therefore, they have been found wider applications.

Keywords SLon magnetic separator; Iron ore; Ilmenite; Quartz; Beneficiation

1 Introduction

Modern mineral processing industry is developing towards larger scale, higher automation. Rich mineral deposits are getting less and less. These require mineral beneficiation equipment and flow sheet developing toward larger scale, higher automation, higher efficiency and lower coast. Till Now, about 3000 SLon vertical ring and pulsating high gradient magnetic separators have been widely applied to beneficiate oxidized iron ore, ilmenite and manganese ore and to purify nonferrous minerals such as quartz, feldspar, kaolin. Through continuous and tremendous creative work and improvement(Xiong, 1998, 2006, 2008), SLon magnetic separators are rapidly developed towards larger scale, more models, higher automation, higher efficiency, high reliability, lower energy consumption and lower coast, they have been found wider applications in mineral processing industry.

2 Developing larger models of SLon magnetic separator

Developing larger models of SLon vertical ring and pulsating high gradient magnetic separators can raise their capacity, save power consumption and space. In recent years, a lot of research work has been done to develop SLon-2500, SLon-3000 and SLon-4000 models.

---

[1] 原载于《第二十六届国际选矿会议论文集》,印度新德里,2012。

2.1 SLon-2500 magnetic separator

Based on the development of SLon-1000, SLon-1250, SLon-1500, SLon-1750, SLon-2000 vertical ring and pulsating high gradient magnetic separators, we developed the SLon-2500 vertical ring and pulsating high gradient magnetic separator since 2005 to 2008. Table 1 and Fig. 1 are its magnetic energizing data and curve.

Table 1 Magnetic energizing data of SLon-2500

Electric current/A	Voltage/V	Power/kW	Background field/T
200	6.6	1.3	0.168
400	12.7	5.1	0.336
600	19.5	11.7	0.495
800	26.1	20.9	0.654
1000	32.7	32.7	0.793
1200	38.6	46.3	0.911
1400	45.0	63.0	1.001
1600	52.0	83.2	1.074
1700	54.5	92.7	1.107

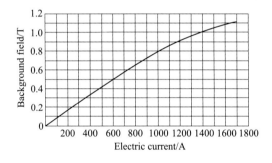

Fig. 1 Energizing curve of SLon-2500 magnetic separator

2.2 SLon-3000 magnetic separator

Since 2009 to 2011 we developed the SLon-3000 vertical ring and pulsating high gradient magnetic separator. Table 2 and Fig. 2 are its magnetic energizing data and curve.

Table 2 Magnetic energizing data of SLon-3000

Electric current/A	Voltage/V	Power/kW	Background field/T
200	8.2	1.6	0.189
400	16.2	6.5	0.375
600	24.6	14.8	0.564
800	33.2	26.6	0.734
1000	42.2	42.2	0.880
1200	51.4	61.7	0.995
1300	55.9	72.7	1.041
1400	60.4	84.6	1.079
1500	64.9	97.4	1.116

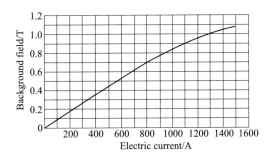

Fig. 2 Energizing curve of SLon-3000 magnetic separator

2.3 SLon-4000 magnetic separator

Since 2010 to now we are developing the SLon-4000 vertical ring and pulsating high gradient magnetic separator. Fig. 3 is the photo of first SLon-4000 magnetic separator Fig. 4 and Table 3. are its magnetic energizing data and curve.

Fig. 3 Photo of SLon-4000 magnetic separator

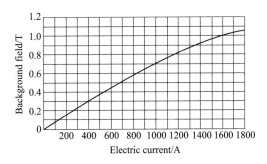

Fig. 4 Energizing curve of SLon-4000 magnetic separator

Table 3 Magnetic energizing data of SLon-4000

Electric current/A	Voltage/V	Power/kW	Background field/T
200	6.7	1.3	0.150
400	13.6	5.4	0.295
600	20.4	12.2	0.442
800	27.1	21.6	0.579
1000	33.5	33.5	0.709
1200	40.8	48.9	0.825
1400	47.5	66.5	0.921
1600	53.9	86.2	0.997
1800	60.6	109.1	1.056

2.4 Technical data of new SLon magnetic separators

Table 4 shows the technical data of SLon-2500, SLon-3000 and SLon-4000 magnetic separators.

Table 4 Technical data of the new models of SLon magnetic separators

Model	SLon-2500	SLon-3000	SLon-4000
Ring diameter/mm	2500	3000	4000
Ore feeding size/mm	0~1.2	0~1.2	0~1.2
Feeding concentration/%	10~40	10~40	10~40
Slurry put through/$m^3 \cdot h^{-1}$	200~400	350~650	750~1400
Capacity/$t \cdot h^{-1}$	100~150	150~250	350~550
Background field/T	1.0	1.0	1.0
Energizing current/A	1400	1300	1700
Energizing voltage/V	45	58	60
Energizing power/kW	63	75	102
Ring power/kW	11	18.5	37
Pulsating power/kW	11	18.5	37
Pulsating stroke/mm	0~30	0~30	0~30
Pulsating frequency/$r \cdot min^{-1}$	0~300	0~300	0~300
Flushing water/$m^3 \cdot h^{-1}$	160~220	280~400	560~800
Cooling water/$m^3 \cdot h^{-1}$	5~6	7~8	8~9
Machine weight/t	105	175	380
Weight of heaviest part/t	15	25	35

2.5 Energy Saving Comparison

Developing larger models of SLon magnetic separators helps to save energy consumption. Through optimizing magnetic energizing system, they realized very low energy consumption in mineral processing industry. Table 5 and Fig. 5 show the energy consumption vs. the ore treating capacities of SLon magnetic separators. Suppose that all SLon models are working at the background field 1.0T, according their ore treating capacity to calculate tonnage ore energy consumption. The larger capacity, the less energy consumption. For example, the tonnage ore energy consumption of SLon-1000 is 3.69kW · h. But the tonnage ore energy consumption of SLon-4000 is only 0.41kW · h, or 88.89% less energy consumption. The energy consumption of SLon-4000 is also 51.76% less than that of SLon-2000, which is now the most popular model in mineral processing industry.

Table 5 Energy consumption vs. the capacities of SLon magnetic separators

Model	Background field/T	Energizing power/kW	Driving power/kW	Total power/kW	Capacity /$t \cdot h^{-1}$	Tonnage power consumption /$kW \cdot h \cdot t^{-1}$
SLon-1000	1.0	17	3.3	20.3	5.5	3.6
SLon-1250	1.0	19	3.7	22.7	14	1.62
SLon-1500	1.0	27	7	34	25	1.36
SLon-1750	1.0	37	8	45	40	1.13
SLon-2000	1.0	42	13	55	65	0.85
SLon-2500	1.0	63	22	85	125	0.68
SLon-3000	1.0	75	37	112	200	0.56
SLon-4000	1.0	102	74	176	430	0.41

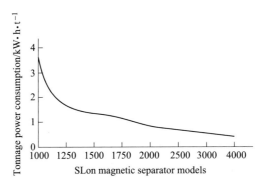

Fig. 5 Tonnage ore energy consumption vs. SLon magnetic separator models

3 Industriral applications

3.1 SLon-2500 magnetic separator applications

SLon-2500 applied in Hainan Iron Mine to recover hematite.

The first SLon-2500 vertical ring-pulsating high gradient magnetic separator was designed and built up in 2006. It was installed at the tails dam of Hai Nan Iron Mining Company to recover hematite from the pumping tails (Xiong, 2010).

The commercial processing flow sheet and the average results are shown in Fig. 6.

3.2 SLon-2500 applied in Essar Steel of India

In 2007, Essar Steel of India did a test on its Kirandul hematite ore with SLon magnetic separator and got very good results. The feed grade is 59.77%Fe, after SLon one roughing and one scavenging, the total iron concentrate grade is 65.00%Fe and iron recovery 93.86%. In 2008-2010, the company bought 10 SLon-2500 magnetic separators to treat such iron ores as shown in Fig. 7.

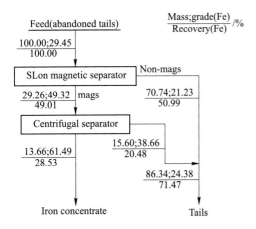

Fig. 6 SLon-2500 applied to recover hematite from abandoned tails

Fig. 7 SLon-2500 applied to recover hematite in Essar Steel of India

3.3 SLon-2500 applied to recover limonite

Beiya Gold Processing Plant in Yunnan Province of China recovers gold from complex ore. The ore consists of gold, magnetite, limonite and gang minerals. At first, the plant only recovered gold by leakage method. The iron minerals were abandoned because particle sizes were very fine. In 2010, 4 units of SLon-2500 magnetic separators were applied to build an iron recovering line as shown in Fig. 8 and Fig. 9. The gold tails enter screens to remove >1.2mm matters and low intensity magnetic separators are used to take out magnetite. Then SLon-2500 magnetic separators are used for one roughing and one scavenging to take out hematite. The flow sheet treat about 1.5 million ton of gold tails and about 1 million tons of iron concentrate are recovered each year.

Fig. 8　Photo of SLon-2500 applied to recover limonite

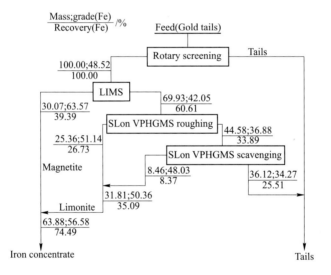

Fig. 9　Flow sheet of SLon-2500 applied to recover limonite

3.4 SLon-2500 applied to purify quartz

In 2009, a SLon-2500 magnetic separator was successfully applied to purify quartz in Shanxi Province of China as shown in Fig. 10. The feed is 0-1mm quartz which contains 0.15%-0.25% Fe_2O_3. The SLon-2500 magnetic separator is used to remove the iron impurities. After one roughing, Its non-mags product contains 0.05%-0.06% Fe_2O_3. Much better than the designed data (non-mags product $Fe_2O_3 \leq 0.08\%$).

3.5 SLon-3000 applications

In 2010 the first SLon-3000 was applied to process ilmenite. The magnetite-ilmenite deposit is located in Panzhihua, Sichuan Province of China. The ore processing flow sheet is: Drum low intensity magnetic separator (LIMS) is used to take out magnetite. The non-mags of the LIMS are the feed of ilmenite processing flow sheet shown as Fig. 11. The SLon-3000 is used to recover ilmenite for roughing. The mags of the SLon-3000 are further ground and cleaned by a SLon-2000 magnetic separator. The mags of the SLon-2000 magnetic separator are condensed and cleaned further by flotation. The SLon-3000 average operational results are: feed grade 9.52% TiO_2, mags grade 17.43% TiO_2, mags mass ratio 41.01%, TiO_2 recovery 75.09%, non-mags grade 4.02% TiO_2, non-mags mass ratio 58.99%.

Fig. 10 SLon-2500 applied
to purify quartz

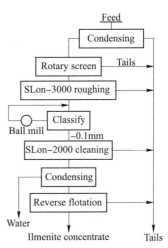

Fig. 11 SLon-3000 applied
to recover ilmenite

3.6 SLon-4000 applications

Panzhihua Iron and Steel Company in Sichuan Province of China produces 6 million tons of iron concentrate and 0.5 million tons of ilmenite concentrate, and discharges 7 million tons of tails to its tails dam annually. The tails still contain ilmenite assaying about 6.5% TiO_2. Laboratory test shows that there is a potential to recover more than 200000 tons of ilmenite concentrate (47% TiO_2) from the tails annually. A site commercial test is now being carried out near the tails dam. By the end of 2011, the first SLon-4000 magnetic separator is installed at the site as shown in Fig. 12. The commercial test flow sheet is shown in Fig. 13.

Fig. 12 SLon-4000 applied to recover ilmenite

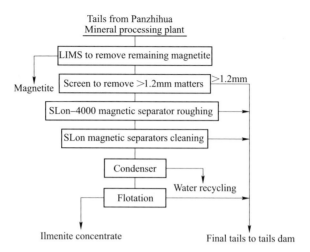

Fig. 13 Commercial test flow sheet of SLon-4000

The SLon-4000 magnetic separator treats half of 7 million tons of tails (450t/h). Because of its operating cost is very low and its capacity is big enough, it can discharge about 75% of the final tails with very low cost. Only about 25% of its mags go to the cleaning procedure. It is the key equipment to efficiently and profitably to recover high quality ilmenite concentrate from the abandoned tails.

4 New flow sheet to process oxidized iron ore

Most Chinese iron ore deposits are very poor. Because SLon-magnetic separators possesses the advantages of high beneficial efficiency and low operating cost, they have been found more and more applications in oxidized iron ore processing industry.

4.1 SLon applied in Sijiaying Iron Mine to recover hematite

Sijiaying is a big oxidized iron mine located in Hebei Province of China. Its oxidized iron ore consists of mainly hematite, magnetite and quartz. In 2005-2010, a plant annually treating 10 million tons of oxidized iron ore was built in which 39 SLon magnetic separators are applied (refer Fig. 14). The total flow sheet results are: Feed grade 30.44%Fe, iron concentrate grade 66%Fe, iron recovery 80%Fe, tails grade 9.65Fe%.

4.2 SLon Applied in Lilou Iron Mine to recover specularite

Lilou Iron Mine in Anhui Province of China built a plant annually processing 5 million tons of iron ore in 2010-2011. The iron minerals are mainly specularite and only about 0.5% of magnetite. 25 SLon-2000 magnetic separators are applied to recover specularite as shown in the flow sheet in Fig. 15. The total flow sheet results are: Feed grade 31.65%Fe, iron concentrate grade 65%Fe, iron recovery 80%Fe, tails grade 10.37%Fe.

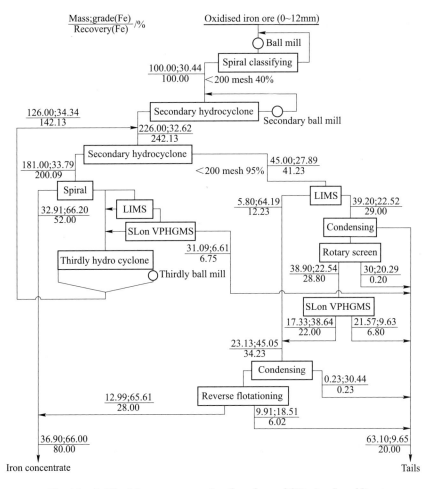

Fig. 14 Oxidized iron ore processing flow sheet of Sijiaying Iron Mine

5 Conclusions

(1) The development of SLon-2500, SLon-3000, SLon-4000 vertical ring and pulsating high gradient magnetic separators represented the trend of larger capacity. They help mineral processing plant to save energy and space, get lower operating cost and easier to raise automation.

(2) The optimization on magnetic system and larger models help SLon magnetic separators to reduce energy consumption dramatically. For example, SLon-4000 consumes 0.41kW · h/t energy at the magnetic background field of 1.0T, but SLon-1000 consumes 3.6kW · h/t at the same field.

(3) The recent developed SLon-3000 and SLon-4000 magnetic separators have been successfully applied in ilmenite processing industry. They are making significant contribution in raising ilmenite recovery.

(4) SLon magnetic separators are found more and more applications in oxidized iron ore processing industry and in non-metallic minerals purification industry.

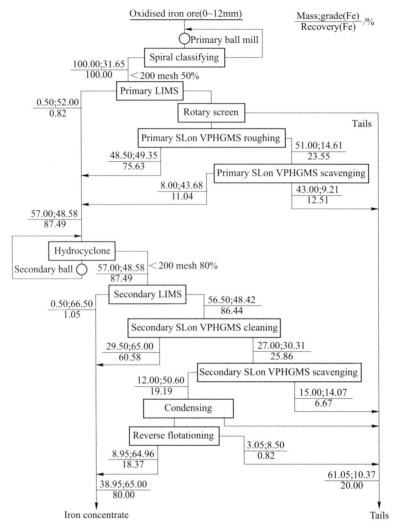

Fig. 15 Specularite processing flow sheet of Lilou Iron Mine

References

[1] Xiong D H, Liu S Y, Chen J. New technology of pulsating high gradient magnetic separation [C]// Int. J. Miner. Process. ,1998, 54:111-127.
[2] Xiong D H. SLon magnetic separator promoting Chinese oxidized iron ore processing industry [C]// Int. Proceedings of XXIII IMPC, Istanbul. 2006:276-281.
[3] Xiong D H. SLon magnetic separators applied to beneficiate low grade oxidized iron ores[C]//Int. Proceedings of XXIV IMPC. 2008:813-818.
[4] Xiong D H. A new technology of applying SLon-2500 magnetic separator to recover iron concentrate from abandoned tails[C]// Int. Proceedings of XXV IMPC, Brisbane. 2010:1403-1412.

写作背景 本文向国外同行介绍了 SLon-2500、SLon-3000、SLon-4000 立环脉动高梯度磁选机的研制、技术性能、设备大型化的节能效果,以它们在褐铁矿、赤铁矿、钛铁矿的选矿和石英砂提纯方面的工业应用。

SLon 磁选机在大型氧化铁矿选矿厂的应用

熊大和[1,2]

(1. 赣州金环磁选设备有限公司；2. 赣州有色冶金研究所)

摘　要　SLon 立环脉动高梯度磁选机利用磁力、脉动流体力和重力的综合力场分选弱磁性矿石，其转环立式旋转、反冲精矿和配有脉动机构松散矿粒群，该机具有富集比大、分选效率高、适应性强、磁介质不易堵塞和运转可靠的优点。该机在鞍钢齐大山选矿厂、调军台选矿厂、胡家庙选矿厂、弓长岭选矿厂、昆钢大红山铁矿、河北司家营铁矿、安徽金日盛铁矿和非洲塞拉利昂等大型氧化铁矿选矿厂得到成功的应用，为提高铁矿的选矿效率做出了重要的贡献。

关键词　SLon 立环脉动高梯度磁选机　氧化铁矿　提高选矿效率

1　SLon 磁选机发展简介

高梯度磁选技术对微细粒弱磁性颗粒具有回收率较高的优点，在实验室周期式高梯度磁选机上做的小型选矿试验研究证明，振动或脉动高梯度磁选可显著提高磁选精矿品位，并保持高梯度磁选对细粒弱磁性矿物回收率较高的优点。例如用振动高梯度磁选分选鞍钢齐大山赤铁矿，与无振动高梯度磁选比较，铁精矿品位可提高 8%~12%，尾矿品位略有降低，铁回收率基本不变。这种高效的选矿方法能否转化为生产力，关键在于研制出适应于工业生产的连续振动或脉动高梯度磁选机。

从 1986~2012 年，赣州有色冶金研究所设备研究室（现为赣州金环磁选设备有限公司）先后与中南工业大学、马钢姑山铁矿、赣州有色冶金机械厂、鞍钢矿山公司、弓长岭选矿厂、海南铁矿、攀钢矿业公司等单位合作研制出了 SLon-1000、SLon-1250、SLon-1500、SLon-1750、SLon-2000、SLon-2500、SLon-3000、SLon-4000 立环脉动高梯度磁选机等机型。该机针对铁矿选矿特点，在技术上采取了如下有力的措施：

(1) 将设备的可靠性和实用性放在第一位，尽可能做到坚固耐用、便于操作与维护。

(2) 将重选理论与磁选理论相结合，设置脉动机构驱动分选区的矿浆脉动，以减少脉石的机械夹杂，提高铁精矿品位。

(3) 转环立式旋转、反冲精矿。对于每一组磁介质而言，其给矿方向与排矿方向相反，粗粒和木渣草屑不必穿过磁介质堆便可冲洗出来，可有效地防止磁介质堵塞。

(4) 扩大分选粒度范围，简化现场筛分作业。该机的给矿粒度上限可达 2mm，其有效粒度下限为 10μm 左右。

(5) SLon 磁选机介质采用导磁不锈钢棒作为磁介质，通过多年的生产应用证明，棒介质具有工作稳定、不易堵塞、使用寿命长、维护工作量小和选矿指标稳定的优点。棒介质的应用为该机的大型化和大规模应用打下了良好的基础。

---

❶ 原载于《2012 年全国选矿前沿技术大会论文集》，武汉。

由于SLon立环脉动高梯度磁选机具有运转可靠、分选效率高、磁介质不易堵塞、分选粒度范围宽、适应性强的优点，该机首先在马钢姑山铁矿、宝钢梅山铁矿、鞍钢弓长岭选矿厂和昆钢罗次铁矿等地应用于分选赤铁矿和褐铁矿获得成功。该机大规模推广应用始于2001年。当时余永富院士提出铁精矿要提高品位，降低硅含量，以提高冶炼效益。我国的多个铁矿选厂纷纷进行技术改造提高铁精矿质量和提高选矿效率。自2001~2012年，SLon立环脉动高梯度磁选机有2000多台在我国的多个氧化铁矿选矿厂应用成功，为各厂矿取得了良好的技术经济指标，成为新一代优质高效的强磁选设备（图1）。

图1　SLon立环脉动高梯度磁选机照片

2　SLon磁选机在齐大山选矿厂的应用

齐大山选矿厂是鞍钢主要铁精矿供应基地，2001年前后每年处理铁品位30%左右的鞍山式氧化铁矿800万吨。该厂原矿经粗破和中破后筛分成0~20mm和20~75mm两个粒级。2001年以前的选矿工艺为：0~20mm粒级进一选车间，采用阶段磨矿—重选—强磁—浮选流程分选；20~75mm粒级进二选车间用竖炉进行还原焙烧成磁铁矿，然后进入磨矿和弱磁选流程。该工艺流程存在着能耗高、污染大等问题。此外该流程的最终铁精矿品位仅63%左右，满足不了冶炼的精料方针。

鞍钢矿业公司决定将齐大山二选车间焙烧—弱磁选流程改为阶段磨矿—重选—强磁—反浮选流程。其中强磁选设备的选型是决定该流程改造成功与否的关键之一。

2001年初，鞍山矿业公司在齐大山选矿厂组织了SLon立环脉动高梯度磁选机、永磁立环高梯度磁选机和平环强磁选机的工业对比试验。通过三个月的工业对比试验，SLon立环脉动高梯度磁选机获得了最好的选矿指标，其设备作业率、水耗、电耗及设备价格均具有较大的优势。因此鞍钢矿业公司决定采用SLon-1750立环脉动高梯度磁选机作为齐大山选矿厂二选车间技术改造的强磁选设备，尔后又用SLon-1500立环脉动中磁机（背景磁感应强度0.4T）取代了滚筒式永磁机。齐大山选矿厂二选车间改造全面完成后的流程如图2所示。

齐大山选矿厂二选车间改造完毕之后，2002年对其一选车间也进行了技术改造，用SLon-1750立环脉动高梯度磁选机取代了原有的永磁立环强磁选机和平环强磁选机。用SLon-1500立环脉动中磁机（背景磁感应强度0.4T）取代了滚筒式永磁机。

齐大山选矿厂的阶段磨矿—重选—强磁—反浮选新流程的改造成功，获得了原矿品位30.15%Fe，铁精矿品位67.50%Fe，铁回收率78.05%的优异指标，将我国氧化铁矿的选矿技术水平推向了一个新的历史高度。使我国的氧化铁矿选矿技术水平从落后于国外的状况转变为达到国际领先水平。从此该流程成了我国鞍山式氧化铁矿选矿流程的样板。

3　SLon磁选机在调军台选矿厂应用

调军台选矿厂是鞍钢于20世纪90年代新建的一座具有现代化生产水平的大型选矿厂，其设计规模为年处理鞍山式氧化铁矿900万吨。该厂于1998年3月投产，采用连续磨矿—弱磁—中磁—强磁—反浮选流程。

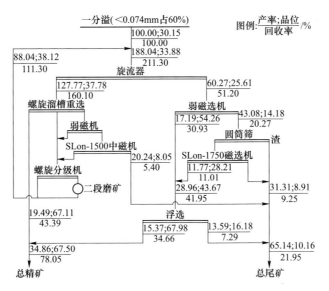

图 2 齐大山选矿厂二选车间数质量工艺流程

阶段磨矿—重选—强磁—反浮选流程具有生产成本较低的优点，2002～2007 年，调军台选矿厂经过几次技术改造和设备更新，将其选矿流程也改造成为阶段磨矿—重选—强磁—反浮选流程，该厂改造后的选矿流程如图 3 所示。其中采用了 15 台 SLon-2000 立环脉

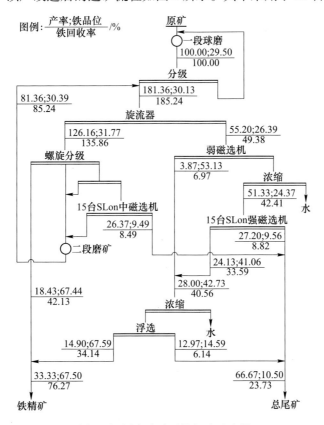

图 3 调军台选矿厂最新选矿流程

动高梯度磁选机（额定背景磁感应强度 1.0T）作为细粒级抛尾设备，15 台 SLon-2000 立环脉动中磁机（额定背景磁感应强度 0.4T）作为粗粒级抛尾设备。目前该厂生产能力达到年处理鞍山式氧化铁矿 1440 万吨。生产上所获得的选矿指标为：原矿品位 29.50%Fe，铁精矿品位 67.50%Fe，尾矿品位 10.50%Fe，铁回收率 76.27%。

4 SLon 磁选机在胡家庙选矿厂的应用

鞍钢胡家庙选矿厂建于 2006 年，年处理鞍山式氧化铁矿 800 万吨。胡家庙铁矿的特点是原矿品位很低，入选品位只有 24%Fe 左右。采用阶段磨矿—分级重选—强磁—反浮选的选矿流程（见图4）。其中采用了 8 台 SLon-2000 立环脉动高梯度磁选机（额定背景磁感应强度 1.0T）作为细粒级抛尾设备，8 台 SLon-2000 立环脉动中磁机（额定背景磁感应强度 0.4T）作为粗粒级抛尾设备。由于 SLon 磁选机的作业成本非常低，因此用它们尽可能提前抛尾是降低生产成本的关键。它们抛弃的尾矿产率合计占原矿的 68.63%，占总尾矿量的 88.55%，反浮选的给矿量只占原矿产率 20.99%。全流程的选矿指标为：原矿品位 23.10%Fe，铁精矿品位 67.54%Fe，尾矿品位 10.20%Fe，铁回收率 65.78%。

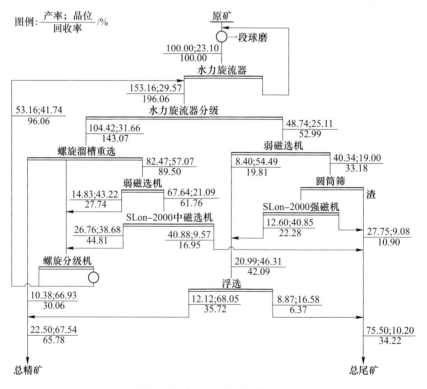

图 4 胡家庙选矿厂选矿流程

5 SLon 磁选机在弓长岭选矿厂的应用

鞍钢弓长岭选矿厂建于 1975 年，其二选车间年处理氧化铁矿 300 万吨，最初采用两段磨矿—离心选矿机—粗—精—扫的选矿流程,所获得的选矿指标为:原矿品位31.80%Fe,铁精矿品位60.50%Fe,铁回收率66.35%。

由于1970~1985年强磁选技术的快速发展,原选矿流程逐步转化为阶段磨矿—强磁选—螺旋溜槽—离心选矿机的选矿流程,所获得的选矿指标为:原矿品位29.20%Fe,铁精矿品位63.90%Fe,铁回收率67.51%。1995年因矿业行情低迷该流程停产。

2005年弓长岭选矿厂恢复了氧化铁矿的选矿,建设了第三选矿车间,年处理300万吨氧化铁矿,采用了阶段磨矿—螺旋溜槽—SLon磁选机—反浮选流程如图5所示。其中采用5台SLon-2000磁选机(1.0T)控制细粒级尾矿品位,5台SLon-2000中磁机(0.6T)控制粗粒级尾矿品位。全流程的选矿指标为:原矿品位28.78%Fe,铁精矿品位67.19%Fe,尾矿品位10.13%Fe,铁回收率76.29%。弓长岭氧化铁矿三个阶段不同选矿流程的指标见表1。

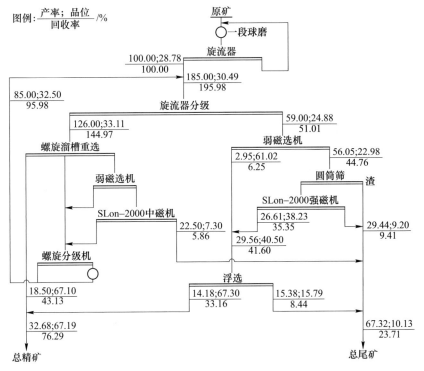

图5 弓长岭选矿厂三选车间选矿流程

表1 弓长岭氧化铁矿三个阶段不同选矿流程的指标对比 (%)

选矿流程	原矿铁品位	精矿铁品位	尾矿铁品位	精矿产率	铁回收率
连续磨矿—离心机	31.80	60.50	16.43	34.88	66.35
阶段磨矿—强磁—螺旋溜槽—离心机	29.20	63.90	13.72	30.85	67.51
阶段磨矿—螺旋溜槽—SLon磁选机—反浮选	28.78	67.19	10.13	32.68	76.29

从表1可见,弓长岭新流程与以前两种老流程比较,原矿品位下降了,而铁精矿品位和回收率都有大幅度的提高。

6 SLon 磁选机在司家营氧化铁矿的应用

唐山司家营铁矿是我国大型铁矿山之一，浅部以氧化铁矿为主，深部以磁铁矿为主。其氧化铁矿类似于鞍山式氧化铁矿，但是结晶粒度比鞍山式氧化铁矿更细，含铁矿物以赤铁矿为主并含少量的褐铁矿，脉石矿物以石英为主，属低品位难选氧化铁矿。自1958年以来，我国多个科研机构对司家营铁矿石进行了大量的研究工作。近年，氧化铁矿选矿技术水平因为选矿新技术、新药剂和新设备的应用得到了飞速的发展，司家营氧化铁矿的选矿技术难题得到解决。2005年至今该矿分两期建成了年处理1000万吨氧化铁矿的选矿厂，其中采用了39台SLon立环脉动高梯度磁选机，该厂第一期600万吨/a设计的选矿流程如图6所示。

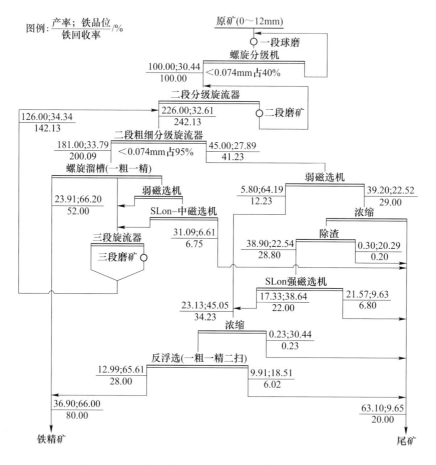

图 6 司家营第一期 600 万吨/a 氧化铁矿设计选矿流程

该流程的特点为：采用阶段磨矿，螺旋溜槽拿出了大部分结晶粒度较粗并已单体解离的铁矿物，SLon中磁机提前抛弃产率占原矿31.09%、粒度较粗品位仅6.61%Fe的尾矿，实现了能收早收，能丢早丢的原则，减少了磨矿量，节约了选矿成本；SLon强磁机抛弃了产率占原矿21.57%、品位仅9.63%Fe的细粒尾矿。SLon中磁机和SLon强磁机抛弃的尾矿产率合计占原矿的52.66%，占尾矿总量的83.45%，它们有效地控制全流程的尾矿品位，保证了全流程获得较高的选矿回收率。尽管反浮选的成本较高，但该流程反浮选的给矿产率仅占原矿的 23.13%，浮选药剂消耗量较少，因此全流程选矿成本较低。

该流程综合选矿指标为：给矿品位 30.44%Fe，精矿品位 66.00%Fe，铁精矿产率 36.90%，铁回收率 80.00%，尾矿品位 9.65%Fe。

7 SLon 磁选机在安徽金日盛铁矿的应用

安徽省霍邱县铁矿资源丰富，近年有几个铁矿选矿厂相继建成。安徽金日盛铁矿的矿石由镜铁矿、赤铁矿、磁铁矿和脉石组成。近几年建成了一座年处理 450 万吨原矿的现代化选矿厂，其选矿流程如图 7 所示。该流程的特点为：

（1）原矿经一段磨矿分级至小于 0.074mm 占 55%，用弱磁选机选出部分品位达到 66%Fe 的磁铁矿作为最终铁精矿，弱磁选尾矿再用螺旋溜槽选出部分品位达到 65%Fe 的氧化铁矿（单体解离度较好的镜铁矿和赤铁矿）。

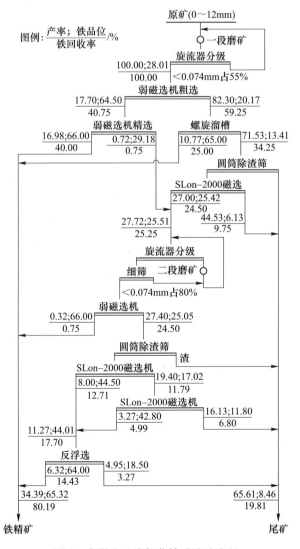

图 7　安徽金日盛氧化铁矿选矿流程

(2) 用 SLon-2000 立环脉动高梯度磁选机作为一段强磁选抛尾设备，该机一次选别抛弃了占原矿产率 44.53%，品位为 6.13%Fe 的尾矿，有效地控制了一段尾矿品位。

(3) 一段强磁选精矿和一段弱磁精选的尾矿合并后进入二段分级和磨矿。由于一段成功地分选出了大部分合格铁精矿和抛弃了大部分低品位尾矿，进入二段分级磨矿的矿量只占原矿产率的 27.72%。

(4) 二段分级磨矿至小于 0.074mm 占 80%，用弱磁选机分选出剩余的磁铁矿作为最终精矿，弱磁选尾矿用 SLon 磁选机一次粗选和一次扫选分选出镜铁矿和赤铁矿。二段强磁选精矿合并后用反浮选精选，得到品位为 64%Fe 的反浮选铁精矿。

该流程综合选矿指标为：原矿品位 28.01%Fe，铁精矿品位 65.32%Fe，铁精矿产率 34.39%，铁回收率 80.19%，尾矿品位 8.46%Fe。整个生产流程稳定，指标优异。

8 SLon 磁选机在昆钢大红山铁矿的应用

昆钢大红山铁矿是磁铁矿和赤铁矿共生的混合矿，原设计采用阶段磨矿—弱磁选—SLon 强磁选—反浮选流程，由于脉石矿物复杂多变，反浮选作业无法稳定而未应用成功，前几年用摇床取代了反浮选。但是，摇床存在处理量较小，对细粒铁矿回收率较低的缺点。大红山铁矿采用赣州金环磁选设备有限公司研制的 SLon-ϕ2400 离心选矿机取代反浮选作业，该流程于 2010 年顺利投产。新改造的选矿流程如图 8 所示，全流程铁精矿品位从改造前的 62%~63% 提高至目前 64%~65%。

该流程的主要特点为：SLon-2000 磁选机对细粒氧化铁矿回收率较高，可有效地控制强磁选作业的尾矿品位；采用 SLon-ϕ2400 离心选矿机取代反浮选作业，对强磁选精矿进行精选，2009 年在大红山铁矿的工业试验表明，离心选矿机与螺旋溜槽或摇床比较具有富集比较大，对细粒铁矿回收率较高的优点。这些新技术和新设备的应用，使该流程既可运行顺畅，又能获得良好的选矿指标。全流程的选矿指标为：原矿品位 35.61%，铁精矿品位 64.95%，铁回收率 74.06%。

9 SLon 磁选机在非洲塞拉利昂分选氧化铁矿的应用

2011 年，11 台 SLon-2500 磁选机出口到非洲塞拉利昂，应用于分选镜铁矿（图 9），第一期年处理 500 万吨原矿。采用 SLon-2500 磁选机一次粗选和一次精选简单的选矿流程，其入选原矿品位为 34.94%Fe，综合铁精矿品位为 67.10%Fe，综合铁回收率为 74.74%，获得了优异的选矿指标。2012 年该矿又购买 11 台 SLon-2500 磁选机进行第二期工程建设，建设成功后该矿的原矿处理量将达到每年 1000 万吨。

10 结论

(1) SLon 立环脉动高梯度磁选机的研制进行了一系列的技术创新，该机利用磁力、脉动流体力和重力的综合力场分选弱磁性矿石，其转环立式旋转、反冲精矿和配有脉动机构松散矿粒群，具有富集比大、分选效率高、适应性强、磁介质不易堵塞和运转可靠的优点。

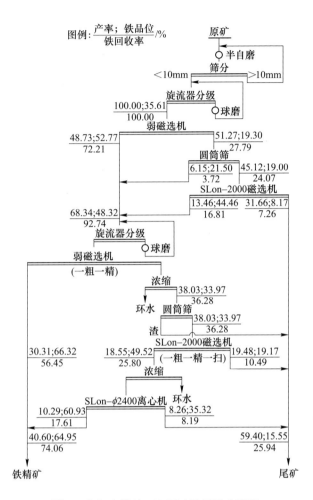

图 8 大红山铁矿 400 万吨选厂选矿流程

（2）2001~2012 年的 12 年中，在余永富院士的倡导下，我国许多大型氧化铁矿选矿厂进行了提铁降硅的技术改造或新厂建设。SLon 立环脉动高梯度磁选机在鞍钢齐大山选矿厂、东鞍山烧结厂、调军台选矿厂、弓长岭选矿厂、胡家庙选矿厂、司家营铁矿、安徽金日盛铁矿、李楼铁矿、昆钢大红山铁矿、非洲塞拉利昂铁矿等地成功地应用于分选氧化铁矿，协助这些厂矿获得了优异的技术经济指标，该机为氧化铁矿的选矿技术进步作出了重要贡献。

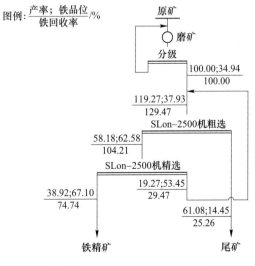

图 9 塞拉利昂镜铁矿选矿流程

写作背景 本文介绍了 SLon 立环脉动高梯度磁选机在我国一些大型选矿厂分选氧化铁矿的应用，该机为提高这些厂矿的选矿效率作出了重大贡献。其中 SLon 磁选机在非洲塞拉利昂应用于分选镜铁矿的为最新公布的内容。

SLon 立环脉动高梯度磁选机研制与应用新进展[1]

熊大和[1,2]

(1. 赣州金环磁选设备有限公司;2. 赣州有色冶金研究所)

摘　要　经济的高速发展和矿产资源的日益贫化促使了选矿设备朝着大型化、高效率、低成本方向发展。本文介绍 SLon-2500、SLon-3000 和 SLon-4000 立环脉动高梯度磁选机的研制与应用。其中 SLon-4000 磁选机是目前我国自主研发的处理量最大的高梯度磁选机或强磁选机。它们具有处理量大、选矿效率高、设备可靠性高、电耗低和生产成本低的优点,它们已成功地在赤铁矿、褐铁矿、钛铁矿选矿和在石英除铁提纯工业中应用。

关键词　SLon 立环脉动高梯度磁选机　大型化　节能　弱磁性矿石选矿

现代选矿工业的特点是:一方面经济的发展需要更多更好的矿产品作为工业原料,另一方面是复杂难选的原矿增多。这就要求选矿设备和选矿工艺向大型化、自动化、高效率和低成本方向发展,以满足现代选矿工业的需要。目前有 3000 多台 SLon 立环脉动高梯度磁选机在工业上广泛应用于氧化铁矿、钛铁矿、锰矿等弱磁性矿石的选矿及石英、长石、高岭土等非金属矿的提纯。通过多年持续的技术创新与改进,该机在设备大型化、多样化、自动化、高效、节能、提高可靠性等方面得到了快速的发展,并且得到了更为广泛的应用。

1　设备大型化研制

设备大型化有利于提高选矿设备的台时处理能力和提高自动化水平,同时大型化设备具有节能、节约占地面积、操作成本低和设备作业率高等优点,可显著为用户降低生产成本和提高用户的经济效益。

1.1　SLon-2500 磁选机的研制

赣州金环磁选设备有限公司在总结研制 SLon-1000、SLon-1250、SLon-1500、SLon-1750、SLon-2000 立环脉动高梯度磁选机成功经验的基础上,研制了 SLon-2500 立环脉动高梯度磁选机,如图 1 所示,其电磁性能数据见表 1,该机干矿处理量达到 100~150t/h。

图 1　SLon-2500 磁选机照片

---

[1] 原载于《金属材料与冶金工程》,2013,41:134~139。

表1 SLon-2500磁选机电磁性能测定数据

电流/A	电压/V	功率/kW	磁场强度/T
200	6.6	1.3	0.168
400	12.7	5.1	0.336
600	19.5	11.7	0.495
800	26.1	20.9	0.654
1000	32.7	32.7	0.793
1200	38.6	46.3	0.911
1400	45.0	63.0	1.001
1600	52.0	83.2	1.074
1700	54.5	92.7	1.107

1.2 SLon-3000磁选机的研制

2010年研制出了SLon-3000立环脉动高梯度磁选机，图2所示为第一台SLon-3000磁选机现场应用照片，其电磁性能数据见表2，该机干矿处理量达到150~250t/h。

图2 SLon-3000磁选机应用照片

表2 SLon-3000磁选机电磁性能测定数据

电流/A	电压/V	功率/kW	磁场强度/T
200	8.2	1.6	0.189
400	16.2	6.5	0.375
600	24.6	14.8	0.564
800	33.2	26.6	0.734
1000	42.2	42.2	0.880
1200	51.4	61.7	0.995
1300	55.9	72.7	1.041
1400	60.4	84.6	1.079
1500	64.9	97.4	1.116

1.3 SLon-4000磁选机的研制

在完成SLon-2500磁选机和SLon-3000磁选机研制的基础上，2011年又研制出了SLon-

4000 立环脉动高梯度磁选机。图 3 为第一台 SLon-4000 磁选机照片，其电磁性能数据见表 3，该机干矿处理量达到 350~550t/h。

图 3 SLon-4000 磁选机照片

表 3 SLon-4000 磁选机电磁性能测定数据

电流/A	电压/V	功率/kW	磁场强度/T
200	6.7	1.3	0.15
400	13.6	5.4	0.295
600	20.4	12.2	0.442
800	27.1	21.6	0.579
1000	33.5	33.5	0.709
1200	40.8	48.9	0.825
1400	47.5	66.5	0.921
1600	53.9	86.2	0.997
1800	60.6	109.1	1.056

在研制 SLon-2500、SLon-3000 和 SLon-4000 磁选机的过程中，采用了一系列的新技术提升它们的选矿性能和机电性能，主要措施如下：

（1）通过优化磁系设计、减少漏磁、降低电阻和提高电效率。使它们有良好的激磁性能。采取低电压大电流激磁，有利于提高激磁线圈的安全可靠性。激磁线圈采用空心铜管绕制，水内冷散热，冷却水直接贴着铜管内壁流动，散热效率高。此外，由于冷却水的流速较高，微细泥沙不易沉淀，可保证激磁线圈长期稳定工作。

（2）新型磁介质的研究。为了提高 SLon 磁选机的选矿效率，我们针对各种矿石的不同特点，分别研制了 $\phi 1mm$、$\phi 1.5mm$、$\phi 2mm$、$\phi 3mm$、$\phi 4mm$、$\phi 5mm$、$\phi 6mm$ 磁介质组，及各种磁介质的优化组合，如 $\phi 1mm+\phi 2mm$，$\phi 2mm+\phi 3mm$ 等磁介质组合，可根据入选矿石的粒度，比磁化系数选择，以获得最佳的选矿效果。

（3）采用新型数字式脉动冲程箱，具有力量大、平衡性好、可靠性高和能耗低的优点。

（4）采用新型的整流控制柜，激磁电流可根据选矿的需要调节，并稳定在人工设定点，不会随外电压的波动而波动。人机界面友好，自动化高，可远程控制。

SLon-2500、SLon-3000 和 SLon-4000 磁选机的主要技术参数见表 4。

表4 大型SLon磁选机主要技术参数

机型	转环外径/mm	给矿粒度/mm	给矿浓度/%	矿浆通过能力/m³·h⁻¹	干矿处理量/t·h⁻¹	额定背景磁感应强度/T	额定激磁电流/A	额定激磁电压/V	额定激磁功率/kW
SLon-2500	2 500	0~1.2	10~40	200~400	100~150	1.0	1 400	45	63
SLon-3000	3 000	0~1.2	10~40	350~650	150~250	1.0	1 300	58	75
SLon-4000	4 000	0~1.2	10~40	750~1 400	350~550	1.0	1 700	60	102

机型	传动功率/kW	脉动冲程/mm	脉动冲次/r·min⁻¹	供水压力/MPa	冲洗水用量/m³·h⁻¹	冷却水用量/m³·h⁻¹	主机质量/t	最大部件质量/t
SLon-2500	11+11	0~30	0~300	0.2~0.4	160~220	5~6	105	15
SLon-3000	18.5+18.5	0~30	0~300	0.2~0.4	240~400	7~8	175	25
SLon-4000	37+37	0~30	0~300	0.2~0.4	560~800	8~9	380	35

2 设备节能研究

SLon立环脉动高梯度磁选机的大型化有利于降低处理每吨矿石的能耗，通过优化磁系结构，减少漏磁，采用优质铜管、增加铜的用量和降低激磁电流密度，使电耗有了显著的降低。

图4和表5为几种型号SLon磁选机能耗对比。各种机型在背景磁感应强度达到1.0T时，按设备处理量的平均值计算吨矿消耗功率，设备越大，能耗越低，例如SLon-1000磁选机单位能耗为3.69kW·h/t，而SLon-4000磁选机仅为0.41kW·h/t。后者比前者节能88.89%。SLon-4000磁选机与目前工业上大量使用的SLon-2000磁选机比较也节能51.76%。

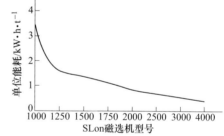

图4 SLon磁选机处理每吨矿石能耗对比

表5 SLon磁选机处理量与能耗对比

机型	额定背景磁感强度/T	额定激磁功率/kW	传动功率/kW	总功率/kW	干矿平均处理量/t·h⁻¹	每吨矿石电耗/kW·h·t⁻¹
SLon-1000	1.0	17	3.3	20.3	5.5	3.69
SLon-1250	1.0	19	3.7	22.7	14	1.62
SLon-1500	1.0	27	7	34	25	1.36
SLon-1750	1.0	37	8	45	40	1.13
SLon-2000	1.0	42	13	55	65	0.85
SLon-2500	1.0	63	22	85	125	0.68
SLon-3000	1.0	75	37	112	200	0.56
SLon-4000	1.0	102	74	176	430	0.41

3 工业应用

3.1 SLon-2500 磁选机在海钢红旗尾矿库的应用

首台 SLon-2500 磁选机在海南钢铁公司（现已更名为海南矿业联合有限公司，下同）应用于从尾矿中回收铁精矿。

海南钢铁公司有三个选矿车间分选氧化铁矿，选矿流程为强磁选—粗—精—扫，或强磁选—粗—扫—反浮选。每年入选原矿约 230 万吨，目前每年大约有 120 万吨尾矿排入到红旗尾矿库。其总尾矿的平均品位为 28%Fe 左右，其中的铁主要是存在于 0~19μm 粒级中。由于该尾矿氧化程度较深，含泥量大、且含有较多的连生体，因此要从该尾矿中再回收铁精矿技术难度较大。

SLon-2500 磁选机安装于该尾矿输送管的排矿口，从该排矿口截取尾矿入选。每年处理大约 72 万吨尾矿。该尾矿再回收系统的选矿工艺流程如图 5 所示。首先用圆筒除渣筛筛出大于 0.8mm 的粗粒。接着用 SLon-2500 磁选机粗选，其粗选精矿经浓缩后用离心选矿机精选。离心机精矿作为最终精矿。

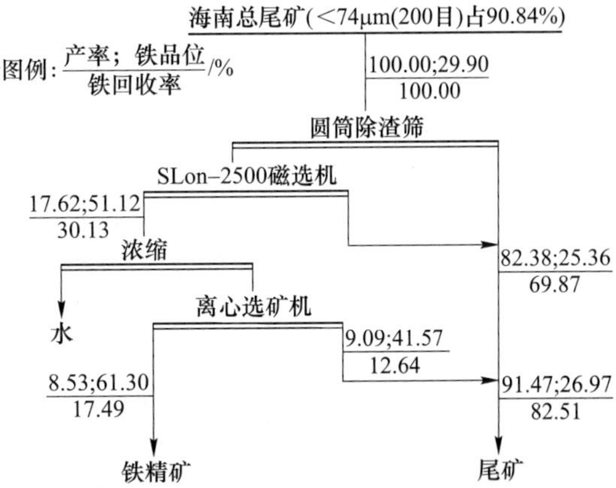

该流程综合选矿指标为：给矿 Fe 品位 29.90%，精矿 Fe 品位 61.30%，铁精矿产率 8.53%，铁回收率 17.49%，尾矿 Fe 品位 26.97%。

图 5 SLon-2500 磁选机分选海钢尾矿流程

该流程至今已运转 6 年。该机的单机作业率达到 99%以上。整个生产流程稳定，指标先进。

目前，海南钢铁公司新建了一个年处理 200 万吨低品位氧化铁矿的选矿车间，其中采用了 11 台 SLon-2500 磁选机用于强磁选作业，从第一台研制到大规模应用，证明了 SLon-2500 磁选机的优越性。

3.2 SLon-2500 磁选机在氧化铁矿选矿的应用

SLon-2500 磁选机研制成功以后，很快在氧化铁矿的选矿工业中得到大规模应用。2008~2010 年，印度一家大型钢铁公司采购了 10 台 SLon-2500 磁选机分选赤铁矿，采用 SLon-2500 磁选机一次粗选和一次扫选的选矿流程，其入选原矿 Fe 品位为 59.77%，综合铁精矿 Fe 品位为 65.00%，综合铁回收率为 93.86%，选矿指标优异。

2011 年，11 台 SLon-2500 磁选机出口到非洲塞拉利昂，分选镜铁矿获得成功应用（图 6，图 7），采用 SLon-2500 磁选机一次粗选和一次精选的选矿流程，其入选原矿 Fe 品位为 34.94%。综合铁精矿 Fe 品位为 67.10%，综合铁回收率为 74.74%，获得了优异的选矿指标。2012 年该选厂又订购了 11 台 SLon-2500 磁选机，计划将原矿处理量由每年 500 万吨扩产到 1000 万吨。

3.3 SLon-2500 磁选机分选褐铁矿的工业应用

云南北衙黄金选矿厂的原矿为金矿与铁矿共生的矿石，其早期的生产仅用浸出法回收金，而选金后的尾矿因粒度很细末得到回收。通过探索试验表明，该选金尾矿中含有磁铁矿和褐铁矿。2006 年该厂建成了一条年处理 50 万吨的选铁生产线，从选金尾矿中回收铁矿物，其中采用 2 台 SLon-2000 磁选机回收褐铁矿。2010 年至 2012 年。该厂又建设了一条年处理 150 万吨原矿的选矿生产线，其选金后的尾矿用弱磁选机选出磁铁矿，然后用 6 台 SLon-2500 磁选机分选褐铁矿（粗选和扫选各 3 台，图 8），该生产线选铁部分的流程如图 9 所示。目前该厂每年可回收 100 多万吨铁精矿，并可根据市场价格安排磁铁矿和褐铁矿分别销售或合并销售。

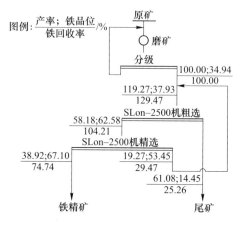

图 6 塞拉利昂镜铁矿选矿原则流程

图 7 SLon-2500 磁选机在塞拉利昂应用照片

图 8 SLon-2500 磁选机在北衙黄金选矿厂应用照片

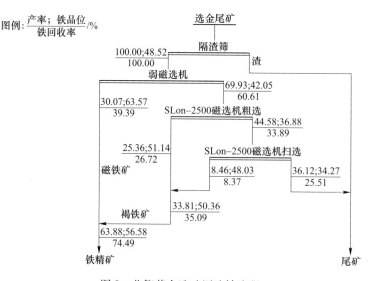

图 9 北衙黄金选矿厂选铁流程

3.4 SLon-2500 磁选机在石英提纯工业中的应用

SLon-2500 磁选机于 2009 年 7 月首先在陕西汉中石英砂提纯工业中应用成功。尔后 SLon-2500 磁选机在凤阳高纯石英砂提纯方面又获得成功应用。凤阳县位于安徽省东北部，处于淮河中游南岸，因石英砂资源丰富，因而享有"中国石英砂之乡"的美誉。SLon-2500 磁选机于 2011 年 7 月在凤阳县远东石英砂有限公司安装完毕，10 月份投入生产（图10）。原石英砂矿中 Fe_2O_3 的含量在 0.01%～0.02% 之间，经过 SLon-2500 磁选机磁选后，精矿产率达到了 95% 以上，精矿中 $Fe_2O_3 \leqslant 0.007\%$、$SiO_2 >$ 99.8%，可达到太阳能、光纤通信等高尖端原料的要求。

图 10　SLon-2500 磁选机在安徽凤阳提纯石英砂

3.5 SLon-3000 磁选机在攀枝花的应用

2010 年首台 SLon-3000 磁选机在攀枝花安宁铁钛有限公司应用于钛铁矿选矿获得成功。攀枝花地区的钒钛磁铁矿储量丰富，目前该地区的选矿厂的选矿流程普遍是先用弱磁选机选出磁铁矿精矿，选完磁铁矿的尾矿即作为选钛作业的原矿，选钛作业普遍用强磁选机预先抛弃大部分尾矿，强磁选精矿再用浮选或重选精选。由于选钛流程中入选原料含钛品位低，要求强磁选机处理量大，生产成本低，尽可能多抛弃尾矿，SLon-3000 磁选机正好具备这些优势，该厂选矿流程如图 11 所示。

SLon-3000 磁选机在该厂用于钛铁矿的强磁粗选作业，其给矿粒度为小于 74μm（200 目）占 40%，台时处理量为 200t/h 左右。该机的选矿指标为：给矿 TiO_2 品位 9.52%，精矿 TiO_2 品位 17.43%，精矿产率 41.01%，TiO_2 回收率 75.09%，尾矿 TiO_2 品位 4.02%，抛弃的尾矿产率为 58.99%。该机使用的背景磁感强度为

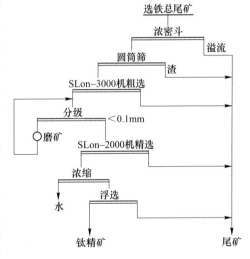

图 11　安宁铁钛有限公司选钛流程

0.7T，处理每吨给矿的电耗仅 0.3kW·h。目前该公司共订购了 9 台 SLon-3000 磁选机用于分选钛铁矿。从首台 SLon-3000 磁选机的工业试验到大规模的工业应用，证明了该机具有选矿效率高、设备可靠性高和生产成本低的优点。

3.6 SLon-3000 磁选机在太和铁矿选钛工业中的应用

重庆钢铁公司太和铁矿近年将原矿处理量由 220 万吨/a 扩建至 630 万吨/a，扩建过程中共采用了 5 台 SLon-3000 磁选机用于钛铁矿分选（图12）。目前该生产流程已正常投产。

3.7 首台 SLon-4000 磁选机在攀钢的选钛工业试验

攀钢密地选矿厂从钒钛磁铁矿中回收铁精矿和钛精矿，其综合尾矿每年约 700 万吨（每小时约 885 吨），全部排往尾矿库。但是其综合尾矿中还含有少量的可回收铁矿和钛铁矿。攀钢选矿综合厂主要从攀钢排放的总尾矿中回收铁精矿和钛精矿，最早的选钛流程采用螺旋溜槽粗选，摇床精选的单一重选流程。由于螺旋溜槽和摇床对细粒级回收率都很低，全流程的选钛 TiO_2 回收率

图 12　SLon-3000 磁选机在太和铁矿分选钛铁矿

仅为 7%左右。2002 年该厂采用了三台 SLon-1500 磁选机建立新的选钛流程（图 13）。该流程的选钛 TiO_2 回收率由原重选作业的 7%提高到 27%左右，但是由于用于粗选作业的 2 台 SLon-1500 磁选机合计只能处理 60t/h 给矿，该厂钛精矿产量只能达到每年 2 万吨左右。随着攀钢密地选钛厂选钛技术水平的提高，目前排往尾矿库的尾矿 TiO_2 品位由原来的 7.68%左右下降至 6.5%左右，因此，如果不改造，该厂的钛精矿产量有进一步下降的趋势。2011 年年底，首台 SLon-4000 立环脉动高梯度磁选机在该厂应用于工业试验（图 14），取代原有粗选作业的 SLon-1500 磁选机，该机每小时给矿量为 460 吨（总尾矿的 50%）左右，至今已运转 14 个月，已取得的工业试验指标为（SLon-4000 单机作业指标）：给矿 TiO_2 品位 6.41%，精矿 TiO_2 品位 13.38%，尾矿 TiO_2 品位 3.64%，TiO_2 作业回收率 59.36%的良好指标。全流程的选钛指标为：给矿 TiO_2 品位 6.41%，钛精矿 TiO_2 品位 47.02%，TiO_2 回收率 28.16%的良好指标。目前该厂又订购了一台 SLon-4000 磁选机进行扩产改造，预计改造完成后，该流程年处理 700 万吨综合尾矿，年产 TiO_2 品位为 47%的钛精矿 20 万吨左右，年新增销售收入 4 亿元左右。年新增利税 2.5 亿元左右。

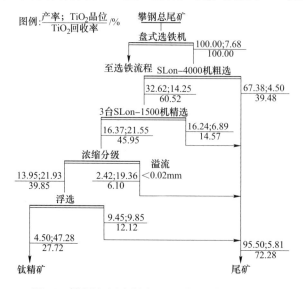

图 13　攀钢综合厂采用 SLon 磁选机的选钛流程

图 14　SLon-4000 磁选机在攀钢从总尾矿中回收钛铁矿

4　结论

（1）SLon-2500、SLon-3000 和 SLon-4000 磁选机的研制代表了该机型大型化发展方向。设备的大型化有利于降低生产成本，节约占地面积，提高选矿厂自动化生产水平。

（2）对磁系的优化设计使 SLon 磁选机的激磁功率有了大幅度的下降，设备的大型化更有利于降低单位处理量的能耗。以 SLon-4000 磁选机为例，处理每吨矿石的额定电耗为 0.41kW·h（按工作场强 1.0T 计算），处理每吨钛铁矿的实际电耗仅为 0.25kW·h（按工作场强 0.8T 计算）。

（3）SLon-2500、SLon-3000 和 SLon-4000 磁选机具有处理量大、选矿效率高、设备可靠性高、电耗低和生产成本低的优点。它们已成功地在赤铁矿、褐铁矿、钛铁矿选矿和在石英除铁提纯工业中应用。

写作背景　本文介绍了 SLon-2500、SLon-3000、SLon-4000 大型磁选机研制与应用新进展，其中包括非洲塞拉利昂分选镜铁矿的 SLon-2500 磁选机由 11 台增加到 22 台，原矿处理量由每年 500 万吨增加到 1000 万吨；SLon-2500 磁选机在安徽凤阳分选高档石英砂；攀枝花安宁铁钛有限公司分选钛铁矿由一台 SLon-3000 磁选机增加至 9 台 SLon-3000 磁选机；SLon-4000 磁选机在攀钢尾矿库分选钛铁矿工业试验的试验数据。

SLon 磁选机提高弱磁性金属矿回收率的研究与应用

熊大和[1,2]

(1. 赣州金环磁选设备有限公司；2. 赣州有色冶金研究所)

摘 要 SLon 立环脉动高梯度磁选机的发展显著地提高了弱磁性矿石的选矿效率，其中包括设备本身的技术创新和选矿流程的改进。SLon-2000、SLon-2500、SLon-3000 和 SLon-4000 立环脉动高梯度磁选等大型设备具有处理量大、选矿效率高、设备可靠性高、电耗低、投资成本低和生产成本低的优点，在氧化铁矿和钛铁矿等弱磁性金属矿的选矿流程中适当增加强磁选作业、优化选矿流程，可显著地提高弱磁性矿物的选矿回收率，使矿产资源的利用率进一步提高。

关键词 SLon 立环脉动高梯度磁选机　氧化铁矿　钛铁矿　提高选矿回收率

Applications of SLon Magnetic Separators in Raising the Recoveries of Feebly Magnetic Minerals

Xiong Dahe[1,2]

(1. *SLon Magnetic Separator Ltd.*;
2. *Ganzhou Nonferrous Metallurgy Research Institute*)

Abstract The development of SLon vertical ring and pulsating high gradient magnetic separators greatly raised the beneficiation efficiency of feebly magnetic minerals. The measures include equipment revolution and mineral processing flow sheet improvement. Fully utilizing the advantages of large capacity, high beneficiation efficiency, high reliability, low energy consumption, low invest cost and low operating cost of SLon-2000, SLon-2500, SLon-3000 and SLon-4000 magnetic separators, properly adding high intensity magnetic operation stages and optimizing processing flow sheet, The recoveries of feebly magnetic minerals can be greatly raised. Thus the utilization ratio of mineral resources can also be raised further.

Keywords SLon magnetic separator; Oxidized iron ore; Ilmenite; Raise recovery

在过去的 20 多年中，我国强磁选机技术得到了快速发展，通过持续的技术创新和设备的大型化研制，SLon 立环脉动高梯度磁选机具有优异的选矿性能，目前有 3000 多台 SLon 立环脉动高梯度磁选机得到了广泛应用。通过设备选型和优化选矿流程，可显著地提高弱磁性矿物的选矿回收率，使矿产资源的利用率进一步提高。

1 强磁选机价格和生产成本变化趋势

在过去的 20 多年中，我国强磁选机技术得到了快速发展，由于持续的技术创新和设备的大型化，以台时处理每吨矿石计算的强磁选机的销售价格大约为 2.5 万~3 万元，这个价格维持了 20 年左右并有所下降。而同期的住房价格上涨了十倍左右，铁矿石的价格上涨了 3 倍左右，同期 SLon 磁选机的作业成本降低 50% 左右。图 1 为过去 20 年这几种因素的变化曲线。

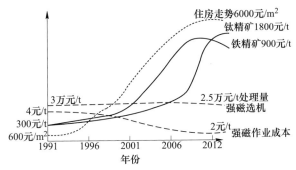

图 1 过去 20 年几种因素的变化曲线

由于强磁选机的相对价格和选矿成本的大幅度降低，使选矿厂可采用较多的强磁选机优化选矿流程，提高选矿回收率和选矿效率。

2 设备节能研究

SLon 立环脉动高梯度磁选机的大型化有利于降低处理每吨矿石的能耗，通过优化磁系结构，减少漏磁，采用优质铜管、增加铜的用量和降低激磁电流密度等措施，使电耗有了显著的降低。

如图 2 所示为几种型号 SLon 磁选机能耗对比情况。各种机型在背景磁感应强度达到 1.0T 时，按设备处理量的平均值计算吨矿消耗功率，设备越大，能耗越低。

图 3 为最新研制的 SLon-4000 立环脉动高梯度磁选机，该机干矿处理量达到 350~550t/h，

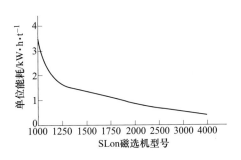

图 2 SLon 磁选机处理每吨矿石能耗对比

图 3 SLon-4000 磁选机照片

处理每吨矿石的单位能耗仅为 0.41kW·h/t，比 SLon-1000 磁选机（单位能耗为 3.69kW·h/t）节能 88.89%，SLon-4000 磁选机与目前工业上大量使用的 SLon-2000 磁选机（单位能耗为 0.85kW·h/t）比较也节能 51.76%。

3 工业应用

由于强磁选机价格下降和作业成本的降低，SLon 立环脉动高梯度磁选机得到了更为广泛的应用，许多氧化铁矿和钛铁矿的选矿流程中增加了强磁选作业。

3.1 SLon 磁选机在安徽金日盛铁矿的应用

安徽省霍邱县铁矿资源丰富，近年有几个铁矿选矿厂相继建成。安徽金日盛铁矿的矿石由镜铁矿、赤铁矿、磁铁矿和脉石组成。近几年建成了一座年处理 450 万吨原矿的现代化选矿厂，其选矿流程如图 4 所示。

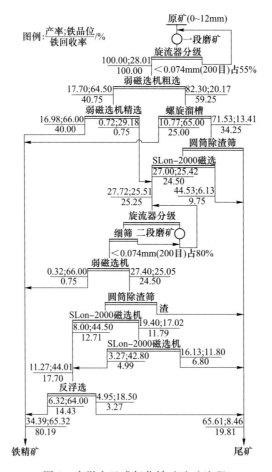

图 4 安徽金日盛氧化铁矿选矿流程

该流程的特点为：

（1）原矿经一段磨矿分级至小于 0.074mm（200 目）占 55%，用弱磁选机选出部分品位达到 66%Fe 的磁铁矿作为最终铁精矿，弱磁选尾矿再用螺旋溜槽选出部分品位达到

65%Fe 的氧化铁矿（单体解离度较好的镜铁矿和赤铁矿）。

（2）用 SLon-2000 立环脉动高梯度磁选机作为一段强磁选抛尾设备，该机一次选别抛弃了占原矿产率 44.53%，品位为 6.13%Fe 的尾矿，有效地控制了一段尾矿品位。

（3）一段强磁选精矿和一段弱磁精选的尾矿合并后进入二段分级和磨矿。由于一段成功地分选出了大部分合格铁精矿和抛弃了大部分低品位尾矿，进入二段分级磨矿的矿量只占原矿产率的 27.72%。

（4）二段分级磨矿至小于 0.074mm（200 目）占 80%，用弱磁选机分选出剩余的磁铁矿作为最终精矿，弱磁选尾矿用 SLon 磁选机一次粗选和一次扫选分选出镜铁矿和赤铁矿。二段强磁选精矿合并后用反浮选精选，得到品位为 64%Fe 的反浮选铁精矿。

该流程综合选矿指标为：原矿品位 28.01%Fe，铁精矿品位 65.32%Fe，铁精矿产率 34.39%，铁回收率 80.19%，尾矿品位 8.46%Fe。整个生产流程稳定，指标优异。

3.2 SLon 磁选机在李楼铁矿的应用

安徽开发矿业有限公司李楼铁矿近几年建成了一座年处理 500 万吨镜铁矿的选矿厂，其设计的选矿流程如图 5 所示。

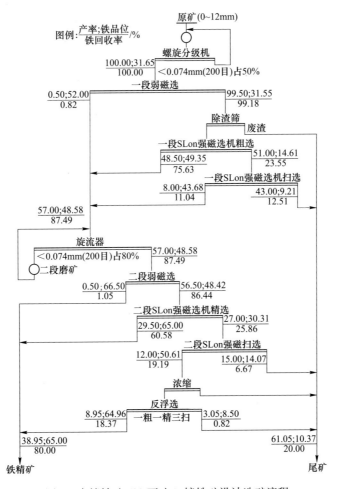

图 5 李楼铁矿 500 万吨/a 镜铁矿设计选矿流程

李楼铁矿镜铁矿的含铁矿物主要是镜铁矿，其他铁矿物含量很少，例如原矿中磁铁矿产率仅为 0.5%左右。其选矿流程的特点为：

（1）原矿铁品位 31.65%Fe，经一段闭路磨矿分级至小于 0.074mm（200 目）占 50%左右，用弱磁选机选出少量的赤铁矿，用强磁选机进行一次粗选和一次扫选提前抛弃产率为 43.00%，品位为 9.21%左右的尾矿，保证了一段强磁尾矿品位处于较低的水平。

（2）弱磁选精矿和强磁选精矿经二段闭路分级磨矿至小于 0.074mm（200 目）占 80%左右，然后用二段弱磁选和二段强磁精选得到一部分品位 65%Fe 以上的铁精矿成品，其中二段强磁精选的铁精矿产率占原矿的 29.5%，占全流程总精矿的 75.74%。

（3）二段强磁精选的尾矿再经二段强磁扫选，二段强磁扫选精矿经浓缩后用反浮选精选，反浮选得到另一部分品位为 64.96%左右的铁精矿成品。

（4）由于强磁选作业成本低，强磁选提前抛弃了大部分尾矿和得出了大部分精矿，仅有占原矿 12%左右的二段强磁扫选精矿进入反浮选作业，由于反浮选的作业成本较高，进入反浮选的矿量越少，全流程的生产成本越低。

（5）该流程中一段强磁选采用一粗一扫，二段强磁选采用一精一扫，保证了全流程获得较高的选矿回收率。

全流程设计的综合选矿指标为：给矿品位 31.65%Fe，精矿品位 65.00%Fe，铁精矿产率 38.95%，铁回收率 80.00%，尾矿品位 10.37%Fe。

3.3 SLon 磁选机在攀钢选钛厂的应用

3.3.1 SLon 磁选机在攀钢选钛厂回收小于 0.045mm 粒级钛铁矿

攀钢选钛厂原先的选钛流程中，小于 0.045mm 粒级作为尾矿排放，其中 TiO_2 的金属占有率为 40%左右，因此，攻克这部分物料的选钛难题，对提高全流程的选钛回收率具有重要的意义。1994 年至 1996 年，攀钢选钛厂采用一台 SLon-1500 立环脉动高梯度磁选机建立了第一条微细粒级选钛试验生产线（图 6）。

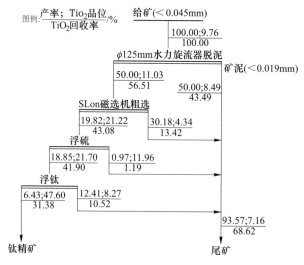

图 6 攀钢选钛厂第一条小于 0.045mm 选钛生产流程

该流程中首先用旋流器脱出小于0.019mm矿泥，然后用一台SLon-1500磁选机粗选，其粗选精矿再用浮选脱硫和选钛铁矿，最终得到品位为47%TiO_2以上的优质钛精矿。

该流程中，SLon磁选机的作业回收率达到76.24%TiO_2，浮选的作业回收率也达到了75%TiO_2左右，该流程的选钛回收率达到了30%TiO_2左右，由于小于0.019mm粒级未入选，这部分TiO_2金属量损失了43.49%，影响了钛铁矿回收率的进一步提高。

上述选钛生产线每年生产2万吨优质钛精矿，随后的几年攀钢选钛厂又分两期建成后八系统和前八系统小于0.045mm粒级选钛生产线，至2004年，攀钢选钛厂小于0.045mm钛精矿的年产量达到15万吨左右，加上粗粒级钛精矿，攀钢选钛厂年产钛精矿达到28万吨左右。

3.3.2 攀钢选钛厂钛铁矿选矿的扩能改造工程

2008年，攀钢选钛厂进行选钛扩能改造工程，具体实施计划有：
（1）将粗粒级选钛流程由原来的重选—电选流程改为强磁选—浮选流程。
（2）优化细粒级选钛流程。
（3）钛精矿产量由改造前的28万吨/a提高到40万吨/a。

攀钢选钛厂委托赣州金环磁选设备有限公司所做的选钛试验流程如图7所示。

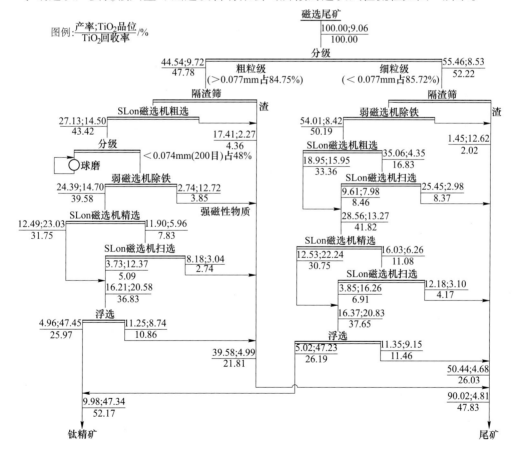

图7 攀钢选钛厂扩能改造试验流程（赣州金环公司试验）

该流程的主要特点为：

（1）磁选尾矿只分成粗粒级和细粒级入选，充分发挥 SLon 磁选机本身的脱泥功能，尽可能减少分级溢流的损失。

（2）细粒级强磁选作业由过去的 SLon 磁选机一次选别改为四次选别，即 SLon 磁选机一粗一扫和一精一扫；粗粒级强磁选作业采用 SLon 磁选机一次粗选，其粗精矿分级再磨后用弱磁选机除铁，其尾矿用 SLon 磁选机一精一扫选。这些措施保证了强磁选作业对钛铁矿有较高的回收率和较高的精矿品位。比较图 6 和图 7 可见，细粒级进入浮选作业的 TiO_2 金属量由 43% 左右提高到（37.65%/52.55% = 72.10%）左右。粗粒级进入浮选作业的 TiO_2 金属量高达该粒级的（36.83%/47.78% = 77.08%）77% 左右。

（3）该选钛流程充分利用了 SLon 磁选机处理量大，生产成本低，脱泥效果较好的优点，大幅度提高强磁选作业的选钛回收率，为浮选创造了较好的条件。在保证钛精矿品位达到 47% TiO_2 以上的前提下，浮选作业 TiO_2 回收率可保证在 70% 左右，全流程的 TiO_2 回收率达到 52.17%。

2008~2009 年，攀钢选钛厂参考各家的选钛试验流程并结合现场的条件进行了选钛扩能改造，2010 年、2011 年和 2012 年该厂分别实现了年产优质钛精矿 48 万吨、50 万吨和 52 万吨，比改造前每年提高 20 余万吨优质钛精矿。目前生产上全流程实际达到的 TiO_2 回收率为 40% 左右。

3.4 SLon 磁选机在重钢太和铁矿分选钛铁矿

3.4.1 重钢太和铁矿早期的重选—浮选选钛流程

重庆钢铁公司太和铁矿位于四川省西昌市，其矿石为钒钛磁铁矿，矿石性质与攀枝花钒钛磁铁矿类似。其选矿流程是原矿经破碎、磨矿、永磁磁选机回收磁铁矿。其选铁尾矿中含 TiO_2 12% 左右。为回收该尾矿中的钛资源，太和铁矿于 1995 年建成了第一条选钛生产线（图 8）。该流程首先将磁选尾矿分级成粗粒级、细粒级及溢流。粗粒级用螺旋溜槽粗选，其粗精矿进行分级磨矿后与细粒级合并，用滚筒永磁机选出残余的强磁性物质，然后用摇床分选，摇床精矿用浮选精选，浮选精矿作为最终钛精矿。

该流程存在的主要问题有：

（1）由于当时选用的水力分级设备分级效率不高，有 32% 左右的 TiO_2 损失在溢流中。

（2）螺旋溜槽和摇床的分选粒度下限只能达到 30μm 左右，对细粒级钛铁矿的回收率很低。螺旋溜槽的作业回收率仅为 59% 左右，摇床的作业回收率仅为 47% 左右。

（3）该流程的重选段包括分级、螺旋溜槽和摇床的 TiO_2 综合回收率仅为 22.42%。

（4）钛铁矿浮选是正浮选，即泡沫产品为钛铁矿精矿，螺旋溜槽和摇床都不能回收细粒级钛铁矿，造成浮选的给矿粒度偏粗，由于钛铁矿密度（4.7~4.8g/cm³）较大，较粗粒级的钛铁矿容易沉槽，因此造成浮选回收率偏低。

上述原因导致了该流程选钛回收率只能达到 10% TiO_2 左右。1995~2000 年，太和铁矿采用该流程每年只能回收 2000~5000t 钛精矿。

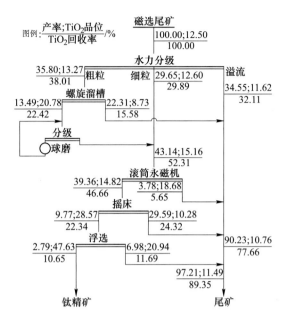

图 8 重钢太和铁矿 2001 年以前采用的选钛流程

3.4.2 重钢太和铁矿强磁—浮选选钛流程

重钢太和铁矿于 2000 年开始采用 SLon 立环脉动高梯度磁选机进行选钛流程的技术改造，经过几次改造后的选钛流程如图 9 所示。

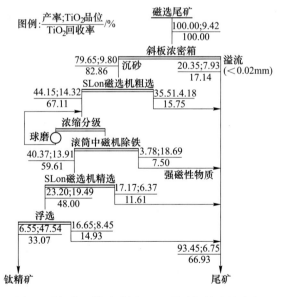

图 9 重钢太和铁矿采用 SLon 磁选机的选钛流程

该流程采用 SLon 磁选机取代了螺旋溜槽和摇床。磁选尾矿首先用水力分级机分成大于 0.02mm（沉砂）和小于 0.02mm（溢流），沉砂用 SLon 磁选机粗选，其精矿分级磨矿，用滚筒永磁中磁机除去磁铁矿等强磁性物质后再用 SLon 磁选机精选，其精矿进入浮选精选。

SLon 磁选机在太和铁矿的成功应用大幅度提高了该矿钛铁矿的回收率，尽管选钛系统入选原矿的 TiO_2 由过去的 12%左右降低至目前的 9%左右，TiO_2 的回收率由过去的 10%左右提高到目前的 33%左右。钛精矿品位保证在 47%TiO_2 以上。至 2007 年，该矿实现年产优质钛精矿 12 万吨左右。

3.4.3 重钢太和铁矿新建钛铁矿选矿项目

重庆钢铁公司太和铁矿前几年计划将原矿处理量由 220 万吨/a 扩建至 630 万吨/a，其选钛流程重新建设。该矿于 2008 年委托赣州金环磁选设备有限公司进行选钛流程优化试验，试验结果如图 10 所示。该选钛流程的特点为（注：磁选部分由赣州金环公司试验，浮选部分由太和铁矿试验）：

（1）磁选尾矿（选钛尾矿）全粒级进入强磁选，第一段强磁选采用 SLon 磁选机一粗一扫选，实现了第一段弱磁选 TiO_2 综合回收率达到 92.12%，磁选尾矿全粒级入选可以避免分级溢流中的 TiO_2 损失，另一个好处是全粒级入选矿浆的流动性更好，细粒级矿物可作为粗粒级矿物的载体，减缓矿浆管道和各种矿斗的磨损（注：攀钢选钛厂粗、细粒级分别入选，粗粒级部分存在流动性较差，矿斗磨损较快的问题）。

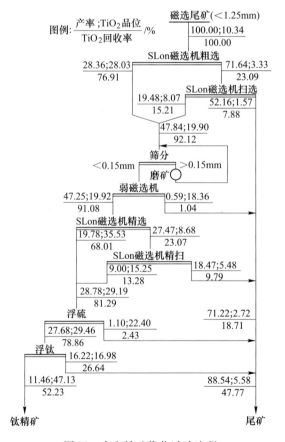

图 10 太和铁矿优化试验流程

(2) 第二段强磁选采用 SLon 磁选机一精一扫选，第二段强磁选 TiO$_2$ 作业回收率达到 (81.29%/91.08%=89.25%) 89.25%。

(3) 两段强磁选 TiO$_2$ 综合回收率达到 (92.12%×89.25%=82.22%) 82.22%，实现了浮选的给矿品位 29.19% TiO$_2$ 和 TiO$_2$ 金属量占选钛原矿的 81.29%。

(4) SLon 磁选机有良好的脱泥效果，两段强磁选精矿粒度组成适合浮选，可以直接作为浮选给料。

(5) 强磁选综合精矿直接进行浮硫和浮钛作业，浮选获得钛精矿品位 47.13%TiO$_2$ 和浮选作业回收率 69.55%的指标。

通过选钛流程的优化，太和铁矿选钛流程试验综合指标为：磁选原矿品位 10.34%TiO$_2$，钛精矿品位 47.13%TiO$_2$，TiO$_2$ 回收率 52.23%。

3.4.4 SLon 磁选机在太和铁矿选钛工业中的应用

重庆钢铁公司太和铁矿近年将原矿处理量由 220 万吨/a 扩建至 630 万吨/a，扩建过程中增添了 6 台 SLon-3000 磁选机作为一段分选钛铁矿的强磁选设备（图 11），二段沿用原有的 SLon-2000 和 SLon-1750 磁选机。该生产流程已于 2012 年全面投产。

图 11 SLon 磁选机在太和铁矿分选钛铁矿

3.5 首台 SLon-4000 磁选机在攀钢的选钛工业试验

攀钢密地选矿厂从钒钛磁铁矿中回收铁精矿和钛精矿，其综合尾矿每年约 700 万吨（每小时约 885t），全部排往尾矿库。但是其综合尾矿中还含有少量的可回收铁矿和钛铁矿。攀钢选矿综合厂主要从攀钢排放的总尾矿中回收铁精矿和钛精矿，最早的选钛流程采用螺旋溜槽粗选，摇床精选的单一重选流程。由于螺旋溜槽和摇床对细粒级回收率都很低，全流程的选钛回收率仅为 7%TiO$_2$ 左右。2002 年该厂采用了 3 台 SLon-1500 磁选机建立新的选钛流程。该流程的选钛回收率由原重选作业的 7%TiO$_2$ 提高到 27%TiO$_2$ 左右。但是由于用于粗选作业的 2 台 SLon-1500 磁选机合计只能处理 60t/h 给矿，该厂钛精矿产量只能达到每年 2 万吨左右。随着攀钢密地选钛厂选钛技术水平的提高，目前排往尾矿库的尾矿 TiO$_2$ 品位由原来的 7.68%左右下降至 6.5%左右，因此，如果不改造，该厂的钛精矿产量有进一步下降的趋势。2011 年年底，首台 SLon-4000 立环脉动高梯度磁选机在该厂应用于工业试验（图 12），取代原有粗选作业的 SLon-1500 磁选机，试验流程（图 13）。该机每小时给矿量为 460t（总尾矿的 50%）左右，至今已运转 14 个月，已取得的工业试验指标为（SLon-4000 单机作业指标）：给矿品位 6.41%TiO$_2$，精矿品位 13.38%TiO$_2$，尾矿品位 3.64%TiO$_2$，TiO$_2$ 作业回收率 59.36%的良好指标。全流程的选钛指标为：给矿品位 6.41%TiO$_2$，钛精矿品位 47.02%TiO$_2$，TiO$_2$ 回收率 28.16%的良好指标。目前该厂又订购了 1 台 SLon-4000 磁选机进行扩产改造，预计改造完成后，该流程年处理 700 万吨综合尾矿，年产品位为 47%TiO$_2$ 的钛精矿 20 万吨左右，年新增销售收入 4 亿元左右。年新增利税 2.5 亿元左右。

 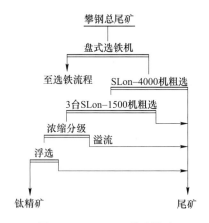

图 12　SLon-4000 磁选机在攀钢从总尾矿中回收钛铁矿　　　图 13　SLon-4000 磁选机选钛的工业试验流程

4　结论

（1）SLon 磁选机的技术进步及大型化显著地提高了强磁选设备的选矿效率，降低了设备采购成本和弱磁性矿石的选矿作业成本，使弱磁性矿石的选矿流程中可以较多的采用强磁选作业和强磁选设备，优化选矿流程，获得更高的选矿回收率。

（2）弱磁性铁矿选矿流程中，为了有效地降低尾矿品位，一段强磁选可采用强磁一粗一扫作业（如李楼氧化铁矿选矿流程），二段强磁选作业可采用一精一扫作业（如金日盛铁矿和李楼铁矿选矿流程），保证全流程获得较高的选矿回收率。

（3）在钛铁矿选矿流程中，过去一段强磁选和二段强磁选分别只采用一次强磁选作业，现在一段强磁选可以采用强磁一粗一扫（如攀钢选钛厂细粒级部分）；二段强磁选可采用强磁一精一扫作业（如攀钢选钛厂选钛流程，太和铁矿选钛流程）。攀钢选钛厂经扩能改造后，钛精矿产量由 28 万吨/a 增加至 50 万吨/a。

写作背景　因学术会议交流需要，这是本人第一次在《昆明理工大学学报（自然科学版）》发表论文，介绍了 SLon 立环脉动高梯度磁选机大型化与节能的优势，SLon 磁选机在安徽金日盛铁矿和李楼铁矿分选氧化铁矿，在攀钢选钛厂和重钢太和铁矿分选钛铁矿的选矿流程和指标。

SLon 立环脉动高梯度磁选机
研制与应用新进展[1]

熊大和[1,2]

(1. 赣州金环磁选设备有限公司；2. 赣州有色冶金研究所)

摘　要　介绍了 SLon 立环脉动高梯度磁选机的工作原理以及大型设备 SLon-2500、SLon-3000 和 SLon-4000 立环脉动高梯度磁选机的研制与应用。其中 SLon-4000 磁选机是目前我国自主研发的处理量最大的高梯度磁选机或强磁选机，具有处理量大、选矿效率高、设备可靠性高、电耗低和生产成本低的优点，在赤铁矿、褐铁矿、钛铁矿选矿和在非金属可除铁提纯工业中得到了广泛的应用。

关键词　SLon 立环脉动高梯度磁选机　大型化　节能　弱磁性矿石选矿

现代选矿工业的特点是：一方面经济的发展需要更多更好的矿产品作为工业原料，另一方面是复杂难选的原矿增多。这就要求选矿设备和选矿工艺向大型化、自动化、高效率和低成本方向发展，以满足现代选矿工业的需要。目前有 3000 多台 SLon 立环脉动高梯度磁选机在工业上广泛应用于氧化铁矿、钛铁矿、锰矿等弱磁性矿石的选矿及石英、长石、高岭土等非金属矿的提纯。通过多年持续的技术创新与改进，该机在设备大型化、多样化、自动化、高效、节能、提高可靠性等方面得到了快速的发展，并且得到了更为广泛的应用。

1　SLon 磁选机工作原理简介

SLon 立环脉动高梯度磁选机利用磁力、脉动流体力和重力的综合力场分选弱磁性矿物，具有富集比大、选矿效率高、选矿成本低、适应面广、设备作业率高、使用寿命长、易安装和检修工作量小的优点。

SLon 磁选机结构如图 1 所示，它主要由脉动机构、激磁线圈、铁轭、转环和各种矿斗、水斗组成。用导磁不锈钢棒作磁介质。该机工作原理如下。

激磁线圈通以直流电，在分选区产生感应磁场，位于分选区的磁介质表面产生高梯度磁场；转环作顺时针旋转，将磁介质不断送入和运出分选区；矿浆从给矿斗给入，进入位于转环下部的分选区，矿浆中的磁性颗粒吸附在磁介质棒表面上，被转环带至顶部无磁场区，被冲洗水冲入精矿斗；非磁性颗粒在重力、脉动流体力的作用下穿过磁介质堆，流入尾矿斗排走，实现与磁性颗粒的分离。

该机的转环采用立式旋转方式，对于每一组磁介质而言，冲洗磁性精矿的方向与给矿方向相反，粗颗粒不必穿过磁介质堆便可冲洗出来。该机的脉动机构驱动矿浆产生脉动，可使位于分选区磁介质堆中的矿粒群保持松散状态，使磁性矿粒更容易被捕获，使非磁性

---

[1] 原载于《2013 年第四届中国矿业科技大会论文集》。

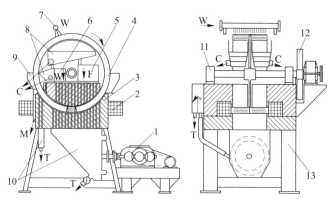

图 1 SLon 立环脉动高梯度磁选机结构

1—脉动机构；2—激磁线圈；3—铁轭；4—转环；5—给矿斗；
6—漂洗水斗；7—精矿冲洗装置；8—精矿斗；9—中矿斗；
10—尾矿斗；11—液位斗；12—转环驱动机构；13—机架；
F—给矿；W—清水；C—精矿；M—中矿；T—尾矿

矿粒尽快穿过磁介质堆到尾矿中去。

显然，反冲精矿和矿浆脉动可防止磁介质堵塞；脉动分选可提高磁性精矿的质量。这些措施保证了该机具有较大的富集比、较高的分选效率和较强的适应能力。

SLon 磁选机的激磁线圈的冷却方式采用水内冷，主要优点为：

（1）节电。水的比热容为 4.2×10^{-3} J/(kg·K)，而变压器油的比热容为 2.06×10^{-3} J/(kg·K)，在相同的激磁功率和相同冷却液流量时，水冷线圈的温升比油冷线圈的温升低得多，例如水温升高 20℃，则油温升为 40℃。

铜导线或铝合金导线的电阻率随温度升高而升高，因此水冷线圈有电阻低、节电的优点（$P=RI^2$）。

（2）散热效率高。激磁线圈采用空心铜管绕制，水内冷散热，冷却水直接贴着铜管内壁流动，散热效率最高，水的温度和导体的温度基本一致。而油冷线圈冷却油和导体之间有绝缘层，导体产生的热量要穿过绝缘层才能传到油里去，导线的温度要比油温高得多。因此油冷线圈散热效率较低。

（3）水内冷是一种最先进的冷却方式，国内外大型发电机的励磁线圈大部分都是采用水内冷。

（4）水内冷是一种有利于环保的冷却方式，万一管路元件漏水，不会污染环境。而油冷线圈一旦管路元件漏油，则会污染环境。

（5）水内冷线圈安全可靠。SLon 磁选机现有 3000 多台在工业上应用，每年线圈的返修率不到 1%。如果用户能按使用手册要求供水（普通生活用水标准），则线圈可做到长期不堵塞。

（6）水内冷更经济。水的购买成本比油低，冷却水可以循环使用，基本上不耗水。例如鞍钢地区有 200 多台 SLon 磁选机，攀枝花地区有 300 多台 SLon 磁选机，绝大部分都做到了冷却水循环使用。

（7）水内冷线圈节约占地面积。SLon 磁选机冷却系统不与操作人员争位置，更有利于操作人员观察和操作设备,易于打扫卫生和设备维护。

(8) 水内冷线圈可获得更高的磁场强度。由于水内冷冷却效率高,与油冷线圈比较,水内冷激磁线圈可获得较高的磁场强度,在高场强下运转更稳定可靠。从而获得较高的选矿回收率。

2 设备大型化研制

设备大型化有利于提高选矿设备的台时处理能力和提高自动化水平,同时大型化设备具有节能、节约占地面积、操作成本低和设备作业率高等优点,可显著为用户降低生产成本和提高用户的经济效益。

2.1 SLon-2500 磁选机的研制

在总结 SLon-1000、SLon-1250、SLon-1500、SLon-1750、SLon-2000 立环脉动高梯度磁选机研制成功经验的基础上,研制了 SLon-2500 立环脉动高梯度磁选机,如图 2 所示,其电磁性能数据见表 1,该机干矿处理量达到 100~150t/h。

图 2　SLon-2500 磁选机照片

表 1　SLon-2500 磁选机电磁性能测定数据

电流/A	电压/V	功率/kW	磁感应强度/T
200	6.6	1.3	0.168
400	12.7	5.1	0.336
600	19.5	11.7	0.495
800	26.1	20.9	0.654
1000	32.7	32.7	0.793
1200	38.6	46.3	0.911
1400	45.0	63.0	1.001
1600	52.0	83.2	1.074
1700	54.5	92.7	1.107

2.2 SLon-3000 磁选机的研制

2010 年研制出了 SLon-3000 立环脉动高梯度磁选机。图 3 所示为第一台 SLon-3000 磁选机现场应用照片,其电磁性能数据见表 2,该机干矿处理量达到 150~250t/h。

图 3　SLon-3000 磁选机应用照片

表 2　SLon-3000 磁选机电磁性能测定数据

电流/A	电压/V	功率/kW	磁感应强度/T
200	8.2	1.6	0.189
400	16.2	6.5	0.375
600	24.6	14.8	0.564
800	33.2	26.6	0.734
1000	42.2	42.2	0.880
1200	51.4	61.7	0.995
1300	55.9	72.7	1.041
1400	60.4	84.6	1.079
1500	64.9	97.4	1.116

2.3 SLon-4000 磁选机的研制

在完成 SLon-2500 磁选机和 SLon-3000 磁选机研制的基础上,2010~2012 年又研制出了 SLon-4000 立环脉动高梯度磁选机。图 4 所示为第一台 SLon-4000 磁选机照片,其电磁性能数据见表 3。该机干矿处理量达到 350~550t/h。

图 4 SLon-4000 磁选机照片

表 3 SLon-4000 磁选机电磁性能测定数据

电流/A	电压/V	功率/kW	磁场强度/T
200	6.7	1.3	0.15
400	13.6	5.4	0.295
600	20.4	12.2	0.442
800	27.1	21.6	0.579
1000	33.5	33.5	0.709
1200	40.8	48.9	0.825
1400	47.5	66.5	0.921
1600	53.9	86.2	0.997
1800	60.6	109.1	1.056

在研制 SLon-2500、SLon-3000 和 SLon-4000 磁选机的过程中,采用了以下新技术提升它们的选矿性能和机电性能:

(1) 通过优化磁系设计、减少漏磁、降低电阻和提高电效率,使其有良好的激磁性能。采取低电压大电流激磁,有利于提高激磁线圈的安全可靠性。激磁线圈采用空心铜管绕制,水内冷散热,冷却水直接贴着铜管内壁流动,散热效率高。此外,由于冷却水的流速较高,微细泥沙不易沉淀,可保证激磁线圈长期稳定工作。

(2) 新型磁介质的研究。为了提高 SLon 磁选机的选矿效率,针对各种矿石的不同特点,分别研制了 $\phi1mm$、$\phi1.5mm$、$\phi2mm$、$\phi3mm$、$\phi4mm$、$\phi5mm$、$\phi6mm$ 磁介质组,及各种磁介质的优化组合,如 $\phi1mm+\phi2mm$,$\phi2mm+\phi3mm$ 等磁介质组合,可根据入选矿石的粒度,比磁化系数选择,获得最佳的选矿效果。

(3) 采用新型数字式脉动冲程箱,具有力量大、平衡性好、可靠性高和能耗低的优点。

(4) 采用新型的整流控制柜,激磁电流可根据选矿的需要调节,并稳定在人工设定点,不会随外电压的波动而波动。人机界面友好,自动化程度高,可远程控制。

SLon-2500、SLon-3000 和 SLon-4000 磁选机的主要技术参数见表 4。

3 设备节能研究

SLon 立环脉动高梯度磁选机的大型化有利于降低处理矿石的能耗,通过优化磁系结构,减少漏磁,采用优质铜管、增加铜的用量和降低激磁电流密度,使电耗显著降低。

表 5 和图 5 为几种型号 SLon 磁选机能耗对比情况。各种机型在背景磁感应强度达到 1.0 T 时,按设备处理量的平均值计算吨矿消耗功率,设备越大,能耗越低,例如 SLon-1000 磁选机单位能耗为 3.69kW·h/t,而 SLon-4000 磁选机仅为 0.41kW·h/t。后者比前者节能 88.89%。SLon-4000 磁选机与目前工业上大量使用的 SLon-2000 磁选机比较也节能 51.76%。

表 4 大型 SLon 磁选机主要技术参数

机 型	转环外径 /mm	给矿粒度 /mm	给矿浓度 /%	矿浆通过能力 /m³·h⁻¹	干矿处理量 /t·h⁻¹	额定背景磁感应强度 /T	额定激磁电流 /A	额定激磁电压 /V
SLon-2500	2500	0~1.2	10~40	200~400	100~150	1.0	1400	45
SLon-3000	3000	0~1.2	10~40	350~650	150~250	1.0	1300	58
SLon-4000	4000	0~1.2	10~40	750~1400	350~550	1.0	1700	60

机 型	额定激磁功率 /kW	传动功率 /kW	脉动冲程 /mm	脉动冲次 /r·min⁻¹	供水压力 /MPa	冲洗水用量 /m³·h⁻¹	冷却水用量 /m³·h⁻¹	主机重量 /t	最大部件质量 /t
SLon-2500	63	11+11	0~30	0~300	0.2~0.4	160~220	5~6	105	15
SLon-3000	75	18.5+18.5	0~30	0~300	0.2~0.4	240~400	7~8	175	25
SLon-4000	102	37+37	0~30	0~300	0.2~0.4	560~800	8~9	380	35

表 5 SLon 磁选机处理量与能耗对比

机 型	额定背景磁感强度 /T	额定激磁功率 /kW	传动功率 /kW	总功率 /kW	干矿平均处理量 /t·h⁻¹	矿石电耗 /kW·h·t⁻¹
SLon-1000	1.0	17	3.3	20.3	5.5	3.69
SLon-1250	1.0	19	3.7	22.7	14	1.62
SLon-1500	1.0	27	7	34	25	1.36
SLon-1750	1.0	37	8	45	40	1.13
SLon-2000	1.0	42	13	55	65	0.85
SLon-2500	1.0	63	22	85	125	0.68
SLon-3000	1.0	75	37	112	200	0.56
SLon-4000	1.0	102	74	176	430	0.41

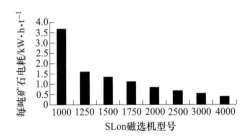

图 5 SLon 磁选机处理每吨矿石能耗对比

4 工业应用

4.1 SLon-2500 磁选机在海南矿业分选氧化铁矿

海南矿业联合有限公司 2010~2013 年新建成一座年处理量 200 万吨氧化铁矿选矿厂，11 台 SLon-2500 磁选机和 80 台 SLon-2400 离心选矿机获得成功应用，在分选低品位氧化铁矿技术上取得了新的突破。

海南矿业联合有限公司（前称为海南钢铁公司）拥有一座储量较大的氧化铁矿，过去长期开采铁品位50%左右的富矿，大约有1000多万吨铁品位40%左右已采出的贫矿（未采出的贫矿还有1亿多吨），因存在选矿技术难题未得到利用。

该矿的选矿技术难点在于：

（1）矿石结晶粒度很细，磨矿至小于43μm（325目）仍然得不到充分解离。

（2）脉石复杂多变，工业试验表明，浮选药剂难以适应，浮选作业难以稳定。

（3）脉石如绿泥石、橄榄石等含铁，尾矿难以降低。

（4）微量磁铁矿和石英的贫连生体较多，强磁选难以选出高品位的铁精矿。

近年有关单位对这部分氧化铁进行了系统的选矿试验研究，实践证明，若仅用强磁选流程只能得到品位为60%左右的铁精矿。而市场上铁品位为63%以上的铁精矿好销且价格较高。因此，选矿试验和生产实践都要求铁精矿铁品位达到63%以上。

经过多次试验，获得如图6所示的SLon磁选机与SLon离心选矿机的组合流程。目前该流程已投产6个月。其入选原矿铁品位为39.15%，含铁矿物主要是赤铁矿及占原矿产率13%左右的磁铁矿。首先磨矿至小于74μm（200目）占90%，然后用弱磁选机选出磁铁矿，弱磁选机的尾矿用SLon-2500磁选机1次粗选和1次扫选，磁选精矿合并，浓缩后用离心选矿机精选出部分铁品位为63.5%左右的铁精矿，离心机尾矿经浓缩后用旋流器分级，旋流器沉砂进入二段球磨。旋流器溢流用弱磁选机和SLon磁选机分别选出磁铁矿和赤铁矿。二者的混合精矿再用离心机精选得出部分铁品位为62%左右的铁精矿。

由图6可知，采用该流程在给矿铁品位为39.15%的条件下，获得了铁品位为63.20%、回收率为63.26%的精矿，尾矿铁品位为23.65%。

4.2 SLon-2500磁选机在印度分选氧化铁矿选矿

SLon-2500磁选机研制成功以后，很快在氧化铁矿的选矿工业中得到大规模应用。2008～2013年，印度一家大型钢铁公

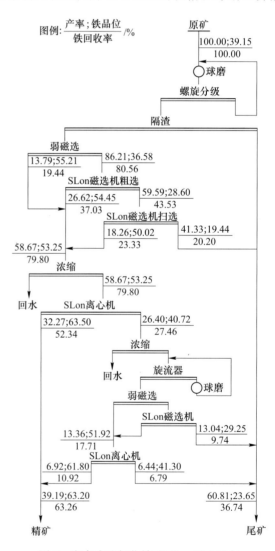

图6 海南难选氧化铁矿SLon磁选机与SLon离心选矿机的组合分选流程

司采购了 20 台 SLon-2500 磁选机分选赤铁矿,采用 SLon-2500 磁选机一次粗选和一次扫选的选矿流程,其入选原矿铁品位为 59.77%,综合铁精矿铁品位为 65.00%,铁回收率为 93.86%,选矿指标优异。

4.3 SLon-2500 磁选机在塞拉利昂分选镜铁矿

2011~2013 年,22 台 SLon-2500 磁选机出口到非洲塞拉利昂,分选镜铁矿获得成功应用,选矿工艺流程如图 7 所示,现场应用照片如图 8 所示,采用 SLon-2500 磁选机 1 次粗选、1 次精选的工艺流程,在每年原矿处理量达到 1000 万吨,入选原矿铁品位为 34.94% 的条件下,获得了综合铁品位为 67.10% 的铁精矿,铁回收率为 74.74%,获得了优异的选矿指标。

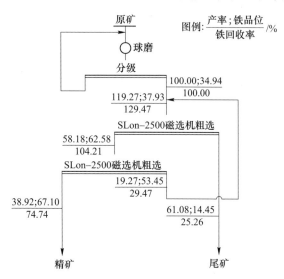

图 7 塞拉利昂镜铁矿选矿原则流程

图 8 SLon-2500 磁选机在塞拉利昂应用照片

4.4 SLon-2500 磁选机分选褐铁矿的工业应用

云南北衙黄金选矿厂的原矿为金矿与铁矿共生的矿石,其早期的生产仅用浸出法回收金,而选金后的尾矿因粒度很细未得到利用。通过探索试验表明,该选金尾矿中含有磁铁矿和褐铁矿。2006 年该厂建成了一条年处理 50 万吨的选铁生产线,从选金尾矿中回收铁矿物,其中采用 2 台 SLon-2000 磁选机回收褐铁矿。2010~2012 年,该厂又建设了一条年处理 150 万吨原矿的选矿生产线,其选金后的尾矿用弱磁选机选出磁铁矿,然后用 6 台 SLon-2500 磁选机分选褐铁矿(粗选和扫选各用 3 台,现场照片如图 9 所示),该生产工艺流程如图 10 所示。目前该厂每年可回收 100 多万吨铁精矿,并可根据市场价格安排磁铁矿和褐铁矿分别销售或合并销售。

图 9 SLon-2500 磁选机在北衙黄金选矿厂应用照片

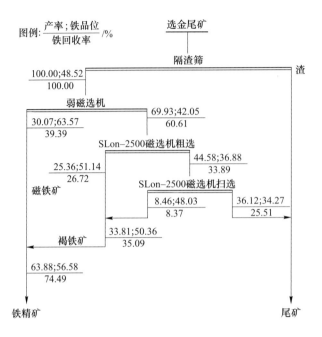

图 10　北衙黄金选矿厂选铁流程

4.5　SLon-2500 磁选机在石英提纯工业中的应用

SLon-2500 磁选机于 2009 年 7 月首先在陕西汉中石英砂提纯工业中应用成功。之后 SLon-2500 磁选机在凤阳高纯石英砂提纯方面又获得成功应用。凤阳县位于安徽省东北部，处于淮河中游南岸，因石英砂资源丰富，因而享有"中国石英砂之乡"的美誉。SLon-2500 磁选机于 2011 年 7 月在凤阳县远东石英砂有限公司安装完毕，10 月投入生产，现场照片见图 11。原石英砂矿中 Fe_2O_3 的含量在 0.01%～0.02%，经过 SLon-2500 磁选机磁选后，精矿产率达到了 95% 以上，精矿中 Fe_2O_3 含量小于 0.007%、SiO_2 含量大于 99.8%，可达到太阳能、光纤通信等高尖端原料的要求。

4.6　SLon-3000 磁选机在攀枝花的应用

2010 年首台 SLon-3000 磁选机在攀枝花安宁铁钛有限公司应用于钛铁矿选矿获得成功。攀枝花地区的钒钛磁铁矿储量丰富，目前该地区的选矿厂的选矿流程普遍是先用弱磁选机选出磁铁矿精矿，选完磁铁矿的尾矿即作为选钛作业的原矿。选钛作业普遍用强磁选机预先抛弃大部分尾矿，强磁精矿再用浮选或重选精选。由于选钛流程中入选原料含钛品位低，要求强磁选机处理量大，生产成本低，尽可能多抛弃尾矿，SLon-3000 磁选机正好具备这些优势，其选矿流程如图 12 所示。

SLon-3000 磁选机在该厂用于钛铁矿的强磁粗选作业，其给矿粒度为小于 74μm（200目）占 40%，台时处理量为 150t/h 左右。该机的选矿指标为：给矿 TiO_2 品位为 9.52%；精矿 TiO_2 品位为 17.43%、回收率为 75.09%，尾矿 TiO_2 品位为 4.02%，抛弃的尾矿产率为 58.99%。该机使用的背景磁感强度为 0.7T，处理每吨给矿的电耗仅 0.3kW·h。目前

图 11 SLon-2500 磁选机在安徽凤阳提纯石英砂

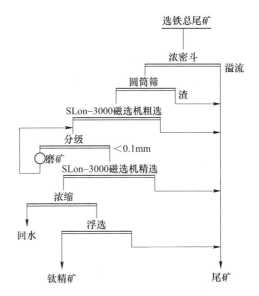

图 12 安宁铁钛有限公司选钛流程

该公司共订购了 9 台 SLon-3000 磁选机用于分选钛铁矿。从首台 SLon-3000 磁选机的工业试验到大规模的工业应用，证明了该机具有选矿效率高、设备可靠性高和生产成本低的优点。

4.7 SLon-3000 磁选机在太和铁矿选钛工业中的应用

重庆钢铁公司太和铁矿近年将原矿处理量由 220 万吨/a 扩建至 630 万吨/a，扩建过程中共采用了 6 台 SLon-3000 磁选机用于钛铁矿分选，目前该生产流程已正常投产，现场应用照片如图 13 所示。

4.8 首台 SLon-4000 磁选机在攀钢的选钛工业试验

攀钢密地选矿厂从钒钛磁铁矿中回收铁精矿和钛精矿，每年产生约 700 万吨尾矿，（每小时即产生约

图 13 SLon-3000 磁选机在太和铁矿分选钛铁矿

885t），全部排往尾矿库。但是其尾矿中还含有少量的可回收铁矿和钛铁矿。攀钢选矿综合厂主要从攀钢排放的总尾矿中回收铁精矿和钛精矿，最早的选钛流程采用螺旋溜槽粗选、摇床精选的单一重选流程。由于螺旋溜槽和摇床对细粒级含钛矿物回收率都很低，全流程的 TiO_2 回收率仅 7% 左右。2002 年该厂采用了 3 台 SLon-1500 磁选机建立新的选钛流程，工艺流程图如图 14 所示。该流程的 TiO_2 回收率由原重选作业的 7% 提高到 27% 左右。但是由于用于粗选作业的 2 台 SLon-1500 磁选机合计处理量仅 60t/h，该厂钛精矿产量仅 2 万吨/a 左右。随着攀钢密地选钛厂选钛技术水平的提高，目前排往尾矿库的尾矿 TiO_2 品位由原来的 7.68% 左右下降至 6.5% 左右，因此，如果不改造，该厂的钛精矿产量有进一步下降的趋势。2011 年年底，首台 SLon-4000 立环脉动高梯度磁选机在该厂应用于工业试

验，现场照片如图 15 所示，取代原有粗选作业的 SLon-1500 磁选机，该机给矿量为 460t/h（总尾矿的 50%）左右，至今已运转 18 个月，已取得的工业试验指标为（SLon-4000 单机作业指标）：给矿 TiO_2 品位为 6.41%，精矿 TiO_2 品位为 13.38%，作业回收率为 59.36%，尾矿 TiO_2 品位为 3.64%。全流程的选钛指标为：给矿 TiO_2 品位为 6.41%，钛精矿 TiO_2 品位为 47.02%、回收率为 28.16%，指标良好。目前该厂又订购了 1 台 SLon-4000 磁选机进行扩产改造，预计改造完成后，该流程年处理 700 万吨综合尾矿，年产 TiO_2 品位为 47% 的钛精矿 20 万吨左右，年新增销售收入 4 亿元左右，年新增利税 2.5 亿元左右。

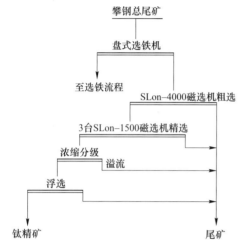

图 14　攀钢综合厂采用 SLon 磁选机的选钛工业试验流程

图 15　SLon-4000 磁选机在攀钢从总尾矿中回收钛铁矿

5　结论

（1）SLon 立环脉动高梯度磁选机利用磁力、脉动流体力和重力的综合力场分选弱磁性矿石，具有富集比大、选矿效率高、选矿成本低、适应面广、设备作业率高、使用寿命长、易安装和检修工作量小的优点。

（2）SLon 磁选机采取低电压大电流激磁，激磁线圈采用水内冷散热，冷却水可以循环使用，具有节电、散热效率高、先进、清洁环保、安全可靠、成本低、占地面积小、磁感应强度高和选矿回收率高的优点。

（3）SLon-2500、SLon-3000 和 SLon-4000 磁选机的研制代表了该机型大型化发展方向。设备的大型化有利于降低生产成本，节约占地面积，提高选矿厂自动化生产水平。

（4）对磁系的优化设计使 SLon 磁选机的激磁功率有了大幅度的下降，设备的大型化更有利于降低单位处理量的能耗。以 SLon-4000 磁选机为例，处理每吨矿石的额定电耗为 $0.41 kW \cdot h$（按工作场强 1.0T 计算），处理每吨钛铁矿的实际电耗仅为 $0.25 kW \cdot h$（按实际工作场强 0.8T 计算）。

（5）SLon-2500、SLon-3000 和 SLon-4000 磁选机具有处理量大、选矿效率高、设备可靠性高、电耗低和生产成本低的优点。它们已成功地在赤铁矿、褐铁矿、钛铁矿选矿和在石英除铁提纯工业中应用。

写作背景 本文介绍了 SLon-2500、SLon-3000、SLon-4000 立环脉动高梯度磁选机的研制与节能原理，强调了 SLon 磁选机采用水内冷的 8 个优点；SLon 磁选机在工业上应用于分选氧化铁矿、钛铁矿和石英砂提纯，其中海南铁矿应用 11 台 SLon-2500 磁选机和 80 台 SLon-ϕ2400 离心机建成年处理 200 万吨难选氧化铁矿流程为新项目。

SLon 立环脉动高梯度磁选机在多种金属矿选矿中的应用[1]

熊大和[1,2]

(1. 赣州金环磁选设备有限公司；2. 赣州有色冶金研究所)

摘　要　SLon 立环脉动高梯度磁选机利用磁力、脉动流体力和重力的综合力场分选弱磁性矿石，具有富集比大、选矿效率高、设备可靠性高、选矿生产成本低、适应面广、设备作业率高、使用寿命长、易安装和检修工作量小的优点。该机在氧化铁矿、钛铁矿、铬铁矿、锰矿、钨矿选矿和在非金属除铁提纯工业中得到了广泛的应用。本文重点介绍该机在多种金属矿方面的应用。

关键词　SLon 立环脉动高梯度磁选机　氧化铁矿　锰矿　铬铁矿　钨矿选矿

The Application of SLon VPHGMS for Processing Several Metallic Ores

Xiong Dahe[1,2]

(1. Ganzhou SLon Magnetic Separator Ltd.,
2. Ganzhou Non-ferrous Metallurgy Research Institute)

Abstract　SLon vertical ring and pulsating high gradient magnetic separator (SLon VPHGMS) utilizes the combined force field of magnetism, pulsating fluid and gravity to continuously separate weakly magnetic minerals. It is equipped with unique pulsating mechanism and possesses the advantages of large beneficiation ratio, high concentration efficiency, high equipment reliability, low production cost, wide adaptation, high operation efficiency, long service life, easy installation and small maintenance workload. It has been widely used for processing iron oxides, ilmenite, chromite, manganese ores, tungsten ores and removing iron impurities from nonmetallic ores. In this paper, the application of SLon VPHGMS for processing several metallic ores was introduced.

Keywords　SLon VPHGMS; Iron oxides; Manganese ore; Chromite; Tungsten ores magnetic separation

　　SLon 立环脉动高梯度磁选机是新一代优质高效强磁选设备。通过多年持续的技术创新与改进，该机在设备大型化、多样化、自动化、高效、节能、提高可靠性等方面得到了快速的发展，并且得到了更为广泛的应用。目前有 3000 多台 SLon 磁选机在工业上广泛用于氧化铁矿、钛铁矿、锰矿、铬铁矿、钨矿等弱磁性矿石的选矿及石英、长石、高岭土等非金属矿的除杂提纯。

---

[1] 原载于《矿产保护与利用》，2013 (06)：51~56。

1 SLon 磁选机工作原理简介

SLon 立环脉动高梯度磁选机利用磁力、脉动流体力和重力的综合力场分选弱磁性矿石,具有富集比大、选矿效率高、选矿成本低、适应面广、设备作业率高、使用寿命长、易安装和检修工作量小等优点。

SLon 磁选机结构如图 1 和图 2 所示,主要由脉动机构、激磁线圈、铁轭、转环和各种矿斗、水斗组成。用导磁不锈钢棒作磁介质。该机工作原理如下。

激磁线圈通以直流电,在分选区产生感应磁场,位于分选区的磁介质表面产生高梯度磁场;转环作顺时针旋转,将磁介质不断送入和运出分选区;矿浆从给矿斗给入,进入位于转环下部的分选区,矿浆中的磁性颗粒吸附在磁介质棒表面上,被转环带至顶部无磁场区,被冲洗水冲入磁性矿斗;非磁性颗粒在重力、脉动流体力的作用下穿过磁介质堆,流入非磁性矿斗排走,实现与磁性颗粒的分离。

图 1 SLon 立环脉动高梯度磁选机照片

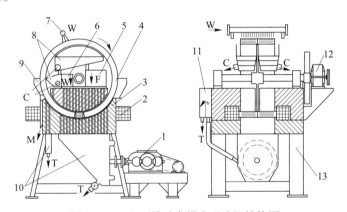

图 2 SLon 立环脉动高梯度磁选机结构图

1—脉动机构;2—激磁线圈;3—铁轭;4—转环;5—给矿斗;6—漂洗水斗;7—精矿冲洗装置;
8—磁性矿斗;9—中矿斗;10—非磁性矿斗;11—液位斗;12—转环驱动机构;13—机架;
F—给矿;W—清水;C—磁性矿;M—中矿;T—非磁性矿

该机的转环采用立式旋转方式,对于每一组磁介质而言,冲洗磁性精矿的方向与给矿方向相反,粗颗粒不必穿过磁介质堆便可冲洗出来。该机的脉动机构驱动矿浆产生脉动,可使位于分选区磁介质堆中的矿粒群保持松散状态,使磁性矿粒更容易被捕获,使非磁性矿粒尽快穿过磁介质堆到非磁性矿中去。

显然,反冲精矿和矿浆脉动可防止磁介质堵塞;脉动分选可提高磁性精矿的质量。这些措施保证了该机具有较大的富集比、较高的分选效率和较强的适应能力。

SLon 磁选机的激磁线圈的冷却方式采用水内冷,由于水内冷冷却效率高,与油冷线圈比较,水内冷激磁线圈可获得较高的磁场强度,在高场强下运转更稳定可靠。从而获得较高的设备作业率和选矿回收率。

2 工业应用

由于 SLon 磁选机具有优异的选矿性能,该机在弱磁性金属矿的选矿领域得到了广泛的应用。

2.1 SLon 磁选机在齐大山选矿厂的应用

齐大山选矿厂是鞍钢主要铁精矿供应基地,每年处理铁品位 30%左右的鞍山式氧化铁矿 800 万吨。其矿物组成主要是赤铁矿、磁铁矿和石英。采用阶段磨矿—重选—强磁—反浮选流程(图 3)。该流程中,原矿首先磨矿至小于 0.074mm(200 目)占 60%左右,用旋流器分级,旋流器的沉砂用螺旋流槽选出一部分较粗粒级且单体解离的铁精矿,螺旋流槽的尾矿用滚筒弱磁选机和 SLon 中磁机扫选,抛弃产率为 20.24%,品位为 8.05%Fe 的尾矿,螺旋流槽的中矿和磁选精矿进入二段磨矿,二段磨矿产品返回到旋流器;旋流器的溢流(小于 0.074mm(200 目)占 90%左右)用滚筒弱磁选机选出磁铁矿,用 SLon 磁选机选出赤铁矿并抛弃产率为 31.31%,品位为 8.91%的尾矿,二者的磁选精矿再用反浮选精选得到高品位的铁精矿。

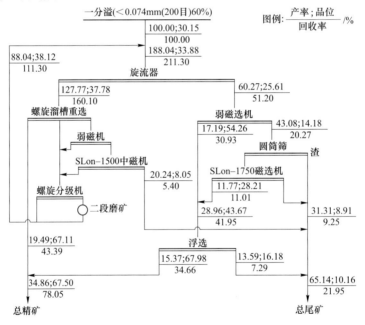

图 3 齐大山选矿厂二选车间数质量原则流程

该流程中 SLon 磁选机抛弃了占原矿产率 51.55%(占总尾矿产率的 79.14%)的低品位尾矿,对控制尾矿品位和提高全流程的选矿回收率起到了关键的作用。

全流程的选矿指标为:原矿品位 30.15% Fe,铁精矿品位 67.50% Fe,铁回收率 78.05%,该选矿指标代表了我国氧化铁矿选矿技术的最高水平,使我国的氧化铁矿选矿技术水平从落后于国外转变为达到国际领先的水平。该流程成为我国鞍山式氧化铁矿选矿流程的样板。

2.2 SLon 磁选机在攀钢选钛厂的应用

攀钢选钛厂原先的选钛流程中小于 0.045mm 粒级作为尾矿排放,其中 TiO_2 的金属占有率为 40% 左右,因此,攻克这部分物料的选钛难题,对提高全流程的选钛回收率具有重要的意义。

1994~1996 年,攀钢选钛厂采用一台 SLon-1500 立环脉动高梯度磁选机建立了第一条微细粒级选钛试验生产线,获得成功,每年生产 2 万吨优质钛精矿。随后的几年攀钢选钛厂又分两期建成后八系统和前八系统小于 0.045mm 粒级选钛生产线,至 2004 年,攀钢选钛厂小于 0.045mm 钛精矿的年产量达到 15 万吨左右,加上粗粒级钛精矿,攀钢选钛厂年产钛精矿达到 28 万吨左右。

2008 年,攀钢选钛厂进行选钛扩能改造工程,具体实施计划有:

(1) 将粗粒级选钛流程由原来的重选—电选流程改为强磁选—浮选流程。
(2) 优化细粒级选钛流程。
(3) 钛精矿产量由改造前的 28 万吨/a 提高到 40 万吨/a。

攀钢选钛厂委托赣州金环磁选设备有限公司所做的选钛试验流程如图 4 所示。该流程的主要特点为:

(1) 磁选尾矿只分成粗粒级和细粒级入选,充分发挥 SLon 磁选机本身的脱泥功能,

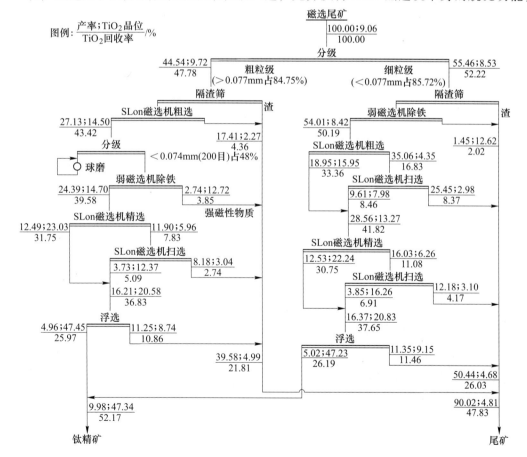

图 4 攀钢选钛厂扩能改造试验流程(赣州金环公司试验)

尽可能减少分级溢流的损失。

（2）细粒级强磁选作业由过去的 SLon 磁选机一次选别改为四次选别，即 SLon 磁选机一粗一扫和一精一扫；粗粒级强磁选作业采用 SLon 磁选机一次粗选，其粗精矿分级再磨后用弱磁选机除铁，其尾矿用 SLon 磁选机一精一扫选。这些措施保证了强磁选作业对钛铁矿有较高的回收率和较高的精矿品位。

（3）该选钛流程充分利用了 SLon 磁选机处理量大，生产成本低，脱泥效果较好的优点，大幅度提高强磁选作业的选钛回收率，为浮选创造了较好的条件。在保证钛精矿品位达到47% TiO_2 以上的前提下，浮选作业 TiO_2 回收率可保证在70%左右，全流程的 TiO_2 回收率达到52.17%。

2008~2009 年，攀钢选钛厂参考各家的选钛试验流程并结合现场的条件进行了选钛扩能改造，2010 年、2011 年和 2012 年该厂分别实现了年产优质钛精矿48万吨、50万吨和52万吨，比改造前每年提高20余万吨优质钛精矿。目前生产上全流程实际达到的 TiO_2 回收率为40%左右。

2.3 SLon 磁选机从选金尾矿中回收褐铁矿的工业应用

云南北衙黄金选矿厂的原矿为金矿与铁矿共生的矿石，其早期的生产仅用浸出法回收金，而选金后的尾矿因粒度很细未得到利用。通过探索试验表明，该选金尾矿中含有磁铁矿和褐铁矿。2006 年该厂建成了一条年处理50万吨的选铁生产线，从选金尾矿中回收铁矿物，其中采用2台 SLon-2000 磁选机回收褐铁矿。2010 年至 2012 年，该厂又建设了一条年处理150万吨尾矿的选矿生产线，其选金后的尾矿用弱磁选机选出磁铁矿，然后用6台 SLon-2500 磁选机分选褐铁矿（粗选和扫选各3台，图5)，该生产线选铁部分的流程如图6。目前该厂每年可回收100多万吨铁精矿，并可根据市场价格安排磁铁矿和褐铁矿分别销售或合并销售。

图 5　SLon-2500 磁选机在北衙黄金选矿厂应用照片

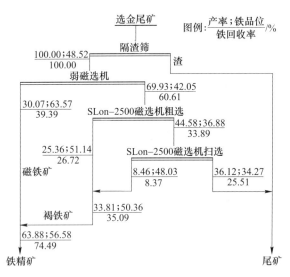

图 6　北衙黄金选矿厂选铁流程

2.4 SLon 磁选机从铅锌尾矿中回收锰矿的应用

南京栖霞山铅锌矿以开采铅锌矿为主，其铅锌选矿的尾矿中含锰12%左右。若从中选出品位为22%以上的锰精矿，则可作为制造硫酸锰的原料。2005年该厂建成了从尾矿中回收锰精矿的选矿生产流程（图7），采用SLon磁选机一次粗选和一次精选，获得的选锰综合指标为：给矿锰品位12.57%，锰精矿品位25.20%，锰精矿产率27.50%，锰回收率55.13%。取得了良好的经济效益。

2.5 SLon 磁选机分选铬铁矿的应用

铬是制造不锈钢的重要原料，我国铬的储量很少，大部分要依靠进口。SLon磁选机在甘肃分选铬铁矿取得了成功应用，如图8所示为甘肃景世化工公司铬铁矿选矿流程。该铬铁矿原矿 Cr_2O_3 品位为27.98%，磨矿至小于0.074mm（200目）占90%，通过SLon磁选机一次粗选和一次扫选，获得铬精矿 Cr_2O_3 品位40.07%，回收率96.36%的优良选矿指标。

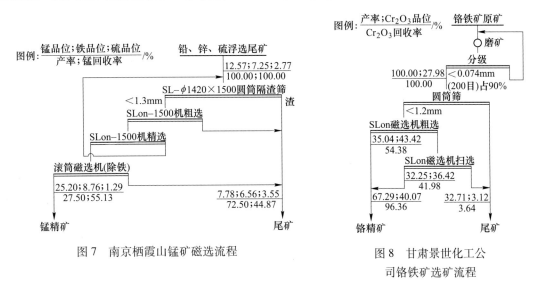

图7 南京栖霞山锰矿磁选流程

图8 甘肃景世化工公司铬铁矿选矿流程

2.6 SLon 磁选机在分选钨矿的应用

黑钨矿是弱磁性矿石，SLon磁选机近年在细粒黑钨矿选矿方面取得了成功的应用。该机可用于：

（1）细粒黑钨矿的粗选作业或预富集作业，抛弃大量的尾矿；

（2）细粒黑钨矿、白钨矿的分离；

（3）细粒黑钨矿与锡石的分离。

2.6.1 SLon 磁选机在荡坪钨矿分选钨细泥的应用

江西荡坪钨矿以黑钨矿为主，其钨细泥选矿流程如图9所示。该流程首先将钨细泥筛分和磨矿至小于0.5mm，用浮选选出硫化矿，然后用SLon磁选机粗选，该机对黑钨矿的富集比为13倍，抛弃的尾矿产率占该流程给矿的92.2%，其粗选精矿 WO_3 品位和产率分别为 7.30% 和 6.93%。这部分粗精矿再用摇床精选。全流程选矿指标为：给矿 WO_3 品位 0.55%，精矿 WO_3 品位 52.82%，WO_3 回收率为 85.01%。

SLon 磁选机在该流程中成功应用，由于该机处理量大，一台设备可代替多台摇床。该机用于粗选作业抛弃了绝大部分尾矿，大幅减轻了后续精选作业的矿量，节约了大量的厂房面积，该机作业回收率高达 92.57%，保证了全流程获得较高的选矿回收率。

2.6.2 SLon 磁选机在从浮铜尾矿中回收钨矿的应用

内蒙古科尔沁旗道伦大坝选厂日处理 3000t 原矿，矿石中有用矿物主要是硫化铜和黑钨矿，其中黑钨矿的嵌布粒度很细。其选矿流程如图 10 所示。该流程首先将原矿磨矿至小于 0.074mm（200 目）90% 左右，用浮选选出铜精矿，然后用 SLon 磁选机选出黑钨矿，该机对黑钨矿的富集比为 26.81，对黑钨矿的作业回收率为 96.43%，抛弃的尾矿产率占选钨流程给矿的 96.43%，其粗选精矿 WO_3 品位和产率分别为 8.31% 和 3.574%。磁性产品再用浮选脱硫，然后用摇床精选得到钨精矿。全流程黑钨矿的选矿指标为：给矿 WO_3 品位 0.31%，钨精矿 WO_3 品位 53.00%，WO_3 回收率为 82.10%。

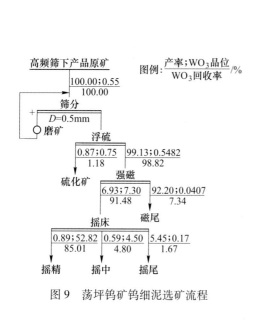

图 9　荡坪钨矿钨细泥选矿流程

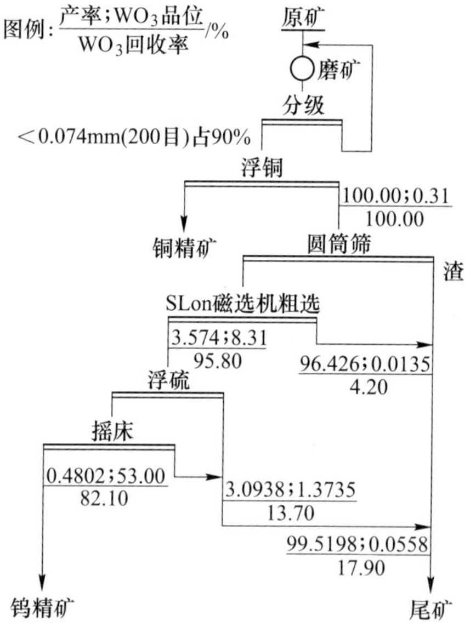

图 10　内蒙古科尔沁旗道伦大坝选厂选矿流程

3　结论

（1）SLon 立环脉动高梯度磁选机利用磁力、脉动流体力和重力的综合力场分选细粒弱磁性矿石，具有富集比大、选矿效率高、选矿成本低、适应面广、设备作业率高、使用寿命长、易安装和检修工作量小等优点。

（2）SLon 磁选机采取低电压大电流激磁，激磁线圈采用水内冷散热，冷却水可以循环使用，具有节电、散热效率高、先进、清洁环保、安全可靠、成本低、占地面积小、磁感应强度高和选矿回收率高的优点。

（3）SLon 磁选机在鞍钢齐大山铁矿的应用，选矿指标创历史最佳，使我国氧化铁矿的选矿达到国际领先水平。

（4）SLon 磁选机在钛铁矿选矿得到广泛应用，为提高我国钛铁矿选矿回收率做出了突出的贡献。

（5）SLon 磁选机在北衙黄金选矿厂从选金尾矿中回收褐铁矿，让过去的尾矿变废为宝。

（6）SLon 磁选机在南京栖霞山从铅锌浮选尾矿中回收锰矿，锰精矿可作为制造硫酸锰的原料。

（7）SLon 磁选机用于甘肃景世化工公司铬铁矿选矿，原矿 Cr_2O_3 品位为 27.98%，获得铬精矿 Cr_2O_3 品位 40.07%，铬回收率 96.36% 的优良选矿指标。

（8）SLon 磁选机在细粒黑钨矿的选矿得到广泛应用。该机用于粗选作业可抛弃绝大部分低品位尾矿，大幅减轻了后续精选作业的给矿量。节约了大量的厂房面积；该机作业回收率高，保证了全流程获得较高的选矿回收率。

写作背景 本文介绍了 SLon 立环脉动高梯度磁选机的优点，该机在多个厂矿分选赤铁矿、褐铁矿、钛铁矿、锰矿、铬铁矿和黑钨矿的应用实例，大幅度提高了这些弱磁性矿的选矿指标，其中 SLon 磁选机应用于分选铬铁矿和黑钨矿的选矿流程为最新公布的内容。

SLon-3000 立环脉动高梯度磁选机的研制与应用[1]

熊大和[1,2]

(1. 赣州金环磁选设备有限公司；2. 赣州有色冶金研究所)

摘 要 赣州金环磁选设备有限公司经过6年的技术创新与改进，研制成功了代表SLon磁选机大型化发展方向的SLon-3000立环脉动高梯度磁选机。该机具有处理量大、选矿效率高、设备可靠性强、电耗和生产成本低的优点，在攀枝花安宁铁钛有限公司和重钢西昌矿业有限公司的选钛流程以及昆钢大红山铁矿的氧化铁矿石分选作业中显示了优良的应用效果。

关键词 SLon-3000立环脉动高梯度磁选机 磁选机大型化 电耗 弱磁性矿石选矿

Development and Application of SLon-3000 Vertical Ring and Pulsating High Gradient Magnetic Separator

Xiong Dahe[1,2]

(1. SLon Magnetic Separator Ltd.；
2. Ganzhou Nonferrous Metallurgy Research Institute)

Abstract Through six-year innovation and improvement, SLon magnetic separator Ltd. has successfully developed SLon-3000 vertical ring pulsating high gradient magnetic separator which is the representative of SLon magnetic separator enlargement. With advantages of large capacity, high beneficial efficiency, high reliability, low energy consumption and low operating cost, it showed an excellent performance in application in Titanium concentration process of Panzhihua Anning Iron-Titanium Co., Ltd. and Xichang Mining Co., Ltd. of Chongqing Iron and Steel Co., Ltd. and iron oxide beneficiation process of Kunsteel Dahongshan Iron Mine.

Keywords SLon-3000 vertical ring pulsating high gradient magnetic separator; Magnetic separator enlargement; Energy saving; Beneficiation of weakly magnetic ore

现代选矿工业的特点是：一方面经济的发展需要更多更好的矿产品作为工业原料，另一方面复杂难选原矿日益增多。这就要求选矿设备和选矿工艺向大型化、自动化、高效率和低成本方向发展[1]，以满足现代选矿工业的需要。SLon立环脉动高梯度磁选机广泛应

---

[1] 原载于《金属矿山》，2013，12：100~104。

用于氧化铁矿、钛铁矿、锰矿等弱磁性矿石的选矿及石英、长石、高岭土等非金属矿的提纯。为了满足大型矿山分选弱磁性矿石的需要，赣州金环磁选设备有限公司从 2007 年开始研制 SLon-3000 立环脉动高梯度磁选机，经过 6 年的技术创新与改进，完成了该机的整个研制过程。该机具有处理量大、自动化程度和选矿效率高、可靠性强、能耗和选矿成本低的优点，已在工业上成功地应用于钛铁矿和氧化铁矿的选矿。

1 SLon-3000 磁选机的研制

1.1 SLon-3000 磁选机结构与工作原理

SLon-3000 立环脉动高梯度磁选机主要由脉动机构、激磁线圈、铁铠、分选转环、给矿斗、磁性产品矿斗和非磁性产品矿斗组成，用导磁不锈钢圆棒作磁介质。如图 1 所示为第 1 台 SLon-3000 立环脉动高梯度磁选机整机照片，如图 2 所示为该机结构图。

图 1 SLon-3000 立环脉动高梯度磁选机整机照片

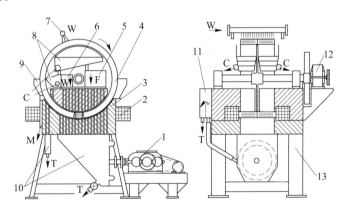

图 2 SLon-3000 立环脉动高梯度磁选机结构

1—脉动机构；2—激磁线圈；3—铁铠；4—转环；5—给矿斗；6—漂洗水斗；7—精矿冲洗装置；8—磁性产品矿斗；9—中矿斗；10—非磁性产品矿斗；11—液位；12—转环驱动机构；13—机架；F—给矿；W—清水；C—精矿；M—中矿；T—尾矿

当激磁线圈通以直流电后，在分选区产生感应磁场，分选转环携带磁介质围绕着其水平轴旋转。矿浆从给矿斗给入，经过位于分选区的磁介质，在磁场力、矿浆脉动力和重力的共同作用下，磁性矿物被吸引至磁介质表面，随转环旋转到顶部的无磁场区时被水冲洗至磁性产品矿斗，非磁性产品则穿过磁介质进入非磁性产品矿斗。

由于转环立式旋转，相对于每一组磁介质而言，给矿方向和磁性产品的冲洗方向相反，因而粗颗粒不必穿过整个磁介质组的深度就可以被冲洗出来。脉动系统驱动分选区的矿浆上下振动，使分选区的矿粒群保持松散状态，磁性矿粒更容易被磁介质捕获，非磁性矿粒更容易穿过磁介质被排入到非磁性产品中去。

反冲磁性产品和矿浆脉动可以防止磁介质堵塞，矿浆脉动还有利于提高磁性产品的品位和非磁性产品的产率。这些措施保证了 SLon 立环脉动高梯度磁选机具有较大的富集比、较高的选矿效率和广泛的适应性。

1.2　SLon-3000 磁选机激磁线圈冷却方式

激磁线圈是电磁强磁选机的最重要部件之一，其结构和技术参数决定磁选机的激磁功率、工作场强和选矿回收率。

直流电磁磁选机所消耗的激磁功率 P 由下式计算：

$$P = VI = RI^2 \tag{1}$$

式中，V 为激磁电压；I 为激磁电流；R 为激磁线圈导线的串联电阻。

电磁强磁选机选矿区的磁感应强度与激磁电流 I 有关系，只要 I 不变，磁感应强度就不会变。当激磁线圈导线长度一定时，其串联电阻越低，激磁功率就越低。例如铜线圈与铝合金线圈比较，铜的电阻率比铝合金的电阻率低 40% 左右，如果采用同样体积的导线绕制同一台强磁选机的激磁线圈，达到同样的磁感应强度时，铜线圈激磁功率比铝合金线圈低 40% 左右。

由于导线存在电阻（超导线圈除外），故由式（1）所计算的激磁功率会全部转化为热量。这使得在狭小空间和密封状态下工作的强磁选机的激磁线圈必须采用水、油、空气或其他方式进行强制冷却，否则，激磁线圈的工作温度就会不断升高直至将线圈烧毁。因此，强磁选机激磁线圈的冷却方式关系到机器是否能长期稳定工作。

SLon-3000 磁选机的激磁线圈采用电阻率低、含铜量为 99.9% 以上、截面为矩形的空心铜管绕制，冷却方式为水内冷，其主要优点如下：

（1）节电。水的比热为 4.2J/(g·℃)，而变压器油的比热约为 2.1J/(g·℃)，因此，在相同的激磁功率和冷却液流量下，水冷线圈的温升比油冷线圈的温升低得多，例如水冷线圈的温升为 20℃时，油冷线圈的温升可达 40℃。另一方面，铜导线或铝合金导线的电阻率随温度升高而升高。例如铜导线的电阻率在 20℃时为 $1.79 \times 10^{-8} \Omega \cdot m$，在 40℃时为 $1.93 \times 10^{-8} \Omega \cdot m$，即温度上升 20℃，磁选机激磁线圈的电阻将提高 7.8%，这样，根据式（1），激磁功率也将提高 7.8%；铝合金导线的激磁线圈也存在同样的规律。可见，水冷线圈具有电阻低、节电的优点。

（2）散热效率高。激磁线圈采用空心铜管绕制，水内冷散热，冷却水直接贴着铜管内壁流动，散热效率最高，水的温度和导体的温度基本一致。而油冷线圈的冷却油与导体之间有绝缘层，导体产生的热量要穿过绝缘层才能被油吸收，因此油冷线圈散热效率较低。

（3）水内冷是一种最先进的冷却方式。国内外大型发电机的励磁线圈大部分都是采用水内冷。

（4）水内冷有利于环保。水内冷方式下，万一管路漏水，不会污染环境。而油冷方式下，一旦管路漏油，则会污染环境。

（5）水内冷线圈安全可靠。SLon 磁选机现有 3000 多台在工业上应用，每年的线圈返修率不到 1%。如果用户能按使用手册要求供水（普通生活用水标准），则线圈可做到长期不堵塞。

（6）水内冷更经济。水的成本比油低，而且冷却水可以循环使用，基本上不耗水。例如鞍钢地区有 200 多台 SLon 磁选机，攀枝花地区有 300 多台 SLon 磁选机，平均已运转 10

年左右，绝大部分冷却水循环使用，未发生过激磁线圈堵塞现象，证明了这种冷却方式具有工作稳定可靠、易于维护的优点。

（7）水内冷线圈节约占地面积。SLon 磁选机冷却系统不与操作人员争位置，更有利于操作人员观察和操作设备，易于打扫卫生和设备维护。

（8）水内冷线圈可获得更高的磁场强度。由于水内冷冷却效率高，故与油冷激磁线圈向比，水内冷激磁线圈可获得更高的磁场强度，从而获得较高的选矿回收率。

1.3 大型 SLon 磁选机的节电原理

下面以圆柱形螺旋管线圈（图3）为例说明大型 SLon 磁选机的节电原理。

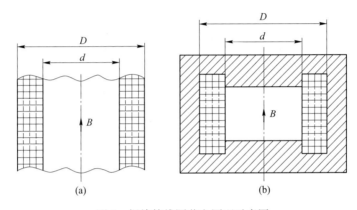

图3 螺旋管线圈节电原理示意图

(a) 无限长螺线管线圈；(b) 带铁铠的螺线管线圈

根据电磁学理论，位于真空中的无限长螺旋管线圈（图3（a））通以直流电后，其轴线上的磁感应强度为

$$B = \mu_0 n I \tag{2}$$

式中，μ_0 为真空磁导率；n 为螺旋管轴线方向单位长度上的线圈匝数。

式（2）说明，无限长螺旋管内部的磁感应强度与其内腔直径 d 和外围直径 D 没有关系。因此，假如线圈内腔直径增加1倍，则在线圈内腔的磁感应强度不变的情况下，对应的磁场空间可扩大至4倍，而 n 匝导线的长度和激磁功率只增加约1倍。

SLon 磁选机的激磁线圈不可能在轴线方向做到无限长，但是，通过采用铁铠（图3（b）），可使激磁线圈内腔（即分选区）的磁感应强度接近于按式（2）计算的预定值，从而实现设备的处理能力扩大至4倍，而激磁线圈的导线长度和激磁功率仅增加1倍。

以上简要分析了大型 SLon 磁选机的节电原理，此外，还可以通过降低激磁电流密度，优化磁系设计和减少漏磁等措施来降低设备的激磁功率。

1.4 SLon-3000 磁选机电磁性能测试

通过激磁系统的优化，使 SLon-3000 立环脉动高梯度磁选机具备了优良的电磁性能。该机设计的额定背景磁感应强度为 1.0T，实测的电磁性能见表1。可见，当背景磁感应强度达到 1.041T 时，相应的激磁电流为 1300A，激磁电压为 55.9V，激磁功率仅 72.7kW。

表 1　SLon-3000 磁选机电磁性能实测结果

激磁电流/A	激磁电压/V	激磁功率/kW	背景磁感应强度/T
200	8.2	1.6	0.189
400	16.2	6.5	0.375
600	24.6	14.8	0.564
800	33.2	26.6	0.734
1000	42.2	42.2	0.880
1200	51.4	61.7	0.995
1300	55.9	72.7	1.041
1400	60.4	84.6	1.079
1500	64.9	97.4	1.116

1.5　SLon 磁选机大型化的节能效果对比

SLon 立环脉动高梯度磁选机的大型化有利于降低单位矿石的能耗，通过优化磁系结构、减少漏磁、采用优质铜管和降低激磁电流密度，使电耗有了显著的降低。

表 2 为几种型号 SLon 磁选机的能耗对比结果。可见，设备越大，能耗越低。例如：SLon-1000 磁选机的单位矿石能耗为 3.69kW·h/t，而 SLon-3000 磁选机的单位矿石能耗仅为 0.56kW·h/t，后者比前者节能 84.82%；与目前工业上大量使用的 SLon-2000 磁选机相比，SLon-3000 磁选机也节能 34.12%。

表 2　不同型号 SLon 磁选机能耗对比

机　型	磁感应强度/T	总功率/kW	平均干矿处理量 /t·h^{-1}	单位矿石电耗 /kW·h·t^{-1}
SLon-1000	1.0	20.3	5.5	3.69
SLon-1250	1.0	22.7	14	1.62
SLon-1500	1.0	34	25	1.36
SLon-1750	1.0	45	40	1.13
SLon-2000	1.0	55	65	0.85
SLon-2500	1.0	85	125	0.68
SLon-3000	1.0	112	200	0.56

1.6　SLon-3000 磁选机整机技术参数

首台 SLon-3000 磁选机的主要技术参数见表 3。

表 3　SLon-3000 磁选机主要技术参数

转环外径/mm	给矿粒度/mm	给矿浓度/%	矿浆通过能力/m^3·h^{-1}	干矿处理量/t·h^{-1}	额定背景磁感应强度/T	额定激磁电流/A	额定激磁电压/V
3000	0~1.2	10~40	350~650	150~250	1.0	1300	58

额定激磁功率/kW	传动功率/kW	脉动冲程/mm	脉动冲次/r·min^{-1}	供水压力/MPa	冲洗水用量/m^3·h^{-1}	冷却水用量/m^3·h^{-1}	主机质量/t
75	18.5+18.5	0~30	0~300	0.2~0.4	240~400	7~8	175

2 工业试验与应用

2.1 首台 SLon-3000 磁选机的工业试验

2010年,首台 SLon-3000 磁选机在攀枝花安宁铁钛有限公司进行了钛铁矿选矿的工业试验。攀枝花地区钒钛磁铁矿储量丰富,目前该地区的选矿厂普遍采用先通过弱磁选机选出磁铁矿精矿,再对选铁尾矿进行选钛的原则流程。选钛时一般先用强磁选机抛弃大部分尾矿,强磁选精矿再进行浮选或重选精选[2]。由于选钛作业入选原料钛品位低,因此要求强磁选机处理量和尾矿抛弃量大、生产成本低,SLon-3000 磁选机正好具备这些优势。工业试验期间安宁铁钛有限公司选钛流程如图4所示。

首台 SLon-3000 磁选机在流程中用于钛铁矿的强磁粗选作业,在给矿粒度为小于 0.074mm(200目)占 40%、TiO_2 品位为 9.52% 的情况下,所获粗精矿的产率为 41.01%、TiO_2 品位为 17.43%、TiO_2 回收率为 75.09%,抛弃的尾矿产率为 58.99%、TiO_2 品位为 4.02%。试验中 SLon-3000 磁选机的台时处理量为 150t/h、背景磁感应强度为 0.7T、单位电耗仅 0.3kW·h/t。

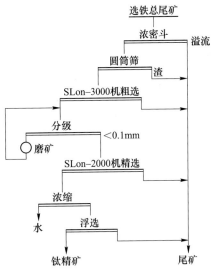

图 4 首台 SLon-3000 磁选机选钛工业试验流程

SLon-3000 磁选机所获粗精矿经分级、磨矿至小于 0.1mm 后,再用 SLon-2000 磁选机精选 1 次,然后经浮选得到 TiO_2 品位在 47% 以上的钛精矿,可作为生产钛白粉或钛合金的优质原料。

首台 SLon-3000 磁选机在工业试验获得成功后随即投入了生产,通过几年的生产应用,证明该机具有选矿效率高、设备可靠性强和生产成本低的优点。至今攀枝花安宁铁钛有限公司共购买了 9 台 SLon-3000 磁选机用于钛铁矿选矿。

2.2 SLon-3000 磁选机在重钢西昌矿业有限公司的工业应用

重钢西昌矿业有限公司拥有 1 座较大的钒钛磁铁矿矿床,近年来正逐步将原矿处理量由 220 万吨/a 扩大到 1000 万吨/a(设计年处理铁矿石 1000 万吨,选出 333 万吨磁铁矿精矿,667 万吨铁尾矿作为选钛原矿)。扩建过程中新增了 7 台 SLon-3000 磁选机用于钛铁矿分选,并将原选钛流程中的 SLon-2000 磁选机和 SLon-1750 磁选机全部加以利用。新的选钛流程如图5所示,SLon-3000 磁选机在该流程中用于浓缩脱泥后选铁尾矿的粗选,获得的粗精矿经浓缩、分级、磨矿至小于 0.074mm(200 目)占 50% 后用 SLon-2000 和 SLon-1750 磁选机精选,所获磁选精矿再经浮选,最终得到 TiO_2 品位在 47% 以上的优质钛精矿。

目前重钢西昌矿业有限公司采出原矿量已达到 700 万吨/a 左右,新的选钛流程已投产 1 年,年产优质钛精矿 30 万吨左右,待项目全部达产,钛精矿产量可望由扩建前的 10 万吨/a 增加至 45 万吨/a。

2.3 SLon-3000 磁选机在昆钢大红山铁矿的工业应用

大红山铁矿是昆钢的主要原料基地，其铁矿石为磁铁矿和氧化铁矿（氧化铁矿包括赤铁矿和褐铁矿，下同）的混合矿石。该矿在第一期 50 万吨/a、第二期 400 万吨/a 选矿车间已分别采用了 SLon-1500 和 SLon-2000 立环脉动高梯度磁选机来分选氧化铁矿[3]。近年来该矿又建成了第三期 600 万吨/a 选矿车间，其工艺流程如图 6 所示，流程中采用 6 台 SLon-3000 磁选机作为氧化铁矿的选别设备，有效地控制了尾矿品位，使全流程获得了优良的选矿指标。

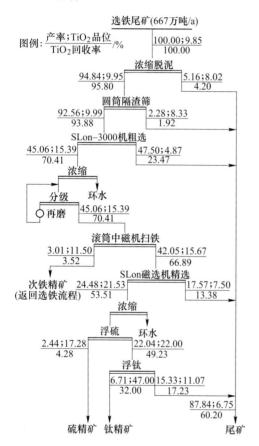

图 5　重钢西昌矿业有限公司选钛设计流程

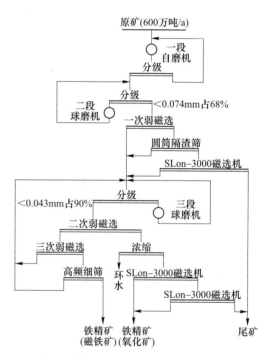

图 6　昆钢大红山铁矿第三期选矿流程

3　结论

（1）SLon-3000 磁选机代表了 SLon-磁选机大型化的发展方向。设备的大型化有利于降低生产成本，节约占地面积，提高选矿厂自动化生产水平。

（2）该机采取低电压大电流激磁，激磁线圈采用水内冷散热，冷却水可以循环使用，具有节电、环保、可靠、成本低、占地面积小、选矿回收率高的优点。

（3）该机的额定单位矿石电耗为 0.56kW·h/t（按额定磁感应强度 1.0T 计算），处理单位钛铁矿的实际电耗仅为 0.3kW·h/t（按工作磁感应强度 0.7T 计算）。

（4）该机已成功地在氧化铁矿和钛铁矿的选矿实践中大规模应用，为弱磁性矿石的选矿工业提供了一种新型高效的强磁选设备。

参 考 文 献

[1] 熊大和.SLon-2500立环脉动高梯度磁选机的研制与应用［J］.金属矿山，2010（6）：133~136.
[2] 熊大和.应用SLon磁选机提高选钛回收率的回顾与展望［J］.金属矿山，2011（10）：1~8.
[3] 熊大和.SLon立环脉动高梯度磁选机分选红矿的研究与应用［J］.金属矿山，2005（8）：24~29.

写作背景　从设计SLon-3000立环脉动高梯度磁选机到该机成熟的工业应用大约花了6年的时间。此文介绍了该机的结构、技术参数、工作原理、水内冷优点、节电原理；首台SLon-3000磁选机在攀枝花安宁铁钛有限公司分选钛铁矿的工业试验，在重钢西昌矿业有限公司（原名：太和铁矿）应用于分选钛铁矿，在昆钢大红山铁矿第三期选矿流程中分选氧化铁矿的工业应用。

SLon 磁选机与离心机分选氧化铁矿新技术

熊大和[1,2]

(1. 赣州金环磁选设备有限公司；2. 赣州有色冶金研究所)

摘 要 SLon立环脉动高梯度磁选机具有优异的选矿性能，分选氧化铁矿具有处理量大、抛弃的尾矿品位低和生产成本低的优点，SLon离心选矿机用于强磁精矿的精选作业，可有效地剔除含少量磁铁矿的石英等脉石，获得较高的铁精矿品位，这两种设备的优势互补，用它们的组合流程分选某些氧化铁矿可获得良好的选矿指标。本文介绍该技术的最新研究与应用。

关键词 SLon立环脉动高梯度磁选机 SLon离心选矿机 氧化铁矿 选矿

The New Development of the Cooperational Flowsheet of SLon Magnetic Separator and Centrifugal Separator in Processing Oxidized Iron Ores

Xiong Dahe[1,2]

(1. *SLon Magnetic Separator Ltd.*；2. *Ganzhou Nonferrous Metallurgy Research Institute*)

Abstract SLon vertical ring and pulsating high gradient magnetic separators possess excellent mineral processing ability. They can discharge low grade tails with big capacity and low operating cost when process oxidized iron ores. SLon centrifugal separators are applied to clean the SLon magnetic concentrate. They can remove quartz and other gangue minerals which associated with a small portion of magnetite and get high grage iron cocentrate. Cooperating the advantages of the two kinds of equipments to process some oxidized iron ores, good beneficial results can be achieved. This paper introduces the latest research and application of this technology.

Keywords SLon magnetic separator; SLon centrifugal separator; Oxidized iron ore; Process

1 SLon 磁选机与离心机组合技术的特点

SLon立环脉动高梯度磁选机广泛应用于分选弱磁性铁矿，在分选鞍山式氧化铁矿的生产流程中，主要是用于粗选作业。精选作业一般是用螺旋溜槽分选粗粒级和用反浮选分选细粒级。为什么分选鞍山式铁矿的流程中强磁选很少用于精选作业呢？原因是鞍山式氧化铁矿是由磁铁矿、假象赤铁矿、赤铁矿、镜铁矿组成，由表1可知，磁铁矿的比磁化率

---

❶ 原载于《2014年全国选矿前沿技术大会论文集》。

是赤铁矿和镜铁矿的 100 倍以上。

表 1　几种铁矿的比磁化率

矿石名称	磁铁矿	假象赤铁矿	赤铁矿	镜铁矿
比磁化率 $\chi/\text{m}^3 \cdot \text{kg}^{-1}$	$(6.25 \sim 11.6) \times 10^{-4}$	$(6.2 \sim 13.5) \times 10^{-6}$	$(0.6 \sim 2.16) \times 10^{-6}$	3.7×10^{-6}

如图 1 所示，鞍山式氧化铁矿中含有较多的磁铁矿和石英的连生体，这些连生体在磁场中受到的磁力很大，表 2 为磁铁矿与石英连生体的视在比磁化率。例如，一颗连生体含 1% 质量的磁铁矿和 99% 质量的石英，它的视在比磁化率达到了 $6.25 \times 10^{-6} \text{ m}^3/\text{kg}$，

图 1　磁铁矿和石英的连生体

已远远大于赤铁矿单体的比磁化率，它在磁场中所受到的磁力就比相同质量的赤铁矿要大。它很容易被强磁机捕捉到铁精矿中去，从而降低铁精矿品位。

表 2　磁铁矿和石英连生体的视在比磁化率

连生体中磁铁矿质量/%	连生体中石英的质量/%	连生体的视在比磁化率 $\chi/\text{m}^3 \cdot \text{kg}^{-1}$
0	100	0
1	99	6.25×10^{-6}
5	95	3.125×10^{-5}
10	90	6.250×10^{-5}
20	80	1.25×10^{-4}
50	50	3.125×10^{-4}
100	0	6.25×10^{-4}

离心选矿机是一种较好的细粒矿物重选设备，它可提供 $30g \sim 120g$（g 表示重力加速度）的离心力，能将细粒矿物按密度分选。石英的密度是 $2.6\text{g}/\text{cm}^3$，磁铁矿和赤铁矿的密度为 $5.0\text{g}/\text{cm}^3$，表 3 为磁铁矿和石英的连生体的视在密度。

表 3　磁铁矿和石英连生体的视在密度

连生体中磁铁矿质量/%	连生体中石英的质量/%	连生体的视在密度/$\text{g} \cdot \text{cm}^{-3}$
0	100	2.60
10	90	2.73
20	80	2.88
30	70	3.04
40	60	3.22
50	50	3.42
60	40	3.65
70	30	3.92
80	20	4.22
90	10	4.58
100	0	5.00

根据生产经验，含 30% 的石英与 70% 的磁铁矿的连生体（视在密度为 $3.92\text{g}/\text{cm}^3$，石英与磁铁矿的体积比为 45∶55），能被离心机排入尾矿中去。因此离心机的精选能力要高于强磁选机的精选能力。SLon 立环脉动高梯度磁选机处理量大，用于粗选作业具有富集

比大、选矿效率高的优点，而用于精选作业则存在含少量磁铁矿和大部分石英的贫连生体难以剔除的制约因素，而离心选矿机用于精选作业则可较好地解决这个问题，这两种设备相结合分选某些氧化铁矿可获得较好的选矿指标。

2 新设备的研究

2.1 SLon 立环脉动高梯度磁选机的研制

SLon 立环脉动高梯度磁选机是新一代优质高效强磁选设备（图2），该机利用磁力、脉动流体力和重力的综合力场分选弱磁性矿石，具有富集比大、选矿效率高、选矿成本低、适应面广、设备作业率高、使用寿命长、易安装和检修工作量小的优点。

通过多年持续的技术创新与改进，多种型号的机型，在设备大型化、多样化、自动化、高效、节能、提高可靠性等方面得到了快速的发展，并且得到了更为广泛的应用。目前有3000多台 SLon 磁选机在工业上广泛用于氧化铁矿、钛铁矿、锰矿、铬铁矿、钨矿等弱磁性矿石的选矿及石英、长石、高岭土等非金属矿的提纯。

2.2 SLon-离心选矿机的研制

针对铁矿选矿要求设备处理量大的特点，我们近年研制出了 SLon-ϕ1600、SLon-ϕ2400 离心选矿机（图3）。通过生产应用证明，用它们对强磁粗选的铁精矿进行精选可获得较高的铁精矿品位。表4为 SLon 磁选机和 SLon 离心机应用于分选海钢强磁选精矿的对比指标，离心机的精矿品位比强磁选精矿品位高2.77%，尾矿品位低2.51%。因此离心机的精选指标明显优于强磁机的精选指标。

图2 SLon 立环脉动高梯度磁选机照片

图3 SLon-ϕ2400 离心选矿机

表4 SLon 磁选机和 SLon 离心机应用于精选作业的对比指标 （%）

名称	给矿		铁精矿			尾矿铁品位
	<0.074mm（200目）	铁品位	铁品位	产率	回收率	
SLon 磁选机	90.55	55.45	61.75	60.38	67.24	45.85
SLon 离心机	90.55	55.45	64.52	57.18	66.53	43.34
离心—磁选差值	0	0	+2.77	-3.20	-0.71	-2.51

3 工业应用

3.1 在昆钢大红山铁矿的应用

昆钢大红山铁矿是磁铁矿和赤铁矿共生的混合矿，原设计采用阶段磨矿—弱磁—SLon强磁选—反浮选流程，由于脉石矿物复杂多变，反浮选作业无法稳定而未用成功，前几年用摇床取代了反浮选。但是，摇床存在处理量较小，对细粒铁矿回收率较低的缺点。2010年大红山铁矿进行了技术改造，采用72台赣州金环磁选设备有限公司研制的SLon-ϕ2400离心选矿机取代反浮选作业。工业试验指标及新改造的原则流程如图4所示。该流程从2010年投产至今已成功运转近4年，全流程铁精矿品位从62%~63%提高至64%~65%。

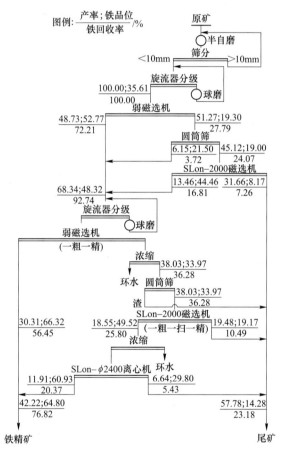

图4 大红山铁矿400万吨选厂工业试验流程

该流程的主要特点为：SLon-2000磁选机对细粒氧化铁矿回收率较高，可有效地控制强磁选作业的尾矿品位；采用SLon-ϕ2400离心选矿机取代反浮选作业，对强磁选精矿进行精选，2009年在大红山铁矿的工业试验表明，离心选矿机具有富集比较大，对细粒铁矿回收率较高的优点。这些新技术和新设备的应用，使该流程既可运行顺畅，又能获得良好的选矿指标。全流程的选矿指标为：原矿品位35.61%，铁精矿品位64.80%，铁回收率76.82%。

3.2 分选海南难选氧化铁矿

海南矿业联合有限公司 2010~2013 年新建成一座年处理 200 万吨氧化铁矿选矿厂，采用 11 台 SLon-2500 磁选机和 80 台 SLon-2400 离心选矿机，在分选低品位氧化铁矿技术上取得新的突破。

该矿拥有一座储量较大的氧化铁矿，过去长期开采品位 50%Fe 左右的富矿，目前可采富矿日益减少，大约有 1000 多万吨品位 40%Fe 左右已采出的贫矿（未采出的贫矿还有 1 亿多吨）过去因无合适的选矿技术未得到利用。

该矿的选矿技术难点在于：

矿石结晶粒度很细，磨矿至小于 0.043mm（325 目）仍然得不到充分解离。

脉石复杂多变，工业试验表明，浮选药剂难以适应，浮选作业难以稳定。

脉石如绿泥石、橄榄石等含铁，尾矿难以降低。

微量磁铁矿和石英的贫连生体较多，强磁选难以选出高品位的铁精矿。

几年前有关单位对这部分氧化铁矿进行了系统的选矿试验研究，通过工业试验证明，若用强磁—反浮选流程，因脉石矿物复杂多变导致反浮选作业无法稳定。若仅用强磁选流程又只能得到品位为 60% 左右的铁精矿。而市场上品位为 63% 以上的铁精矿好销且价格较高。因此，选矿试验和生产实践都要求铁精矿品位达到 63% 以上。通过工业试验证明，SLon 磁选机和离心机的组合流程可获得 63% 以上的铁精矿品位，选矿回收率与强磁—反浮选试验流程或单一磁选流程相当，且流程运行稳定，环境友好，生产成本较低。

该矿年处理 200 万吨原矿的选矿厂采用如图 5 所示的选矿流程，其入选原矿品位为 39.15%，含铁矿物主要是赤铁矿及占原矿产率 13% 左右的磁铁矿。首先磨矿至小于 0.074mm（200 目）占 90%，然后用弱磁选机选出磁铁矿，弱磁选机的尾矿用 SLon-2500 磁选机一次粗选和一次扫选，磁选精矿合并，浓缩后用离心选矿机精选拿出大部分品位为 63.5%Fe 左右的铁精矿，离心机尾矿经浓缩后用旋流器分级，旋流器沉砂进入二段球磨。旋流器溢流用弱磁选机和 SLon 磁选机分别选出磁铁矿和赤铁矿。二者的混合精矿再用离心机精选得出小部分品位为 62% 左右的铁精矿。

该流程综合选矿指标为：给矿品位 39.15%Fe，精矿品位 63.20%Fe，铁精矿产率 39.19%，铁回收率 63.26%，尾矿品位 23.65%Fe。

该流程于 2013 年初投产，至今已成功运转 16 个月，生产流程稳定，生产指标良好，选矿成本较低，且无药剂污染。

目前海南昌江县另有两家民营企业也采用上述流程分选同类矿石，其年处理海南难选氧化铁矿合计 100 万吨左右。

3.3 在低品位镜铁矿中的应用

江西省境内有一部分低品位镜铁矿，原矿品位为 22%Fe 左右，含铁矿物主要是镜铁矿和少量磁铁矿，脉石矿物主要是石英、云母、绿泥石、石榴石和磷灰石。

如图 6 所示为一座日处理 500t 低品位镜铁矿的选厂生产流程图。其原矿品位为 22.63%TFe，铁矿物主要是镜铁矿和少量的磁铁矿，磁铁矿产率占原矿的 3% 左右。

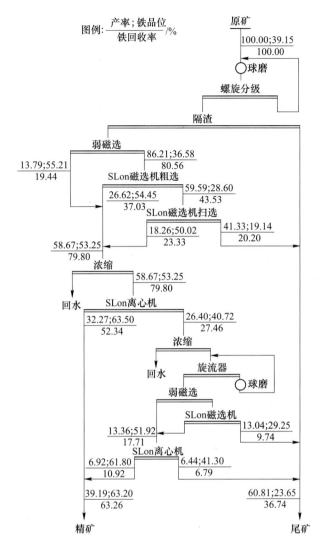

图 5 海南难选氧化铁矿 SLon 磁选机与 SLon 离心选矿机的组合分选流程

采用阶段磨矿，SLon 强磁抛尾—离心机精选流程。一段磨矿将矿石磨至小于 0.074mm（200 目）60% 左右，用弱磁选机选出磁铁矿精矿，用 SLon 磁选机分选抛去一部分低品位尾矿，其强磁精矿用旋流器分级，旋流器沉砂进二段磨矿，二段磨矿的排矿返回到旋流器，旋流器的溢流粒度为小于 0.074mm（200 目）占 90% 左右。旋流器溢流进入二段 SLon 强磁选机分选，二段强磁精矿浓缩后用离心机精选，二段强磁尾矿和离心机尾矿作为最终尾矿。该流程获得最终综合精矿品位为 62.05%，铁回收率 65.41%。

该流程的特点是：

（1）一段 SLon 强磁选可抛弃大量的低品位尾矿，较大幅度地节约了二段磨矿的生产成本。

（2）离心选矿机用于二段强磁精矿的精选作业，可有效地剔除含磁铁矿的石英连生

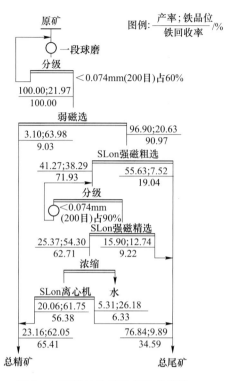

图 6 江西低品位镜铁矿选矿流程

体,其精矿品位提高幅度较大,从 54.30%TFe 提高到 61.75%TFe。

(3) 整个流程为开路分选流程,选矿作业不存在循环负荷,生产上很好控制。

上述生产流程具有流程较简单,选矿指标较好,生产成本较低的优点。

3.4 在鞍山式氧化铁矿的应用

目前我国鞍山式氧化铁矿大多数都采用阶段磨矿—分级重选—强磁—反浮选流程。该流程具有节能、生产成本较低及选矿指标较好的优点。但是,对于一些中小型选矿厂来说,反浮选作业存在技术复杂、环保审批难等问题,因此,采用离心选矿机代替反浮选作业在一些中小型选厂得到应用。如图 7 所示为 SLon 磁选机与 SLon 离心机的组合流程在辽宁省保国铁矿的应用,其矿石为鞍山式氧化铁矿,含铁矿物以赤铁矿为主,含有一部分磁铁矿和假象赤铁矿,脉石矿物以石英为主。选矿流程为:一段磨矿后用水力旋流器分级,旋流器沉砂用螺旋溜槽选出一部分粒度较粗已经单体解离了的铁精矿,螺旋溜槽尾矿用 SLon 立环脉动中磁机分选,抛弃一部分粗粒级尾矿。旋流器溢流用弱磁选机选出磁铁矿精矿,弱磁选机尾矿经浓缩后用 SLon 立环脉动强磁机分选,该机抛弃一部分细粒尾矿。其强磁精矿浓缩后用 SLon 离心机进行一次精选,离心机取得一部分的细粒级铁精矿。该流程综合选矿指标为:原矿品位 30.09%TFe,综合铁精矿品位为 62.43%,综合铁回收率 68.95%。综合尾矿品位 14.00%。该流程全部采用磁选和重选作业,具有节能、环保、生产成本较低的优点。

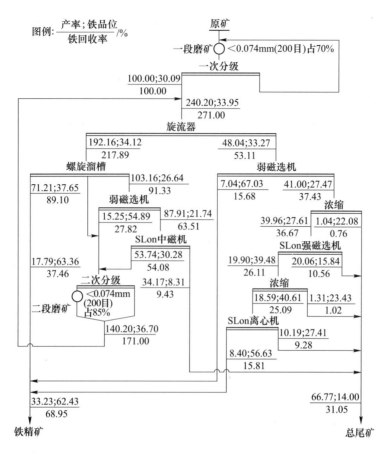

图 7 在辽宁保国铁矿分选鞍山式氧化铁矿的选矿流程

3.5 从尾矿中回收铁精矿的应用

我国分选磁铁矿的小厂很多，有的小厂的入选原矿中含有一部分赤铁矿或镜铁矿，这些小选厂用弱磁选机选完磁铁矿后，弱磁选尾矿直接排入尾矿库。采用 SLon 磁选机和 SLon 离心选矿机的组合流程对这种尾矿进行再回收，往往可以获得较好的技术经济指标。如图 8 所示为 SLon 磁选机与 SLon 离心机的组合流程分选某尾矿库的堆存尾矿的试验指标。该尾矿中含铁矿物主要是镜铁矿和少量的磁铁矿。该流程的特点是：先搅拌，分级磨矿至小于 0.074mm（200 目）占 95%，利用弱磁选机分选出少量的磁铁矿，然后利用 SLon 立环脉动高梯度磁选机处理量大，作业成本低的特点，一次粗选抛弃产率 62.17%，品位 6.17%TFe 的低品位尾矿，SLon 磁选机的粗选精矿品位已达到 36.42%TFe，而产率只占原矿的 33.68%，这部分粗精矿进入二段磨矿分级，二段分级溢流粒度为小于 0.048mm（300 目）占 95%，浓缩后再用 SLon 磁选机精选一次，其精矿用离心选矿机精选。该流程的综合选矿指标为：给矿品位 18.80%TFe，铁精矿品位为 60.15%TFe，铁精矿产率 15.40%，铁回收率 49.26%，综合尾矿品位为 11.28%。该流程目前已在生产中应用，实现了低成本从尾矿中回收铁精矿的目的，使二次资源得到利用。

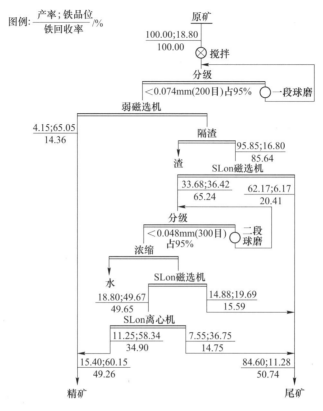

图 8 SLon 磁选机与离心机组合流程从尾矿中回收铁精矿

4 结论

（1）SLon 立环脉动高梯度磁选机具有处理量大、生产成本低，它用于低品位氧化铁矿的粗选作业具有富集比大、选矿效率高的优点。

（2）强磁粗选精矿中，若含有磁铁矿和石英的贫连生体，其视在比磁化率远高于赤铁矿的比磁化率，则再利用强磁精选作用不大，例如鞍山式的氧化铁矿是由磁铁矿、假象赤铁矿和赤铁矿组成，这种矿石用强磁粗选后，后续作业一般不再用强磁精选。

（3）SLon 离心选矿机是利用矿石密度差异进行分选的，可产生 $30g \sim 120g$（g 为重力加速度）的离心力，可有效地回收微细粒铁精矿，并可有效地剔除含少量磁铁矿的贫连生体脉石，强磁选精矿用离心机再进行精选往往可获得较高的铁精矿品位。

（4）SLon 立环脉动高梯度磁选机和 SLon 离心选矿机的组合流程具有生产成本低、环境友好、易于操作管理的特点。这种流程已在多个选矿厂推广应用。

写作背景 本文总结了近几年 SLon 立环脉动高梯度磁选机与离心选矿机组合技术分选氧化铁矿的研究与应用新进展。此时 SLon-ϕ2400 离心机在工业上已得到大规模应用。文章中列举了昆钢大红山铁矿采用 72 台 SLon-ϕ2400 离心机取代反浮选作业；海南铁矿采用 11 台 SLon-2500 磁选机和 80 台 SLon-ϕ2400 离心机建成年处理 200 万吨难选氧化铁矿流程；辽宁保国铁矿采用 SLon-2000 磁选机和 48 台 SLon-ϕ1600 磁选机分选鞍山式氧化铁矿。

SLon 磁选机在安徽分选弱磁性矿石的研究与应用

熊大和[1,2]

（1. 赣州金环磁选设备有限公司；2. 赣州有色冶金研究所）

摘　要　SLon 立环脉动高梯度磁选机利用磁力、脉动流体力和重力的综合力场分选弱磁性矿石，具有富集比大、选矿效率高、设备可靠性高、选矿生产成本低、适应面广、设备作业率高、使用寿命长、易安装、检修工作量小等优点。介绍了 SLon 磁选机在安徽省境内的发展和在氧化铁矿、长石、石英和高岭土选矿方面的应用。SLon 磁选机在安徽省的试验与应用，完成了该设备原始创新的重要阶段，为该设备的广泛应用打下了良好的基础，也为安徽省经济发展作出了一定的贡献。

关键词　SLon 立环脉动高梯度磁选机　氧化铁矿选矿　石英　长石　高岭土提纯

我国拥有丰富的氧化铁矿、钛铁矿、锰矿、黑钨矿等弱磁性矿石资源，但多数矿床的原矿品位低、嵌布粒度细，而原有的选矿设备和工艺不能满足选矿工业的要求。

SLon 立环脉动高梯度磁选机是新一代优质高效强磁选设备。通过多年持续的技术创新与改进，该机在设备大型化、多样化、自动化、高效、节能、提高可靠性等方面得到了快速发展，目前已有 3500 多台 SLon 磁选机在工业上广泛应用于氧化铁矿、钛铁矿、锰矿、铬铁矿、钨矿等弱磁性矿石的选矿及石英、长石、高岭土等非金属矿的提纯。

SLon 磁选机在安徽省境内应用实现了如下九个"第一"：

（1）第一台 SLon-1000 磁选机在姑山铁矿分选赤铁矿。

（2）第一台 SLon-1500 磁选机在姑山铁矿分选赤铁矿。

（3）第一台 SLon-1750 磁选机在姑山铁矿分选赤铁矿。

（4）第一台 SLon-1250 磁选机在来安县分选长石。

（5）SLon 磁选机在来安县分选长石用量第一。

（6）SLon-1500 磁选机第一次在煤系高岭土除铁应用。

（7）第一台 SLon-1000 干式振动高梯度磁选机在煤系高岭土应用。

（8）SLon 磁选机第一次在大型镜铁矿选矿厂应用。

（9）SLon 磁选机在凤阳石英砂除铁用量第一。

SLon 磁选机在安徽省境内完成了设备原始创新的重要阶段，为该设备的大规模应用打下了基础，也为安徽省的经济发展作出了一定的贡献。

1　SLon 磁选机工作原理简介

SLon 立环脉动高梯度磁选机利用磁力、脉动流体力和重力的综合力场分选弱磁性矿

---

❶ 原载于《2014 年中国矿业科技大会论文集》，合肥。

石，具有富集比大、选矿效率高、选矿成本低、适应面广、设备作业率高、使用寿命长、易安装和检修工作量小的优点。

SLon 磁选机结构见图 1 和图 2，主要由脉动机构、激磁线圈、铁轭、转环和各种矿斗、水斗组成，用导磁不锈钢棒作磁介质。该机工作原理为，激磁线圈通以直流电，在分选区产生感应磁场，位于分选区的磁介质表面产生高梯度磁场；转环作顺时针旋转，将磁介质不断送入和运出分选区；矿浆从给矿斗给入，进入转环下部的分选区，矿浆中的磁性颗粒吸附在磁介质棒表面上，随转环运转带至顶部无磁场区，被冲洗水冲入磁性产品矿斗；非磁性颗粒在重力、脉动流体力的作用下穿过磁介质堆，流入非磁性产品矿斗排走，实现与磁性颗粒的分离。

图 1　SLon 磁选机照片

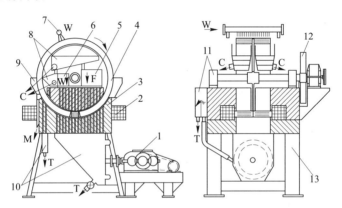

图 2　SLon 立环脉动高梯度磁选机结构

1—脉动机构；2—激磁线圈；3—铁轭；4—转环；5—给矿斗；6—漂洗水斗；7—磁性矿冲洗装置；8—磁性矿斗；9—中矿斗；10—非磁性矿斗；11—液位斗；12—转环驱动机构；13—机架；F—给矿；W—清水；C—磁性矿；M—中矿；T—非磁性矿

该机的转环采用立式旋转方式，对于每一组磁介质堆而言，冲洗磁精矿的方向与给矿方向相反，粗颗粒不必穿过磁介质堆便可冲洗出来。该机的脉动机构驱动矿浆产生脉动，可使位于分选区磁介质堆中的矿粒群保持松散状态，使磁性矿粒更容易被捕获，使非磁性矿粒尽快穿过磁介质堆到非磁性矿斗中去。

反冲精矿和矿浆脉动可防止磁介质堆堵塞；脉动分选可提高磁性精矿的质量。这些措施保证了该机具有较大的富集比、较高的分选效率和较强的适应能力。

2　首台 SLon-1000 磁选机在姑山铁矿的工业试验

为了提高弱磁性矿石的选矿指标，1986~1987 年赣州有色冶金研究所与中南工业大学联合研制开发了首台 SLon-1000 立环脉动高梯度磁选机。

这台 SLon-1000 立环脉动高梯度磁选机于 1987 年 8 月在赣州有色冶金研究所制造出来，然后运至马钢姑山铁矿进行工业考核试验，当时姑山铁矿有 10 台 $\phi1600mm \times 900mm$

离心机分选细粒赤铁矿，由于离心机用于粗选作业选矿效率较低，姑山铁矿决定拆去1台离心机装上SLon-1000磁选机，经过3000余小时的工业试验，该机获得了远优于离心机的选矿指标，见表1。与同段离心机选矿指标比较，该机产生铁精矿品位提高了4.42%，尾矿品位降低了7.48%，精矿铁回收率提高了19.57%。

表1 SLon-1000磁选机与ϕ1600mm×900mm离心机对比指标　　（%）

设备	给矿铁品位	精矿铁品位	精矿产率	精矿铁回收率	尾矿铁品位
SLon-1000磁选机	33.26	56.78	37.22	63.56	19.31
ϕ1600mm×900mm离心机	34.12	52.36	28.67	43.99	26.79
差值	-0.86	+4.42	+8.55	+19.57	-7.48

SLon-1000磁选机在马钢姑山铁矿的成功试验，为提高我国细粒弱磁性矿石的选矿水平开创了一条新的途径，也为该机的进一步发展奠定了基础。

3 首台SLon-1500磁选机在姑山铁矿的工业试验

由于SLon-1000磁选机的处理量较小，该机处理细粒氧化铁矿每小时处理量约4~7t，远远满足不了大规模选矿工业的要求。为了满足大中型选矿厂的需要，赣州有色冶金研究所于1989年研制出首台SLon-1500立环脉动高梯度磁选机，同年6月在马钢姑山铁矿投入工业试验，该机处理量是SLon-1000磁选机处理量的4~5倍。用于分选洗矿溢流、SQC强磁选机尾矿经ϕ350mm旋流器分级的溢流等细粒级赤铁矿，这些细粒赤铁矿之前是作为尾矿排放。工业试验共运转5000余小时，所获得的平均选矿指标见表2。

表2 SLon-1500磁选机从细粒尾矿中回收赤铁矿的选矿指标　　（%）

给矿粒度（-200目）	给矿铁品位	精矿铁品位	精矿产率	精矿铁回收率	尾矿铁品位
71	28.13	55.65	28.28	55.94	17.28

由表2可知，SLon-1500磁选机可从铁品位为28.13%的给矿中，一次分选得到铁品位为55.56%的合格铁精矿（当时姑山选厂合格铁精矿品位为54.5%），作业回收率为55.94%。

SLon-1500磁选机在姑山铁矿试验成功，使立环脉动高梯度磁选机在选矿工业设备大型化方面前进了一大步。

4 SLon-1750磁选机在姑山铁矿的工业试验与应用

姑山铁矿是马钢的主要原料基地之一，分选矿物主要为赤铁矿，矿石硬度大、嵌布粒度细且不均匀，属难选红铁矿。该矿选矿车间自1978年投产以来，选矿工艺流程历经多次改造，生产技术指标逐步提高，但随着矿床开采深度增加，矿石含磷增高是影响铁精矿质量的主要因素之一。

1995~1999年，姑山选厂主厂房采用阶段磨矿、强磁—脉动高梯度磁选流程选别细粒赤铁矿。其中一段磨矿分级溢流采用3台SQC-6-2770强磁选机抛尾，由于给矿粒度较粗（小于0.074mm（200目）占42%左右），该机齿板经常堵塞，精矿品位较低，尾矿品位较高，使全厂选矿指标受到较大影响。

为使姑山铁矿选矿指标进一步提高，赣州有色冶金研究所于1999年研制出首台SLon-

1750立环脉动高梯度磁选机,到姑山铁矿取代了1台SQC-6-2770强磁选机,同年4月1日至5月31日开展了为期2个月的工业对比试验,该机与平行的另一台SQC-6-2770强磁选机对比的平均选矿指标见表3。该机在给矿品位低0.93%的情况下,获得的铁精矿品位高1.28%、铁回收率高7.64%。其选矿指标远优于SQC-6-2770强磁选机。

2000年姑山铁矿又引进1台SLon-1750磁选机取代原SQC-6-2770强磁选机,并于同年6月完成了一个磨选系列的优化改造,试验流程和选矿指标如图3所示。其中粗选作业和精选作业各采用1台SLon-1750立环脉动高梯度磁选机,扫选作业采用原有的1台SLon-1500磁选机。这3台磁选机与1台球磨机构成了完整的工业试验流程。该系统每小时处理干矿量33t左右,与其他系统处理量相同。

表3 SLon-1750磁选机与SQC-6-2770强磁选机对比指标 （%）

设备	给矿铁品位	精矿铁品位	精矿产率	精矿铁回收率	尾矿铁品位
SLon-1750磁选机	41.20	51.33	71.50	89.09	15.78
SQC-6-2770强磁机	42.13	50.05	68.56	81.45	24.86
差值	-0.93	+1.28	+2.94	+7.64	-9.08

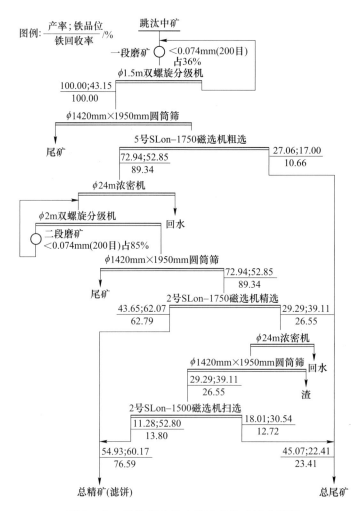

图3 SLon磁选机在姑山铁矿分选赤铁矿流程

该流程工业试验实现了精矿铁品位达到 60%、铁回收率 76%的指标，与其他系统比较，铁精矿品位提高约 5%，铁回收率提高约 10%。

姑山铁矿在随后的两年中，共采用 6 台 SLon-1750 磁选机和 4 台 SLon-1500 磁选机将另外 3 个磨选系统全部改为上述流程。

SLon 磁选机在姑山铁矿的试验与应用，解决了平环强磁机齿板易堵塞和选矿效率较低的问题，选矿指标创姑山铁矿历史新高，同时为该机在全国其他的细粒氧化铁矿选矿的应用打下了良好的基础。

5 SLon 磁选机在来安县长石除铁中的应用

长石是制造陶瓷和玻璃的主要原料，安徽省来安县境内有丰富的长石资源，铁、钛等是长石中的有害元素。过去采用筛洗的方法将相对较纯的矿石直接开采，作为低档产品出售，剩余部分作为废矿排弃，大量排弃的废石不仅浪费了国家资源，而且严重污染了环境。

1998 年来安县皖东长石有限公司建成了一座年产 3 万吨长石精矿的选矿厂。首台 SLon-1250 立环脉动高梯度磁选机在该厂用于长石除铁获得成功，该厂的选矿流程如图 4 所示。棒磨机和高频细筛组成的闭合回路将矿石磨至小于 0.8mm，用滚筒中磁机除去强磁性铁物质，然后用一台 SLon-1250 磁选机除去弱磁性铁矿物。含 1.48%Fe_2O_3 的原矿经该流程分选后，得到长石精矿产品含 0.26% Fe_2O_3，满足了玻璃工业和陶瓷工业 Fe_2O_3 小于 0.30%的要求。

SLon 磁选机在长石除铁工业中的成功应用，为提高我国长石资源的利用率做出了重要贡献。目前来安县境内有十多家长石加工厂采用 SLon 磁选机用于长石除铁。该设备和相应的长石除铁技术已在全国各地得到广泛应用。

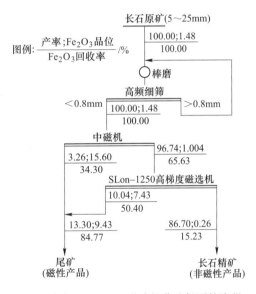

图 4 首台 SLon-1250 磁选机分选长石的流程

6 SLon-1500 磁选机在淮北煤系高岭土提纯中的应用

高岭土是一种非金属矿产资源，在工业上广泛应用于造纸涂料、油漆填料、橡胶填料和陶瓷行业。用于造纸涂料的为高等级高岭土，其技术要求为白度大于 90%，粒度小于 2μm 大于 86%。煤系高岭土的储量占我国高岭土储量的 98%，安徽淮北市是我国重要的煤炭生产基地之一，其煤矸石中存在着大量的煤系高岭土。过去，因煤系高岭土含铁较高和颜色呈灰黑状而很少得到利用，自 20 世纪 80 年代以来，随着高岭土开发技术的发展和我国工业对高岭土需求量的增大，煤系高岭土的开发越来越受到重视。实践证明，经深加工后的煤系高岭土具有白度高，质量好等优点。

2002~2003 年，淮北金岩高岭土公司建成了年产 1 万吨精矿的生产线，采用一台 SLon-1500 立环脉动高梯度磁选机作为除铁设备，其生产流程如图 5 所示。原矿含 0.7%~1.0% Fe_2O_3，经破碎、磨矿至小于 74μm（200 目）占 100%，调浆至浓度 30% 进入 SLon-1500 立环脉动高梯度磁选机除铁，使非磁性产品 Fe_2O_3 含量小于 0.5%。非磁性产品用旋流器脱水，脱水后用立式搅拌磨研磨至小于 2μm 占 90%，然后用煅烧炉煅烧，高岭土中的有机物质经煅烧后变成灰分。煅烧后的高岭土白度达到 93%，粒度为小于 2μm 占 90%，可作为造纸的优质涂料。

SLon 磁选机在煤系高岭土除铁提纯工业中首次获得成功应用，标志着我国煤系高岭土加工技术达到了一个新的高度。

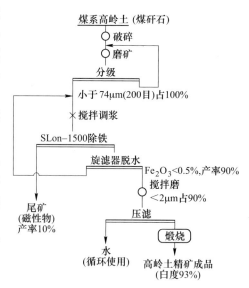

图 5　淮北煤系高岭土加工流程

7　首台 SLon-1000 干式振动高梯度磁选机在淮北煤系高岭土应用

淮北煤系高岭土的另一种加工方法是将高岭土原矿先煅烧，干法研磨至一定的粒度，然后用强磁选除铁。这种方法的优点是节省了湿法调浆和最后干燥的过程，具有流程短、节能和生产成本较低的优点。

2012 年，首台 SLon-1000 干式振动高梯度磁选机在淮北成功应用于煤系高岭土干法除铁，该机的作业指标见表 4。由于该机具有场强高、磁场梯度高、对微细粒弱磁性铁矿物捕收力大的优点，应用于煤系高岭土干法除铁获得了良好的选矿指标。

表 4　SLon 磁选机用于煤系煅烧高岭土干法除铁指标

粒度/目	原矿 Fe_2O_3/%	精矿 Fe_2O_3/%	除铁率/%
10~16	0.98	0.78	20.4
16~30	0.96	0.63	34.2
30~60	1.02	0.59	42.5
60~80	1.10	0.54	51.1
80~120	1.25	0.47	62.3
120~200	1.25	0.62	50.2
200~300	1.29	1.06	18.1

8　SLon 磁选机在李楼铁矿的应用

安徽省霍邱县铁矿资源丰富，近年有几个铁矿选矿厂相继建成。安徽开发矿业有限公司李楼铁矿建成了一座年处理 500 万吨镜铁矿的选矿厂，其设计的选矿流程如图 6 所示。该流程中采用了 25 台 SLon-2000 立环脉动高梯度磁选机作为强磁选设备。

李楼铁矿镜铁矿的含铁矿物主要是镜铁矿，其他铁矿物含量很少，例如原矿中磁铁矿

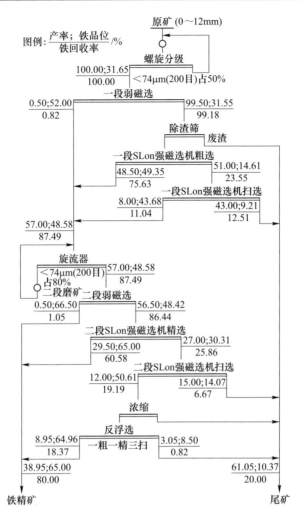

图 6 李楼铁矿 500 万吨/a 镜铁矿设计选矿流程

产率仅为 0.5% 左右。其选矿流程的特点为：

（1）该流程中一段强磁选采用一粗一扫，二段强磁选采用一精一扫，保证了全流程获得较高的选矿回收率。

（2）由于强磁选作业成本低，强磁选提前抛弃了大部分尾矿和得出了大部分精矿，仅有占原矿 12% 左右的二段强磁扫选精矿进入反浮选作业，由于反浮选的作业成本较高，进入反浮选的矿量越少，全流程的生产成本越低。

全流程设计的综合选矿指标为：给矿铁品位 31.65%，精矿铁品位 65.00%、回收率 80.00%，尾矿铁品位 10.37%。

9 SLon 磁选机在安徽金日盛铁矿的应用

安徽金日盛铁矿的矿石由镜铁矿、赤铁矿、磁铁矿和脉石组成。近几年建成了一座年处理量 450 万吨的现代化选矿厂，其选矿流程如图 7 所示。

该流程中采用了 22 台 SLon-2000 和 6 台 SLon-2500 立环脉动高梯度磁选机作为强磁选

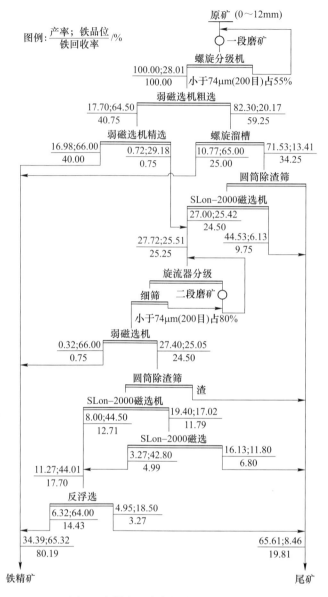

图 7 安徽金日盛氧化铁矿选矿流程

设备,该流程具有以下特点:

(1) 原矿经一段磨矿分级至小于 74μm(200 目) 占 55%,用弱磁选机选出部分铁品位达到 66% 的磁铁矿作为最终铁精矿,弱磁选尾矿再用螺旋溜槽选出部分铁品位达到 65% 的氧化铁矿(单体解离度较好的镜铁矿和赤铁矿)。

(2) 用 SLon-2000 立环脉动高梯度磁选机作为一段强磁选抛尾设备,该机一次选别抛弃了产率为 44.53%,铁品位为 6.13% 的尾矿,有效地控制了一段尾矿品位。

(3) 一段强磁选精矿和一段弱磁精选的尾矿合并后进入二段分级和磨矿。由于一段成功地分选出了大部分合格铁精矿和抛弃了大部分低品位尾矿,进入二段分级磨矿的矿量只占原矿产率的 27.72%。

（4）二段分级磨矿至小于 74μm（200 目）占 80%，用弱磁选机分选出剩余的磁铁矿作为最终精矿，弱磁选尾矿用 SLon 磁选机一次粗选和一次扫选分选出镜铁矿和赤铁矿。二段强磁选精矿合并后用反浮选精选，得到品位为 64% 的反浮选铁精矿。该流程综合选矿指标为：原矿铁品位 28.01%，精矿铁品位 65.32%、回收率 80.19%，尾矿铁品位 8.46%。整个生产流程稳定，指标优异。

10 SLon 磁选机在凤阳石英提纯工业中的应用

安徽凤阳县石英砂资源丰富，目前已探明储量为 50 亿吨，远景储量 100 亿吨以上，享有"中国石英砂之乡"的美誉。当地的石英砂含硅高，SiO_2>99.5%，天然品质较好，但是含铁量偏高，Fe_2O_3 含量在 0.03% 左右，而光伏行业高品质石英砂要求 Fe_2O_3 含量在 0.008% 以下。SLon 立环脉动高梯度磁选机应用于凤阳石英砂除铁，具有除铁效果好，设备运转稳定可靠的优点。如图 8 所示为 SLon-2500 磁选机在凤阳县远东石英砂有限公司应用于石英砂除铁的生产流程。原石英砂矿中 Fe_2O_3 的含量在 0.01%~0.02%，经过 SLon-2500 磁选机磁选除铁后，精矿产率达到了 95% 以上，精矿中 Fe_2O_3 含量小于 0.007%、SiO_2 含量大于 99.8%，达到了太阳能、光纤通信等高尖端原料的要求。

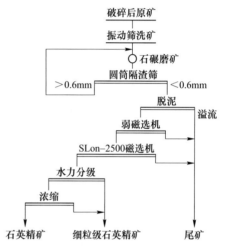

图 8 SLon-2500 磁选机在安徽凤阳提纯石英砂流程

目前，凤阳地区已有几十家生产企业采用 SLon 立环脉动高梯度磁选机用于石英砂除铁。

11 结论

（1）SLon 立环脉动高梯度磁选机利用磁力、脉动流体力和重力的综合力场分选弱磁性矿石，具有富集比大、选矿效率高、选矿成本低、适应面广、设备作业率高、使用寿命长、易安装和检修工作量小的优点。

（2）SLon-1000、SLon-1500、SLon-1750 磁选机在马钢姑山铁矿应用于分选细粒赤铁矿，大幅度地提高了姑山铁矿细粒氧化铁矿的选矿指标和经济效益。

（3）SLon 磁选机在来安县成功地应用于长石除铁，使该地区大量的废石变成了宝贵的矿产资源。

（4）SLon 磁选机在淮北煤系高岭土除铁中的成功应用，为我国煤系高岭土的开发和利用发展了一条新的途径。

（5）SLon 磁选机在李楼铁矿和金日盛铁矿得到了大规模应用，分选氧化铁矿获得了良好的选矿指标。

（6）SLon 磁选机在凤阳石英砂除铁工业中应用，具有除铁效果好、产品质量高的优点，将低档次的石英砂加工成满足光伏行业要求的优质原料。

（7）SLon 磁选机在安徽省的试验与应用，完成了该设备原始创新的重要阶段，为该设备的广泛应用打下了良好的基础，也为安徽省经济发展作出了一定的贡献。

写作背景 为感谢安徽省境内各个厂矿和同行人士对 SLon 立环脉动高梯度磁选机的发展所作的贡献而写本文。文章中介绍了 SLon 磁选机在安徽省境内创新的 9 个第一；SLon 磁选机在安徽省境内应用于分选氧化铁矿、应用于煤系高岭土、长石和石英砂提纯。该机在安徽省境内完成了原始创新的重要阶段，也为安徽省经济的发展作出了一定的贡献。

SLon 磁选机提高选钛回收率的研究与应用[❶]

熊大和[1,2]

(1. 赣州金环磁选设备有限公司; 2. 赣州有色冶金研究所)

摘 要 我国攀枝花和承德等地区拥有丰富的钒钛磁铁矿资源,早期的选钛以重选为主,TiO_2 的选矿回收率只能达到 10% 左右,大量的细粒钛铁矿损失在尾矿中。自从 1994 年以后,SLon 立环脉动高梯度磁选机开始在钛铁矿选矿工业中应用,使细粒级和微细粒级钛铁矿得到了较好的回收。SLon 磁选机和浮选新技术的发展使我国钛铁矿选矿技术水平得到了迅速提高。目前我国钛铁矿选矿回收率普遍达到 20%~40% TiO_2。本文介绍 SLon 磁选机提高钛铁矿选矿回收率的研究与应用。

关键词 SLon 立环脉动高梯度磁选机 钛铁矿 回收率

我国四川省的攀枝花、西昌,河北省的承德及山东、云南等地拥有丰富的钒钛磁铁矿资源,这些地区的选矿厂普遍是通过破碎、磨矿后用弱磁选机选出磁铁矿。弱磁选尾矿中含有 3%~15% TiO_2 的钛铁矿,在 1980 年以前,大部分选厂将弱磁选尾矿作为最终尾矿排放,其钛铁矿基本上没有回收。1980 年攀枝花钢铁公司建成了我国第一条从钒钛磁铁矿磁选尾矿中回收钛铁矿的生产线[1]。生产采用的工艺流程为重选—浮选(脱硫)—电选流程,主要回收 0.25~0.074mm 粒级钛铁矿,每年生产 5 万吨钛精矿,钛精矿品位 46.2%~46.5% TiO_2 (注:生产钛白粉的钛精矿品位要求达到 46% TiO_2 以上),TiO_2 回收率占磁选尾矿的 8.72% 左右[2]。

1980 年至 1995 年期间,我国选钛技术主要是将螺旋溜槽、摇床等重选设备用于粗选作业回收较粗粒级的钛铁矿[3],其精矿再用电选或浮选的方法精选。选钛回收率一般占磁选尾矿的 10% TiO_2 左右。由于当时选矿设备和选矿技术的限制,细粒级钛铁矿绝大部分当尾矿排放。

SLon 立环脉动高梯度磁选机是我国自行研制的新一代高效强磁选设备,该机利用磁力、脉动流体力和重力的综合力场分选弱磁性矿石,该机转环立式旋转、反冲精矿、配有矿浆脉动机构、具有选矿效率高、磁介质不易堵塞、分选粒度范围较宽、可靠性高和能耗低的优点。该机从 1994 年开始在国内外的钛铁矿选矿中试验和应用,对提高我国的选钛回收率做出了重要的贡献。

1 攀钢选钛厂的选钛技术进步

1.1 攀钢选钛厂早期的选钛流程

攀钢选钛厂早期的选钛流程始建于 1980 年,后经过多次技术改造形成如图 1 所示的

[❶] 原载于《矿冶工程》,2014,34:57~62。

流程,该流程将磁选尾矿(选磁铁矿的尾矿,即选钛原矿,选钛回收率从这里开始计算,下同)除渣,弱磁选除铁后,分成三个粒级,即粗粒级(>0.1mm)、细粒级(0.1~0.045mm)和微细粒级(<0.045mm)。其中粗粒级和细粒级分别用强磁选、螺旋溜槽和摇床重选,最后干燥后用电选精选,得到品位为46.5%TiO_2左右的钛精矿。而微细粒级因受当时选矿设备和技术的限制作为尾矿排放。

该流程在1991年以前每年可生产合格的钛精矿5万吨,1992年增加了一条生产线以后达到年产钛精矿10万吨。该流程的建设成功,实现了我国钒钛磁铁矿选钛工业从无到有的转变。

上述流程存在的主要问题有:

(1)小于0.045mm微细粒级直接排入尾矿,虽然流程图上注明的微细粒级是小于0.045mm,但是由于水力分级设备分级效率不高,该粒级中仍有较多的大于0.045mm钛铁矿直接排入尾矿。

(2)该流程的分级、浓缩作业环节多,其溢流中损失的钛铁矿较多。

(3)螺旋溜槽和摇床选钛的作业回收率较低。

上述原因导致了该流程选钛回收率较低,以年处理500万吨磁选尾矿,其中TiO_2品位为10%,年产10万吨品位为47% TiO_2的铁精矿计算,TiO_2的回收率为9.40%左右。

1.2 SLon磁选机在攀钢选钛厂回收小于0.045mm粒级钛铁矿

攀钢选钛厂原先的选钛流程中,小于0.045mm粒级作为尾矿排放(图1),其中TiO_2的金属占有率为40%左右,因此,攻克这部分物料的选钛难题,对提高全流程的选钛回收

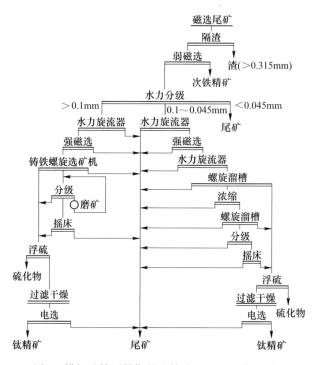

图1 攀钢选钛厂早期的选钛流程(1995年以前)

率具有重要的意义。1994~1996 年，攀钢选钛厂联合赣州有色冶金研究所等单位开展了从这部分物料中回收钛铁矿的小型试验和工业试验，取得了突破性的研究成果，并于 1996 年在攀钢选钛厂采用一台 SLon-1500 立环脉动高梯度磁选机建立了第一条微细粒级选钛试验生产线（图 2）。

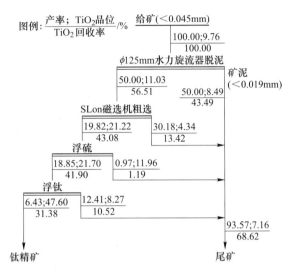

图 2　攀钢选钛厂第一条小于 0.045mm 选钛生产流程

由于浮选作业要求脱出小于 0.019mm 粒级矿泥，否则会造成浮选作业泡沫过多跑槽，该流程中首先用旋流器脱出小于 0.019mm 矿泥，然后用一台 SLon-1500 磁选机粗选，其粗选精矿再用浮选脱硫和浮选精选钛铁矿，最终得到品位为 47%TiO₂ 以上的优质钛精矿。

该流程中，SLon 磁选机的作业回收率达到 76.24% TiO_2，浮选的作业回收率也达到了 75% TiO_2 左右，全流程的选钛回收率达到了 30% TiO_2 左右。但是由于小于 0.019mm 粒级未入选，这部分 TiO_2 金属量损失了 43.49%，严重影响了钛铁矿回收率的进一步提高。

上述选钛生产线一年生产 2 万吨优质钛精矿。随后的几年该厂又分两期建成后八系统和前八系统小于 0.045mm 粒级选钛生产线，至 2004 年，该厂小于 0.045mm 钛精矿的年产量达到 15 万吨左右，加上粗粒级钛精矿，该厂年产钛精矿达到 28 万吨左右。

1.3　攀钢选钛厂钛铁矿选矿的扩能改造工程

经过多年的生产实践证明，强磁选—浮选的选钛流程与重选—电选流程比较具有回收率较高、生产成本较低、生产环节较易控制的优点。2008 年，攀钢选钛厂进行选钛扩能改造工程，具体实施计划有：

（1）将粗粒级选钛流程由原来的重选—电选流程改为强磁选—浮选流程。
（2）优化细粒级选钛流程。
（3）钛精矿产量由改造前的 28 万吨/a 提高到 40 万吨/a。

攀钢选钛厂委托赣州金环磁选设备有限公司所做的选钛流程如图 3 所示。
该流程的主要特点为：
（1）磁选尾矿只分成粗粒级和细粒级入选，充分发挥 SLon 磁选机本身的脱泥功能，尽可能减少分级溢流的损失。

（2）细粒级强磁选作业由过去的 SLon 磁选机一次选别改为四次选别，即 SLon 磁选机一粗一扫选和一精一扫选；粗粒级强磁选作业采用 SLon 磁选机一次粗选，其粗精矿分级再磨后用弱磁选机除铁，其尾矿用 SLon 磁选机一精一扫选。这些措施保证了强磁选作业对钛铁矿有较高的回收率和较高的精矿品位。比较图 2 和图 3 可见，细粒级进入浮选作业的 TiO_2 金属量由 43% 左右提高到 72% 左右（37.65%/52.55% = 72.10%）。粗粒级进入浮选作业的 TiO_2 金属量高达该粒级的 77% 左右（36.83%/47.78% = 77.08%）。

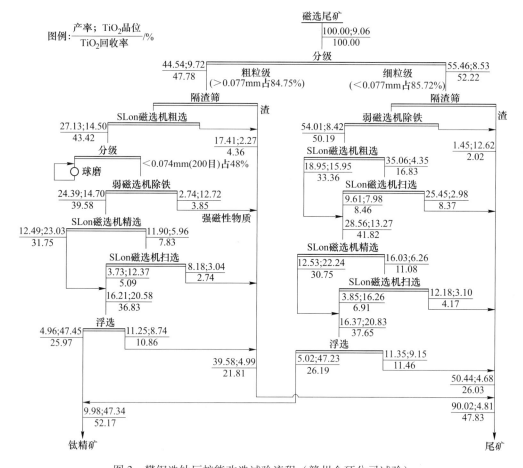

图 3　攀钢选钛厂扩能改造试验流程（赣州金环公司试验）

（3）该选钛流程充分利用了 SLon 磁选机处理量大、生产成本低、脱泥效果较好的优点，大幅度提高强磁选作业的选钛回收率，为浮选创造了较好的条件。在保证钛精矿品位达到 47% TiO_2 以上的前提下，浮选作业 TiO_2 回收率可保证在 70% 左右，全流程的 TiO_2 回收率达到 52.17%。

2008~2009 年，攀钢选钛厂参考各家的选钛试验流程并结合现场的条件进行了选钛扩能改造，2010~2012 年该厂实现年产优质钛精矿分别为 48 万吨、50 万吨和 52 万吨，目前生产上全流程实际达到的 TiO_2 回收率为 40% 左右，虽然选钛回收率比以前提高了很多，由于生产条件的限制，许多作业环节还没有达到最佳状况。选钛回收率的提高还有很大的潜力。

2 SLon 磁选机在攀钢综合厂的应用

2.1 攀钢综合厂早期的选钛流程

攀钢综合厂为攀钢附属集体企业，主要从攀钢排放的总尾矿中分流一部分回收铁精矿和钛精矿，原选钛流程采用螺旋溜槽粗选，摇床精选的重选流程（图4）。由于螺旋溜槽和摇床对细粒级回收率都很低，因此全流程的选钛回收率仅为 7% TiO_2 左右。

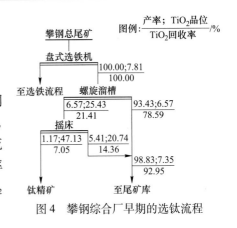

图 4　攀钢综合厂早期的选钛流程

2.2 攀钢综合厂早期采用 SLon 磁选机的选钛流程

攀钢综合厂 2002 年采用 SLon 磁选机建立了新的选钛流程（图 5）。攀钢的总尾矿用盘式选铁机选出残余的磁铁矿等强磁性物质。剩余的作为选钛原矿。该流程的特点是：选钛原矿全粒级进入 SLon 磁选机作一次粗选一次精选，避免了水力分级机溢流中 TiO_2 损失（注：由于水力分级机的分级效率不高，溢流中往往有 15% 左右大于 20μm 的钛铁矿适合于浮选回收）。由于 SLon 磁选机本身有脱泥的作用，它可将绝大部分 10μm 以下的矿泥排入尾矿中，在 SLon 磁选机的精选作业中，采用较低的背景磁感应强度还可将大部分小于 20μm 的矿泥排入尾矿中（注：小于 20μm 的矿泥包括该粒级的钛铁矿在进入浮选之前要尽可能脱除，否则会对浮选造成不利影响）。SLon 磁选机精选作业的精矿产率只有 16% 左右，再用小型的高效浓缩机脱除小于 20μm 的残余矿泥，然后用浮选得到合格钛精矿。通过技术改造，该流程的选钛回收率由原重选作业的 7%TiO_2 提高到 27%TiO_2 左右。

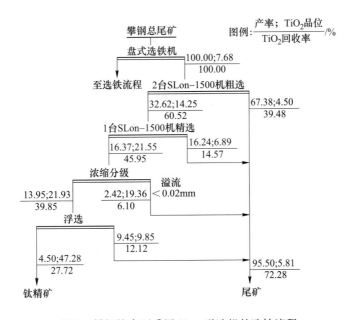

图 5　攀钢综合厂采用 SLon 磁选机的选钛流程

2.3 SLon-4000 磁选机在攀钢综合厂的应用

攀钢密地选矿厂从钒钛磁铁矿中回收铁精矿和钛精矿，其综合尾矿每年约 700 万吨，全部排往尾矿库。攀钢选矿综合厂 2002 年采用了 3 台 SLon-1500 磁选机建立新的选钛流程（图 5）。但是用于粗选作业的 2 台 SLon-1500 磁选机合计只能处理 60t/h 给矿，该厂钛精矿产量只能达到每年 2 万吨左右。随着攀钢密地选钛厂选钛技术水平的提高，目前排往尾矿库的尾矿 TiO_2 品位由原来的 7.68% 左右下降至 6.5% 左右，因此，如果不改造，该厂的钛精矿产量有进一步下降的趋势。2011 年年底，首台 SLon-4000 立环脉动高梯度磁选机在该厂应用（进行工业试验），取代原有粗选作业的 SLon-1500 磁选机，该机每小时给矿量为 460t（约占总尾矿的 50%）左右，一年多的工业试验，取得的工业试验指标为（SLon-4000 单机作业指标）：给矿品位 6.41% TiO_2，精矿品位 13.38% TiO_2，尾矿品位 3.64% TiO_2，TiO_2 作业回收率 59.36% 的良好指标。全流程的选钛指标为：给矿品位 6.41% TiO_2，钛精矿品位 47.02% TiO_2，TiO_2 回收率 28.16%。2013 年该厂又采购了 1 台 SLon-4000 磁选机进行扩产改造，2 台 SLon-4000 磁选机均已投入生产（图 6），该流程

图 6 2 台 SLon-4000 磁选机在攀钢从总尾矿中回收钛铁矿

年处理攀钢密地选厂 700 万吨综合尾矿，目前已达到年产品位为 47% TiO_2 的优质钛精矿 18 万吨左右。

3 重钢西昌矿业有限公司选钛流程的进步

3.1 重钢西昌矿业有限公司早期的选钛流程

重钢西昌矿业有限公司（原名为重钢太和铁矿）位于四川省西昌市，该公司拥有一座较大的钒钛磁铁矿矿床，矿石性质与攀枝花钒钛磁铁矿类似。其选矿流程是原矿经破碎、磨矿、弱磁磁选机回收磁铁矿。其选铁尾矿中含 TiO_2 12% 左右。为回收该尾矿中的钛资源，该公司于 1995 年建成了第一条选钛生产线（图 7）。该流程首先将磁选尾矿分级成粗粒级、细粒级及溢流。粗粒级用螺旋溜槽粗选，其粗精矿进行分级磨矿后与细粒级合并，用滚筒中磁机选出残余的强磁性物质，然后用摇床分选，摇床精矿再用浮选精选，浮选精矿作为最终钛精矿。

该重—浮流程存在的主要问题有：

(1) 由于当时选用的水力分级设备（斜板浓密箱）分级效率不高，有 30% 左右的 TiO_2 损失在溢流中。

(2) 螺旋溜槽和摇床的分选粒度下限只能达到 30μm 左右，对细粒级钛铁矿的回收率很低。螺旋溜槽的作业回收率仅为 59% 左右，摇床的作业回收率仅为 47% 左右。

(3) 该流程的重选段包括分级、螺旋溜槽和摇床的 TiO_2 综合回收率仅为 22.42%。

(4) 钛铁矿浮选是正浮选，即泡沫产品为钛铁矿精矿，螺旋溜槽和摇床都不能回收细

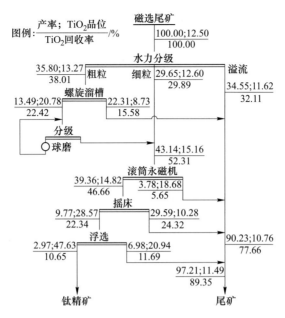

图 7 重钢太和铁矿 2000 年以前采用的选钛流程

粒级钛铁矿，造成浮选的给矿粒度偏粗，由于钛铁矿密度（$4.7\sim4.8\mathrm{g/cm^3}$）较大，较粗粒级的钛铁矿容易沉槽，因此造成浮选作业回收率偏低。

上述原因导致了该流程选钛回收率只能达到 10% TiO_2 左右。1995~2000 年，该公司采用该流程每年只能回收 2000 余吨钛精矿。

3.2 SLon 磁选机早期在西昌矿业公司分选钛铁矿的流程

重钢西昌矿业有限公司于 2000 年开始采用 SLon 立环脉动高梯度磁选机进行选钛流程的技术改造，经过改造后的选钛流程如图 8 所示。

该流程采用 SLon 磁选机取代了螺旋溜槽和摇床。磁选尾矿首先用水力分级机分成大于 0.02mm（沉砂）和小于 0.02mm（溢流），沉砂用 SLon 磁选机粗选，其精矿分级磨矿，用滚筒永磁中磁机除去磁铁矿等强磁性物质后再用 SLon 磁选机精选，其精矿进入浮选精选。

上述流程的主要优点有：

（1）由于 SLon 磁选机的选矿粒度下限可达到 $20\sim10\mu m$，磁选尾矿用水力分级机分级时可控制溢流部分少跑一些。让更多的细粒级钛铁矿进入 SLon 磁选机的选别作业。对比图 7 和图 8 可知，溢流部分流失的 TiO_2 金属量由 32% 左右降低至 17% 左右。

（2）由于 SLon 磁选机分选粒度范围较宽，其入选原矿不再分粗粒级和细粒级，简化了选矿流程。

（3）SLon 磁选机取代螺旋溜槽和摇床后作业回收率大幅度提高。尽管原矿品位大幅度降低了（经过几年开采后，太和铁矿的原矿 TiO_2 品位大幅度降低了），但 SLon 磁选机的作业回收率分别比螺旋溜槽和摇床高 22% 左右和 32% 左右。进入浮选作业的 TiO_2 金属量由原重选作业的 22% 左右提高到 48% 左右。

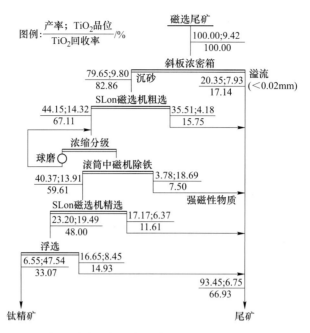

图 8 重钢西昌矿业有限公司采用 SLon 磁选机的选钛流程

（4）浮选作业的给矿粒度组成更好了。由于 SLon 磁选机对 0.15~0.02mm 钛铁矿粒级回收率高达 85% 以上，这一粒度范围正好是浮选的最佳粒度。因此浮选的作业回收率由过去的 48%TiO_2 左右提高到目前的 69%TiO_2 左右。

SLon 磁选机在西昌矿业公司的成功应用大幅度提高了该矿钛铁矿的回收率，尽管入选原矿品位降低了，TiO_2 的回收率由过去的 10% 左右提高到 33% 左右。钛精矿品位保证在 47%TiO_2 以上。至 2007 年，该矿实现年产优质钛精矿 12 万吨。

3.3 SLon 磁选机在西昌矿业公司分选钛铁矿的扩建流程

近年，重钢西昌矿业有限公司计划逐步将原矿处理量由 220 万吨/a 扩建至 1000 万吨/a（设计年处理铁矿石 1000 万吨，选出 333 万吨磁铁矿精矿，选完磁铁矿的尾矿 667 万吨作为选钛原矿），扩建过程中新增了 7 台 SLon-3000 和 1 台 SLon-2500 磁选机用于钛铁矿分选，原选钛流程中的 SLon-2000 磁选机和 SLon-1750 磁选机也部分得到利用。其新的选钛流程如图 9 所示。该流程中，选钛原矿（即选铁尾矿）经浓缩脱泥后用 SLon-3000 磁选机粗选，其粗精矿经浓缩、分级磨矿至 -0.074mm 粒级占 50%，然后用 SLon-2000 和 SLon-1750 磁选机精选，其磁选精矿再用浮选脱硫并精选钛精矿，最终得到品位为 47%TiO_2 以上的优质钛精矿。

目前该生产流程已投产 18 个月，采出原矿处理量达到 700 万吨/a 左右，年产优质钛精矿 30 万吨左右，待项目全部达产，其钛精矿产量可望由扩建前的 12 万吨/a 增加至 45 万吨/a 左右。

4 结论

（1）我国钒钛磁铁矿的选钛技术在 1980 年以前几乎是一片空白，1980~1990 年发展

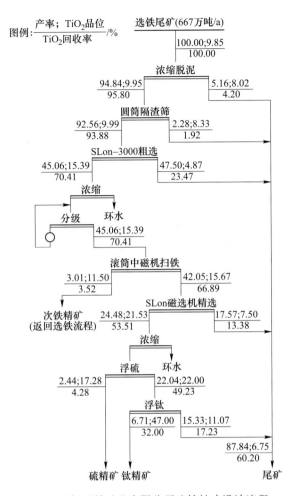

图 9 重钢西昌矿业有限公司选钛扩建设计流程

了以重选和电选为主的选钛流程，实现了我国钒钛磁铁矿选钛技术从无到有的转变，但是，由于当时细粒级钛铁矿不能回收等原因，TiO_2 回收率只能达到占选钛原矿的 $10\% TiO_2$ 左右。

（2）从 1994 年以后，SLon 立环脉动高梯度磁选机在钒钛磁铁矿的选钛流程中得到应用，使 $45\sim20\mu m$ 的钛铁矿得到较好的回收，TiO_2 回收率普遍达到占选钛原矿的 $20\%\sim40\% TiO_2$，钛精矿品位达到 $47\% TiO_2$ 以上。

（3）通过对选钛流程的优化，充分利用强磁选机处理量大和作业成本低的优点，在选钛流程的一段强磁选和二段强磁选作业中增加扫选作业，可较大幅度地提高强磁选钛作业回收率。通过优化的选钛流程，在选钛原矿品位 $10\% TiO_2$ 左右的前提下，目前生产上已经达到全流程 TiO_2 回收率占选钛原矿的 40% 左右，钛精矿品位达到 $47\% TiO_2$ 以上。

参 考 文 献

[1] 孟长春，刘胜华. 攀钢选钛厂细粒钛铁矿浮选工艺技术研究与发展探讨 [J]. 矿冶工程，2010（8）：11~15.

[2] 邹建新，杨成，彭富昌，等. 攀西地区钒钛磁铁矿提钛工艺与技术进展 [J]. 金属矿山, 2007 (7): 7~9.
[3] 袁国红. 重钢太和铁矿选钛流程技术改造 [J]. 金属矿山, 2001 (6): 39~40.

写作背景　本文介绍了 SLon 立环脉动高梯度磁选机在钛铁矿选矿方面的应用，以及近几年取得的技术进步，例如 2 台 SLon-4000 磁选机在攀钢尾矿库回收钛铁矿已达到年产优质铁精矿 18 万吨；重钢西昌矿业公司新扩建的选钛流程，铁精矿年产量由扩建前的 12 万吨增加至 30 万吨。

新型 SLon 立环脉动高梯度磁选机应用实践[1]

熊大和[1,2]

(1. 赣州金环磁选设备有限公司；2. 赣州有色冶金研究所)

摘　要　矿产资源的日益贫化和矿业市场的激烈竞争迫使选矿企业努力降低选矿成本，提高选矿效率。为此，赣州金环磁选设备有限公司研发了几种新型 SLon 立环脉动高梯度磁选机，并在工业生产上得到应用。SLon-2250 磁选机在太钢袁家村铁矿应用，强磁选尾矿品位降低1.7%，作业铁回收率提高 2.67%；SLon-2000-1.3T 磁选机在安徽开发矿业李楼铁矿和鞍钢东鞍山烧结厂分选氧化铁矿，铁的回收率显著提高；新型 SLon-2500 磁选机在宝钢梅山铁矿应用，使强磁扫选的尾矿铁品位降低 1.63%，作业铁回收率提高 3.78%；SLon-2000-1.5T、SLon-2500-1.5T、SLon-2000-1.8T 和 SLon-2500-1.8T 磁选机已成功应用在长石、石英等非金属矿的除铁提纯中，取得了良好的技术经济指标。

关键词　SLon 立环脉动高梯度磁选机　提铁　非金属矿　除铁提纯

在当前矿产品需求不旺、价格持续走低的形势下，选矿厂必须不断降低生产成本、提高选矿效率才有生存和持续发展的空间。SLon 立环脉动高梯度磁选机利用磁力、脉动流体力和重力的综合力场分选弱磁性矿石，具有富集比大、选矿效率高、选矿成本低、适应面广、设备作业率高、使用寿命长、易安装和检修工作量小的优点。SLon 立环脉动高梯度磁选机在工业上广泛应用于氧化铁矿、钛铁矿、锰矿等弱磁性矿石的选矿及石英、长石、高岭土等非金属矿的除铁提纯。经过多年持续的技术创新与改进，该磁选机在设备大型化、多样化、自动化和高效节能及可靠性等方面得到了快速的发展，并得到了广泛的应用。SLon 立环脉动高梯度磁选机如图 1 所示。

为了进一步提高选矿技术指标、降低选矿成本，多种新型 SLon 立环脉动高梯度磁选机研制成功，背景磁感强度为 1.0~1.8T，并在分选氧化铁矿、石英、长石等领域得到工业应用，获得了良好的选矿指标。

图 1　SLon 立环脉动高梯度磁选机

1　SLon-2250 磁选机在太钢袁家村铁矿的应用

太钢袁家村铁矿选矿厂是我国最大的铁矿氧化矿石选矿厂之一，设计年处理氧化铁矿 2200 万吨，采用两段磨矿—弱磁选—强磁选—脱泥—反浮选处理流程，数质量流程见图 2。

实际生产中由于强磁选给矿铁品位较高、含泥量较大，造成强磁选尾矿铁品位偏高。

---

[1] 原载于《中国矿业科技文汇—2015》。

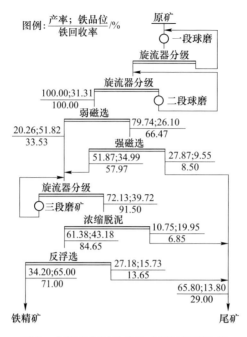

图 2 太钢袁家村铁矿选矿设计原则流程

为降低强磁选尾矿铁品位、提高强磁选作业回收率，赣州金环磁选设备有限公司研制了 SLon-2250 立环脉动高梯度磁选机，最高背景磁感强度为 1.051T，电磁性能见表 1。

表 1 SLon-2250 磁选机电磁性能

电流/A	电压/V	功率/kW	场强/T
200	8.0	1.60	0.203
400	16.6	6.6	0.390
600	24.7	14.8	0.576
800	33.2	26.6	0.745
1000	41.3	41.3	0.873
1200	48.8	58.6	0.970
1400	57.6	80.6	1.051

该设备的特点是结构紧凑，干矿处理量是 SLon-2000 立环脉动高梯度磁选机的 1.4 倍左右，而设备的安装空间与现场 ϕ2000mm 立环强磁选机基本相同。2014 年 9 月将 6 台 SLon-2250 磁选机在太钢袁家村铁矿选矿厂投入生产，与现场 ϕ2000mm 立环强磁选机进行对比试验，结果见表 2。

表 2 SLon-2250 磁选机与现场 ϕ2000 立环强磁选机对比试验结果

设备	给矿体积 /m³·h⁻¹	给矿浓度 /%	干矿量 /t·h⁻¹	<74μm (200目) /%	给矿铁品位/%	精矿铁品位/%	尾矿铁品位/%	精矿产率 /%	铁回收率 /%
SLon-2250 磁选机	125	33.94	55.47	87.47	29.71	37.49	10.99	70.64	89.14
ϕ2000 立环强磁选机	90.00	33.47	43.00	87.47	29.66	37.51	12.69	68.37	86.47

由表2可知，在相同给矿条件下，SLon-2250磁选机比现场φ2000mm立环强磁选机强磁选尾矿铁品位降低1.7%，作业铁回收率提高2.67%，单机处理量高12.47t/h左右。

2 SLon-2000-1.3T磁选机的应用

SLon-2000-1.3T立环脉动高梯度磁选机是为提高非金属矿选矿效果而研制的，近几年在长石和石英等非金属矿选矿中得到了广泛应用，在铁矿石氧化矿的选矿中也得到了较好的应用。

2.1 SLon-2000-1.3T磁选机的电磁性能

SLon-2000-1.3T磁选机电磁性能见表3，设计的额定背景磁感强度为1.3T，实际测定最高背景磁感强度为1.375T。

表3 SLon-2000-1.3T磁选机电磁性能

电流/A	电压/V	功率/kW	场强/T
200	8.2	1.6	0.289
400	16.6	6.6	0.543
600	24.9	14.9	0.796
800	33.6	26.9	1.004
1000	42.3	42.3	1.151
1200	48.8	58.6	1.297
1400	58.2	81.5	1.375

2.2 SLon-2000-1.3T磁选机在安徽李楼铁矿的应用

安徽开发矿业有限公司李楼铁矿选矿厂年处理550万吨镜铁矿，入选含铁矿物主要是镜铁矿，其他铁矿物含量很少，磁铁矿仅0.5%左右，设计选矿数质量流程如图3所示。

选矿流程的特点为：（1）该流程中一段强磁选采用一粗一扫，二段强磁选采用一精一扫，保证了全流程获得较高的选矿回收率；（2）由于强磁选作业成本低，强磁选提前抛除了大部分尾矿并获得大部分精矿，仅有占原矿12%左右的二段强磁扫选精矿进入反浮选作业。反浮选的作业成本较高，进入反浮选的矿量越少，全流程的生产成本越低。全流程设计的综合选矿指标为：给矿铁品位31.65%、精矿铁品位65.00%、铁精矿产率38.95%、铁回收率80.00%、尾矿铁品位10.37%。

经过几年生产，发现二段强磁扫选尾矿铁品位偏高，设计的二段强磁扫选尾矿铁品位为14.07%，而实际为20%左右，影响了全流程的选矿回收率。因此生产上迫切需要降低二段强磁选尾矿铁品位。2014年李楼铁矿采用了3台SLon-2000-1.3T立环脉动高梯度磁选机进行工业试验，位于二段强磁扫选作业，取得了良好的选矿指标，见表4。

表4 SLon-2000-1.3T磁选机在李楼铁矿对比试验指标　　　　（%）

设备	给矿铁品位	精矿铁品位	尾矿铁品位	作业铁回收率
设计指标	30.31	50.31	14.07	74.38
SLon-2000-1.3T	36.15	51.26	15.91	80.83
原φ2000立环强磁选机	36.15	57.36	22.83	60.96

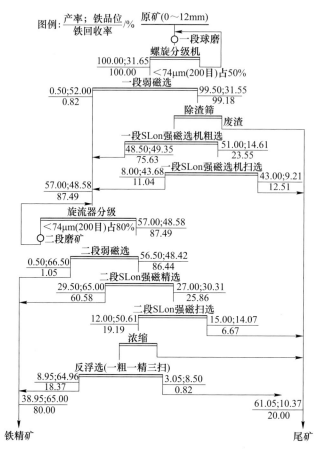

图 3 李楼铁矿 500 万吨/a 镜铁矿设计选矿数质量流程

由表 4 可知，SLon-2000-1.3T 立环脉动高梯度磁选机作业的精矿铁品位和回收率均优于设计指标，与原 φ2000mm 立环强磁选机比较，作业回收率有显著提高。

2.3 SLon-2000-1.3T 立环脉动高梯度磁选机在鞍钢东鞍山烧结厂的应用

鞍钢东鞍山矿区铁矿氧化矿石结晶粒度细、氧化程度深，且含有一定比例的菱铁矿，是鞍山地区最难选的红矿。鞍钢东鞍山烧结厂一选车间年处理氧化铁矿 470 万吨，2002 年开始采用阶段磨矿—分级—螺旋溜槽重选—强磁—反浮选工艺流程，见图 4。

采用 SLon 立环脉动高梯度磁选机（额定背景磁感应强度 1.0T）作为强磁选设备控制细粒级尾矿品位，另外采用 SLon 立环脉动中磁机（额定背景磁感应强度 0.6T）扫选螺旋溜槽尾矿。由于开采深度不断增加，目前原矿选别难度不断增加，现场强磁尾矿铁品位偏高。如何降低强磁尾矿铁品位是生产上亟待解决的问题。

2014 年东鞍山烧结厂采用了 8 台 SLon-2000-1.3T 立环脉动高梯度磁选机取代原有的 SLon-2000-1.0T 立环脉动高梯度磁选机以降低强磁选尾矿品位。现场工业对比试验选矿指标见表 5。

由表 5 可知，使用 SLon-2000-1.3T 立环脉动高梯度磁选机后强磁尾矿铁品位降低 4.28%，作业铁回收率提高 12.92%。

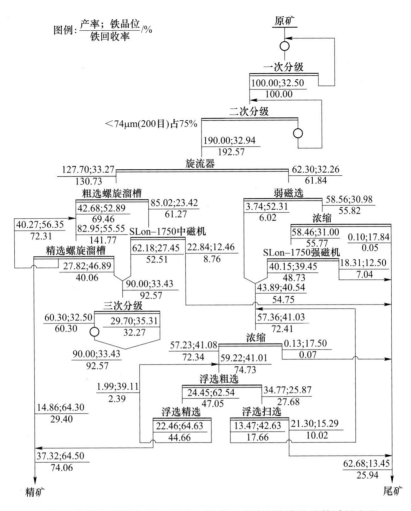

图 4 东鞍山阶段磨矿—重选—强磁—反浮选设计选矿数质量流程

表 5 SLon-2000-1.3T 磁选机与现场 SLon-2000-1.0T 磁选机对比试验选矿指标 （%）

设 备	给矿品位	精矿铁品位	尾矿铁品位	铁精矿产率	作业铁回收率
SLon-2000-1.3T	30.24	47.74	15.40	45.90	72.45
SLon-2000-1.0T	30.39	48.08	19.68	37.64	59.53

3 新型 SLon-2500 磁选机的研制及在宝钢梅山铁矿的应用

宝钢梅山铁矿选矿厂年处理氧化铁矿石 450 万吨，年生产铁精矿 300 万吨，采用重磁预选—磨矿—浮硫—弱磁选—强磁选选矿工艺流程，其中局部弱磁选—强磁选流程如图 5 所示。

为提高选矿指标和处理量，梅山铁矿拟采用较大的强磁选机取代强磁扫选作业的 SLon-1500 立环脉动高梯度磁选机。针对梅山铁矿的条件，赣州金环磁选设备有限公司研制了新型 SLon-2500 立环脉动高梯度磁选机，电磁性能见表 6。

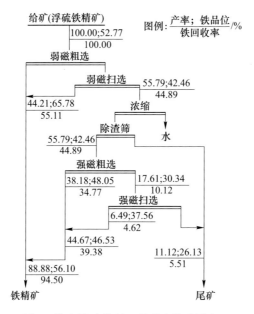

图 5 梅山铁矿弱磁—强磁选数质量流程

表 6 新型 SLon-2500 磁选机电磁性能

电流/A	电压/V	功率/kW	场强/T
200	10.2	2.0	0.183
400	19.3	7.7	0.363
600	28.9	17.3	0.534
800	38.9	31.1	0.700
1000	49.4	49.4	0.832
1200	59.5	71.4	0.930
1400	69.4	97.2	1.009

2014 年 4 台新型 SLon-2500 立环脉动高梯度磁选机取代强磁扫选作业的 8 台 SLon-1500 立环脉动高梯度磁选机，工业对比试验指标见表 7。

表 7 新型 SLon-2500 磁选机与现场 SLon-1500 磁选机对比试验指标　　　（%）

设备	给矿品位	精矿铁品位	尾矿铁品位	铁精矿产率	作业铁回收率
新型 SLon-2500	29.48	39.72	19.02	50.53	68.08
SLon-1500	29.15	37.81	20.65	49.55	64.26

由表 7 可知，新型 SLon-2500 磁选机替代现场 SLon-1500 磁选机作业铁精矿品位提高 1.91%、尾矿品位降低 1.63%、作业铁回收率提高 3.78%，获得了良好的选矿指标。

4 SLon-2000-1.5T 磁选机

2015 年赣州金环磁选设备有限公司研制出 SLon-2000-1.5T 立环脉动高梯度磁选机，最高背景磁感强度达到 1.513T，电磁性能见表 8。目前该机已应用在长石和石英的除铁提纯工业中。

表8 SLon-2000-1.5T磁选机电磁性能

电流/A	电压/V	功率/kW	场强/T
200	7.5	1.5	0.274
400	15.1	6.04	0.543
600	23.1	13.86	0.796
800	31.1	24.88	1.013
1000	39.1	39.1	1.173
1200	47.1	56.52	1.282
1400	55.1	77.14	1.370
1600	63.3	101.3	1.442
1800	71.2	128.2	1.513

5 SLon-2500-1.5T磁选机在长石除铁应用

由于非金属矿除铁提纯的需要，2015年赣州金环磁选设备有限公司研制出SLon-2500-1.5T立环脉动高梯度磁选机，设计的额定背景磁感强度为1.5T，电磁性能见表9。

表9 SLon-2500-1.5T磁选机电磁性能

电流/A	电压/V	功率/kW	场强/T
200	5.3	1.06	0.229
400	10.8	4.32	0.446
600	16.2	9.72	0.674
800	21.6	17.28	0.880
1000	27.2	27.20	1.058
1200	33.2	39.84	1.182
1400	38.7	54.18	1.286
1600	44.3	70.88	1.362
1800	50.0	90.0	1.42
2000	54.8	109.6	1.468
2200	60.8	133.8	1.511

SLon-2500-1.5T立环脉动高梯度磁选机已在长石除铁工业中得到应用，生产工艺流程如图6所示。

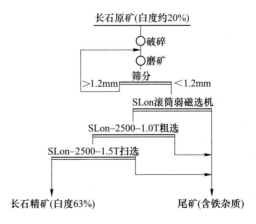

图6 SLon磁选机应用于长石除铁的工艺流程

长石原矿经破碎、磨矿后用滚筒永磁磁选机选出机械铁,然后分别用 SLon-2500-1.0T 和 SLon-2500-1.5T 立环脉动高梯度磁选机进行一次粗选和一次扫选。白度为 20% 左右的长石原矿通过该流程除铁,长石精矿白度达到 63%,除铁效果良好。

6 SLon-2500-1.8T 磁选机的研制及在石英除铁提纯中应用

建筑行业景气度下降导致大量用于建筑行业的普通石英、长石滞销。但用于做光导纤维、太阳能电池的高纯石英仍然具有较好的市场。为了提高非金属矿除铁提纯的选矿指标,2014 年赣州金环磁选设备有限公司研制出 SLon-2500-1.8T 立环脉动高梯度磁选机,设计的额定场强为 1.8T,电磁性能见表 10。

表 10 SLon-2500-1.8T 磁选机电磁性能

电流/A	电压/V	功率/kW	场强/T
200	7.0	1.4	0.283
400	14.5	5.8	0.558
600	21.4	12.84	0.839
800	28.6	22.88	1.092
1000	35.7	35.7	1.297
1200	43.2	51.84	1.451
1400	50.4	70.56	1.556
1600	57.4	91.84	1.642
1800	65.3	117.5	1.715
2000	71.5	143.0	1.781
2100	73.6	154.6	1.812

目前该机已应用在石英的除铁提纯工业中,选矿流程如图 7 所示。

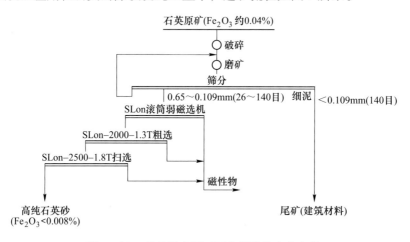

图 7 SLon 磁选机应用于石英提纯的生产流程

石英原矿含 Fe_2O_3 约 $400×10^{-6}$,经破碎、磨矿、脱泥后用滚筒弱磁选机选出机械铁,然后分别用 SLon-2000-1.3T 和 SLon-2500-1.8T 磁选机进行一次粗选和一次扫选。非磁性产品为高纯石英砂($Fe_2O_3<8×10^{-5}$),小于 0.109mm(140 目)细泥和磁性物为尾矿,可分别作为普通建筑材料出售。

7 结论

(1) 为了提高选矿效率、降低选矿企业的生产成本，赣州金环磁选设备有限公司近几年研制了多种型号的 SLon 立环脉动高梯度磁选机，具有磁场强度高、选矿效率高、选矿成本低、适应面广、设备作业率高、使用寿命长、易安装和检修工作量小的优点。

(2) SLon-2250 磁选机在太钢袁家村铁矿应用，强磁选尾矿品位降低 1.7%，作业铁回收率提高 2.67%；SLon-2000-1.3T 磁选机在安徽开发矿业李楼铁矿和鞍钢东鞍山烧结厂分选氧化铁矿石，铁的回收率显著提高；新型 SLon-2500 磁选机在宝钢梅山铁矿应用，使强磁扫选的尾矿铁品位降低 1.63%，作业铁回收率提高 3.78%。

(3) SLon-2000-1.5T、SLon-2500-1.5T、SLon-2000-1.8T 和 SLon-2500-1.8T 磁选机均已研制成功，它们已成功地应用在长石、石英等非金属矿的除铁提纯中，取得了良好的技术经济指标。

写作背景 本文介绍了近几年研制的新型 SLon 立环脉动高梯度磁选机及其工业应用，例如 SLon-2250 磁选机在太钢袁家村铁矿分选赤铁矿，SLon-2000-1.3T 磁选机在安徽李楼铁矿和鞍钢东鞍山烧结厂分选氧化铁矿，SLon-2500 磁选机在宝钢梅山铁矿分选氧化铁矿，SLon-2500-1.5T 磁选机应用于长石除铁，SLon-2500-1.8T 磁选机应用于石英提纯。